COVER-UP

COLLUSION IN THE HALLS OF ACADEMIA

Second Edition

HELENE Z. HILL, PH.D.

As told to Amy Waters Yarsinke

Paperback: 978-1-968519-13-1
Ebook: 978-1-968519-14-8

Ordering Information:

Books to Life Marketing Ltd
128 City Road, London, EC1V 2NX, UK

Printed in the United States of America

Dedication

To all whistleblowers who had the strength and passion to put their reputations on the line and to accept the risks of blowing the whistle

Table of Contents

Preface to the Second Edition

After my appeal to my case for *Qui tam* failed and there were no additional avenues to pursue my charge of fraud at the NJ Medical School, I knew I had to write a book. I had kept a careful diary starting in March of 2001. I entered my first record on March 8 of that year and my last on September 13, 2011. By that time, I had filled 76 pages of WORD. In addition, I had at my disposal several thousand documents emanating from the complete subpoenaed lab notebooks covering the period three years before and three years after the arrival of the key post-doctoral fellow at the laboratory. My lawyer and I had sat through eight depositions, three by expert witnesses, present at my expense. All in all, my cost was more than $200,000.

My time after the case ended was filled with my efforts to keep the shreds of my scientific career alive. I taught core medical courses to freshman and sophomore medical students. I interviewed medical school applicants, and I joined forces with Dr. Karen Hubbard at CCNY in New York City. She had been a post-doctoral fellow in my laboratory in earlier days and now was a professor with her own laboratory and research support. For two years or so, I commuted several days a week into the City to avail myself of her hospitality working at the bench with her students and post-doctoral fellows.

So, I started to work on "the book" and I got bogged down. I turned to Dorothy Okie Beach, my former classmate at day school in Bryn Mawr, PA who was now in business to unite people who had stories with writers who had the literary gift. Dorothy matched me with Amy Waters Yarsinske who had recently finished her own book *An American in the Basement*. This book, *Cover-Up: Collusion in the Halls of Academia*, is the result of our collaboration. Amy has accomplished far more than I expected. She not only turned my notes into a cogent story, but she did an enormous amount of research on her own to fill in the background about scientific fraud and reviewing other cases, many of which were lost, as was mine.

If Amy were my student instead of my writer, she would certainly merit the grade of A+.

Introduction

I n pursuit of information on the fraud, fabrication and misconduct that has plagued scientific and medical research in recent years and has too frequently made negative headlines for over a decade, Dr. Helene Z. Hill's story stood out and in the pursuit of further information on her case, it was clear that what she had done by bringing a *qui tam* lawsuit was the centerpiece of the greater story – it was the unique twist that set apart her experience from a desktop stacked with hundreds of cases, transgressors unpunished, and the hopes of cure and relief dashed for tens of thousands of terminally and chronically ill. All of us know the latter, someone who wants to believe that a cure will come in time to heal them because it is promised in a nightly news health segment or has stared at them as a hopeful story from the cover of a newsstand magazine. Following the light that Hill was shining brightly in chaotic lab space gone awry, it is not enough to know what she knows about the broken science. As you read what follows, it will become clear that what is so badly broken has to be fixed.

While she would shed no tears publicly in her fight to expose a medical cover up at her university, privately it took its toll. Hill's employer at the time of the *qui tam* filing was the University of Medicine and Dentistry of New Jersey (UMDNJ), which ceased operation by that name on July 1, 2013, and Rutgers University took over seven of the UMDNJ's schools, including two of its medical schools as part of a statewide higher education restructuring. The reconstituted Rutgers University/New Jersey Medical School, and Hill's institution, inherited several major whistleblower cases from the UMDNJ merger, including Hill's claims of wrongdoing. UMDNJ was a troubled medical university.

If the history books *get it right*, as Paul S. Brookes, Ph.D., a professor of anesthesiology at the University of Rochester Medical Center, School of Medicine and Dentistry, Rochester, New York, wrote of Helene Z. Hill, Ph.D.'s case in August 2015, she will be remembered as a scientist ahead of her time. "Typically such accolades are afforded to Nobel Prize winners and others who make ground-breaking discoveries, [but] in Dr. Hill's case, her efforts to shine light onto the dark underbelly of scientific misconduct began at a time when nobody else was even considering such things could happen."[1] Helene Hill's story to expose the fraud, fabrication and scientific misconduct that too often puts the public health at immediate risk, started in the late 1990s when the number of publicly known misconduct cases, according to Brookes and many others, could be counted on the digits of one hand and it ends in the present and includes more than a decade of fighting the establishment to get the story out in the open, to tell the tale of how science commits and subsequently deals with the misconduct that occurs in today's laboratories all too frequently. Brookes noted and it has been documented in this narrative that in the past five years, the pace of scientific misconduct has quickened further, but with the help of the Internet there are fewer and fewer opportunities for less-than-honorable scientists to continue operating undiscovered.

Retractions of scientific papers have increased ten-fold worldwide over the past decade. Turnitin.com and iThenticate.com are popular vendors of plagiarism-detection software. The latter estimates on its website that there were seven million researchers competing to publish nearly two million manuscripts in 31,758 scholarly journals throughout the world in 2010. They report that one leading journal published by Taylor and Francis rejected about twenty-three percent of its submissions because of plagiarism. Also, nearly two percent of scientists questioned in a recent study admitted to having falsified data at least once. As many as seventy-two percent of those polled had engaged in some form of questionable research practice. iThenticate estimates that a single case of scientific misconduct costs the affected university/research institution about $525,000. In 2010, the total cost came to about $110 million for investigations throughout the United States. Of even greater concern, more than 70,000 patients had been treated or otherwise participated in 859 retracted clinical studies. The iThenticate report concludes that "time spent on research based on fraudulent work is wasted effort."[2] But in Dr. Hill's experience, the hidden cost of time expended on research based on big science gone wrong is several million dollars.

Hill has observed that while plagiarism software programs are valuable tools for its detection, the Department of Health and Human Services (HHS) Office of Research Integrity (ORI) has downloadable software that utilizes Photoshop for the analysis and detection of manipulated images. For statistical data there is no patent solution. Programs accessible to investigators, she cautioned, are needed that detect numbers that should be but are not random, variances, standard deviations and confidence limits that are too small or too uniform, and patterns that deviate from expectation.

Yet all the detection software on the market has failed to tamp down the growing plethora of retractions, ethics reviews and court cases as the result of fraud, fabrication and misconduct that now involve millions of dollars in federal grants, and that is largely kept secret by the scientific community establishment has continued in earnest, mostly unknown to the public and, to a greater degree than first believed, to those in the allied health professions. This narrative turns Hill's shining light on the growing list of scientific protocols, bad medicine and pharmaceuticals that trickle down and become that immediate threat to the health of the general population.

Further, as Hill and Brookes and a litany of supportive documentation confirm, the increasingly litigious environment surrounding scientific misconduct has overwhelmingly been on the defensive side, with a far greater willingness of accused scientists to fight back in recent years. This has included scientists threatening to sue bloggers for defamation, trying to subpoena websites to gain identifying information on anonymous commenters who criticized their work, and university professors suing their institutions for wrongful dismissal or withdrawal of job offers when misconduct accusations come to light, according to Brookes. While such legal threats often amount to little more than chest-thumping, there can be little doubt that they have a dampening effect on the desire of whistleblowers to speak out, he would observe. Very often, legal threats from allegedly defamed individuals fall under the umbrella term "SLAPP" (Strategic Lawsuit Against Public Participation), with the goal being not to find the truth, but rather to scare the accuser into silence. Thankfully,

Brookes reported, several states have anti-SLAPP statutes, and a number of lawyers committed to First Amendment rights have taken active roles in defending bloggers and other online correspondents. Exemplary among these is Ken White of the popular legal blog *PopeHat*.[3] On the site, White wrote "[W]inning in court generally requires competent representation, which is ruinously expensive for normal people. It's not fair, it's not right, but it's true." Another key feature of such cases in the Internet age, added Brookes, is the so-called "Streisand Effect," named after an attempt by Barbara Streisand to prevent a website from publishing photos of her Malibu home ended up drawing far more attention to the photographs than they otherwise would have received. In several recent cases, among them Bharat Aggarwal, Rakesh Kumar and Falzul Sarkar, attempts by accused scientists to bring legal action have resulted in widespread reporting, often spilling over into mainstream media outlets. "Scientists who pursue such legal retribution would do well to remember the adage," continued Brookes, "that scientific facts are determined empirically, not in the courts."[4]

The other side of the legal landscape for those in science and medicine is when researchers sue to get to the truth, and there are far fewer of them in this category. Helene Hill's case serves as a biting narrative, Brookes would emphasize, on the inadequacies of the current policing and legal systems that scientists have at their disposal to fight misconduct. The specific example of *qui tam* – in which reporters of misappropriation of government funds are entitled to a cut of any funds recovered – is touted as a readily accessible option; however, as Hill's story famously reveals, this is far from an easy road since the bar that must be crossed to prove intentional misappropriation of funds is incredibly high. For a *qui tam* case to stick, the whistleblower must prove that the accused knew what they were doing was wrong but did it anyway. Placing such a burden of proof on the accuser is a likely reason why there have been so few successful cases brought via this method. On this basis alone, Hill deserves credit for having the strength and will, observed Brookes, to actually bring a *qui tam* case all the way to court. Thus, in addition to the law, another central theme that underlies this story is one of perseverance. The sheer amount of mental, physical and financial effort required to bring such cases to light is often a significant deterrent to anyone considering playing the role, he cautioned, of whistleblower.

As Hill's story illustrates, once the decision has been made to inform on fraud, fabrication and misconduct, there is no turning back, and often the person who is already exposed as a tipster must go even further still, to have any hope of bringing a case to conclusion. As readers of the popular blog *Retraction Watch*, or the post-publication peer review site *PubPeer* are sadly aware, opined Brookes, "there are numerous cases in which significant evidence regarding alleged misconduct simply sits in public view for years or even decades, before anything is done at the government or administrative level. In many cases, the *modus operandi* on the part of journals and institutions appears to be to outlast the whistleblowers – literally."[5]

As Hill would quickly discover, being a whistleblower is not easy. The accused are more often than not friends or colleagues, and one or all of them has friends and colleagues who come to the accused's defense – lining up against the tipster. The accused's motives are suspect, she would later note. Are they in it for the money? "I think rarely," she observed. "We are a breed that believe in truth – truth at all cost – and we know that we are right. To be a whistleblower is to stick your neck out," she continued, "to endanger your own position, possibly your livelihood, even your own future."[6] But some whistleblowers give up their science careers and this was certainly the case for Kathryn Milam, Ph.D., who ultimately lost her *qui tam* case predicated on the False Claims Act[7] in 1995 and rather than give up entirely, went to law school. But in her case, at least, scientists had an ally in the judge, who wrote "the legal process is not suited to resolving scientific disputes or identifying scientific misconduct.[8] "I read of one whistleblower," Hill noted, "who lost his home, his wife, [and] his job yet when asked if he would do it again, he replied 'yes.'" Hill's own case is just one of a number of failures intended to set the record straight. In its November 2014 newsletter, the Office of Research Integrity (ORI) reported that in the year prior it had closed seventy-one cases of scientific misconduct but found only twelve of the accused – just seventeen percent – guilty. "Does this mean that fifty-nine whistleblowers barked up the wrong tree?" asked Hill.

The case that ultimately drew long-overdue attention to scientific misconduct is popularly called the David Baltimore Affair (also known as the Baltimore Affair and Imanishi-Kari case). Though discussed in some detail in the chapters that follow (among several other landmark cases new and old), this well publicized case has affected ethics in biomedical research since post-doctoral fellow Margot O'Toole, working in her laboratory at Massachusetts Institute of Technology (MIT) of junior faculty member Thereza Imanishi-Kari[9], reported to her department superiors that she believed Imanishi-Kari had fabricated data that supported a seminal paper published in the journal *Cell* in 1986. Though O'Toole was certainly admired by some for exposing the misconduct, she was vilified by others. Imanishi-Kari, finding out what O'Toole had done, dismissed her from the laboratory. Since that time, O'Toole has worked in the pharmaceutical industry, raised a family, and became a consultant.

Beyond the personal effect, the Imanishi-Kari case has also been the subject of several college and university ethics courses. Notably, the University of Miami provides a ten-year timeline of the case, prefaced with this cautionary: "Civilization expects a lot from scientists, perhaps especially from those in the biological sciences. Some expectations are straightforward. Citizens in the mature democracies spend large amounts of tax money to support research. They expect," it continues, "to sponsor work that is free of intellectual dishonesty. But the expectations are less clear when it is realized that misconduct can come in many forms, not all of them as clear-cut as falsifying data."[10] Beyond all of this, the David Baltimore Affair has provided what the Miami ethics program has called a "lattice on which to develop an analysis of key issues and controversies in scientific practice." In the responsible conduct of research, Miami students are challenged to ask: What is the difference between fraud and misconduct? What are the responsibilities of researchers and team leaders? Who should police scientific practice? What is the proper role of whistleblowers?

More recently, there was another troubling case, won but lost by the whistleblowers; it is a case that still bothers Hill. Six graduate students in the laboratory of Elizabeth Goodwin, Ph.D., an associate professor of genetics and medical genetics at the University of Wisconsin-Madison (UW-M) College of Agriculture and Life Sciences and the School of Medicine and Public Health, suspected that she had fabricated results in two grant applications. Goodwin had been awarded over $1.8 million in federal grants from the National Institutes of Health (NIH) and the United States Department of Agriculture (USDA). After being turned in by her students, Goodwin resigned her position in February 2006. While Goodwin's case was investigated by UW-M and the Office of Research Integrity, it was also investigated by the Federal Bureau of Investigation (FBI) and the United States Department of Health and Human Services (HHS)' Office of the Inspector General, with the university's full cooperation.[11]

Based on the combined investigations, Goodwin was found guilty of engaging in scientific misconduct while her research supported by National Institute of General Medical Sciences (NIGMS), National Institutes of Health grants R01 GM051836 and R01 GMO73183. The United States Public Health Service (PHS) found that she engaged in misconduct in science by falsifying and fabricating data that she also included in additional grant applications 2 Ro1 GM051836-13 and 1 R01 GM073183-01. Goodwin was also sentenced to two years of probation and a $500 fine after pleading guilty in federal court on June 25, 2010, to one count of making a false statement. United States District Court judge William Conley further demanded that Goodwin pay $50,000 in restitution to the HHS and $50,000 to a UW-M scholarship program. "She was sanctioned by the ORI and received their usual slap on the wrist," Hill opined. That "slap" boiled down to voluntary abstention from any government contract work for three years or service on any Public Health Service advisory or peer review committee during the period of suspension. Hill was struck by how similar Goodwin's background was to her own. Goodwin graduated from Smith College in 1981, and Hill in 1950. The former UW-M geneticist spent her junior year of college abroad, as did Hill, and Goodwin earned her doctoral degree from Brandeis University in 1990 and Hill in 1964. But while Goodwin seemed to land on her feet in the private sector after probation, restitution and ORI's slap on the wrist, the graduate student whistleblowers who turned her in – not so much. "My biggest worry was what if we didn't turn her in…and different grad students got stuck in our position," said Mary Allen, one of the six, at the time.[12] While the university praised the students for having done the right thing, an update to the case published in the June 28, 2010 *Science*, much of it reprising information from the magazine's article published four years before, indicates the outcome for several students was far from a happy ending. The UW-M essentially told the whistleblowers they had to restart their doctoral programs. One, Chantal Ly, had gone through seven years of graduate school and was told that much of her work was unusable and that she had to start a new project for her Ph.D. to complete the program. The reason was not so much due to falsified data introduced by Goodwin but because, so Ly and the others believed, Goodwin had stuck by questionable results. Ly and two others quit graduate school. Allen moved to a school in Colorado. Just two students chose to stay at UW-M. Reflecting on what happened, one student, Sarah LaMartina, was stung by the outcome. "Are we just stupid [to have turned in Goodwin]?" she asked. "Sure, it's the right thing to do, but right for who? Who is going to benefit from this? Nobody."[13] In this case, the students were held penalized for the sins of their principal investigator and it most of them their careers in science.

For whistleblower Robert Bauchwitz, M.D., Ph.D.[14], who filed a *qui tam*[15] in the United States District Court for the Eastern District of Pennsylvania against William K. Holloman, Ph.D., Eric B. Kmiec, Ph.D., Cornell University Medical College (Cornell University), and Thomas Jefferson University in the summer of 2004, more than a decade later, the case is still in appeal. Bauchwitz's most recent filing against Holloman *et al.* was recorded in the United States Court of Appeals for the Third Circuit (Case No. 16-cv-01669) on March 25, 2016, pertaining to statutory actions. Central to Bauchwitz's original suit is scientific research performed by Holloman and Kmiec involving the identity of the gene, protein and activity within the cell of the fungus *Ustilago Maydis* (*U. maydis*), research that represented a promising gene-therapy breakthrough for the direct correction of debilitating genetic defects and which was subsequently given the catchy name chimeraplasty. Great promise would eventually be followed by incredible disappointment.

Holloman was a scientific researcher in Cornell University's Department of Microbiology when Bauchwitz filed his original case. Kmiec had collaborated with Holloman as a graduate student from 1979 to 1984 at the University of Florida, where he worked under Holloman's supervision, and as a scientific researcher in Thomas Jefferson University's Department of Pharmacology and at the Kimmel Cancer Center, where he assisted Holloman on research partly funded by NIH grants from 1991 to 1999. When he was a student in the early 1980s, Kmiec and Holloman coauthored several articles in which they claimed to be the first to isolate a protein associated with nucleated cells from the eukaryotic fungus *U. Maydis*. They called the protein Rec1 and claimed that the Rec1 protein was derived from REC1 gene and had activity that allowed for the specific recombination of strands of DNA, called recombinase. In the DNA research field, this was a significant revelation.[16]

Bauchwitz, according to the original court filing, worked in Holloman's laboratory from 1987 to 1990 as a graduate student. During that period, he and his colleagues at Holloman's lab isolated the genes for REC1 and REC2 and obtained their sequences. The results were that the REC1 DNA sequence would not specify the protein elements common to known DNA recombinases, but the REC2 gene had a sequence consistent with a recombinase. After Kmiec left Holloman's lab, they continued to collaborate and coauthor articles. One of those articles, "The REC2 gene encodes the homologous pairing protein of *Ustilago Maydis* ("the 1994 article"), was published in the November 1994 issue of *Molecular and Cellular Biology*. The authors claimed that the protein they had isolated from *Ustilago Maydis* was a Rec2 protein, not a Rec1 protein, which was derived from the REC2 gene.

After reading the 1994 article, sometime around November 28, 1994, Bauchwitz had suspicions about the reported findings. He had already been dubious of Holloman and Kmiec's research. His doubts could be dated to when he had worked in Holloman's lab in the late 1980s. At that time, Bauchwitz suspected, for several reasons, that Holloman was engaging in scientific misconduct. First, no one else in Holloman's lab was able to replicate Kmiec and Holloman's findings that the Rec1 protein was derived from the REC1 gene and had recombinase activity. Second, when Holloman was having difficulty obtaining approval of grant applications, another graduate student in his lab was able to achieve results that were contradicted by prior results in other labs as well as Bauchwitz's initial findings, which enabled Holloman to obtain funding. Third, cited in the filing, Bauchwitz's own sequencing results obtained in the late 1980s on the REC1 and REC2 genes were inconsistent with Holloman and Kmiec's earlier findings linking the Rec1 protein activity to the REC1 gene, and instead supported a link between the Rec1 protein and the REC2 gene.

Bauchwitz's suspicions were, in the words of the filing, "so profound" that after he left the lab, he continued to follow Holloman's work. In late 1990, he informed Dr. Kenneth Berns[17], then chairman of the Cornell University Graduate School of Medical Sciences' Department of Microbiology, that, over his objections, Holloman had removed unfavorable data from the manuscript that they had coauthored, and that Holloman had forged Bauchwitz's initials indicating that he had approved the manuscript. Further, in 1990, he urged the Office of Scientific Integrity (OSI) to investigate the accuracy of Holloman's reported findings regarding the Rec1 protein activity. After reading the 1994 article, in which the individual defendants claimed that the protein they had isolated from *U. maydis* was derived from the REC2, not REC1 gene, Bauchwitz suspected, even more strongly than he had previously, that Holloman and Kmiec had falsified their findings. Bauchwitz believed that the irreproducibility of the Rec1 protein research in the 1980s, as well as his own sequencing results, which were inconsistent with Holloman and Kmiec's findings linking the Rec1 protein activity to the REC1 gene, motivated Holloman and Kmiec to find a way to transform the disputed Rec1 protein activity into Rec2 protein activity. Bauchwitz thus alleged that the 1994 article contained two false statements to support the authors' new "Rec1 to Rec2" theory. Between December 1994 and February 1995, he pursued his own investigation by contacting current and former colleagues of Holloman, and current and former graduate students who worked in Holloman's lab.

As part of Bauchwitz's investigation, he contacted the Office of Research Integrity (ORI)[18] on February 6, 1995, to learn the status of the government's investigation of the defendants instigated by his first call to OSI in 1990, and to give ORI additional information based on his December 1994 telephone call with Brian Rubin, a graduate student who succeeded Bauchwitz at Holloman's lab. The parties do not agree whether Bauchwitz identified himself on this call and whether he provided ORI with any details regarding alleged fraud at Holloman's laboratory. Bauchwitz's own research, which began in 1991 and completed in 1999, failed to replicate Kmiec and Holloman's results. He concluded that the "Rec1 to Rec2" *U. maydis* research was false and the defendants used it to obtain NIH grants. But he argued that he did not reach this conclusion until 2002, after he read another article authored by Holloman, cited in the filing as "the 2001 article," that he believed falsified the rec2-1 mutant gene sequence to appear capable of producing protein. Bauchwitz maintains that he did not learn that the defendants obtained any of the specific grant money at issue in the case until September 23 and November 8, 2002, when he received responses to his Freedom of Information Act (FOIA) requests that his counsel had submitted on June 10, 2002.

Bauchwitz alleged that grant applications, progress reports and financial status reports (FSRs) submitted to NIH by the defendants between October 31, 1991, and September 19, 2006, contained false or misleading statements pertaining to three categories of false statements. Specifically, he accused the defendants of (1) fabrication and/or falsification of the amino acid sequence of the Rec1 protein; (2) falsification of the DNA sequence of the rec2-1 mutant gene; and (3) falsification of data images relating to Rec2 protein activity. With respect to all defendants, he contended that the false statements were published in, or based in substantial part, on conclusions or findings reported in the 1994 article. With respect to the Cornell defendants only, he contends that grant 2 RO1 GM42482-12A2 contains false statements by citing the 2001 article. These false statements, according to Bauchwitz, were material to the NIH grants at issue in the case.

From 2000 through 2002, Bauchwitz assisted science journalist Gary Taubes with an article on various phases of concern in the scientific community over the work of Holloman and his former graduate student Kmiec.[19] Taubes then recommended to the federal Office of Research Integrity (ORI) that they contact Bauchwitz regarding additional potential research misconduct associated with Holloman and Kmiec. ORI told Bauchwitz that they had already been investigating Kmiec's "chimeraplasty" claims, which had received international scrutiny from other scientists.

Bauchwitz's original *qui tam* died in a sea of procedural motions, orders and judicial rulings before any evidence of possible wrongdoing was presented to a jury. Bauchwitz had filed his original complaint under seal on June 30, 2004, in the federal district court in the Eastern District of Pennsylvania, which he would later note was the location recommended by the ORI, and after the ORI agreed to proceed with an attempt to recover government funds against the defendants if he would act as the relator. The government investigated the case while the complaint remained under seal. At the request of the United States Attorney's Office, ORI conducted a scientific review of the allegations set forth in Bauchwitz's complaint. The ORI was to produce a report for the Department of Justice based on the science called into question. Because the research at issue had taken place so many years earlier and because it did not view the statements at issue as intentionally false, ORI concluded that it did "not believe that evidence is available" to prove that any of the three claims alleged by Bauchwitz are true. But in its summary ORI stated:

"Dr. Bauchwitz' complaint identifies three false claims, as identified above. *ORI notes that these false claims deal with only a very small portion of the much larger scope of possible misconduct issues that have been linked to Drs. Kmiec and Holloman.* The reason for this is that Dr. Bauchwitz has limited his claims to issues that he has direct knowledge of. He has made a solid case that the 'story' on *Ustilago maydis* recombination genes, their associated proteins and their enzymatic properties has shifted dramatically over the past 20 years. Many scientists working in this area appear to have believed that erroneous claims have been consistently published by Drs. Holloman and Kmiec. [Bauchwitz emphasis added.][20]

On August 31, 2005, after its fourth request for an extension was denied, the government elected not to intervene. On April 19, 2006, because Bauchwitz did not prosecute the action for over seven months after the government had declined to intervene, the action was dismissed in accordance with legal precedent. The case ultimately settled without going to trial in significant part because the judge reduced the time for discovery to less than the parties had agreed and then rejected a motion to extend it. The volume, complexity, and poor readability of evidence of one of the three claims against the defendants would later take as long to assess as the total, reduced, time which had been provided for discovery and expert reports, observed Bauchwitz.[21] There were no rulings on any of the evidence obtained during the limited discovery which was carried out. There were no protective orders, or any confidentiality agreements made during settlement.

Among the most clear-cut pieces of evidence obtained by subpoena from Harvard University's microchemistry laboratory pertained to amino acid sequencing work that the defendants had claimed was performed by that laboratory and which was foundational to the claims that the defendants had made in scientific publications and grants. Bauchwitz pulled these threads from that report (emphasis added):

> "*These sequences are not consistent with the data we provided.*"

"... none of the sequence data we obtained agrees with the data they claimed was from our lab."

"I am confident that there is no other data."[22]

Although the judge did not permit what was believed to be sufficient time to complete discovery and production of expert reports, nevertheless, after the case, two experts, who had earlier presented comments to the court, reviewed the evidence for two of the three claims, including the one involving the statements by Harvard's Microchemistry Laboratory, reported Bauchwitz via his research site (emphasis added). They concluded:

a. Based on the evidence and the standards of intent provided in this document, *do the allegations in your judgment have merit?* In other words, *does the preponderance of the evidence suggest that the defendants have fabricated or falsified scientific claims?*

Expert 1: Yes.
Expert 2: Clearly the data presented includes a demonstration of data fabrication and falsification of scientific claims. I would suggest that there is a dangerous mix going on here: Out and out fraud together with ignorance of the truth and selective use of facts for the expressed purpose of substantiating a story(-ies) that allowed the perpetrators to secure tangible assets (e.g., grant funding) as well intangible assets (e.g., standing in the scientific community).

b. If you were a member of an NIH scientific review committee (study section) and you were made aware of any of the information presented here as a grant reviewer, *would it have had a significant negative impact on your scoring of the grants* at issue from the applicants/defendants, H (Holloman) and K (Kmiec)?

Expert 1: Yes.
Expert 2: YES. This level of evidence would sway me not to even score such a grant in the current funding climate. That is, the level

of apparent malfeasance and/or open questions regarding the data and prospective conclusion would immediately disqualify this grant for further review.

A third expert reviewer then concluded with respect to all the evidence obtained:

> In every instance the evidence is strong and, in many instances, it is airtight. The fact that there has been no serious investigation to date shows major problems in the system.

Subsequent to conclusion of the case, which ended without any real judgment on the merits, attempts were made to obtain corrections of the scientific literature based on evidence obtained during discovery from Harvard University, according to Bauchwitz; however, the responsible journal, *Molecular and Cellular Biology*, and its parent organization, the American Society for Microbiology, refused to take any action to have the responsible institution investigate or do so themselves, as would have been appropriate under the Committee on Publication Ethics (COPE) guidelines they claimed to follow.[23]

"The responsible former defendant institution, Weill Cornell Medical College, continued its longstanding resistance to initiating a formal inquiry, the first step of record towards an investigation, and if appropriate, retraction of the affected publications," writes Bauchwitz. "Indeed, Cornell has stated that it had *never* initiated a faculty inquiry or investigation on any allegation related to the senior investigator involved in the case."[24] The importance of that remarkably common claim as it relates to the handling of allegation of biomedical research misconduct in the United States has been addressed in a paper[25] coauthored by Bauchwitz, in which he writes that nearly ninety percent of allegations of biomedical research misconduct in the United States are dismissed by responsible institutions without any faculty assessment or auditable record. "Recently, members of the U.S. Congress have complained that the penalties for those against whom findings of research misconduct are made are too light and that too few grant funds associated with research misconduct have been recovered for use by other researchers and taxpayers." Furthermore, the responsible oversight and funding agencies in the United States, primarily the ORI and NIH, also took no known, meaningful action based on the information discovered despite there being no basis in law which would have prevented their doing so.[26]

Bauchwitz has iterated that the all-too-frequent gulf between what the institutions and journals have been claiming about an interest in scientific integrity and their actual behavior does not appear to be specific to this case, or even to science in the United States. The severity and widespread nature of an unwillingness to deal effectively with scientific misconduct has been presented repeatedly by other scientists. Among the most recent of these, he continued, is a report that alleged:

> A secret dossier that warns that fraud in biomedical research is even more prolific than feared is being considered by Jo Johnson, the [UK] universities and sciences minister, documents passed to *Times Higher Education*…conclude that some research institutes, university administrators, funders, journals and science leaders have been covering up malpractice.[27]

The *Times* exposé further warns that "the past three decades have seen an 'alarming' increase in paper retractions, mainly due to misconduct. It catalogues a series of high-profile misconduct scandals involving senior scientists in the [United Kingdom] and abroad." The dossier also points out that although the number of retracted papers is small compared with the huge number published, only a small proportion of articles are genuinely scrutinized; it calculates that of papers that are closely checked, "as many as one in 20 contain errors or falsifications." But there is another reason, reveals the *Times*, that the problem may be larger than thought: "scientists and journals are extremely reluctant to retract their papers, even in the face of damning evidence," meaning that misconduct may go completely unreported.

Back to Bauchwitz's case. On April 19, 2007, the dismissal was vacated on Bauchwitz's motion. An amended complaint was filed on May 16, 2007, adding claims based on allegedly false statements made in connection with a third federal grant submitted by the Thomas Jefferson defendants. It also "identified...the specific statements made by each...defendant that the Relator [Bauchwitz, the whistleblower] contends were false or fraudulent, as well as the specific grant applications and progress reports in which such false or fraudulent statements were made." Though this case will be examined more fully in the narrative that follows, Bauchwitz's reporting exemplifies the legal maze that whistleblowers navigate to get their cases before a court of law. But even if they succeed in getting evidence before a jury, there is no guarantee – and a very poor judicial track record proves it – of courts adjudicating scientific misconduct and, most especially, those predicated on the False Claims Act[28] (FCA) 31 U.S.C. § 3729. In Bauchwitz's case, to be clear, no court ever decided if the defendants did or did not submit grant applications containing false scientific information.[29] Settlement of the case, according to Bauchwitz's counsel, was not decided because of evidence submitted by either side to the court.

"You filed a *qui tam* action under the False Claims Act, seeking to hold Cornell University, and certain defendants employed by it, and Thomas Jefferson University and certain defendants employed by it, responsible for having falsified information in grant applications submitted to the United States government," they wrote on November 13, 2013. "On December 1, 2009, Judge Timothy Savage granted summary judgment on all claims against the Jefferson defendants for failing to file them within the required statute of limitations period.[30] In that opinion, Judge Savage further ruled that all of the claims against the Cornell defendants, except those relating to grant 2 R01 GM 42482-12A2 were dismissed for failing to file them within the required statute of limitations period. The dismissal for failing to file claims within the required statute of limitations is not, to be clear, a dismissal on the merits, and does not resolve the issue of whether the claims were or were not meritorious. The judge's ruling was predicated solely on the basis of the statute of limitations. The docket of the district court shows that on December 16, 2009, the court issued an order requiring that all discovery be completed by April 9, 2010. This, in the opinion of Bauchwitz's legal counsel, was an inadequate amount of time to make the necessary discovery to back his claims. Subsequent motions to extend the discovery were denied by the court. On April 1, 2010, the judge approved a stipulation of the parties that called for a voluntary dismissal pursuant to

Local Rule 41.1(b) and issued an order dismissing the case with prejudice; the same claims by the United States government were also dismissed with prejudice. By operation of the law, a dismissal with prejudice is a technical adjudication on the merits even if the underlying claims were never submitted before the court for determination by a judge or jury. Such a finding of the court is generally used to prevent the plaintiff from filing a new complaint for the same claims. Further, the court was never asked to review evidence for the purpose of determining whether or not the grant applications did or did not contain false information; it was never judicially determined by the court if the grant applications did or did not contain false information.

Thus far, Bauchwitz's case against Holloman and Kmiec has had little impact on either defendant scientists' careers. Holloman is a professor of microbiology and immunology at the Weill Cornell Medical College, and Kmiec is the chairman of the chemistry department at Delaware State University. The NIH has provided millions of dollars in grant funding to Kmiec since 1999, and Holloman since 1991. As for Bauchwitz, he went to Widener University Delaware Law School, where he received a certificate from the school's Legal Education Institute (LEI), a program centered on the theory and philosophy of law and ethical responsibility. He is the director of research and development at Amerandus Research (AmR), which he founded in 2011. Bauchwitz's motivation in forming Amerandus Research, from his website, "was to expand his academic research into the area of handling research misconduct, and to provide practical assistance to others who were involved in conducting scientific fraud investigations." These new goals grew out of his own extensive experiences with the *United States v. Holloman et al. qui tam*, as well as the increasing realization by the scientific community of the significant financial and social harm from such misconduct and fraud.

Kenneth James Jones[31], Ed.D., a statistician and now professor emeritus at Brandeis University, filed a *qui tam* action against the defendants Brigham and Women's Hospital, Massachusetts General Hospital, Marilyn S. Albert, Ph.D., and Ronald J. Killiany, Ph.D., collectively the defendants, under the False Claims Act on June 14, 2006, in the United States District Court for the District of Massachusetts. The case first arose out of alleged false statements contained in a program project grant application to the

National Institute on Aging (NIA), an organization under the National Institutes of Health. Jones, the *ex-rel.* (relator) in the federal government's case, alleged that certain statements contained in the application were false because they had been derived from falsified data. Jones claimed that the defendants falsely certified that the application was in compliance with all relevant statutes and regulations. Thus, he contended that the defendants caused the government to fund the grant in violation of the False Claims Act by asserting false statements in the application and falsely certifying compliance with relevant statutes.[32] The background of the case is a matter of court record, but the particulars focused on the aforementioned multi-million-dollar, multi-year grant, led by Albert. At the time of the alleged violations, research was being done into early detection of Alzheimer's through longitudinal studies of certain regions of the brain. The research aimed to characterize the early phase of Alzheimer's disease and to differentiate it from changes related to normal aging, thus enabling prediction of who will develop Alzheimer's years before the individual displays diagnosable dementia. Albert's work was considered just then as work in the fight against one of the world's most wretched diseases, a neurodegenerative illness associated with aging that is without a good early detection method or a cure.

The research at issue, documented in court proceedings, was conducted under a grant entitled "Age-related changes of cognition in health disease," which the NIA and the NIH first funded in 1980 and continued to fund through 2007, according to Albert's deposition. The defendants, Albert and Killiany, along with Jones, were part of a team of scientists working for several decades on the grant, which consisted of four projects and four cores. The projects were interrelated and comprised: neuropsychological assessment (Project 1), single photon emission computed tomography (SPECT) (Project 2), structural magnetic resonance imaging (MRI) (Project 3), and functional magnetic resonance imaging (fMRI) (Project 4). The cores provided support to the projects and included: the administrative and clinical core (Core A), the data management and statistical core (Core B), the genetics core (Core C), and the neuropathology core (Core D).

There is a grant process that all must follow to receive federal money to continue their research. The NIA systematically reviews submitted applications. Institutions seeking funding are required to submit applications to the Center for Scientific Review and the NIH; the applications on the Albert projects were, per the requirement, then submitted to the NIA for funding consideration. The NIH grants policy statement[33] shows the initial level of review is a peer-review conducted by a committee of experts to assess several factors. Some of the factors include the significance of the proposed study, the approach taken, innovation, whether the investigator is appropriately trained, and whether the environment where research will be conducted will likely contribute to the probability of success. For the purpose of the case filed against the defendants in this case, NIH's grants policy statement dated March 2001 applied. After the scientific review administrator prepares a summary statement (called the "pink sheets") with peer reviewers' comments, including a précis of the strengths and weaknesses of the proposed project and a priority score, a decision is made to move forward or reject the proposal. If recommended for further consideration, the pink sheets, in this case, were presented to the National Advisory Council on Aging for a second level of review. The NIA director has the authority to approve payment of applications reviewed favorably where primary weight is given to the perceived scientific quality of the application.

Albert served as the principal investigator and program director of the grant in question. Jones was the core leader of Core B, the data management and statistical core. As the leader of Core B, his responsibilities included supervising data management, reviewing project progress, carrying out complex analyses pertaining to individual projects, and developing new analytic approaches for the data set. Mary Hyde, Ph.D., worked with Jones as the data manager and programmer for Core B. Hyde's responsibilities included communicating with project leaders, assisting project leaders with data entry, and reviewing the contents of data sets for accuracy and completeness. According to court documentation, Project 3, the structural MRI study, involved analysis of MRI images of certain regions of the brain. In the mid-1990s, with the advent of advanced MRI techniques, a potential method of early detection of Alzheimer's evolved. Measurements of the volumes of certain regions of interest (ROIs) have been shown to be especially useful indicators. Two regions of the brain, the entorhinal cortex and the hippocampus have been shown to be indicators of the atrophy associated with Alzheimer's. While MRI is a useful marker for degenerative changes in the brain, alone it is not a diagnostic for clinical Alzheimer's.

In Project 3, participants were observed for a period of years to track the progression of cognitive development in the prodromal phases of Alzheimer's. Participants in the longitudinal study using MRI data were divided into three groups on the basis of their group status after several years of follow-up: Controls (subjects who remained constant for three follow-up evaluations); questionables (non-demented subjects with memory problems who did not progress to Alzheimer's); and converters (non-demented subjects with memory problems who progressed and eventually were diagnosed with probable Alzheimer's).

Ronald J. Killiany[34], Ph.D. was the project leader of Project 3, the structural MRI project and in this capacity, he was responsible for using MRI scans to trace the boundaries of certain regions of the brain that interested the scientists; that is, he was the primary neuroanatomist tasked with tracing the boundaries of the entorhinal cortex and subsequently sending volumetric data to Hyde in the statistical core. The ultimate objective of Project 3 was to determine whether structural MRI data could be used to predict which non-demented subjects with memory problems would decline into Alzheimer's. The manual outlining of the boundaries of various regions of interest was done using a computer, a track-ball driven mouse, and a software program called Neuroview, information gleaned from Killiany's first set of interrogatories and included in the public record of the proceedings. To trace the boundaries of the entorhinal cortex using the protocol developed by Killiany and other members of the grant, the operator would begin the outline of the region at the angle formed by the junction of the rhinal sulcus and the surface of the brain. The operator would then transect this angle, cutting across the gray matter to the level of the white matter. Next, the operator would follow the edge of the white matter to the inferior surface of the hippocampus. Finally, to complete the outline, the operator would trace the surface of the brain back to the starting point.[35] The minutiae of this case matters.

Jones alleged that Killiany falsified data pertaining to manually drawn boundaries of the entorhinal cortex, producing a second set of data in which the volumes of twelve subjects in the "normal" grouping were enlarged to make data statistically significant. Jones claimed to have learned of this supposed second set of data through discussions with Keith A. Johnson[36], M.D., leader of Project 2. This data, Jones contended, enabled the grant to claim that the entorhinal cortex was a region that could be used to predict conversion to Alzheimer's. Jones additionally claimed that he compared Killiany's first and second sets of data and performed an analysis which showed that the changes Killiany made were responsible for the statistical significance of the reported results. Jones reported his concern over a discrepancy in Killiany's two sets of data to Albert, the principal investigator and overall lead. Albert initiated an inquiry. Mark Moss[37], Ph.D., an eminent neuroanatomist, evaluated both sets of data from twenty-three of Killiany's measurements to determine their accuracy; the cases Moss examined were picked by Jones.

In total, Jones alleged four ways in which the defendants violated the False Claims Act. First, he claimed that the results generated by Killiany's altered MRI data served as the centerpiece of the grant application, resulting in false representations to the NIA. To support this allegation, Jones cited several statements made in the application. According to information from the court's November 10, 2010 memorandum, Jones found that the defendants stated they "achieved a 'major finding' that measures of the [entorhinal cortex] were 'highly predictive' for the course of prodromal AD [Alzheimer's disease]." The application states:

> Our major finding is that measures of memory and executive function, or SPECT and MRI measures of brain regions related to these domains (such as the entorhinal cortex, the hippocampus, and the caudal portion of the anterior cingulate) are highly predictive of subsequent development of dementia among non-demented individuals with memory problems.

But the court memorandum also indicated in the notes section that throughout Jones's complaint, the alleged false statements are never clearly established. Despite claims sounding in fraud, he never articulated the false statements in the complaint. In order to locate the alleged false statements, the court has had to comb through Jones's pleadings and motions. This complicated Jones's case from the start.

Jones also claimed that the defendants "stated prediction of conversion to [Alzheimer's] as one of the primary findings of the MRI data, with entorhinal cortex studies proving to be the most 'discriminating measurements.'"[38] The application, however, read: "The most discriminating MRI measures pertain to atrophy of the medial temporal lobe (particularly the entorhinal cortex), and the volume of anterior and posterior cingulate."[39] Further, Jones claimed that the defendants wrote that they had "identified a selected group of brain regions, primarily the entorhinal cortex, which paralleled the neuropsychological changes during preclinical [Alzheimer's]." But, again, that is not what the application stated when filed for federal agency review:

> [A] selected group of brain regions develop neuropathology during preclinical [Alzheimer's] which, in turn, influence the cognitive deficits of the individuals. Based on combined analyses of the data it appears that problems within a memory circuit (involving the entorhinal cortex and the hippocampus) are essential but not sufficient for a diagnosis of [Alzheimer's]...[40]

These statements together account for the allegedly false statements, predicated on "falsified" data, which comprised Jones's first claim.

Jones also alleged, in the second part of the complaint, that the defendants violated the False Claims Act by falsely stating that Killiany followed blinded methodologies when manually tracing the entorhinal cortex. The application states in relevant part: "In order to prevent possible bias in the drawing of the manually drawn regions, all operators are blinded to the groupings of the subjects." From Jones's deposition, he admitted that he had no evidence that Killiany had not followed proper, blinded methodologies when retracing entorhinal cortex boundaries.[41] The third basis for Jones's claim was that the defendants violated the Act by making false representations that they had conducted a reliability study on the underlying data. Section D 4.3 of the application stated, "The procedures in place for generating the manually drawn image maps have been demonstrated to have high reliability. Inter-rater reliability for these ROIs ranges between $r=0.94-0.99$." Jones contended that the reliability study represented in the application was based on the first set of data, not the second, "falsified" set of data.[42] Contemporaneous emails show that Killiany submitted the second set of data to Hyde for reliability testing, from the affidavit of counsel Lisa A. Tenerowicz, as cited in the memorandum. In addition, Killiany testified that he had no knowledge of the statistical significance of the MRI data nor what happened to the entorhinal cortex tracings once the measurements were emailed to Hyde.

From November 10, 2010, it is known that Jones also raised a fourth novel claim in his motion for summary judgment that was not originally pled in his second amended complaint. In the motion, he alleged that the defendants violated the Act by both expressly and impliedly certifying compliance with the "Responsibilities of Awardee and Applicant Institutions for Dealing with and Reporting Possible Misconduct in Science" ("Responsibilities of Applicants"), 42 C.F.R. (Code of Federal Regulations) Part 50, Subpart A (2001) (replaced by 42 C.F.R. Part 93). Additionally, he filed a motion for sanctions for spoliation of evidence on September 28, 2010, asserting yet another theory of liability for violating 45 C.F.R. § 74.53, a separate regulation regarding post-award requirements.

Jones was losing his *qui tam*. Despite this, he insisted that if the court did not hold the defendants in violation of the Act under an express certification theory, then the defendants must be found liable under a theory of implied certification for falsely certifying compliance with the responsibilities of applicants for the federal grant. Allegations regarding the responsibilities of applicants regulation were raised for the first time in Jones's summary judgment motion. While the second amended complaint referenced an express certification theory for false statements, no mention of an implied certification theory of liability was made until Jones filed his motion for summary judgment on September 15, 2010. The claim under an implied certification theory is based on allegations that the defendants did not follow an investigating procedure under the responsibilities of applicants, whereas the original claims in the second amended complaint were based on allegations of falsified data and submissions of false statements predicated on that data. While related, the factual record for these new claims would be substantially different.[43]

Jones made new allegations in 2010 regarding conduct that may have occurred in 2001. As the relator, his original complaint was filed on

June 14, 2006, just within the six-year statute of limitations period for a claim under the False Claims Act. 31 U.S.C. § 3731(b)(1). To allow him to effectively amend his complaint almost four years later to include an allegation dependent on different evidence would be unduly prejudicial to the defendants – and there was ample precedent to support this ruling. Moreover, even if the court allowed this claim to go forward, it would not survive summary judgment. In *Hutcheson v. Blackstone Medical, Inc.*, the court observed that a claim may be legally false when "a claimant makes no express statement about compliance with a statute or regulation, but by submitting a claim for payment implies that it has complied with any preconditions to payment." Arguably, as recipients of NIH funding in the past, from Albert's deposition, the defendants should have been aware of any regulations with which they were required to comply as a prerequisite for payment of funds. Per *Hutcheson's* requirement that the regulation explicitly state compliance as a precondition for payment, this regulation states: "An institution's failure to comply with its assurances and requirements of this subpart may result in enforcement action against the institution, including loss of funding..." 42 C.F.R. § 50.105 (2001).

According to the regulation, once scientific misconduct is suspected or alleged, an applicant institution must take "immediate and appropriate action" in accordance with 42 C.F.R. § 50.103(c)(3) (2001). An inquiry into any allegation of scientific misconduct must be completed within sixty calendar days; "[a] written report shall be prepared that states what evidence was reviewed, summarizes relevant interviews, and includes the conclusions of the inquiry per 42 C.F.R. § 50.103(d)(1). The regulation further stipulates that the director of the Office of Scientific Integrity (OSI) shall be notified only when "on the basis of the initial inquiry, the institution determines that an investigation is warranted" following 42 C.F.R. § 50.103(d)(4); *see also id.* § 50.104(a)(1). Finally, the regulation demands secure maintenance of sufficiently detailed documentation of inquiries for at least three years after the inquiry's termination per *id.* § 50.103(d)(6).

According to Jones, he first made Albert aware of his concerns about the data on March 15, 2001, from Jones's declaration before the court. Notably, he did not accuse Killiany of scientific misconduct; rather, he voiced concern that there was a discrepancy between the two sets of data that he could not resolve and that it was a serious matter requiring action, which is known per his deposition. At that point, Jones suggested Albert secure an independent evaluation of the circumstances leading up to the second set of data. He then provided Albert with a list of the aforementioned twenty-three cases that he believed to be suspect. Albert engaged Moss to reevaluate the data. Thus, it is undisputed that an inquiry was performed regarding the discrepancy between the two sets of data. Jones introduced his own and an expert declaration suggesting that the inquiry was inadequate because it was not done by an independent evaluator. Neither declaration identifies any part of the regulation that requires the inquiry be done by an outside, independent evaluator. That he was not satisfied with the person chosen by Albert, as the principal investigator, to conduct the inquiry was of no consequence to the court because it is not supported by a requirement in the regulation. The only evidence available regarding the next steps taken was testimonial; documentary evidence no longer existed. Dr. Johnson was satisfied with the results of the inquiry, according to his deposition in the court record. Albert, also satisfied with Moss' evaluation, considered the issue resolved because it was no longer discussed. Jones did not produce any objective evidence which would suggest that a formal investigation was warranted beyond the initial inquiry. The record does not contain sufficient evidence to show that the defendants failed to comply with the responsibilities of applicants; therefore, the claim did not survive summary judgment. Jones's first federal court fight was effectively over because a *qui tam* case requires proof of false or fraudulent claims or statements to the government and that proof must be clear and supported by ample evidence to succeed. The defendants argued that a dispute over scientific methods did not meet the letter of the law. Jones could not demonstrate that the hospitals – Brigham and Women's Hospital and Massachusetts General Hospital – knowingly presented a false or fraudulent claim for payment or approval to the federal government nor that the defendants knowingly made or used a false record, not even a statement, to compel a false or fraudulent claim to be paid or approved via the NIH. Jones did also not prove to the court's satisfaction that any allegedly false record or statement the defendants made to the NIH was material to its decision to fund the defendants' project that focused on Alzheimer's disease research. Further, there was no evidence that Killiany was not blinded to the diagnostic group status of the study participants in the Alzheimer's study in question. A close reading of the trial's July 2009 hearing suggested quite clearly that had the court been aware that Jones was unable to prove what his counsel represented he could prove, "the court would have dismissed the case and denied leave to amend."[44] Taking these statements at face value, Jones's case was done. But *qui tam* cases hardly die quiet deaths.

Despite the court's dismissal, Jones and his wife, Priscilla Pitt Jones, weren't done with the defendants, including Harvard University[45], which had been removed from the original *qui tam* case in March 2009. On May 7, 2012, Jones won an appeal of the case in the United States Court of Appeals for the First Circuit. "After careful review of the record, we conclude that the district court abused its discretion by excluding or failing to consider certain expert testimony," according to the First Circuit opinion. "It then committed an error of law by failing to consider statements of the parties and experts in a manner required by the summary judgment standard. When properly considered, those statements generate genuine issues of material fact concerning some of the relator's [Jones's] FCA claims." In vacating a summary judgment from the federal district court, the appeals court held that Jones presented sufficient evidence of fraud to require a jury trial. Jones renewed his claims that Albert and Killiany falsified brain scans in the Alzheimer's disease study to support continued federal funding of their research. The Albert *et al.* project, at $12 million, was one of the largest Alzheimer's disease research grants awarded by the National Institutes of Health. After the study failed to show positive results, Killiany retraced certain areas of selected brain scans to get the positive results the project team had sought. Further, they reported that their data exceeded reliability measures without disclosing that those measures applied to the negative results, not the positive results that came from the retracings. Ultimately, Jones would argue that the defendants made a false certification that they complied with the regulations governing investigations into scientific misconduct (42 C.F.R. § 50.103(c)(3)). The published opinion of the First Circuit thus concludes that "the essential dispute is about whether Killiany falsified scientific data by intentionally exaggerating the re-measurements of the EC [entorhinal cortex] to cause proof of a particular scientific hypothesis to emerge from the data, and whether statements made in the application about having used blinded, reliable methods to produce these results were true."

Allowing the appeal was, in the opinion of Michael D. Kohn, one of Jones's lead attorneys and the president of the National Whistleblowers Center, "a major breakthrough holding universities accountable for the integrity of reported research results. Fraud committed to obtaining NIH funding not only robs taxpayers but also sets back long-term medical research goals. The facts of this case indicate the report of false data misdirected research efforts at other institutions."[46] Kohn would also observe that "[...] bullying and blacklisting scientists by the likes of Harvard to cover-up research fraud represents a terrible disservice to society as a whole." Jones's case thereafter became the first lawsuit alleging scientific fraud to be allowed to progress to trial under the False Claims Act. This was a vindication for Jones, whose prior case had been dismissed in the lower court just three days before it was headed to trial in Boston. The stakes were enormous for all parties named in the appeal. If guilty, the defendants were more than aware that they could end up paying more than $45 million to the federal government and, by law, whistleblower Jones stood to receive fifteen to thirty percent of recovered funds as the relator of the *qui tam*.

A trial now afforded the plaintiff and defendants to put expert testimony in front of a jury. The defense had previously elicited the help of Andrew J. Saykin, Psy.D., a professor of radiology at the Indiana University School of Medicine and director of the Indiana University Center for Neuroimaging, with a research focus on "the integration of neuroimaging and genomic data with emphasis on early detection of Alzheimer's disease."[47] In his report to the court[48], Saykin stated that outlining "ROIs on brain scans inevitably involves a learning curve for the neuroanatomic rater and some degree of trial and error, especially as methods are refined and solidified." Saykin did "not believe that it was unusual or inappropriate for Killiany to perform re-measurements to improve his accuracy of measurement, as long as he remained blinded to the clinical status of the participants." Saykin stated that re-measurements like those that Killiany made are normal and need not be reported "unless the re-measurements were undertaken as part of a formal reliability assessment after a change in the protocol for measuring the EC [entorhinal cortex]." He also stated that he believed that "Killiany took great pains to refine and update his measurements, following the protocol or procedural guidelines he had developed, in an attempt to make the measurements more accurate." Saykin concluded that Killiany was blinded at the time of the re-measurements.

What about the allegation of misconduct? In Saykin's opinion there was none. He stated that it was "appropriate for [Albert] to ask a prominent neuroanatomist to review Killiany's measurements and provide his analysis demonstrating whether the re-measurements were done in an attempt at greater accuracy or simply to prove the hypothesis. Since Albert concluded the former had occurred, there was no need to report what had happened to anyone. Saykin stated that if, in fact, "Killiany's first set of measurements did not demonstrate the predictive value of the volume of the EC, those measurements were inconsistent with the findings from many other scientific studies," as "[i]t is now scientifically accepted that the volume of the EC is a good predictor of conversion to AD [Alzheimer's disease]."

Not surprisingly, Jones's expert, Norbert Schuff, Ph.D., a professor of radiology at the University of California at San Francisco, an investigator at the Veterans Administration (VA) Medical Center in San Francisco, lead physicist at the Center for Imaging of Neurodegenerative Diseases, also at the San Francisco VA Medical Center, and a researcher focusing on the development of new MRI methods and concepts to identify markers of neurodegenerative diseases, including Alzheimer's disease, offered a contrasting conclusion.[49] He stated that while Killiany and the other rater, M. Teresa Gomez-Isla, M.D., Ph.D., had reached an "initial conceptual understanding [of] how to trace the [entorhinal cortex]" and had achieved "high inter-rater reliability" using that protocol for the first set of measurements. Killiany's re-measurements included "extensive" revisions, according to Schuff's report to the court, that were "frequently inconsistent with the initially adapted protocol and seemed to introduce greater rater variability." Schuff further noted that any changes to the protocol should have been discussed with Gomez-Isla, after which Gomez-Isla would have revised her tracings and the statistics core would have retested reliability. Killiany, because he failed to take these steps, led Schuff to conclude that the only "explanation for the highly selective revisions of the entorhinal cortex tracings in [the normal group]" was that such "revisions were made with knowledge of the [participants'] diagnoses." Schuff iterated that the objectivity of the entorhinal cortex measurements was material to the NIH's and the peer reviewers' assessments of the merit of the proposed project.

The district court found Schuff unqualified to "testify as to the materiality of a statement regarding the NIH review process" because he did not "list any qualifications regarding the NIH application review process or the peer editing process." To the contrary, Schuff's curriculum vitae – submitted with his report – listed four NIA/NIH grant proposals on which he was a reviewer between 2006 and 2009 and numerous other experiences as a grant and peer reviewer for other institutions as well. Schuff's curriculum vitae also stated that he had specialized knowledge and training in relevant topics such as neuroimaging and neurodegeneration, has published more than 150 peer-reviewed articles in those and related fields, and had acted as the principal investigator on several clinical trials that utilized MRI to examine ROIs including the entorhinal cortex. Considering the information in Schuff's curriculum vitae, the appellate court concluded that the district court abused its discretion by excluding Schuff's opinions regarding the materiality of the application. Thus, it treated those statements as if they had been admitted and considered them in its *de novo* review of the district court's summary judgment determination.

Jones also elicited the help Martha Isabel Dávila-Garcia, Ph.D., associate professor at Howard University College of Medicine[50], who explained the NIH application consideration process based on her knowledge as a past application reviewer. She stated, among other things, that the application contained "a number of statements that were material to the government's funding decision, including the preliminary data and progress report of the scientists' ongoing research project, as well as the reliability of the blinded methodologies the scientists claimed they followed in validating that data." Dávila-Garcia further suggested that "each of these statements was required to be made in the grant application, and each is fundamental to the peer review ranking of the application." She also stated that Albert's inquiry into the alleged misconduct was insufficient per 42 C.F.R. § 50.103(d)(8)-(9) (2001). Subsection 50.103(d) describes the various assurances that subject institutions must make to ensure that they are capable of dealing with and reporting possible scientific misconduct, including "securing necessary and appropriate expertise to carry out a thorough and authoritative evaluation of the relevant evidence in any inquiry or investigation" (§ 50.103(d)(8)), and "taking precautions against real or apparent conflicts of interest on the part of those involved in the inquiry or investigation (§ 50.103(d)(9)). This regulation has since been replaced by 42 C.F.R. § 93.100-319.

The district court had similarly excluded Dávila-Garcia's statement "that the reliability study was material to NIH's decision to fund the grant." Although the court stated that Dávila-Garcia "appears qualified to opine" on the materiality of statements in the application concerning the reliability study, it rejected her report because it found that she "[did] not support her opinion with any evidence from the record." Specifically, the district court found that although Dávila-Garcia stated that the reliability analysis was material because it was a required element of the application, it excluded her opinion because she "[did] not provide any support for [her] statement from a statute, regulation, instruction manual or personal experience […] [and did not] cite any of the reviewers' comments from the pink sheets regarding the strengths and weaknesses of the application." Further, the district court found that the record contradicted Dávila-Garcia's testimony, because the pink sheets stated that "the use of the Pearson correlation coefficients and student t-tests to assess reliability, as proposed, is inadequate." But the appellate court found that the district court abused its discretion by concluding that Dávila-Garcia did not sufficiently rely on personal experience in formulating her opinion about the materiality of the reliability study. Dávila-Garcia had personal experience as a peer reviewer and was familiar with the NIH grant application process. According to her curriculum vitae, Dávila-Garcia reviewed four articles between 1999 and 2009, and acted as a grant reviewer on six committees between 2001 and 2010, including one NIH committee. Dávila-Garcia also acted as a grant consultant on two projects, one with the NIH from 2007-2012. In her report, Dávila–García expressly stated that "each opinion [in the report] is based upon a reasonable degree of certainty, in light of [her] experience and background." She explained the peer-review process that grant applications undergo at the NIH:

> In making its ranking decision, NIH peer reviewers care-
> fully consider several factors, as reflected in the NIH grant

application. Instructions require applicants to include in the proposal a detailed research plan, describing the specific aims of the scientific project, preliminary data and progress reports on the work performed prior to the grant's submission, the procedures used and proposed to be used during the course of the project, and anticipated problems for the project with proposed solutions. These disclosures are essential to the peer-review process, as it is the responsibility of the reviewers to provide their best assessment of the chances for success and significance of that success on any given project."[51]

She then applied her experience and knowledge of the peer review process to the facts of the case as she understood them:

Within the grant proposal at issue in this case, the applicants make a number of statements that were material to the government's funding decision. Such material information undisputedly included the preliminary data and progress report of the scientists' ongoing research project, as well as the reliability of the blinded methodologies the scientists claimed they followed in validating that data. Defendants represented that reliability studies had confirmed a high correlation between and among blinded raters (measuring in ranges between 0.94 and 0.99 for Pearson r correlation coefficients). Each of these statements was required to be made in the grant application, and each is fundamental to the peer review ranking of the application.[52]

Moreover, the appellate court noted that the pink sheet statement cited by the defendants and the district court appeared in section D of the application, entitled "Research and Design Methods." Section D outlined the methodologies that would be employed in future studies to be conducted if grant funds were awarded. In contrast, reliability study results from past studies were presented in Section C, "Progress Report/Preliminary

Studies." Although the appeal's court agreed with the district court that the methodology concerns described in the pink sheets indicate that the defendants' proposed method of future reliability testing was immaterial to the reviewers' decision, that conclusion does not satisfy the materiality determination about the results of the reliability study conducted in previous research projects. The alleged falsity in this case does not rest upon the adequacy of the reliability method to be employed in the future, but rather on results allegedly already obtained in a past reliability study and relied upon by the applicants in their proposal. In sum, the evidence brought forth by Jones on the reliability issue generates an issue of fact regarding materiality, specifically, whether providing a reliability coefficient of 0.54, or stating that no reliability study was conducted on the measurements that gave rise to the scientifically significant results, would be capable of influencing the reviewers' decision.

When Jones filed the original claim in 2006, the False Claims Act imposed civil liability on any person who either "knowingly presents, or causes to be presented to an officer or employee of the United States government a false or fraudulent claim for payment or approval" per 31 U.S.C. § 3729(a)(1) or "knowingly makes, uses, or causes to be made or used, a false record or statement to get a false or fraudulent claim paid or approved by the government" *id.* § 3729(a)(2). A person acts "knowingly" if he or she "(1) had actual knowledge of the information; (2) acts in deliberate ignorance of the truth or falsity of the information; or (3) acts in reckless disregard of the truth or falsity of the information" reference *id.* § 3729(b). The relator need not show any "proof of specific intent to defraud" *id.* The court has also long held that the False Claims Act is subject to a judicially imposed requirement that the "allegedly false claim or statement be material" per *United States ex rel. Loughren v. Unum Group,* 613 F.3d 300, 307 (1ˢᵗ Cir. 2010).

The district court found that Jones had not generated genuine issues of material fact on any of the elements at issue-falsity, materiality, and knowledge. Jones argued to the contrary. He noted that the application relied on Killiany's research and claims that Killiany "fraudulently altered the MRI study data prior to 1998 to produce false results of a statistically significant correlation between conversion to Alzheimer's disease and volume of the EC [entorhinal cortex]," and did so "after the scientists had conducted a reliability study showing consistency in the manual drawings Jones claims that the alterations led to results of EC boundaries." Jones claimed that the alterations led to results that were incompatible with the results of the reliability study that was conducted and with the stated protocols. He further alleged that "Albert and the defendant hospitals submitted the subject application for NIH funding on the entire program project grant, including a separate MRI project, without disclosing to NIH the second set of MRI data, its statistical impact or the allegations and appearance of Killiany's misconduct. Jones also came back to the claim that the district court abused its discretion by not finding that the defendants had spoliated evidence, including scientific notebooks, MRI images, data files, e-mails, and any record of the internal inquiry into the allegations of scientific misconduct. Jones argued that such evidence should have been preserved under 45 C.F.R. § 74.53(b) "Post-Award Requirements, Reports and Records." Under § 74.53(b), financial records, supporting documents, statistical records, and all other records pertinent to an award are to be retained for a period of three years from the date of submission of the final expenditure report. The district court found – as the name of the regulation suggests – that the post-award requirements are forward-looking and apply only once a grant has been awarded. Thus, because the grant in question in the Jones case was not funded until 2002, the post-award requirements did not apply to records pre-dating 2002. The appellate court agreed with the district court that Jones's spoliation claimed failed on this basis. The district court also iterated that applicant institutions are required to retain documents related to misconduct inquiries for three years, a time span that terminated in 2004 in this case. Jones provided no basis to doubt this determination. Of note, he did not bring his first suit until 2006, two years too late to include critical data pertinent to his case.

The case went back to the district court for a jury trial. At the close of evidence presented in the appeal, the defendants moved for judgment as a matter of law on all claims, and Jones moved for a ruling on damages. The jury returned a verdict in favor of the defendants, and Jones subsequently filed a motion for judgment as a matter of law under Rule 50(b) and incorporated an alternative request for a new trial under Rule 59. Among other grounds, Jones contended that "undisputed, substantial evidence" existed as to each element of the False Claims Act claim. In that motion Jones also attempted to "renew [...] his Rule 56 motion for summary judgment," conceding that he made no "separate, formal written motion under Rule 50(a)." The district court denied the motion and Jones's timely appeal soon followed. But finding no reason to upset the jury's considered verdict, a panel of First Circuit judges tossed the second appeal on March 16, 2015. Judge Jeffrey Howard wrote that the jury made a reasonable decision against Jones and that the judge did not make mistakes in the handling of the trial. In this case, Jones plainly failed to preserve his Rule 50(b) arguments. Jones made no Rule 50(a) motion challenging the sufficiency of evidence to support a verdict in favor of the defendants. Rather, it is undisputed that Jones made an oral motion regarding a singular issue of damages, alone, at the close of evidence. Having reviewed the record, the appellate court found that this motion did not encompass or necessarily include an argument that Jones was entitled to judgment as a matter of law. "Jones has had the opportunity to present his claims in court before a jury," the opinion read. "That jury ultimately concluded that Killiany did not intentionally falsify scientific data and that the application's statement that the study used blinded, reliable methods was not false."

Jones tried to argue before the appeal's panel that the trial judge should have reweighed the evidence in deciding whether to grant a new trial in 2014 and that the weight of the evidence should favor him based on his uncontested expert witnesses at trial, uncontested simply because the defendants did not call their own expert witnesses during proceedings. Judge Howard was unconvinced. He wrote that the court did not have to reweigh all the evidence, and that considerable findings were gleaned from the witnesses Jones put on the stand on cross-examination. For example, Schuff testified that Killiany's "original measurements were with major error according to the protocol," that Killiany's later changes "deviated substantially from that initial protocol" and that there was "no scientific justification" to make such revisions to some, but not other, scans. Yet, the defense elicited a multitude of damaging concessions from Schuff, including that he was not a neuroanatomist; he had never attempted to employ Killiany's method to measure the entorhinal cortex; he used a different protocol in his own lab that consistently produced a much larger volume of entorhinal cortex; and he had only measured entorhinal cortex on an MRI on fifty prior occasions and, when he did so, had consistently drawn the entorhinal cortex too short and too small. The jury concluded that his testimony did nothing to counteract the defendants' theory of the case. Schuff conceded that he had no basis to determine whether Killiany's original or revised measurements were more or less accurate. Jones's case was over.

Whistleblower *qui tam* cases are growing in number, evidenced by recent trends being reported by legal and health care watchdogs. The *National Law Review* reported on April 13, 2016, that 42 health care–related *qui tam* cases had recently been unsealed in whole or in part. Included in that count was one proceeding (the "consolidated case") in which three separate actions had been consolidated into a single case. A substantial majority of the unsealed cases had been under seal for periods well more than the required statutory period. Of the 44 complaints filed in those 42 cases, three quarters of the complaints (33 of 44) were filed before 2015, with one unsealed complaint dating back to November 2008 and two others dating back to 2010 and 2011, respectively. Of the remaining complaints reported in the *Review*, five were filed in 2012, fourteen in 2013, eleven in 2014 and eleven in 2015. As these cases illustrate, extensions of the seal on *qui tam* actions continue to be routine. The cases identified were filed in federal district courts in fifteen states and the District of Columbia, including California (6), New York (6), Florida (4), New Jersey (4), Ohio (4), Tennessee (3), Texas (3), Indiana (2), South Carolina (2), Wisconsin (2), Alabama (1), the District of Columbia (1), Georgia (1), Oregon (1), Pennsylvania (1), and Utah (1).[53] The federal government, it observed, declined to intervene in 25 of the 42 *qui tam* cases and although the government did not become involved in the three individual cases comprising the consolidated case, it did intervene on the overarching consolidated case. Accordingly, the three consolidated individual actions were omitted from the 25 cases in which the government declined to intervene, but the consolidated case accounts for one of the seven cases in which the government did get involved, in whole or in part. In another case, the state governments of New York and New Jersey declined to intervene while the federal government's intervention decision was not made clear by the unsealed filings. Six more cases were voluntarily dismissed before any action was taken by the government. In the remaining cases, the government's intervention status could not be discerned from the unsealed filings. Claims for relief under state or federal anti-whistleblower retaliation provisions appeared in 19 of the 42 recently unsealed cases. In over two-thirds of the unsealed cases (30 of 42), relators were current or former employees of the defendant. In two cases, the relator's relationship with the defendant was not revealed by the unsealed filings.[54]

Prior reporting indicates that *qui tam* filings are on the increase and have been for several years. A June 1, 2015, report included 33 recently unsealed health care *qui tam* cases. In one of those cases the plaintiff informed the reporter that it was her intent to withdraw the action and submit a redacted version of the original complaint. Of the *qui tam* cases, four were filed within a year of the report. One case had been filed in 2010 but was only unsealed in April 2015; three were filed in late 2011 but only unsealed in March 2015; eight of the cases in the newly unsealed batch were filed in 2012; and the balance were filed in 2013 and 2014.[55] Claims for relief under state or federal whistleblower retaliation provisions appeared in nine of the cases in this reporting period.

Sixty-five health care *qui tam* cases[56] were released between April and June 2014. The federal government partially intervened in ten cases and fully intervened in seven. Nineteen cases were dismissed prior to settlement, nine of which were dismissed voluntarily by the relator, and twelve cases were settled or dismissed pursuant to a settlement agreement. The government declined to intervene in 34 cases and filed notices of non-intervention in ten. Remarkably, a majority of the unsealed cases had been filed within the prior two years but about one-third of them predated 2012. According to the report, while the cases were filed in a wide variety of federal courts, historically active jurisdictions for False Claims Act enforcement saw a significantly higher number of the cases filed, including C.D. California (C.D. Cal.); M.D. Florida (M.D. Fla.); S.D. Florida (S.D. Fla.), E.D. New York (E.D. N.Y.); and E.D. Pennsylvania (E.D. Pa.).[57]

Faked research results have been on the rise for over two decades. Beyond the *qui tam* cases, allegations of misconduct by American researchers started spiking in 2004, when the Department of Health and Human Services received 274 complaints, fifty percent higher than 2003 and the most since 1989 when the federal government established a program to deal with scientific misconduct – the Office of Research Integrity (ORI). The

ORI could not keep pace with the number of reports it received, closing only 23 cases within the year. Of those 23, eight individuals were found guilty of research misconduct. From 1990 to 2005 – a period of fifteen years – the ORI confirmed roughly 185 cases of scientific misconduct, meaning those found guilty. But research suggests that this is but a small fraction of all incidents of fabrication, falsification and misconduct.[58] A survey published on June 9, 2005, *Nature* reported by the Associated Press revealed that 1.5 percent of 3.2 researchers who responded admitted to falsification or plagiarism. One in three admitted to some type of professional misconduct.

Years before Kenneth Jones, Ed.D., filed his *qui tam* against, in part, Brigham and Women's Hospital and Harvard Medical School, there was the case of Andrew Friedman, M.D., a brilliant surgeon and researcher, documented in the Associated Press[59] as a remarkable example of scientific misconduct. Friedman's department chair was coming closer to the truth, that for three years prior to discovery, Friedman had been faking – making up – data in some of the respected, peer-reviewed studies he had published in top medical journals. "It is difficult for me to describe the degree of panic and irrational thought that I was going through," he later told the Harvard University inquiry panel.[60] On the night of March 13, 1995, he was ordered in writing by his department chair to fix his suspect data. But he didn't do it. Rather, he did the just the opposite – dug deeper. "I did something which was the worst possible thing I could have done," he testified. From the seven-foot-high-stack of documents on his case at the Massachusetts Board of Registration in Medicine, it is known that he went to the medical room, and for the next three to four hours, pulled out permanent medical files of a handful of patients. Friedman then covered up his lies, writing in information he needed to support his study. "I created data. I made it up. I also made up patients that were fictitious," he testified. He continued to lie for another ten days until he broke under the weight of his deception and what has been often described as "the crushing pressure of academia."[61] The Associated Press reported that although Friedman ultimately confessed, retracted his articles, apologized to his department colleagues and was punished, he managed to recover his career, eventually becoming senior director of clinical research at Ortho-McNeil Pharmaceutical, a division of Johnson and Johnson.

There were many other cases that made the news at the time Andrew Friedman's case broke. In November 2004, from reporting, federal officials found that Ali A. Sultan, M.D., Ph.D., an award-winning malaria researcher at the Harvard School of Public Health, had plagiarized text and figures, and falsified his data – substituting results from one type of malaria for another – on a grant application for federal funds to study malaria drugs. When brought before an inquiry committee, Sultan tried to pin the blame on a postdoctoral student. Sultan resigned and is now a faculty member at Weill Cornell Medical College in Qatar, according to a spokeswoman there.[62] The truth of the matter was – and is – that the public only hears about the high-profile scientific misconduct cases when scientists have been cheating for decades. But why?

In 1974, William Summerlin, M.D., a top-ranked researcher working under immunologist Robert A. Good, M.D., Ph.D., at New York's Sloan-Kettering Cancer Institute, one of the world's leading biomedical research centers, claimed to have shown that success of skin transplants between genetically unrelated animals was enhanced by culturing the skin in special medium for several weeks. The experimental method involved transplantation of skin from black mice (with black melanocyte pigment cells) to white mice (without melanocytes). Over time, the melanocytes would naturally tend to migrate out of the transplanted tissue, so as to produce a grayish patch rather than a distinctly black patch. In the incident that became notorious, Summerlin was called to a meeting with Good, and took with him the single experimental animal that was purported to be the best evidence of transplant success. Noting that the patch had somewhat grayed, Summerlin by his own subsequent admission darkened it with a black permanent marker. The mouse was not produced at the meeting with Good. Summerlin's action was discovered when he returned the mouse to animal care technicians, who immediately noticed that the patch could be removed with alcohol. Senior staff and Dr. Good were notified within minutes and Summerlin's deception began to unravel. In the subsequent investigation, it became apparent that the original transplantation experiments were poorly controlled, and that other experiments from Summerlin's lab at Sloan-Kettering were misrepresented in reports and to colleagues. Regrettably, had Summerlin's experiment worked, it would have had major implications as a method to suppress immunological rejection of transplanted tissue.

Summerlin's misconduct continued unchecked until it could be ignored no longer. At the end of it all, his transgressions were attributed to mental and physical exhaustion, and three words that ring loudly in the hallowed halls of academia: "publish or perish." More recent media stories document that it was the Summerlin case that prompted Al Gore, at that time a Democratic congressman from Tennessee, to hold the first congressional hearings on the issue of scientific misconduct. "At the base of our involvement in research lies the trust of American people and the integrity of the scientific exercise," said Gore at the time. As a result of their hearings, Congress passed a law in 1985 requiring institutions that receive federal money for scientific research to have some system to report rule breakers.[63] "Often we're confronted with people who are brilliant, absolutely incredible researchers, but that's not what makes them great scientists. It's the character," said Debbi Gilad, a research compliance and integrity officer at the University of California, Davis, which has taken a lead on handling scientific misconduct, just then.[64]

David Wright, Ph.D., a Michigan State University professor (and a consultant just then of the United States Department of Health and Human Services' Office of Research Integrity) who researched why scientists cheat, said there are four basic reasons: some sort of mental disorder; foreign nationals who learned somewhat different scientific standards; inadequate mentoring; and, most commonly, tremendous and increasing professional pressure to publish studies.[65] From his testimony, Freidman's inability to handle that pressure, was his downfall. "And it was almost as though you're on a treadmill that starts out slowly and gradually increases in speed. And it happens so gradually you don't realize that eventually you're just hoping you don't fall off," he told a magistrate during a state hearing in 1995. "You're sprinting near the end and taking it all you can not to fall off."[66] Friedman started cheating when he was in his late thirties, married and the father of two young children. To some degree, the pressure on Friedman was complicated by the fact he was trying to fill several pairs of shoes, following his father, grandfather and uncle, who were all physicians and medical researchers. He was an associate professor of obstetrics, gynecology and reproductive biology at Harvard Medical School and chief of the department of reproductive endocrinology at Brigham and Women's Hospital.

Media reports indicate that Friedman's reputation was "tremendous" and his work "groundbreaking." Friedman's thirty-page curriculum vitae was peppered with awards and honors, lectures he'd delivered around the world, and an impressive listing of more than 150 articles, book chapters, reviews and abstracts. Of those, 58 were original research articles, where he had designed studies, conducted clinical trials, enrolled patients, collected and analyzed data and made conclusions. In the end, investigators found – and Friedman confessed – to making up information for three separate journal articles (one of them never published) involving hormonal treatment of gynecological conditions. So why did he do it? He testified that he was working 80 to 90 hours a week, seeing patients two days a week, doing surgery one day a week, supervising medical residents, serving on as many as ten different committees at the hospital and the medical school and putting on national medical conferences.[67] While Friedman sought help from a psychiatrist who advised he cut back his schedule, including research projects and patient load, his department chair refused to accommodate either request and Friedman continued careening down the path of self-destruction. As great a physician and researcher as Friedman was purported to be, he was dubbed "a terrible cheater." On the documents he fudged to make his study data work, the initials he used for fictitious patients were the same as those of residents and faculty members in his program. Unlike many scientists who file immediate lawsuits when they're caught, Friedman was repentant, resigning from his positions at both Brigham and Women's, and Harvard, and in 1996, he voluntarily agreed to be excluded from working on federally funded research for three years. During the next three years he consulted with drug companies; he paid a $10,000 fine to the commonwealth of Massachusetts and surrendered his medical license for a year; became very active with the American Red Cross, donating more than 500 hours, and attended several lectures on ethics and record-keeping.[68] "Andy can never undo the damage that his actions have caused. However, he has paid the price – his academic career is ruined, his reputation sullied, and his personal shame unremitting," wrote Charles Lockwood, M.D., M.H.C.M., then chair of obstetrics and gynecology at New York University School of Medicine, in a letter on Friedman's behalf.[69]

But there is more to the story. After successfully petitioning to get his license reinstated in 1999, Friedman went to work as director of women's health care at Ortho-McNeil Pharmaceuticals. The job involved designing and reviewing clinical trials for hormonal birth control, writing package insert labels and lecturing to doctors. He had even appeared on television and in newspaper articles responding to concerns about the safety of the birth control patch. Mary Anne Wyatt, a retired biochemist in Natick, Massachusetts, was one of several former patients. When asked by the Associated Press for a comment, she responded: "I think it's not at all surprising that a drug company would hire somebody who is very comfortable with hiding the effects of very dangerous drugs," said Wyatt, who unsuccessfully sued him. More than a decade removed from the media exposé that dredged his misconduct case to the surface for all to read, Friedman is today head of the Global Labeling Center of Excellence at Janssen, Pharmaceutical Companies of Johnson and Johnson. In this capacity, Friedman is responsible for providing strategic leadership and oversight of pharmaceutical product labeling development, execution and compliance for all Janssen products globally.

Rutgers University Biomedical and Health Sciences (RBHS), aligned with Rutgers University–New Brunswick and collaborating university wide, RBHS includes eight schools, a behavioral health network, and five centers and institutes that focus on cancer treatment and research, neuroscience, advanced biotechnology and medicine, environmental and occupational health, and health care policy and aging research.[70] Posted to the school's Robert Wood Johnson Library of the Health Sciences website is a page[71] devoted to the ins and outs of scientific misconduct. In the "quick facts" section, it lists the more than two thousand retractions of articles in scientific journals that have occurred over the span of forty years, give or take. Journal articles pertaining to anesthesia account for the most retractions. Are more people doing wrong or are more people speaking up? This was the question the journal *Nature* posed in the fall of 2013. To clarify the retraction issue, it was also noted that retractions of scientific papers, the ones Rutgers puts at a little over two thousand, have increased about tenfold during the past decade, with many studies crumbling in cases of high-profile research misconduct that ranges from plagiarism to image manipulation to outright data fabrication.[72] Nine scientists (and counting) have more than twenty retractions each. The Rutgers site further informs us that scientific misconduct is old, noting that Claudius Ptolemy, an Egyptian astronomer, mathematician, and geographer of Greek descent who flourished in Alexandria during the 2nd century A.D., is highly likely guilty of having copied his now famous sky charts from another astronomer. The site provides a few examples of the better-known scientific misconduct cases that have been the focus of intense scrutiny from 1950 to present, skirting a few of its own.

For Summerlin and Friedman, what motivated them to commit fraud varied from mental and physical exhaustion, work overload to the pressure within academia to bring in substantive research projects and publish the outcomes. Rutgers suggests that what motivates a scientist to risk fame and fortune are exactly the sources of the problem: an excessive need for name recognition, ever larger grants, rapid promotion within a research institute or university, old fashioned rivalry with a colleague, and plain old ego stroking. What nearly all scientists gone bad have in common falls into one or more of these motivators. Even the apologies for their wrongdoing share a familiar ring, often sliding quickly between real or portended mortification to hasty promises of corrections, denial, and the one most often found as a matter of record when a whistleblower files a *qui tam*: an admission of responsibility but the blame is shifted to someone else, and in many cases – too many – it is the postdoctoral research fellow, a lab assistant and even the one who blew the whistle on the wrongdoer.

When worries about a scientist's work reaches a critical point, it falls to a colleague, supervisor, junior partner or uninvolved bystander to decide whether to keep quiet or step up and blow the whistle, something that takes much soul searching to do.[73] Doing the latter – blowing the whistle – comes at significant risk, and the path is rarely simple and what has been observed many times over is that the whistleblower will follow one of two paths: make the case and move on, or never give up. In what seems to be a growing trend, anonymous watchdogs are airing their concerns through e-mail and public forums.[74] When *Nature* profiled Helene Hill, it subtitled her story "the quixotic," inferring her fourteen-year effort to hold accountable junior colleague Anupam Bishayee for misconduct resembled Don Quixote tilting at windmills. Though university panels (at least three of them), the Office of Research Integrity (twice), and two courts of law evaluated and dismissed her concerns, she didn't relent. The cost of blowing the whistle was emotional and financial. Hill spent thousands of dollars – well over two hundred thousand – in legal fees. She pored over more than thirty thousand documents, consulted experts and prepared for each legal action required. "A person has an obligation to do the right thing if they can," she said just then.[75] But the "right thing" could have cost her her job as a professor of radiation biology at the University of Medicine and Dentistry of New Jersey[76] (UMDNJ), now Rutgers New Jersey Medical School, part of the university's Biomedical and Health Sciences. Hill has been at New Jersey university since 1981. The "right thing" also consumed her.

In the whistleblowing game, it is abundantly clear that staying power is an absolute must, and Hill, as Paul Brookes wrote of her, has it in spades. Sadly, the non-legal options available to discoverers of misconduct are no easier than the legal route. In Hill's case, it should be clear that she is a pioneer in the field of science ethics. "By sticking her neck out," Brookes wrote, "and bringing a *qui tam* case, she serves as a rare example of someone fully committed to seeing something through, rather than simply quitting due to being shouted down or countersued into submission, as so many others have done. She has been ostracized from her chosen career path and subjected to discrimination on a level reminiscent of the pre-civil-rights era and yet has bounced back at every level with fresh evidence that cannot be refuted." She has remained civil, polite, and composed, he would further observe and should be an example to anyone wanting a role model in how to conduct oneself in science.

Hill's case is also rare because it allows access to a side of the scientific process that few members of the public ever see. The "face" of science to most people, we are reminded by Brookes, is the popular media, with puffed-up reports claiming a cure for cancer in five years, or a new epilepsy drug that works in rats and will be in humans within a decade. The middle ground is the scientific literature – unintelligible to most people, but still a polished and carefully crafted set of stories about what really went on in a laboratory. The *dark side of the moon* in science, Brookes and others have emphasized repeatedly, the side never seen by the public, is the chaos of a modern laboratory.

Brookes wrote that it is those piles of laboratory books on the graduate students' desks; the critical details scribbled on post-it notes waiting to be formally documented; the Excel spreadsheets scattered across USB thumb-drives; the masses of online folders containing data and images, all with similar but ever-so-slightly different names; the millions of Petri dishes sitting in freezers and incubators, their barely legible labels scrawled in Sharpie marker gradually wearing off. It the mass of intricate experimental details in the head of every laboratory scientist that simply never ever gets written down. Yet somehow, Brookes would tell anyone who might listen, it is a miracle that all this chaos can be distilled down into a reliable narrative that can be published and understood by other scientists within the context of the rest of the literature. "Is it any wonder then, that the process of going from the chaos of the lab to the neat thirty-second snippet on the evening news 'Health Watch' section, is littered with opportunities for mistakes, and perhaps more scarily, with opportunities for perversion of information."[77]

This book offers a unique window for the professional to the layman to understand what happens when it all goes wrong, when that path to which Brookes refers is corrupted. Perhaps the scariest message of all, between these lines, is how easy it would be for such things to happen again – and again. "This is not an isolated incident for filing away in the history books," he continued. "Rather, it is a wake-up call to fix the process of science, before it is too late," and the public trust is lost." Beyond eroding public trust, millions of taxpayer dollars are lost to fraud, fabrication and scientists behaving badly, and beyond that – and worse yet – the promise of a cure here or a miracle drug there dashes the hopes of the terminally and chronically ill.

The long view of this story begins with a woman of science who earned her degrees from prestigious colleges and universities after World War II and started, like other women of her education and ambition, her career in what Hill has described as "the dark times when we [women of science and medicine] were not welcome and had to fight every step of the way. The story," she has opined, "is one of the struggles for recognition, for respect and for trust." Moreover, it is the story of one person who stood by her principles and refused to give up. Above all, it is the story of the quest for truth in science despite being marginalized and ostracized. Hill would later confess that she was ashamed for being a whistleblower, but she was more ashamed that her colleagues held back what they knew, that they would not acknowledge what they knew about the fraud, fabrication and misconduct that she uncovered in her department. Instead of telling the truth, they left her alone to fight for the truth for more than fourteen years.

Hill, to be clear, never wanted to be a whistleblower. But that is what she is. It has defined the last quarter of a lengthy, brilliant scientific career. If you're reading this and are already asking why not give up, it wasn't in her character to do that, to surrender truth for expediency, to go along to get along. "I want other scientists but especially women to know that you have to keep fighting for what you believe in, no matter what the odds," she would write after she lost her *qui tam*, which demonstrated to Hill and others that a whistleblower's fate is rarely a happy one. Despite this loss, and while it is important for Hill and others like her to tell their story, for her it is far more important that as a result of the outcome of her case, the system for adjudicating and punishing fraud, fabrication and misconduct in scientific and medical research, most especially, must change. Hill needs to be heard and we need to listen.

Chapter 1 –
The Ueno Phenomenon

elene Hill thought hard about retirement as she approached her seventieth birthday in the spring of 1999, invoking what she called the Ueno phenomenon. "I had friends, colleagues, in Fort Collins, Colorado, Charles A. Waldren [Ph.D.] and his wife, Diane B. Vannais [Ph.D.]," she opined. Waldren was a professor in the Department of Radiological and Environmental Health Sciences at Colorado State University and Vannais ran his laboratory. "I visited them several times in the 1990s to learn the techniques that they had developed to assess mutations in Chinese hamster cells in tissue culture." Waldren and Vannais' laboratory was particularly productive, she recalled later, not only because the two scientists were intelligent and creative, but because they had been able to collaborate with Akiko Ueno, Ph.D.

"Unassuming and hard-working, Dr. Akiko Ueno doesn't seem like a radical figure," declared the Colorado State University College of Veterinary and Biological Sciences page[78] devoted to her. "But as a Japanese woman working in the field of radiation biology, she was just that. Dr. Ueno, now retired, pursued her passion for science around the globe, including working in some of the most prestigious laboratories researching basic questions in radiation biology and cytogenetics analysis. She was one of a handful of Japanese women who doggedly pursued a research career in radiation biology." Ueno worked for many years alongside and collaborated on research projects including tracking survivors of the Hiroshima bombing. "During her time at Colorado State University, Dr. Ueno worked closely with Dr. Waldren to advance basic understanding of radiation biology, particularly cytogenetics," said Jac A. Nickoloff, Ph.D., head of the Department of Environmental and Radiological Health Sciences. "She worked at the National Institute of Radiological Sciences, as well as the University of Tokyo and Harvard University. She was truly a pioneer in her field."[79]

After working in the United States, Ueno went back to Japan, where the mandatory retirement age is sixty. Faced with no option to continue her research there, Ueno found herself bored and returned to Colorado State to continue her collaboration with co-investigators Waldren, Vannais and others on new projects. Ueno is now fully retired and living in Japan. But what she and Hill shared, beyond the science, was age. Hill and Ueno were roughly the same age in 1994. "But she did not want to stop working [speaking of Ueno] [...] and Charles and Diane were the lucky ones [...]," explained Hill. "She did beautiful, meticulous work that culminated in several papers for them and certainly were a help in getting funding to continue their research. I was very impressed with Dr. Ueno and vowed that when it came to the waning years of my career, I would like to offer myself up to some researcher who needed help and could use my talents." Hill did not like writing grants, discouraged just then by a ten percent or less success rate at applying for and being awarded money to do her research.

As her seventieth birthday approached in the spring of 1999, Hill was ready to put the "Ueno phenomenon" into practice. "I was about out of funds, and I had made a wrong turn," she said. "I had established a fair name for myself in pigment cell research, studying the effect of the skin pigment melanin on the photobiological responses of mammalian cells in tissue culture." She wrote an oft quoted review of the same, comparing the effects of melanin that are numerous and varied to the parable of the blind men and the elephant. In this well-known tale, a group of blind men (or men in the dark) touch an elephant to learn what it is like. Each one feels a different part, but only one part, such as the side or the tusk. They then compare notes and learn that they are in complete disagreement. The parable has been used to illustrate a range of truths and falsities; broadly, it implies that one's subjective experience can be true, but that such experience is inherently limited by its failure to account for other truths or a totality of truth. At various times the parable has provided insight into the relativism, opaqueness or inexpressible nature of truth, the behavior of experts in fields where there is a deficit or inaccessibility of information, the need for communication, and respect for different perspectives. The parable of blind men and an elephant suited – even foreshadowed – the circumstance in which Hill would soon find herself.

"I had obtained a three-year National Institutes of Health (NIH) grant to study melanin and had only been moderately successful in the research," Hill later observed. "Three years were simply not enough time to accomplish much. I had written a renewal, which failed to get funded; however, in the meantime, I had also discovered a factor elaborated by mouse melanoma cells in tissue culture that protected the cells from the lethal effects of radiation and of several chemotherapeutic agents." Though she'd received funding for the latter studies, funding was running, and she found herself at a crossroads. What research should she pursue? Believing that she could only do justice to one, Hill opted to study the so-called multifactor resistance factor (MTRF), also the best chance for funding. When the renewal application came back for her melanin studies came back, at roughly the same time as her dilemma over "what next?" Hill realized she should have rewritten it and kept trying to get the funds to continue her work. The wrong turn came when she put it aside and decided to go with the MTRF study. But it, too, failed.

Though Hill, a tenured professor with a salary covered by the state of New Jersey, could continue at the university, she no longer had funds to support anyone in her laboratory but herself. This predicament came with a price. "I was told that I would now have to share space […] with Roger Howell[80], a younger member of my department who also was running out of funds and trying to get new funding." Though the loss of funding for both meant shared space and a certain amount of downsizing, it was also the perfect opportunity, she thought, to put the Ueno phenomenon into effect. Perhaps it was time to retire.

Hill packed up what was left of her laboratory and moved it up one floor, where she was assigned an office space just opposite Howell's, and shown a large laboratory that she would have to share with him. A smaller lab was also provided for her, one that was exclusively hers and hers only. As fortune dictated, Hill and Howell were both applying for research grants from the NIH at the same time and both applications were being reviewed by the same radiation study section. "I went to a meeting of the Radiation Research Society just after the time that the study section met and I asked the study section staff person, Paul Strudler[81], how our grant applications had fared," she remembered. "He told me that mine would not make the pay line and that Roger's wouldn't either but that it was close and would probably succeed on the next go around with some needed modifications."

When Hill returned to New Jersey, she told Howell the good news, also informing him that she had decided not to pursue her own research but would be willing to help him with his study to get Howell's work funded through NIH. Howell agreed to let her help, but she quickly sensed that he had some reservations about it. "My expertise is in DNA damage, its repair and in the induction of mutations; Roger is a physicist with little knowledge of biology," she explained, "so my expertise would have done well to complement his own." In the fall of 1999, Howell rewrote his grant. Feedback from NIH pink sheets advised that Howell's goal was too narrow and that he should examine additional end points than just survival of cells after irradiation. "That was where I came in because I could measure other end points such as mutation and various types of DNA damage," she continued. "[Howell] asked me to write up the aims directed at these determinations, and I agreed but made it clear that I expected to be a co-investigator on the grant." But it was here that Howell drew the line in the sand and said no. "At that point I said 'too bad' and ended the collaboration," Hill offered. Several days later, Howell changed his mind. She speculated that further discussion with others was key in his reversal. Hill would be his co-investigator. She wrote up the experiments she would perform and turned them over to Howell to submit with the application.

The devil is in this detail: As it turned out, had Hill not been the co-investigator on the grant (NIH Grant RO1#CA83838) that would subsequently be called into question, she would not have had the rights to the data that she later used to file the *qui tam*. While she also obtained all of the data involved in the fraud she reported, she would still need to call upon her rights as a collaborator to publish, discuss or share in any form her findings.

Howell, who earned his Ph.D. in physics from the University of Massachusetts-Amherst in 1987, was mentored by Kandula S. R. Sastry[82], Ph.D., who had also advised Dandamudi V. Rao[83], Ph.D., another future

4

professor and department chairman in Hill's radiology department. When Sastry died on November 23, 2001, Howell and Rao memorialized him, noting that the Society of Nuclear Medicine had lost an influential force in radiation dosimetry.[84] Rao's research was relevant to nuclear medicine, Hill recalled later, while her own focused on radiation therapy. It was Rao who recruited Howell to the University of Medicine and Dentistry of New Jersey (UMDNJ), again, now Rutgers New Jersey Medical School, soon after Howell received his doctorate. Rao and Howell worked together, rarely interacting with Hill. Rao, who had been successful obtaining grant funding with Howell, retired just as Hill's lab was moved to shared quarters with his protégé. Howell was thus on his own to obtain funding for his research. Edouard I. Azzam[85], Ph.D., another figure who would play prominently in the inquiry to come, joined the faculty in 2000 in the division of radiation research as an assistant professor. Anupam Bishayee[86], Ph.D., who Hill trained and shared lab space but was then compelled to accuse him of misconduct, came to UMDNJ in February 1997 as a research and teaching specialist.

The NIH application for the grant that would soon be called into question listed those involved along with a level of effort expected of each contributor (the percentage is noted after each participant's name): Howell as the principal investigator (25 percent); Hill as co-investigator (5 percent); Rao, co-investigator (5 percent); Bishayee, research and teaching specialist (100 percent), and an unnamed postdoctoral fellow (100 percent). Thus, the research workload fell on Bishayee and a "to be determined" unnamed postdoctoral fellow, a funding line in the grant budget. Also of note in the grant budget is an equipment request for a Packard automatic gamma counter, Cobra Model 5003, for $25,000; it is a piece of information to keep in mind as the story unfolds.

Howell's grant application to the NIH National Cancer Institute indicated that the primary focus of the proposed research was the role of "bystander effect," in the biological response of mammalian cells to ionizing radiation, a process, in simplest terms, whereby cells exposed to an agent, in this case radiation, influence the behavior of unexposed neighbors. "It has long been believed that the principal genetic effects of ionizing radiation on mammalian cells are the direct result of DNA damage in irradiated cells that has not been repaired adequately," he submitted. "Therefore, when cells are exposed to non-uniform distributions of radioactivity, only those cells which receive 'hits' from the emitted radiations would be damaged. No effects would be observed in cells that were not 'hit.' These cells are referred to as bystanders."[87] Howell submitted the application for the November 1, 1999 deadline. "Roger's research goal, at least as it pertained to our collaboration and the events [that subsequently unfolded], was to develop a model system to measure the effects of non-uniform distributions of radioactivity that would be encountered in nuclear medicine during the course of diagnostic or therapeutic procedures," Hill later recalled. "Nuclear medicine involves the use of radioactive isotopes in various ways. In diagnostic procedures, radioactive isotopes – that is radioactive forms of elements attached to various carrier molecules – are injected in the body with the goal of seeking out certain organs. For example," she continued, "in a bone scan, an isotope of technetium – ^{99m}Tc – incorporated into a phosphate compound seeks out actively metabolizing bone which is said to light up. Such areas can be indicative of arthritis, cancer and infection."

Hill explained that the radiation dose received by various tissues in a nuclear medicine procedure is non-uniform and therefore problematic, so it is the overall dose to particular structures that is computed. It can be shown that this is not an adequate measure of dose because the radioactive compounds distribute themselves in tissues in such a way that there may be high doses in some areas and small doses in others. "It was to get at this non-uniform distribution of radionuclides that Roger designed his model." Howell's grant was funded by the Radiology Study Section at the National Institutes of Health. Hill would further note only that Howell is a physicist, not a biologist and there were elements of the grant – biological in nature – that he missed, as did the NIH study section referees. The model predicted how the results of the experimental manipulations might come out. "In radiation biology and in many other branches of basic science, we often develop theories that can be translated into mathematical models resulting in theoretical lines on a graph close to which, if the model is a good one," observed Hill, "the experimental data will fall. Roger's publications with his post-doctoral fellow, Anupam Bishayee, did a fine job of fitting the data to his mathematical model.

Bishayee had been working in the lab since February 1997 just then and it was his research that formed the experiments reported in several research papers published between 1998 and 2002. Hill's expertise was in DNA damage and mutagenesis. Howell thus asked that she work with Bishayee on an experiment that he devised to determine whether the experimental model that he had designed would provide adequate oxygen to the cells that he was studying. The model involved incubating several million mammalian cells in loose pellets that Howell dubbed "clusters." Howell then passed dishes containing mammalian cells to Hill to study for mutations. "My results showed that it was likely that the cells were not getting adequate oxygen," she noted. Howell then had Bishayee repeat both parts of the experiment by himself. "During the course of his experiment I examined dishes that I was quite certain were his from that experiment," Hill continued, "I believed those dishes were empty, that is, they did not contain the colonies that were expected and were later reported to contain. I was suspicious of the results and related my concerns to Howell. Bishayee's results supported Howell's hypothesis regarding oxygen whereas my results did not support that hypothesis. Howell used Bishayee's results in his grant application, although he knew – because I had told him – that my outcome conflicted with that of Bishayee."

Hill kept her notes, and it is from those that it is clear there was a problem that predated the filing of the November 1, 1999 NIH grant application. "To my knowledge, that experiment has never been repeated although good scientific practice would require that the conflict between two [different] outcomes be resolved. Scientific misconduct was not of much concern in those days. I wrote a memo to myself," she about my suspicions but did nothing further." A copy of that "note to herself" dated October 23, 1999, went as follows and bears inclusion:

> I write this because I suspect that Dr. Anupam Bishayee, a post-doc working for Dr. Roger Howell, is fudging data and I do not believe that Roger is taking what I have told him seriously.

> [The rest is a day-to-day log of Hill's observations:]

Monday, October 11, 1999: Anupam had started an experiment involving mutagenesis which would end today. If there had been no contamination, there should be 50 P100s ready to stain. I ask him during the day if he will stain them today and he replies that he is busy with other things but that he will stain them later. When I leave around 6 o'clock, I stop in to see if they have been stained and he says that they have not but [that] he will work late and do them later.

Tuesday, October 12, 1999: When I come in this morning, Anupam has not arrived. I go into the lab to look for the plates. There is no sign of them around the lab or in the trash. I look in the incubator and there is a tray containing 10 stacks of 5 p100s each and marked consecutively from 1 to 10. This is the expected numbering that would result from the experiment. I look at several plates by I-ball and under the microscope. I do not see any colonies. I know from my experience that cells that are not going to form colonies will lyse on about day 4 after plating leaving nuclei – this should be day 7. There are very few nuclei. I worry that these are the plates from the experiment and that Anupam miscounted resulting in too low cell numbers to get colonies. I also figure that he decided not to fix and stain them the night before because it had gotten too late. An extra day would not make any difference. He may also have become aware that they didn't have many, if any, colonies on them. When he comes in, I ask him how the experiment went, and he says it went very well and he will give me the data later. I ask him what the plates are that are in the incubator, and he replies that that is another experiment that he is working on. I know that the other experiments that he does do not utilize the p100s, they use the p60s. In fact, I have supplied the p100s for these experiments. He gives me the data later on and it is as predicted: the data for cells placed in aerobic conditions indicate more mutants than data for cells placed in hypoxic conditions. During the course of the experiment, cells must be plated and replated and the numbers must be carefully recorded in order to calculate the number of cell divisions the cells have undergone before they are challenged with the selection agent 6-TG. I ask Anupam for these data and he tells me he has them and will give them to me later.

Wednesday, October 13, 1999: Again I come in earlier than Anupam. I go to the incubator only to find the mysterious set of 50 p100s is no longer there. Again I search through the trash and around the lab and there is no sign of the plates. Anupam does not give me the missing data.

Tuesday, October 19, 1999: Roger [Howell] has been sick but is now back at work. I relate to him the above events and tell him of my suspicions. He agrees that things sound pretty fishy. He says that he will ask Anupam for the missing data referred to above. I hear him do this later and I hear Anupam reply that he cannot lay his hands on the data because they are at home. Roger scolds him for taking data home.

From now on, I check when I can as to what is going on in Anupam's incubator. There is a set of p60s marked appropriately for the sorts of survival experiments that Anupam does. I don't see anything on these plates but they may be newly plated.

Thursday, October 21, 1999: I discuss the situation with Roger and he [dismisses me] saying that I never really did like Anupam in the first place. This is not true. I don't dislike him and I don't have it in for him. I do not know if Roger ever got the missing data from Anupam. Roger is pleased with Anupam's work because he faithfully turns out data regularly. He also thinks that Anupam is not smart enough to make up data. I decide that, as much as possible, I will distance myself from Anupam's work. Until there is a logical explanation for the mysterious set of 50 p100s that should not have been in the incubator, I believe that Anupam made up the data for the mutagenesis arm of the experiment and I do not want to be associated with him.

Friday, October 22, 1999: Roger is analyzing some old data and finds that the RBE for 3H is at least three times greater than it should be. He will repeat these experiments himself to determine whether the data are real or not.

Saturday, October 23, 1999: I look in the incubator for the above mentioned p60s. They are gone. I check the trash. There are plates in the trash that still have pink medium in them which looks a bit cloudy and may be contaminated.

Howell's grant application was successful and funded for $1.42 million over 5 years; the money started flowing on July 1, 2000. Howell hired a new postdoctoral fellow, Marek Lenarczyk, and by the fall of 2000 he was working closely with Hill and though she later documented Lenarczyk's concern over Bishayee's work, Lenarczyk's statements when subsequently deposed would become a source of confusion to the point of contradicting prior testimony and physical evidence. "One evening in March of 2001," Hill documented later, "Lenarczyk came to me because he was concerned about an experiment that Bishayee was carrying out. We determined," she continued, "to follow that experiment which would last for another 10 to 12 days, believing the experiment was contaminated with mold which would show up in the end." But it was two other events that drew Hill's (and purportedly Lenarczyk's) suspicions. Lenarczyk photographed samples in the incubator that they believed should not have been there because the protocol called for them to be processed in another laboratory.

Further, Bishayee had obtained fresh cultures from Lenarczyk, and she and Lenarczyk believed he used those to simulate the desired results.

When Howell became aware of Hill shadowing Bishayee's work, he went to department chairman Stephen Baker, Ph.D., documenting their meeting the same day in a memorandum dated April 6, 2001. The subject line was "Integrity of data." Howell wrote:

> This memo is to follow up on our discussion this morning regarding Dr. Hill's accusations of falsification of data by my Research Associate, Anupam Bishayee, Ph.D. As I mentioned to you, I first became aware of Dr. Hill's concerns about six to nine months ago when she brought this to my attention. Having worked closely with Anupam on experimental design and analysis of the data, and after further reviewing his data, I told her at that time that I did not believe there was any problem with the data and I did not take any further action at that time. This past Wednesday, April 4, 2001, I was informed by a member of another department in NJMS that Dr. Hill had recently spoken to a few persons regarding her suspicions. Knowing that I already had a meeting scheduled with you for today, April 6, 2001, I decided to ponder the situation for a day or two and discuss it with you at that time. As I told you this morning, I have requested that my post-doctoral fellow Marek Lenarczyk, Ph.D. repeat some of Dr. Bishayee's experiments as a check of the validity of the data. Dr. Bishayee will work closely with Dr. Lenarczyk to ensure that the experiments are performed in an identical manner. I will also personally sign all of Dr. Bishayee's data sheets and culture dishes to ensure that they are not tampered with. I believe that these steps will be sufficient to document the validity of Dr. Bishayee's data. Dr. Bishayee has been a faithful and hard-worker and I fully believe that his results will be reproducible.

He continues,

> There are a number of factors that support my confidence that Dr. Bishayee's data is indeed valid. First and foremost was that Dr. Bishayee's original findings which led to the observation of "bystander effects" in our experimental model were serendipitous. About three years ago when Dr. Bishayee first joined my laboratory, I had planned to use tritiated thymidine as a control so that a pelleted population of cells consisting of heavily radiolabeled and unlabeled cells in a 1:1 ratio would ultimately lead to the death of all of the radiolabeled cells and none of the unlabeled cells. Thus, I fully anticipated a surviving fraction of 50 [percent]. However, Dr. Bishayee found that a large fraction of the unlabeled cells unexpectedly died. Initially, I was upset with Dr. Bishayee because I believed that some mistake had been made in the experiments. However, I had Dr. Bishayee repeat his study at least three times and the same phenomenon was observed. Ultimately, I attributed this phenomenon to bystander effects and we have subsequently shown that the signaling responsible for the death of the unlabeled bystander cells is mediated through gap junctions. Another factor in my confidence is that the mathematical gymnastics required to analyze the data are not trivial and it would be difficult to manipulate them without my knowledge. These are strong arguments that I discussed with Dr. Hill when she first raised her concerns several months ago. However, despite them, her suspicions have persisted and have led to this situation. I will take all steps necessary to prove that the data is valid beginning with those that I have outlined at the end of the first paragraph.

"The memo indicates that I had made 'accusations of falsification of data' although I was not to do so until the following week. It also states that ' I [indicating Howell] have requested that my post-doctoral fellow Marek

Lenarczyk, Ph.D. repeat some of Dr. Bishayee's experiments as a check of the validity of the data,' when, at this time, Lenarczyk had already performed eleven experiments that had failed to reproduce Bishayee's results (non-replication experiments)," Hill documented later. There were to be, in all, 22 experiments that precisely followed the same two protocols and that failed to reproduce Bishayee's findings. Howell would subsequently acknowledge in his deposition that the failure to reproduce was never reported to Baker, Karen Putterman, M.D., M.P.H., UMDNJ vice president for academic affairs, the NIH or to the journal *Radiation Research* that had published two papers based on Bishayee's data. It would not become clear until after Howell's subsequent two-part deposition as part of Hill's *qui tam* case, that he did not perceive an obligation to do so either. Lenarczyk took a number of pictures of the contents of various incubators, but the most significant were of the 10.5-degree-centigrade incubator that housed the slender Helena tubes set to be processed on Friday, March 30, 2001, and "we [to include Lenarczyk] were certain, should have been gone that day," Hill continued. Bishayee's Helena tubes remained there until the following Thursday.

The episode involving what Hill called the "mysterious new cell line (NCL) that comes and goes," was documented in the minutes of the April 27, 2001, Newark Campus Committee on Research Integrity (CCRI) hearing to sort out what was going on in Howell's laboratory. Present were Neil S. Cherniak, M.D., professor of medicine and physiology, vice chair of research and director of clinical affairs; Daniel H. Fine, D.M.D., professor of oral biology; Anthony Forrester, Ph.D., R.N., A.N.E.F., F.A.A.N., professor in the Division of Nursing Science; Teresa Marsico, C.N.M., M.Ed., F.A.C.N.M., director of the university's nurse midwifery education program (by phone); Elizabeth Raveché, Ph.D., professor, Department of Pathology and Laboratory Medicine; Putterman, and Sheila Eder, Ph.D., director of Institutional Research. Raveché presided. None of them, of note, were qualified in radiation biology. The committee interviewed Hill, Howell, Lenarczyk, and Bishayee.

Lenarczyk first explained how he came to work in Howell's laboratory as a postdoctoral fellow before describing for the panel the general nature of the experiments performed in the lab on the survival rates of cells exposed to radionuclides and of bystander cells. He told Raveché and the others that the experiments measuring cell survival rates cannot be validly completed if carried out with contaminated cell material. In the case of the experiment started by Bishayee on Monday, March 26, 2001, Lenarczyk believed that Bishayee had used contaminated cells.

Lenarczyk was clear in testimony, at least in the beginning, and before he was asked to explain the photographs he took of Bishayee's material. He went on to state that by Friday, March 30, 2001, he was certain that the experiment was contaminated. Since he had no reason to check Bishayee's cells before that, he couldn't say for sure that the experiment was begun with contaminated material. But on Friday, March 30, Lenarczyk observed that Bishayee's cells were still in Helena tubes in the 10.5-degree-centigrade incubator when, according to protocol, they should have been taken out by that time. The committee questioned Lenarczyk as to how he came to know that the experiment had been started on Monday, rather than Tuesday or Wednesday. Lenarczyk replied that the protocol for the experiment was more conveniently carried out by beginning on a Monday. Asked at what point he began to observe Bishayee's conduct of this experiment, Lenarczyk wasn't certain how to answer the question. Fine asked Lenarczyk why he had put his own cells in the 37-degree incubator if the incubator was suspected as a source of contamination to which Lenarczyk responded that there were only three places to propagate, and that he himself only put some of his cells in the 37-degree incubator shared with Bishayee.

Lenarczyk told the committee that Bishayee had asked him for cells on Thursday, March 29, 2001, and that this aroused his suspicion because of the long-standing problem of contamination in the lab. He wondered why Bishayee was asking for cells on Thursday, when the cells for the experiment should be removed from the tubes on Friday. When the committee asked whether Bishayee might not have been following a different protocol, Lenarczyk answered that he believed the fact that the cells were in Helena tubes indicated that the experiment was looking for bystander effect and was using the same protocol. He went on to explain that when he went to the 10.5-degree-centigrade incubator on Friday, March 30, to remove his own tubes, he observed Bishayee's tubes still there with one tube missing. He indicated that he counted the tubes to be sure he had the right number of tubes that were his own. Earlier in the week he'd noticed that Bishayee had seven tubes, the expected number, in the incubator. Lenarczyk had witnessed Bishayee sitting in the laminar flow hood that Friday morning, perhaps around 10 or 11, and that while he didn't check with Bishayee just then, he assumed that he was processing cells. "My end point was to seed dishes and put them back into an incubator, which I did around 4 or 5, and then I see his dishes in there. He had already plated cells which should have come from the tubes."

Lenarczyk explained that he double-checked that Bishayee's six were still in the 10.5-degree-centigrade incubator. Further, he couldn't understand why Bishayee would take one tube when all the tubes should have been taken out for the experiment. Lenarczyk began to think "something was going wrong" and took samples of the tubes belonging to Bishayee. He sampled the tubes on Friday because he believed Bishayee had already concluded the experiment when he saw him working in the hood on Friday morning. When asked why he hadn't asked Bishayee directly about what was going on, he replied that it was his choice not to speak to Bishayee, but to talk to Dr. Hill since he was living in her house. At this point the committee questioned him about when in that week he first spoke to Hill. Lenarczyk admitted that the two had spoken about contamination problems in the lab, to which Hill had told him "we have to observe very carefully." The committee asked if Hill had suggested taking samples from Bishayee's six tubes remaining in the incubator, to which he replied, "probably, yes."

Lenarczyk described for the committee how he had removed samples from each of the tubes by popping the top of the tube, putting media into a clean six-well Petri dish, splitting the samples to measure radioactivity, and doing both radioactive count and contamination testing. When prompted for more details, he told the panel that he sampled only five tubes, putting just media into the sixth well. He put a drop in each dish, put the dishes into an incubator in a separate room, and after one day, he observed the six wells by holding them up to the light. One of the wells was still clear and the others were cloudy. He also put the dishes under a microscope and all five of the wells were contaminated. Lenarczyk didn't reply when asked what he did with the wells. He went on to say that he put another drop from the tubes into a small vial designed to measure radioactivity. Raveché pressed Lenarczyk on whether he tested the supernatant or the cells for radioactivity. But he seemed uncertain about whether he had tested supernatant or cells; the exact date when he had run the test for radioactivity, and why he had felt the necessity to test for radioactive counts at all.

Raveché asked whether both Lenarczyk and Hill had gone through the trash looking for the seventh tube, to which he replied that they both looked through the trash on Saturday, March 31, and found it in a non-radioactive trash can. The tube should have been radioactive, although Lenarczyk admitted he didn't test it for radioactivity. He described checking his own plate cells on March 31, and determined that all were fine. He also saw Bishayee's two sets of dishes (dyed/radioactive and undyed/bystander) marked A and B, and one of the sets was fully contaminated. Raveché asked what happened to the dishes – in truth, critical evidence of wrongdoing – and Lenarczyk told her that they had disappeared by Tuesday or Wednesday, April 2 or 3, 2001. He had looked for them in the trash but couldn't locate them. The committee members wanted to know when Lenarczyk started taking pictures. But he didn't remember. After being prompted, he would admit taking pictures of Helena tubes and stated that the date should be on the picture. Fine reminded Lenarczyk that the first time he used the camera, there was no date. The second time there was a date. The camera was new to Lenarczyk, he explained, and he had to learn how to set it to record the date. Raveché was fixed on why the order in which the tubes appeared in the rack differed from one photograph to the other. He told them he had to take the tubes out of the rack because it was difficult to sample from them without doing so. The committee, at that point, became concerned that there were inconsistencies in Lenarczyk's remarks.

Certainly, Lenarczyk's responses when asked to explain the difference in tube order – the inconsistencies – between the photographs in Exhibit A and Exhibit B, both dated April 1, 2001, gave them pause. He answered that he thought the pictures were in reverse order. When Raveché reminded him that he had previously stated that he took the samples from the six tubes on Friday, March 30, he answered that he didn't want to disturb the cells, so he took samples only once, "maybe Sunday." But when asked when he took the photograph in Exhibit D, he told the committee that it "might have been Friday" because there were only six tubes in the photograph. Raveché asked about a photograph Exhibit K dated April 7,

2001. She pointed out that the picture was taken a full nine days after the sample was cultured, and that the plate should be solidified by that time if contaminated. But it wasn't solidified in the picture. She asked if he was certain that the picture was taken on April 7, and he said "yes."

Lenarczyk's actions and his responses clouded the hearing. By the time it was over, the committee was concerned that Hill and Lenarczyk's actions might have interfered with Bishayee's experiment. If cultures from the sampled six tubes were allowed to grow for a day, then the samples must have been drawn on Thursday, March 29. If so, this seemed to the panel that it would have interfered with Bishayee's work. Then Lenarczyk admitted that he might have taken samples on Thursday.

Raveché wondered whether there could have been scientific misconduct if Bishayee used contaminated cells but reported in his lab book that half of the Petri dishes were contaminated and half were not. The Petri dishes were in the committee's possession and demonstrated the pattern of contamination reported. She asked Lenarczyk if he was aware of Bishayee's results, and he responded that he never saw the results. Fine suggested that the dishes plated on March 30 could have been from a different experiment. But the latter would be a best guess with no basis in fact.

Lenarczyk was asked if Hill came to him with suspicions or whether he went to Hill. He explained that he went to Hill first. Cherniak asked if Lenarczyk had ever told Howell about his problems in the lab, and he responded that "of course" he had told Howell. Cherniak followed up by asking why he had not told Howell about his suspicions regarding the March 26 experiment to which Lenarczyk replied that he had only spoken to Hill, and admitted that that may have been a mistake, but that Hill told him it was better for her to tell Howell. As a junior postdoctoral fellow in the lab, Lenarczyk felt at a disadvantage in bringing forward a complaint of possible scientific misconduct to Howell. In truth, there were considerable reasons for Lenarczyk to be uncomfortable with CCRI and any other legal proceedings regarding the Bishayee complaint. As a Polish national, English was not his first language and that language barrier – of the idioms English speakers take for granted – spotted his testimony. He also had, at least when he first came to the United States, a family to support at a relatively low salary as a research assistant in a university laboratory; he couldn't afford to displease his employers and lose his job. Lenarczyk's subsequent deposition would support this finding. Certainly, Lenarczyk's hesitation, confusion and reluctance to be involved is palpable throughout this case. Despite whatever reservations he may have kept to himself, when Raveché asked if he felt unimpeded in making any statement to the committee or if he feared reprisals for appearing before the panel, he said "no." He had not been coerced.

Bishayee testified next. If the protocols had been followed (as recorded), there should have been one rack of seven Helena tubes from Tuesday, March 27 until Friday, March 30; no tubes (or rack) from that Friday until Tuesday, April 3, then one rack of seven tubes until Friday, April 6. There was, in fact, as expected, on rack of seven tubes from Tuesday, March 27 to Friday, March 30. The number of tubes dropped from seven to six between 3 and 5 PM and this mystery rack of six tubes – the NCL – remained until Thursday, April 5. A new rack appeared in front of the mystery rack on Tuesday, April 3. But let the extrapolated exchange between Bishayee and the committee tell the story:

> *Bishayee*: [The mystery] NCL was put in the incubator between Monday and Friday, [between] March 26 and 30 [in truth, a second rack should have appeared because there was already an ongoing experiment, but it didn't. Information was not in his notes; he told the committee he was preparing for freezing but the tubes got contaminated.]
>
> *Raveché*: When?
>
> *Bishayee*: [first response] After April 5... [second response] Monday after... [third response] April 9 (Monday).
>
> *Raveché*: When were extra cells in the incubator?
>
> *Bishayee*: Between April 4 and 6 [Hill and Lenarczyk observed them between March 30 and April 5].
>
> *Committee*: [Thought it was the week of March 26].
>
> *Bishayee*: Between March 27 and March 30.
>
> *Raveché*: Why didn't you freeze the NCL?
>
> *Bishayee*: [He explained that he was getting ready to go to Puerto Rico (for a meeting), looked at them and threw them out.]
>
> *Committee*: How many Helenas were present on March 30, 2001?

Bishayee: Seven. [Hill noted 8:30 AM, 12:54 and 3 PM, there are seven. At 5 and 6 PM there are six, tube number seven was gone.]

Committee: Where did you throw them out?

Bishayee: In the trash.

Fine: Was there contamination?

Bishayee: [He did not check them. But note the first statement that indicates they were contaminated.]

Committee minutes: Bishayee had two sets of seven on Thursday March 29 [This is incorrect. On that day, he had only one, the second one (in front) appeared on Tuesday, April 3 and is documented in a photograph.]

Committee: What time on Friday, March 30 were there none?

Bishayee: Possibly mid-day [there were never "none"].

Raveché: Was he sure he had seven tubes for the NCL?

Bishayee: "…for the other study…I probably had five."

Committee: Had he previously tested the NCL?

Bishayee: [He stated he tested it a month earlier, a record of which is in his lab binder no. 9, which the committee had in hand.]

Committee: Why throw out the tubes of the NCL without recording the results?

Bishayee: [He explained he wanted to do other things; it was not a crucial experiment.]

Committee: What did he do with the tubes of the March 26 experiment?

Bishayee: Threw them out [There was confusion in the committee about where.]

Committee: What time did he go to the fluorescence-activated cell sorter (FACS)?

Bishayee: [He replied between 2 and 2:30 PM Friday. Sorting takes 4 hours.]

Raveché: Did he look at the NCL before or after going to the sorter?

Bishayee: Probably.

Raveché: What did he do after sorting?

Bishayee: Plated the cells – into the incubator between 7 and 8 PM (Hill and Lenarczyk leave at 6:30, Bishayee is not around).

Committee: Anything else in the 37-degree incubator?

Bishayee: "Other things were on top." [Starts new experiment Monday, April 2 with seven tubes.]

Committee: Was there anything else in the incubator?

Bishayee: Not sure.

Committee: When did he look at his March 26 dishes?

Bishayee: [He told them it takes seven days, so April 6.]

Committee: What did he find for plates?

Bishayee: All dyed contaminated, undyed were okay.

Bishayee to *Raveché*: He told her he didn't look at the tubes from March 27 (Tuesday) to March 30 (Friday). Also, he did not look at the NCL but per committee minutes, he said he had two sets of seven on March 29. [They show him the photos: the rack is his; it contains six tubes. He doesn't know when taken or why location changed.]

Committee: What happened to the seventh tube?

Bishayee: "No idea."

Committee: Did he ever take one tube out and leave the others?

Bishayee: "No." [There should have been seven according to the protocol.]

Committee: Is it possible to tell whether cells are contaminated?

Raveché: It would not be visible in the Helena tubes.

Raveché: Would he continue experiment with contamination?

Bishayee: He would not. [Recall that he received clean cells from Lenarczyk on Thursday evening, March 29 – FACS processing was March 30.]

Bishayee: Yes, he got them, doesn't know what for.

Committee: Could he have used them in the sort?

Bishayee: Why?

In Bishayee's final statement he indicated that he felt that he was the victim of a conspiracy, and he went on to state that he had had problems with Hill for two years. Bishayee believed that the origin of his problems with Hill was that Howell did not want to incorporate Hill's work into the grant. Bishayee added that he fought with Lenarczyk, who he believes had a conflict of interest because Lenarczyk was living in Hill's house.

When Howell was called to testify, he started by explaining that certain details of the experiment neither Hill nor Lenarczyk would have known. According to Howell, Hill and Lenarczyk believed that contamination existed with the two populations of cells at the point of plating; however, this would be hard to decipher until the seven days of growth were complete. In fact, he told the committee, while the plated Petri dishes of dyed cells were found to be contaminated after seven days and couldn't be counted, the undyed plated cells grew and were able to be counted in Howell's presence. The CCRI wanted to know where the cells in the plated Petri dishes came from and whether something improper could have been done to get the results Howell reported. While it was possible, Howell said, if someone was going to fudge, Bishayee (or any other scientist) wouldn't fudge the wrong population of cells. He went on to explain that each experiment focused on either the radioactive dyed cells or the bystander undyed cells. The amount of radioactivity used varied according to the focus of the experiment. The experiment in question focused on the radioactive cells that was different from previous experiments. Hill and Lenarczyk would be unaware of this, Howell reiterated. "It would make no sense," he suggested, "for Dr. Bishayee to substitute cells for the non-radioactive cells because they were not the focus of the experiment." Howell had nothing to say just then about the photographs attributed to Lenarczyk. When asked to comment on Hill's evidence that Bishayee's experiment was contaminated and that Bishayee knew it on Friday, March 30, Howell told the CCRI that Bishayee would have had no way of knowing about the contamination from observation of the Helena tubes. "The only way would have been if he had plated the cells at the beginning of the experiment." From the record of this hearing, it was clear that Howell did not have many answers. But he did try to shift the panel's focus. Asked for comment about Lenarczyk's sampling of the tubes, Howell responded that he didn't understand why Hill and Lenarczyk didn't confront Bishayee directly.

In response to the information in the committee's possession that Hill and Lenarczyk documented that Bishayee took tube number seven out of the incubator to complete the experiment (leaving the others) and then threw out the tube, Howell (also not a radiation biologist) speculated that if the cells were contaminated, he didn't know how well they would take up radioactivity. But contaminated cells would not grow colonies, he suggested, noting further that it's much more difficult to draw material for the experiment from one tube, rather than taking some material from each tube. Howell had no explanation when asked why Bishayee would have left the tubes in the incubator.

In an attempt to account for there being only six tubes in the 10.5-degree-centigrade incubator, Howell speculated, again, that they could have been the new cell line (NCL) tubes. The committee then reminded him that the tubes in the rack shown in the pictures had a radioactive label and that Bishayee had been very adamant that that's the way he could tell the racks apart. It was also pointed out that the tubes were full. Howell responded that the tubes would have had to be taken out of the trash and filled with medium. There is no evidence, however, of this ever happening.

Forrester asked Howell to comment on interpersonal relationships in the lab. "Dr. Bishayee," he told the CCRI, "has a good record of producing work. Dr. Hill has not produced original research in years, and Dr. Lenarczyk has been nonproductive in his nine months as a postdoc." Raveché wanted Howell's reaction to Lenarczyk's "lack of productivity." Howell remarked that the experimental protocol was very difficult. There are a number of steps that are prone to contamination, he observed. Bishayee had "one complete, two failed and one half-contaminated" experiment. "There is pressure to publish. Dr. Lenarczyk has produced no reasonable data and his position is guaranteed for only one year."

Raveché wanted to know if Howell could explain the surprising fact that only half the experimental tubes were contaminated. Howell could only suggest that it could have something to do with the dye. He knew that the dye was sterile, but the phosphate buffer used with the dye could have been tainted. He explained that after thirty minutes in the dye, the cells are washed, mixed with the unlabeled cells and then chilled. The bacteria would expectantly remain dormant and not infect the unlabeled cells. After further discussion of the color and placement of objects in the photographs. Raveché asked Howell if he could tell from the colony counts whether cells had been removed (when Lenarczyk sampled the tubes for radioactivity). Howell believed that if cells had been removed, he would be able to see that in the counts. But the results on page 32 of the Bishayee lab book are reasonable (as though no cells had been removed), although the count from tube seven was the lowest.

At the end of the hearing, it was clear that the greater sticking point involved the photographs. Were they taken as and when stated by Hill and Lenarczyk? Secondary to that the committee want to know "if Bishayee's testimony about his conduct of the experiment was truthful." No other evidence was available to them just then to either prove or disprove Bishayee's statements. There was also no evidence to confirm the validity of the photographs.

Hill had suspected Bishayee of fabrication of data since she began working with him under Howell in October 1999 and she reported this in her original allegation filed with Raveché, documented in the minutes of the Newark CCRI's April 11, 2001, meeting that first put the issue on the committee's radar. She described her observations of a specific experiment conducted by Bishayee in October 1999 involving irradiating mammalian V79 cells with the mutant gene HPRT. She told Raveché that she looked in dishes in which the cells were supposed to be growing colonies that would be counted as the result of the experiment. Under the microscope, she found that these dishes contained no colonies but, significantly, also no dead cells. She asked Bishayee about this at the time and he told her the dishes she looked at were for another experiment; however, Howell said there was no other experiment underway in the lab that would use this kind of dish. In addition, the dishes she examined disappeared from the lab shortly after she questioned Bishayee and she could not find them, not even in the trash. From this, Hill concluded Bishayee had fabricated the data for this experiment. She reported her suspicions to Howell who, she explained, did not believe her. She did not take the issue further at that time because she was not "absolutely certain" she was correct since she was unfamiliar with and had difficulty using the particular microscope with which she examined the dishes in question. She also thought that Bishayee might have been merely sloppy rather than dishonest. Hill provided Raveché materials relating to the October 1999 experiment, including the experimental protocol, copies of pages containing data from Bishayee's lab book, and her own original data from the same protocol.

What was clear to the committee, documented in Hill's statement and supporting evidence, was that Hill didn't want to be associated with Bishayee after the September/October 1999 experiment that she believed had gone wrong. "[A]fter that time, I determined that I would distance myself as much as possible from Dr. Bishayee and from the projects that he was working on. I went on to develop a related but different project that employed an entirely different cell line," she explained. "Between then and [when she reported the second suspected incident of scientific misconduct] I did not have the opportunity to observe Dr. Bishayee closely again. In conversations with Dr. Howell, however, the incident came up a number of times and I reiterated my concerns. I remember one time when I grudgingly admitted the possibility that Dr. Bishayee might just be sloppy but not dishonest." But there was more. "One thing that makes me feel very badly," Hill continued, "is that I allowed myself to be listed as coauthor on a paper with Dr. Bishayee as the senior author. I actually had little to do with the experiments that were reported, but at the time, I did believe that they were true. I admit that I was worried that I had had nothing to publish for some time, so I relaxed my principles and let them use my name. I should also say that I have allowed Dr. Bishayee to live in an apartment that is rented by me in Ivy Hill. I rented this apartment," she told the committee, "for my son who was disabled by a stroke." Hill offered that her son had become very depressed and was admitted to the Essex County hospital in Cedar Grove in the fall of 1999. "He has been there ever since. The rent for the apartment is very low – about $350 per month – and I did not want to let it go because my son has no resources other than his SSDI [Social Security disability insurance] income and I wanted to be prepared for the possibility that my son might go back there to live. This now seems unlikely," she reported to the committee. "Dr. Bishayee does not pay rent to me, but I have asked him to pay the equivalent of the monthly rent to a trust fund set up for my son's daughter."

During the time between the fall of 1999 and the second misconduct report to the Newark CCRI, Hill tried to remain friendly with Bishayee. "I felt that if I acted overly suspicious," she told them, "This would be very disruptive to the functioning and harmony of the laboratory. This became especially important as our numbers grew." On April 1, 2000, a new faculty member joined Hill's division, Edouard I. Azzam. "This is a person," Hill later wrote, "of absolute honor and integrity and with a fine scientific background. He has his own office, but we all share the laboratory space. He was joined in the summer by his wife, Dr. Sonia de Toledo. It was a great pleasure for me to have these two wonderful people working beside us. I have not mentioned to them any of my suspicions regarding Dr. Bishayee," she informed Raveché's panel. "At the end of August 2000, we were joined by Dr. Marek Lenarczyk, who holds the position of postdoctoral fellow [...]. Once the grant had begun, Dr. Bishayee was promoted from part-time to post-doctoral fellow to research associate."

Bishayee was not the only one to whom Hill provided housing. "My husband and I have a small bedroom and kitchenette in the basement of our house and Dr. Lenarczyk has been living there as our guest since his arrival," she disclosed to the committee. "As a result, I have gotten to know him quite well. Within a month or two of his arrival, he voiced to me his suspicions about Dr. Bishayee and his mistrust of the work that Dr. Bishayee had been doing. I shared with him my concerns." This is how Hill and Lenarczyk first came to commiserate about Bishayee.

Lenarczyk had joined Howell's lab specifically to work on Howell's NIH grant. Hill stated that Lenarczyk told her he became suspicious (a detail that matches Lenarczyk's statements before the committee) of Bishayee's work also, and she then shared her concerns with him. This led to their teaming up to observe and investigate the experiment conducted by Bishayee from March 26-30, 2001. Their investigations of Bishayee's experiment were without his knowledge and were also kept secret from Howell. Hill and Lenarczyk secretly tested Bishayee's incubating test tubes for bacterial or yeast contamination and attempted to monitor the number and location of the test tubes during the experiment, documenting and photographing their findings. As a result of this investigation, Hill told Raveché that she and Lenarczyk found gross contamination of the cultures that should have resulted in their being discarded. But according to Hill, the cultures were not discarded, thus calling into question the validity of any experimental results. Additionally, Hill stated that six of the seven tubes were not removed from the incubator on the day for harvesting the cells. As testimony before the April 27, 2001 CCRI revealed and Hill originally reported to Raveché, that she found the seventh tube that had contained radioactive substances, empty in the non-radioactive trash in the lab. Hill and Lenarczyk tested the tubes for radioactivity and concluded that Bishayee had used the contents of the seventh tube to add radioactive aliquots to the other tubes that were then measured in the FACS laboratory run by Thomas N. Denny[88], M.Sc., M.Phil., director, Center for Laboratory Investigations. All the tubes subsequently disappeared, and Hill could not find them anywhere, including the trash.

From these secret investigations, Hill concluded that Bishayee fabricated and/or falsified data from the March 2001 experiment because he could not have obtained any valid results otherwise under the circumstances in which the experiment was observed by Hill and Lenarczyk and should have been conducted. Hill provided Raveché with a copy of the evidence she gathered from her investigation of the March experiment, including a description of her daily activities and observations, photographs she and Lenarczyk had taken of Bishayee's tubes and the cultures made by Hill and Lenarczyk from the tube contents, radioactive counts of the material in the tubes, map of the lab and other supporting information.

Following that initial meeting between Raveché and Hill, accompanied by Howell, Raveché spoke with Putterman, and it was decided to protect the data in question by sequestering all of it. On April 10, 2001, Raveché proceeded to collect Bishayee's notebooks, binders, diskettes and Zip disks with the assistance of Howell, who identified all relevant material. Thirty-eight Petri dishes containing colonies from the experiment were also secured. These materials remained in Raveché's possession and were secured under lock and key, of which there were no duplicates – no one else was permitted access. But Raveché did make the materials available to Howell as required for the continuing conduct of his research project. Raveché was present each time Howell reviewed material. Howell informed her that Bishayee did not need to have access to the material, nor did he require copies. Howell also collected Bishayee's key and identification card and only permitted him to work in the lab under his direct supervision.

Following Raveché's description of the allegation and of the evidence given to her by Hill, the committee had voted unanimously that the allegations met the definition of misconduct in science under Public Health Service (PHS) regulations and university policy; and that there was adequate information for an initial query to proceed. Thus, the committee commenced its initial inquiry, the official start date of which was April 11, 2001, with the expectation of completing the same within sixty days from that date, a report to be filed with the senior vice president for academic affairs. The university's CCRI reviewed the six circumstances under which the ORI must be immediately notified of an allegation of misconduct in science as set forth in the university policy (Section V.H.). The committee decided that none of the conditions pertained to the current case and the ORI did not need to be notified at that time. Upon hearing from Raveché that the data in question were going to be presented at the meeting of the Radiation Research Society in Puerto Rico later in the month, per the April 11, 2001 CCRI minutes, the committee requested that she write to Howell, the study's principal investigator, and tell him to cancel all presentations of any data that might be tainted, and to withdraw immediately all abstracts containing such data.

The Newark CCRI convened again on May 9, 2001, to determine if credible evidence of misconduct in science was sufficient to warrant further investigation, and to decide whether to recommend remedial action be taken by Howell within his lab to address the problems of scientific oversight and interpersonal relationships. The committee reviewed the evidence and the interviews and unanimously voted that there was insufficient credible evidence of scientific misconduct to warrant further investigation. This conclusion was predicated on the following considerations:

> The major physical evidence in this case is the photographs taken by Drs. Hill and Lenarczyk. These photographs could not be dated definitively and could not be related definitively to the experiment that Dr. Bishayee said he performed from March 26-30, 2001. Therefore, the committee was unconvinced that the photographs credibly proved that the experiment Dr. Bishayee carried out was different from that recorded in his lab book.

> The evidence that Dr. Bishayee's March 26-30, 2001 experimental materials were contaminated from the inception of the experiment was insufficiently credible to support the complainant's contention that Dr. Bishayee could not have obtained the data he recorded from the experiment he actually carried out.

> The testimony of the complainant, Dr. Hill, conflicted with that of Dr. Lenarczyk as to dates, results of their

observations of Dr. Bishayee's test tubes, and what they did with Dr. Bishayee's experimental materials in their attempt to collect evidence of misconduct in the March 26-30, 2001 experiment.

Dr. Hill and Dr. Lenarczyk admitted to tampering with Dr. Bishayee's March 26-30, 2001 experiment, possibly before it was completed.

With regard to Dr. Hill's allegation of falsification/fabrication by Dr. Bishayee in the September/October 1999 experiment under the Banbury Protocol, the committee found insufficient evidence to substantiate this allegation from its examination of Dr. Bishayee's notebooks [provided at the April 27, 2001 CCRI meeting], from Dr. Hill's testimony about her observations of plates she found in the incubator, and from her statements following her review of Banbury Report No. 28 [attached to the April 17, 2001 CCRI meeting].

Although the committee discussed the possible motivations for Dr. Bishayee's alleged actions, it could discern no reason for Dr. Bishayee's falsification or fabrication of the data for the experiments of March 26-30, 2001, or of September/October 1999.

For all of the reasons aforementioned, the Newark CCRI found that there was insufficient evidence to reconcile the purported date of the photographs and what Hill believed they demonstrated about Bishayee's experiment of March 2001 with Bishayee's own testimony that he conducted the experiment as recorded in his lab and obtained the results as recorded therein. The committee further recommended Howell take corrective actions to improve the conduct of research and the environment in his lab.

But the committee's work was far from over. A meeting was convened on June 7, 2001, to consider additional material submitted by Hill concerning experiments performed by her and Bishayee in September/October 1999 on V79 HPRT mutants conducted under the Banbury protocol (M. M. Moore et al., "Mammalian cell mutagenesis," Banbury Report Number 28, 1987). Raveché told the panel she had discussed Hill's comments with her at length prior to introducing it to colleagues on the CCRI, and was familiar with the experiments, the techniques involved, Hill's analysis of Bishayee's data, and Hill's conclusion from her analysis. Raveché explained all of this to the committee, using Bishayee's original handwritten lab notebook data, and Hill's analysis and graphing of the data in Excel. There were two objectives to the experiments, a survival arm followed by a mutagenesis arm. On May 22, 2001, Hill met with Raveché and reported that on September 6, 1999, Bishayee began one such experiment jointly with her, with Bishayee performing the survival part and Hill the mutagenesis part. Hill went on to tell Raveché that on September 20, 1999, Bishayee conducted another one of these experiments, this time doing both parts himself. Hill told Raveché she reviewed Bishayee's survival data, including the Coulter cell counts of September 24 and 27, and October 1 and 4, and graphed his survival and mutagenicity results. Hill explained that she believed that the Coulter counts after irradiation do not show the expected difference between the controls and the irradiated cells. For example, the irradiated cells should be expected to have lower counts than the controls due to cell death or damage from the irradiation making it impossible for the cells to divide normally. Hill showed Raveché her own data from the same protocol she had carried out on September 6, 1999, which she said did show this difference. Hill concluded that, with these Coulter readings three days after irradiation, Bishayee could not have gotten the experimental results he did, which appear to be valid and as predicted for this experiment.

The committee documented that during Hill's interview on April 17, 2001, she had brought to their attention her concerns about the mutagenicity part of the Banbury protocol experiment from September 24, 1999. During that interview, Hill had explained that on October 11, 1999, following ten days of incubation, the plated cells were ready to be fixed and stained and the colonies counted. Hill said that Bishayee told her he was going to stain the plates on October 11. The next day, October 12, Hill said she became suspicious when she found a set of dishes of the number and type that would be used under this protocol still in the incubator. These were the plates she examined under the microscope, and found no colonies or dead cells that should have been present. These were also the Petri dishes that disappeared from the lab on October 14 and never recovered. Hill concluded from these occurrences that Bishayee had fabricated the mutation data from this experiment, or that he may have plagiarized the experimental results from the Banbury publication that had also disappeared from the laboratory. When a copy of the Banbury document was obtained from the university's library and shown to Hill, in Eder's presence, and stated she could not find any data that Bishayee had plagiarized.

The Newark CCRI, documented in the June 7, 2001 meeting, reviewed the steps in the protocol followed by Hill and Bishayee in September/ October 1999 and the specific techniques involved. They noted that high variability in counting cells using Coulter methodology is the norm, and that Coulter counts can be thrown off by technical flaws such as failure to adequately disperse the cells, the presence of bubbles and so on. Also noted was the fact that the Coulter counts were not integral to the experiment in question, but are incidental data not analyzed or used in the results; they are used, it was observed, only as a guide to determine how to dilute the cells to get the correct number of cells for the next step and to determine when the cells had undergone a total of ten divisions. But they did agree that the pattern of Coulter counts in Bishayee's experiment showed inconsistent effects of irradiation compared to the non-irradiated controls. While the members considered consulting an expert in the Banbury protocol, they opted instead to recall Bishayee to speak to them again.

Bishayee agreed to appear before the committee and was interviewed the same day (June 7). The minutes of this meeting reveal that he was provided with his original notebook containing the records of the experiment under discussion in order to refresh his recollection since it had taken place eighteen months before. He explained that plating for survival is done on day zero of exposure (irradiation) and the plates are read seven days later. In his experiment, Bishayee stated that September 24 was day zero (irradiation), which he was able to confirm by pointing to his records in the notebook. If September 24 was day zero, Coulter counts were not expected to show any significant difference between controls and irradiated tubes. He reviewed for the committee the Coulter counts for September 27, day three, at which time differences would be expected. Bishayee and the committee noted that except for tubes five and ten whose counts appear too high for the highest radiation-dose tubes, the expected difference in counts was, in fact, observed (tubes three and four had lower counts than tubes one and two, and tubes eight and nine lower than tube six and seven). The committee agree with Bishayee that the counts in tubes five and ten, although not fitting the expected pattern, were within experimental error. Additionally, Bishayee explained to the panel why even on day three one might not necessarily see survival effects of irradiation (for example, because cell death or damage might not occur right away but be delayed and appear later in an exponential fashion). Survival effects are known to occur for certain by day seven, which is why the plates prepared on day zero are read seven days later for survival. Satisfied with Bishayee's explanations, the committee decided an expert was unnecessary. The committee declined to investigate Hill's allegations of misconduct against Bishayee for the September/October 1999 experiments.

Hill did not accept the Newark CCRI findings. After the final ruling came down, she was subjected to documented retaliation in her department, including being locked out of the shared laboratory and being denigrated by Howell. In a memorandum dated July 2, 2001, to Howell from Stephen R. Baker, M.D., professor and chair of the department of radiology, Howell is appointed chief of the division of radiation research. "If you accept," wrote Baker to Howell, "your leadership responsibilities will encompass the previous division of radiation research and the section of cancer biology [Hill's area of expertise], now to be combined into one division under the appellation of radiation research." Hill's section was being eliminated. In a world in which perception is everything, was the abolishment of Hill's section just bad timing on Baker's part or intentional? Howell accepted Baker's invitation to serve as chief.

Howell's July 30, 2001 memorandum addressed to Azzam, de Toledo (Azzam's wife) and Hill dated July 30, 2001, addressed laboratory space and dispensation thereof; in it, he dictated the space each would have, including offices and laboratory assignment, effective August 3, and at the bottom of the communication, Howell informed them that "laboratories and offices will be rekeyed. Faculty will be provided keys to access their assigned areas only. Postdoctoral fellows will only be provided keys to the laboratories of their mentors." He copied Baker.

Hill questioned Howell about the stipulations of the July 30 memorandum, to which he replied in another memorandum dated the following day and handed directly to her, putting her on notice that "we must protect ourselves from any retaliation by Bishayee:

> Thank you for your inquiry regarding access to MSB F-451 [Hill's former lab space] as a consequence of the recent reorganization of the laboratory space within the division. You have requested clarification of your privileges to access laboratory space within the division. As stated in my memorandum of July 30, 2001, you have been assigned laboratory space MSB F-468. The laboratory MSB F-451 has been assigned to R. W. Howell [himself]. According to the memorandum, you will be denied access to MSB F-451 as of August 3, 2001. By virtue of the layout of the laboratory space, this will also restrict your access to MSB F-451a and MSB F-451b. To facilitate continuation of your collaborative work with Drs. Azzam and de Toledo, I have assigned MSB F-468 as shared laboratory space with Dr. Azzam. Equipment of yours that is necessary for your work will be moved to MSB F-468.

> You have also asked for access to MSB F-451 during my vacation so that you can work closely with Dr. de Toledo during my absence. Drs. Azzam and de Toledo have also informed me that cells will require refeeding during their absence in the latter part of August. Accordingly, access to MSB F-451 is granted during my absence through August 28, 2001. The keys of Drs. Azzam and de Toledo can be used for this purpose.

Howell copied Azzam, de Toledo and Baker. There was no need to copy Lenarczyk or Bishayee. Lenarczyk returned to Poland on July 26, and

Bishayee resigned on July 30, the same day Howell sent his missive to the department. Hill had some idea of what was coming. Azzam had called her over the weekend, on July 29, "very distraught, warning me I will be locked out of the lab. [Bishayee] is gone and has no more access to the lab. The locks were all changed on Friday," she recorded in the timeline. "I will only have access to my office and F-468, the little lab."

Prior to Bishayee's resignation, Howell purportedly had frank discussions with him concerning the ill will in the lab and the "uncomfortable working conditions" due to the dispute between Hill and Bishayee. "Dr. Howell felt that Dr. Hill would not leave Dr. Bishayee along and therefore, it was best for him to leave the lab," according to interrogatories provided by Howell on April 16, 2002, pertaining to Hill's *qui tam* case. Howell, per this document, encouraged Bishayee to find a position outside the university entirely. For this reason, Howell did not give Bishayee a letter of recommendation for any position internal to the campus but would have done so if Bishayee had found a position elsewhere. Bishayee had other ideas. Howell subsequently became aware that Bishayee left his department to work in the department of pharmacology and physiology and even ran into him on occasion. Bishayee told Howell, in truth, that he was applying for a job in radiation safety, and also that he had just gotten married. The rapport of mentor and fellow had not ended with a change of venue.

Hill had met with Bishayee on August 1, the day following his resignation; she later documented that Bishayee was apologetic that he had hurt her. Interestingly, she also noted that Bishayee told her that Howell had handed Bishayee a letter of resignation and told him to sign it. Bishayee was, she continued, "afraid not to sign lest Howell not give him a recommendation." In Hill's chronology of what happened next, prepared for the federal court in her *qui tam* case, Howell told Bishayee not to come [the] campus and took away his keys and identification. Further, Howell reportedly told Bishayee not to apply for research positions on the campus or in radiation biology anywhere. Bishayee's H-1B visa was changed to a visitor visa, good for only three months, renewable one time for three additional months. When Hill reportedly asked Bishayee if he could continue to live in her son's apartment, he told her yes. Bishayee told Hill that his evaluations up to the point of termination had all been good.

Hill: How had the original bystander observations come about?

Bishayee: Some experiments showed it, some did not.

Hill: Did [Howell] pick and choose experiments?

Bishayee: "Yes."

Hill: Were you pressured to produce data?

Bishayee: [No direct answer was provided but Bishayee indicated that Howell was always pleased when data fit his mathematical model.]

Hill reported that she advised Bishayee that rules dictate that he cannot be fired since he was exonerated of scientific misconduct by the Newark CCRI.[89] She recommended just then that he consult legal aid to engage a public defender. Bishayee demurred. Hill characterized Bishayee as afraid of Howell.

Hill did file a grievance with Karen Putterman, M.D., MPH, UMDNJ vice president for academic affairs, on August 13, 2001. The subject line of the confidential memorandum was "Retaliation." She informed Putterman that Mark Schorr, an attorney for the American Association of University Professors (AAUP), suggested she provide the university the opportunity to make things right without going through the formality of the grievance process before exercising her right to file suit. Retaliation against whistleblowers is prohibited by law and the university, he explained, was bound to protect her.

"I presume," Hill wrote to Putterman, "that you will do your best to rectify this matter and assume that I have your assurance that this will be the case." Schorr further suggested, and Hill conveyed to Putterman, that she send Putterman a copy of the grievance that she had prepared in order to properly record the events and her desired remedy. "Please note that I have not included attachments as you have those documents in your possession already." This was a reference to the prior scientific misconduct reporting and the department memorandums that soon followed between Howell and division scientists.

The text of Hill's proposed grievance makes it clear that dysfunction prevailed in the lab in the wake of the Newark CCRI clearing Bishayee of wrongdoing:

On April 21, 2001, I reported to the chairman of the Newark Campus Committee on Research Integrity, Dr. Elizabeth Raveché, acts committed by a fellow in the laboratory of Dr. Roger Howell that I believed satisfied the university definition of scientific misconduct. University guidelines require faculty members to report suspicions of misconduct, and my report was made per these requirements. University policy also protects the reporter of possible scientific misconduct from retaliation. On June 22, 2001, Dr. Raveché reported to me that there was 'insufficient definitive evidence to warrant further misconduct-in-science proceedings in this case.'

On July 30, 2001, Dr. Howell sent me a memo [which she attached], and on July 31, 2001, a second memo [also attached] that have barred my access to the laboratories of the Division of Radiation Research thereby preventing me from effectively carrying out my research in collaboration with Drs. Edouard Azzam and Sonia de Toledo. It needs to be understood that, since I am not allowed to enter MSB-F451 – the door must be locked at all times because of the radioisotopes that are contained therein – I cannot gain access to MSB-F451b where they do most of their work. This will seriously impair our collaborative endeavors. Furthermore, although Dr. Howell has offered to move my equipment (2 large incubators) into the small laboratory assigned to me – MSB-F468 – this is not feasible as there is no more room in that laboratory. There are a number of other shared instruments and fume hoods that I would not be able to use as they would remain in MSB-451. Dr. Howell has not given me any reason for this action. He did, however, on July 6, 2001, say to me, "I do not want to have anything more to do with you." This was followed on July 11 by a copy of a letter to Dr. Howell from my department chairman, Dr. Baker, sent to all the members of my department which was obviously designed to chastise, demean and humiliate me [she attached this communication also]. My name was not mentioned, but the intent of this memo is clear. This letter effectively put Dr. Howell in complete control over my activities in the Division of Radiation Research.

I reported what I believed was misconduct in good faith. It was my understanding that I could do no less as a responsible member of the faculty and the university community. I am deeply saddened that the result has been what appears to be retaliation that impedes my own research.

I believe that Dr. Howell, supported by Dr. Baker, barred me from the division laboratories in retaliation for my report regarding [Howell's] fellow. I know of no other possible explanation.

Hill wanted only a restoration of *status quo* regarding access to division laboratories prior to Howell's July 30 memorandum. That is, she wanted to be issued a key to MSB-F451 so that she could enter and leave that suite of laboratories at will to continue to use instruments and equipment to which she had previously had unfettered access. She also asked that Howell no longer be her supervisor.

Hill met on August 13, the same day she registered her retaliation claim in writing with Putterman, with Baker in the presence of Emanuel Goldman, Ph.D., a professor in the department of microbiology and molecular genetics, and Karel Campbell, Baker's department coordinator. Baker, department chairman, informed Hill that he would instruct Howell to allow her back in the lab. But he declined to remove Howell as Hill's supervisor. At this point, the university had not shown Hill at record of her testimony. Azzam would subsequently confirm what Bishayee had told her: that Howell handed Bishayee the letter of resignation; instructed Bishayee to stay off the campus and not apply for research positions there or in any other radiation biology program. Azzam, it was later learned, encouraged Howell to dismiss Bishayee with the aforementioned conditions. How did Azzam know this? Bishayee was visiting Azzam's house to use his computer to look for employment. Bishayee also spent extended hours in the computer center of the medical school doing the same. Pranela Rameshwar, Ph.D., a professor in the department of medicine, reportedly confirmed to Hill that Bishayee applied for a job in medicine with Stephan K. Schwander, M.D., Ph.D., an associate professor in the Robert Wood Johnson Medical School. When he asked Howell for a recommendation, Howell purportedly told him to "keep looking." From August to November 2001, Bishayee came by Hill's office to pay for the apartment, first calling out "How is the weather?" This was code, Hill explained, for "Has Roger gone home?"

From November 2001 to February 2002, it was clear, by Bishayee's own admission, that he was broke, Hill wrote in the timeline later submitted to the federal court as part of discovery. He told her he couldn't afford the apartment but asked if he could stay anyway. She said yes and didn't deposit Bishayee's November check until March. The relationship between Hill and Bishayee was not acrimonious.

When Hill met with Baker, accompanied by Cynthia J. Stolman, Ph.D., assistant professor of pediatrics, for her annual evaluation on September 4, 2001, she noted that Baker was clearly angered that in her letter to Putterman – the retaliation grievance – Hill implied that Baker was complicit in the lab lockout. Baker asserted his prerogative to appoint division chiefs. He further informed Hill that she should not be locked out of the lab but he wouldn't get involved in "our disputes." Lab notebooks that had been previously retained by the Newark CCRI as part of their inquiry into Hill's allegation of scientific misconduct were returned and Howell sequestered the notebooks in his office where Hill could not gain access to them. A few days later, on September 13, Hill was scheduled to participate in a teleconference in which she recognized none of the other faculty participants. The subject matter was "Conducting Research Responsibly." As a result of this teleconference, Hill learned that it is misconduct once the falsified data is recorded in the notebook. The notebooks in Howell's possession now took on wholly new importance and as Hill was soon to learn, even the Public Health Service's Office of Research Integrity (ORI) didn't seem to be aware of what she had just learned about transposition of falsified data to the notebook.

Chapter 2 –
Nobody Likes a Snitch

Shortly after Roger Howell's July 30, 2001 memo circulated through the division of radiation research, Helene Hill took her suspicions of scientific misconduct to federal investigators at the Office of Research Integrity (ORI). Kay Fields, Ph.D., a scientist investigator and Public Health Service staff fellow, took the case. Hill first spoke to Fields on August 16, 2001, and it was just then that Fields advised her that the university should show Hill a copy of her testimony before the Newark Campus Committee on Research Integrity (CCRI). She also advised that Roger Howell's publications with Anupam Bishayee should be retracted. Hill sent materials pertaining to the case on August 23 and it would take the ORI a year to come to the same conclusions as the first CCRI committee.

As prequel, Hill did not have access, at this point, to any of the CCRI proceedings, including her own testimony and despite the fact that Fields informed her that she should have a transcript of her own appearance before the committee, she had yet to receive it. In a December 11 telephone conversation with Fields, which lasted almost one hour, Fields and Hill conferred on the way ahead, what to do next to determine whether the ORI could build a case for scientific misconduct. Hill and Fields discussed the need to break down the data that went into the papers coauthored by Anupam Bishayee. She wanted to know what figures were false in the two bystander papers. One question that also required a response: Why didn't Roger Howell retract the papers if the data could not be repeated? Fields observed that it is not misconduct to have contamination but it is misconduct to knowingly continue an experiment that is contaminated. She advised Hill, per Hill's notes of the conversation, that Hill would be surprised to learn what Howell told the Newark CCRI. Field informed Hill that Howell had told the committee that the contamination was in the dye, but that would still contaminate everything. Further, she revealed to Hill that Howell should have discussed the falsification issue with his assigned program officer at the NIH and also, she reiterated, retract the papers. Hill wanted to know how she could withdraw as coauthor of the paper published with Bishayee, although it was discussed in this call that he was "a bit player" in a greater drama.

Fields supported Hill's findings but when she reported her results to the ORI board, Fields was overruled by her boss, Alan R. Price, Ph.D., the director, Division of Investigative Oversight [DIO] and associate director of the ORI, in part because he believed that the control data for the analysis, in this case Hill's own work, were statistically questionable.[90] Fields took herself off the case and her report was redacted by Price though it is disputed whether she took herself off Hill's case or was dismissed from it. Both reports became available to Hill during the discovery period of the *qui tam* case, which means that she had no idea what had been reported for years to come. "Just before the ruling by the ORI board [August 2002], Dr. Fields told me about a statistical method for analyzing the rightmost terminal digits of numbers that should be random or uniform as an indication of potential scientific misconduct," Hill wrote later. "These were numbers recorded from the LED screen of the Coulter counter, a [device] used for quantifying the cell numbers in virtually all of the experiments done in Howell's lab. The ORI had analyzed the numbers in that first experiment in 1999 and found that they were highly unlikely to be, as they should have been, random or uniform. Fields told me," she continued, "how to perform the analysis, which I did and found, indeed, that the Coulter counter numbers that Bishayee had produced in that 1999 experiment had a probability of being random on the order of 1 chance in 10 [trillion]. This method is now posted on the ORI website as a forensic tool to use when scientific misconduct is suspected."[91] Both Fields and Price indicated that Hill should return to the university and present the latest analysis to the CCRI. In a letter dated September 5, 2002, Price wrote Hill, in part:

> Since Dr. Kay Fields told me that she encouraged you, when you called her last month, to submit any new allegations or evidence directly to the University of Medicine and Dentistry of New Jersey, the institution responsible for investigating the referenced case, I assume you have done so.
>
> I also assume that Dr. Fields told you to contact me, since she had removed herself from the case. When you called me on August 14, 2002, I told you that I thought Mr. Chris Pascal would be making the decision for ORI in the next couple of weeks, and he did so, before your recent letter [of August 22, 2002] arrived. The closeout letters have been mailed, so you should be receiving one from him soon.

Though Fields was removed from Hill's case [her supervisor Price stated in correspondence that she removed herself], she had already indicated to Hill via a telephone conversation that she believed Hill had a good case for *qui tam*.

Fields' original findings are noteworthy. Later provided to Hill, they provide material Fields included and Price redacted. The case, as indicated, came to ORI in August 2001 as a request that the ORI review the UMDNJ's Newark CCRI decision. ORI verified that the questioned 1999 research data had been included in a grant application submitted to and subsequently funded by the Public Health Service (PHS), and that the questioned 2001 experiment was research funded by the PHS. On September 4, 2001, the Division of Investigative Oversight (DIO) of ORI asked Karen Putterman, M.D., M.P.H., UMDNJ vice president for academic affairs, to send the institution's inquiry report to ORI for review. The PHS had three major issues to sort out:

1. That Dr. Anupam Bishayee fabricated or falsified data in an experiment in September/October 1999 in which he measured the cell survival and induction of mutations following the irradiation of cultured mammalian cells with cesium-137. The

questioned data was included as Figure 7 in grant application 1 RO1 CA83838-01A1.

2. That Dr. Anupam Bishayee falsified data of an experiment done March 26-30, 2001, on the viability of 'bystander cells' incubated for three days in the cold in contact with cells that had incorporated tritiated thymidine into their DNA and the separation of those cells by fluorescence activated cell sorting. This questioned research was supported by PHS grant RO1 CA83838.

3. That Dr. Anupam Bishayee falsified data showing an effect of tritiated thymidine on the survival of bystander cells presented in a PHS grant application (RO1 CA83838-01A1, Figures 2 and 4) and in two published articles, Bishayee et al. (1999)[92], *Radiation Research* 152, 88-97 (Figures 3 and 6) and Bishayee et al. (2001)[93], *Radiation Research* 155, 335-344 (Figure 2A). [Price redacted this issue from the final ORI report on Hill's case.] The questioned data appeared in the PHS grant application CA83838-01A1 and was supported by PHS grants RO1 CA83838 and S10 RR14763-01.

The discussion within ORI focused on the fact [which Price redacted] that the DIO did not concur with the institution's determination that there was insufficient evidence to warrant an investigation, and that DIO request that UMDNJ proceed to an investigation. The basis for this recommendation came about because (1) examination of the 1999 data by DIO suggested that the numbers showed unexpected properties: very striking reproducible replicates and repetitive appearance of certain numbers and digits in purportedly machine generated numbers; (2) the inquiry was incomplete, since it did not verify the experimental basis of the bystander effect on survival, which was crucial to the funded research grant, even though this finding was alleged to be not reproduced by others in this laboratory and, therefore, was falsified (erroneous); (3) the committee apparently was not sufficiently knowledgeable to evaluate these experiments; (4) the inquiry committee appeared to be biased against the complainant; (5) the head of the laboratory, when asked to provide further information, failed to provide experimental data to support the claimed bystander effect (of tritium decay) that was the basis of his funded research.[94]

Hill sought confidential status when she first contacted ORI in August 2001 to discuss her allegations of scientific misconduct going back to September/October 1999 and March 26-30, 2001. The laboratory had published data suggesting that the effect required the formation of gap junctions between irradiated and bystander cells; however, the validity of the bystander effect on survival was challenged in the laboratory in 2000/2001 by the data of a senior postdoctoral associate, Marek Lenarczyk, who was unable to confirm the bystander effect on cell survival. Hill also claimed what she had all along, that Howell himself wasn't able to replicate Bishayee's key experiments, suggesting that Bishayee had falsified data. She also made ORI aware of Howell's retaliation against the respondent, the complainant and the witness after the inquiry was completed, taking actions that included forcing the respondent to resign, not renewing the research appointment of Lenarczyk, a witness to the alleged falsified research, and, with the cooperation of the department chairman, Stephen Baker, abolishing the research section of the complainant (cancer biology) and attempting to exclude her from the department laboratory. Hill considered these actions by Howell indicative of someone attempting to protect and retain his grant funding, even when the bystander effect was not reproducible.

The ORI was able to get specifics of Hill's case that she did not have in the early stages of her effort to press the issue of scientific misconduct against Bishayee, and, later, Howell. ORI reported that UMDNJ Newark CCRI chair Elizabeth Raveché, Ph.D., sequestered "32 binders, 4 notebooks, 46 diskettes, 7 Zip disks and 38 Petri plates." Further, Hill had provided Raveché a binder containing her written allegations, narrative, diaries, photographs, copies of original data from Bishayee's notebook and original data generated by Hill herself. Later, the committee obtained additional materials: the grant application, "all publications on which the grant was based," all publications appearing subsequent to receipt of the grant that reported data developed under the grant, all abstracts pending presentation, and the biographical sketches of Bishayee, Hill and Howell. These materials were stored in Putterman's office and reviewed by the committee as required. Bishayee had been informed in writing on April 12, 2001, that the inquiry committee was considering "questions about whether you falsified or fabricated data for NIH grant RO1 CA83838.

The inquiry, as discussed, included the aforementioned experiments in 1999 and 2001.

The DIO reviewed the university's inquiry report and its attachments. First, the DIO looked at expertise. Four of the inquiry committee members were deans and did not have Medline citation records indicating bench science backgrounds. Raveché, a pathologist, may or may not have had the appropriate experience to evaluate the questions about the performance of these experiments; however, the complainant (Hill) alleged that Raveché suggested that Howell simply terminate Bishayee when Raveché was first informed of the allegation. If Raveché did so, that was clearly inappropriate, according to the DIO finding. Hill later alleged to ORI that Howell forced Bishayee to resign; however, Putterman was not able to confirm this, although she did find that Howell had not written a letter of recommendation for Bishayee. Putterman stated that such a letter was unnecessary because Bishayee's next appointment was within UMDNJ.

The university provided no information about the professional experience of the committee members relevant to the questioned experiments, which involved radiation damage to *in vitro* cultured mammalian cells. Several statements in the inquiry report indicated a lack of such knowledge. For example, according to the ORI findings, Hill had questioned the counts of induced HPRT mutations in assays where colonies of surviving mutant (thioguanine-resistant) cells were evaluated. It was very unsettling to DIO to find that the inquiry report referred to this 1999 experiment as "irradiation of mammalian V79 cells with the mutant gene HPRT." They also considered it harder to substitute new cells than to restart the experiment in 2001, a pretty naïve statement.

Raveché wrote to Hill that she would be given an opportunity to comment on the report per a letter of April/May 2001; however, Hill stated to DIO that she was not provided with a copy of the report nor with that portion of the report that described her views. Perhaps as a consequence, Hill strongly objected to the decision of the inquiry, and the committee had no opportunity to correct factual mistakes in the report. When UMDNJ sent the report to ORI, no comments on the inquiry were included from any of the participants. This is a red flag.

DIO examined the allegation that Bishayee had fabricated or falsified data in the September/October 1999 experiment, largely on material provided by Hill to the inquiry committee and materials sent to the ORI. The material included a set of data and a protocol for the two experiments carried out by Bishayee and Hill in 1999. The protocol for the evaluation of radiation effects on viability and mutation of cells was described in the grant application as the Banbury protocol. The notebook materials included two graphs that showed ORI the induction of mutations (in the HPRT gene) in two sets of cell samples represented by circles and squares on pages dated September 28, 1999. For the experiment done jointly the circles were data for mutant cells obtained after irradiation of resuspended cells (aerobic), the squares were indicated as cells irradiated when the cells are clustered (loose cell pellets created by low speed centrifugation) (hypoxic). Both samples were exposed to variable doses of strong gamma radiation from cesium-137. In the second experiment carried out by Bishayee alone, the results were again graphed, and, as before, circles were resuspended cells, and squares were clusters. In the grant application, this same data was included in the preliminary results section as Figure 7A (survival) and 7B (induction of mutants). The Figure 7 legend noted that all the cells had been incubated (as cell pellets) at 10 degrees centigrade for three days and then exposed with or without resuspension to varied doses of gamma irradiation from cesium-137. This information agrees with the protocol contained in the material provided by Hill.

In the text of the grant, the result of this experiment was described as showing that slightly more mutations and more killing of cells were obtained by irradiation or resuspended cells than by irradiation of clustered cells. This seemed to match the conclusion of the graphed data from Bishayee's experiment number two (and experiment number 1). The data shown in Figure 7 of the grant resembled the graphed data obtained by Bishayee for the experiment (number two) done in October 1999; however, on closer examination, DIO noted that the curves in Figure 7 of the grant application did not accurately represent the result obtained: the sets of samples (curves) had been mixed up. The curves showing more killing and more mutations were drawn with filled squares in Figure 7, and according to the figure legend, these were supposed to be samples of cells irradiated "intact" – for example, as pellets. The second curves were filled circles;

they showed slightly less killing, and fewer mutations. By exclusion, they should have been resuspended, irradiated cell samples. In the figure legend, resuspended cells were supposed to be open squares, but there were no open squares in the graph. DIO concluded that between the notebook and the Figure 7 legend, the curves had been mixed up. Thus, the results recorded by Bishayee in experiment number two were summarized in the text in agreement with the data in the notebook by the statement that "the cells that remained in clusters were somewhat more resistant to killing by acute gamma irradiation relative to those [cells] that had been resuspended," but the graph suggested (and suggests) the opposite result. From Hill's prior statement and supporting documentation, it is known that Bishayee did this October 1999 experiment "completely on his own. It was only after this experiment was said to be complete," she told investigators, "that I found 100-millimeter dishes in the 37-degree incubator with no colonies on them."

It appeared to the DIO that whoever prepared the graphs for the grant application managed to mix up the symbols and even described a symbol in the figure legend that did not appear in the figure itself. The inquiry committee at UMDNJ evidently did not notice, or if they did notice it, did not mention this discrepancy in its final report. But it was clear to the DIO, most especially Fields, that the data Hill criticized as fabricated was used in the grant. The contradiction between actual or stated results and its graphic presentation in the grant is most likely due to honest error or just carelessness, since it clearly did not adequately support the text.

The data for the experiment that Hill had done in September 1999 with Bishayee (experiment number one dated September 6, 1999 to September 28, 1999) was also submitted to the committee by Hill. The results Hill obtained showed no reliable increase with dose of mutants/cell in the cells irradiated under hypoxic conditions, that is as clusters (cell pellets, filled squares) and did show an increase in mutants/cell when the cells were irradiated in suspension (aerobic condition) (filled circles). This experiment supports the statement that cells in suspension were more sensitive to the mutagenic and toxic effects of irradiation than cells left in pellets, as stated in the text of the grant application. Thus, the result obtained by Hill in the mutation arm of experiment one, even with rather erratic values for the (hypoxic) clustered cells, did not contradict the statements in the grant. Hill was not objecting to the results Bishayee claimed to have obtained as a wrong or contradictory result, but her objection was that he had obtained his results by fabrication of this data. This conclusion (by Hill) was predicated on (1) the tissue culture plates full of medium that she observed in the incubator. She considered these plates, which had not been fixed and stained, as evidence that Bishayee had not actually counted surviving and mutant colonies of cells as he claimed. This interpretation was reinforced for Hill by (2) Bishayee's inability to produce the stained tissue culture plates that he claimed to have counted, and (3) Bishayee's claim that he had a second experiment going on involving the plates in the incubator, but had no protocol or data from the experiment, and (4) Howell knew of no second experiment.

Hill noted that Howell went ahead and added this disputed data to his grant application. The DIO found no data on mutant induction in the two published papers mentioned in the allegation. According to Hill, Lenarczyk was carrying out experiments involving the induction of mutants by radiation, but he could not confirm the bystander effect on cell viability. From the summary of Lenarczyk's Newark CCRI interview, this was not discussed with the committee, and the committee may not have known about this issue. No details were provided of his experimental system and results, and he was not asked to evaluate the 1999 data obtained with cesium-137 radiation. Lenarczyk, of course, had not been present in the laboratory in 1999.

In a memorandum and an interview with Raveché, dated May 22, 2001, Hill questioned the reliability of the Coulter counter data recorded by Bishayee for the mutation arm of experiment two. Hill argued that the recorded cell counts in the mutation arm exceeded the expected cell survival by six- to tenfold. She provided comparable data for the samples in the immediately preceding experiment number one, where she herself had carried out the mutation arm. The protocol required that the number of cells be followed at intervals of ten days before plating equal numbers of cells on the selective medium to measure mutation rates.

When the CCRI asked Bishayee for an explanation of the cell counts that he did with a Coulter counter, he admitted that he had not observed an effect on cell survival when he counted cells, but stated during his interview that the effects on the survival and growth of irradiated cells might be delayed and not evident when the cells were counted on day three after irradiation. [The protocol called for counts on days 0, 3, 7 and 10 days after radiation, and his dated Excel sheets showed counts on days 0 (September 24, 1999), 5, 7 and 10.] The CCRI report indicated that the members were satisfied with Bishayee's explanation. DIO reviewed the counts recorded by Bishayee for this experiment. Although the university's committee considered Coulter counting to be subject to variations, Bishayee's counts in this experiment were remarkably close for the replicate samples. The counts for three samples from each culture barely showed the variation expected from recounting the same sample (square root of N) according to a simple analysis done by DIO. When Howell was asked about this variation, he claimed to Putterman that he also obtained Coulter counts that were in close agreement. In contrast, Hill obtained quite highly variable Coulter counts in experiment one; in fact, she switched to counting cells with a hemocytometer. Hill's cell counts on September 20, 1999, appeared to show a twofold decrease in total cells due to radiation (comparing sample four versus sample two); however, Hill stated that she expected a tenfold reduction in cell count in Bishayee's experiment (number two). Hill would have had to provide more explanation of these differences for the DIO to make a determination. DIO was most impressed by the small variation in Bishayee's recorded Coulter counts, and not in a positive way.

The DIO finding reported that Howell ignored the objections of his senior colleague and used data from Bishayee's experiment in the grant application in the absence of verifying counted tissue culture plates. Hill was listed as an experienced co-investigator, and she was the only person on the grant application who had experience with mutagenesis. Howell, on the other hand, was trained as a physicist, and his experience with cell culture was minimal, if judged by his publications, which the DIO obtained during its investigation. It seemed extraordinary to DIO that the Newark CCRI dismissed the testimony and judgment of Hill in this matter, since she had twenty years of experience and any publications in the field of research in question. The committee also accepted Coulter count data that may have been too precise to be (1) likely and (2) have been accurately reported.

According to Hill's written allegation, she reacted to the first incident where she suspected that Bishayee had fabricated data (in October 1999) by informing Howell. Apparently, he took no action except to inform Bishayee (who then disposed of the plates in the incubator) and to retain the preliminary data (inaccurately graphed) in his grant application. The Newark CCRI did not appear to take Hill's allegation about the 1999 experiment seriously. The DIO found that Bishayee's explanation that the plates were in the incubator for a second experiment was not supported by any evidence, and the committee did not pursue that question with Bishayee or Howell. Howell was not asked if the disputed experiment on mutations induced by cesium-137 irradiation of aerobic versus hypoxic cells was ever repeated. If so, the primary data of the later experiment could have been compared to what was in the grant and could further have established to what extent Bishayee's Coulter counts usually varied. In the grant application, further experiments were proposed that involved mutagenesis.

According to Hill, Lenarczyk was measuring mutagenesis in this laboratory, but he was not asked what results he obtained, and whether he was able to replicate Bishayee's experimental data. Lenarczyk's data and notebooks, in the DIO finding, were presumably left with Howell and should have been available to the university's CCRI. Lenarczyk, who had returned briefly to Poland after leaving UMDNJ, had come back to the United States to work in Colorado at the time of the DIO investigation.

More concretely, the DIO noted, it looked at the numbers recorded by Bishayee as his Coulter count data for four dates in 1999 – October 1, 4 and September 24 and 29. Those numbers were entered into a spreadsheet. DIO observed a reuse of two numbers – 72 and 56 – and the high frequency of 1, 2 and 9s in the terminal place of these three digits. The numbers do not appear to be statistically compatible with an origin in an unbiased counting device, such as the Coulter counter.

DIO noticed, too, that the inquiry conducted by the UMDNJ committee ignored or dismissed much of the evidence brought forward by Hill. The source of the dismissive attitude was not clear in the material sent to the DIO. But the DIO did determine the following:

1. The committee did not comment on Dr. Hill's allegation that the recorded data of colony formation did not match the graphic rep-

resentation of the data. The institution did not provide materials that would have allowed DIO to review this claim.

2. The report did not comment on the significance of Dr. Bishayee's lack of any notes to support his claim that he was performing a second experiment in October 1999, even though he had claimed, according to Dr. Hill, that the dishes that she observed and questioned were part of a second experiment he was doing. Dr. Hill reported that the plates disappeared promptly from the incubator and from the laboratory as soon as she questioned Dr. Bishayee. The committee report did not comment on this, nor did they appear to have asked Dr. Bishayee whether he had discarded the plates that raised Dr. Hill's concerns.

3. Dr. Howell had been involved in writing and revising his grant application around the time that the 1999 questioned experiment was done. He signed the grant face page on October 21, 1999, and the application was due at NIH on November 1. Dr. Hill had complained specifically that Dr. Howell had gone ahead and used data whose veracity she questioned. However, it is not clear from the report that the committee examined the data that was incorporated into the grant application 1 RO1 CA 83838-01A1 as Figure 7. DIO has found that the reporting of Dr. Bishayee's data from this experiment was not accurate in the grant, where the data for clusters and resuspended cells have been mislabeled in the figure. The committee did not examine the data closely, did not notice this, or noticed it and failed to comment on it.

The committee failed to determine whether Dr. Howell, given the question about the mutation data, had instructed Dr. Bishayee to repeat the experiment, and whether the claim in the grant application was supported by later data.

The DIO determined that there should have been a discussion at university level as to whether the issue should have proceeded to a review by an investigation committee, especially concerning the Coulter count numbers recorded by Dr. Bishayee as aforementioned.

The second issue the DIO covered was whether Bishayee had falsified data of the experiment done March 26-30, 2001, on the viability of the bystander cells incubated for three days in the cold in contact with cells that had incorporated tritiated thymidine into their DNA and the separation of those cells by fluorescent activated cell sorting. The experiment that provoked Hill's allegation in 2001 involved radiation from tritiated thymidine incorporated into cellular DNA. The radiation of tritiated thymidine (H-3) is much less penetrating (beta decay) than the gamma irradiation of cesium-137 and is not expected to penetrate far enough from its location in DNA in one cell's nucleus to cause mutations in adjacent cells' nuclei.

According to Hill, Lenarczyk attempted to observe the bystander effect on survival and had not replicated the published effect. He knew that Hill had accused Bishayee of fabricating data in 1999, before he joined the laboratory. Lenarczyk mentioned his concerns about Bishayee's ongoing experiment to Hill and provided a fresh cell culture to Bishayee on March 29, 2001, as confirmed by Bishayee. Lenarczyk had observed that Bishayee's cultures in the warm incubator were contaminated with yeast or bacteria (appeared cloudy), but he also noted that Bishayee appeared to be continuing with an experiment that he had started earlier in the week. He and Hill decided to monitor carefully what Bishayee did with the experiment that was under way, and this is when they observed the flasks in the tissue culture room and samples in the incubators. Hill took notes on what they saw, and Lenarczyk, of course, took the photographs that the Newark CCRI found inconclusive. Hill compared what they had observed with the results Bishayee recorded in his notebook. She then wrote out her concerns and went to speak, again, with Howell, the head of the research project, and the principal investigator on the grant supporting both Bishayee and Lenarczyk.

Howell and Hill went together to speak to Stephen Baker, their department chairman, who sent them to discuss the matter with Raveché, chairman of the university's committee on misconduct in research (COMIR). All of this seemed to be straightforward enough, although the DIO found it unusual to have firsthand observations of this sort. The

Newark CCRI was skeptical of Lenarczyk's photographs. They cited the lack of definitive dating of the photographs as a reason to doubt that the plates or tubes that were photographed were relevant to the experiment that Bishayee performed; however, Bishayee had not claimed to have any other plates in the incubator or any other experiment using plates underway, so it is really the set of small centrifuge tubes that are the issue. Bishayee did not claim that he was using any alternative incubator, nor did he have any notes to indicate that he had prepared more than one set of tubes, or anything to show that he was doing a second experiment that Hill and Lenarczyk could have mistaken for his bystander effect experiment. The committee appeared to be suggesting, concluded the DIO, that the photographs were falsified, rather than just interpreted incorrectly by Hill. The veracity, however, of the photographs did not seem to be disputed by Bishayee; he was exceedingly vague about what the other experiment would have looked like (since it was not recorded) or how he was noting growth or survival of cells simply put into the cold incubator. "Frankly, the observations by Lenarczyk supported the idea that Bishayee was trying to grow the new, human cell line, and that they were not growing well," concluded the DIO investigation, "certainly not well enough to generate a set of cell pellets, that were then simply discarded." Again, evidentiary material was thrown away and neither Bishayee or Howell had an explanation of the when, where and manner of disposal.

Bishayee's explanation of a second set of tubes could not be confirmed because he had no record of carrying out any experiment on a human cell line using Helena tubes nor did he explain what he might have been observing about the growth of cells in a second set of pelleted cells in the cold, according to the DIO report, which also remarked that mammalian cells do not grow at 10 degrees Celsius, and their growth, unlike bacteria, could not have been observed without some kind of quantitative measurement with a microscope or spectrophotometer. Bishayee stated that he was evaluating the growth of the cells, but did not explain how he evaluated growth without any recorded measurements. Bishayee had no protocol, no coherent explanation of what he might have been measuring in pellets of the other cell line, and provided no information that would account for the second set of tubes. Nor did he describe what he intended to measure and why he had not done so. To DIO scientists, based on considerable experience with mammalian cell culture, Bishayee's claim of a second experiment appears to be unsupported and to not be credible. By comparison, the photographs provided by Hill were tangible evidence that supported what she and Lenarczyk claimed to have observed, and they were accompanied by authentic-appearing notes taken by Hill at the time of their observations.

The UMDNJ committee failed to explain why they chose Bishayee's account of his experiment rather than the observations and photographs of other members of the same laboratory. They were contradictory, observed the DIO, and Bishayee's account seemed weak. Stripped of the extraneous details, Bishayee claimed to have harvested seven tubes of cold-incubated cells, subjecting them to FACS (Fluorescence-activated cell sorting) sorting[95] on Friday, March 30, 2001, while Hill and Lenarczyk stated that six of the tubes were still in and remained in the cold incubator until they talked Howell and Bishayee the next week. Neither Hill nor Lenarczyk observed two sets of Helena tubes in the cold incubator, and, in fact, DIO could not find Bishayee resolutely claiming that he ever had two sets of tubes in the incubator at the same time. Bishayee did not show the committee at UMDNJ what he had done with the cells he got from Lenarczyk.

The Newark CCRI accused Hill and Lenarczyk of "tampering" with Bishayee's experiment, clearly disapproved of their secret investigation and appeared to be accusing them of producing falsified photographs, per the DIO report. If Hill and Lenarczyk sampled the tubes as they claimed, after the cell sorting had begun, DIO found no evidence that the sampling they performed would have affected Bishayee's bystander experiment nor was the sampling intended to affect his experiment. In truth, Hill and Lenarczyk's claim that they observed radioactivity in the Helena tubes seemed to DIO to strongly favor their correct identification of those tubes as belonging to the "bystander"/tritiated thymidine experiment, since Bishayee implied that the second experiment did not involve radiolabeled cells.

The major evidence concerning the March 2001 experiment was the set of photographs taken by Hill and Lenarczyk, at least to the Newark CCRI. The DIO would disagree. The major evidence was the recorded observations of two witnesses, their lack of motive to fabricate evidence such as photographs, and to a minor extent, the photographs themselves, which were not disputed by Bishayee. The university committee considered the dating of the photographs to not be definitive and the photographs to be possibly unrelated to the experiment that Bishayee claimed to have performed March 26-30. The report prepared by the CCRI, again, stated: "The date of the photographs claimed by Dr. Hill could not be reconciled with what Dr. Hill believed they demonstrated about Dr. Bishayee's experiment and Dr. Bishayee's recorded notes and account of what he carried out." DIO understood this to mean that they concluded that the photographs and observations of Hill were incompatible with Bishayee's claim to have done the experiment as planned. It is even less clear whether they thought that the dated (or undated) photographs might have been falsified, or that the evidence simply was not conclusive. The committee was evidently not totally convinced that the photographs and witness testimony proved that the experiment Bishayee actually carried out was different from that recorded in his lab book. That decision probably should have been made after a thorough investigation, not during the inquiry.

DIO was struck by the similar alleged behavior of Bishayee in both the September/October 1999 and March 26-30, 2001 experiments brought to the attention of federal investigators. In both instances, he claimed to be doing another, unrecorded experiment; he is alleged to have substituted samples or data in failed experiments, and to have discarded samples when they were challenged. Had he retained the fixed, counted plates (primary data) in 1999, he could readily have supported his experimental data (since they could be recounted). In the 2001 case, had he not discarded the disputed set of six Helena tubes left in the cold incubator, he could have demonstrated the absence of tritium label or even the human rather than hamster origin of the cells and thus had evidence to disprove the allegation. Bishayee's actions in discarding crucial evidence after being accused of fabricating or falsifying data was not reassuring to the DIO.

It was also striking to the DIO that neither Howell nor Bishayee, nor the laboratory protocol and notes of the experiment, revealed in what way the experiment was primarily concerned with the radioactivity labeled cells, rather than the bystander cells. The FACS separations clearly were carried out on cell suspensions containing mixtures of fluorescent and non-fluorescent cells. How those cell separations might have been affected by the three-day coincubation was not discussed; however, the

DIO found that Howell's proffered explanation for the contamination if only the fluorescent cells by a dye solution was clearly *off the wall* [emphasis added]. Incubation of a mixture of contaminated, fluorescently labeled cells with uncontaminated, non-radioactive cells for three days as mixed cell pellets, could not have yielded uncontaminated separated, non-tritiated cells because there was plenty of opportunity for the contaminating bacteria or yeast to be equally associated with both labeled and unlabeled cells. DIO concurred with Lenarczyk's comment that the substitution of uncontaminated, unirradiated cells for the cells in the Helena tubes would account for the contamination of only half of the plated samples, if Bishayee had not excluded his source of contamination (medium, serum, pipettes and so forth) by March 30, when the cells were plated out.

Finally, the Newark CCRI stated that the members did not see any motive for Bishayee to go forward with this experiment, substituting fresh cells for the contaminated samples. The committee believed that it would have been easier for him to restart the experiment. The committee appeared to DIO to assume, without sufficient foundation, that it would have been possible for Bishayee to sort cells at the FACS facility at short notice, had he restarted the experiment on March 30. They did not discuss whether the arrangement with the FACS facility was flexible or whether Howell would have approved if his grant were charged twice if the experiment was postponed. Either situation could have contributed to pressure on Bishayee to continue the experiment despite the contamination and have further influence Bishayee when he considered the option of "simply" restarting it. On the contrary, carrying on with fresh cells clearly was easier than starting over, but it would not give any information about the bystander effect or the effects of exposure of any of the cells to decay or tritiated thymidine.

The third issue the DIO addressed – that Bishayee falsified data showing an effect of tritiated thymidine on the survival of bystander cells presented in Howell's PHS grant – and which Price struck from amended ORI report 2001-28, was also perhaps the most damning, at least the proverbial "nail in the coffin." Hill claimed to DIO that data showing a bystander effect on survival could not be replicated by Lenarczyk and was likely to have been falsified by Bishayee. She also claimed that Howell was instructed to repeat the experiment by Raveché during the inquiry, and that Howell was unable to repeat the experiment. DIO asked the UMDNJ committee to ask Howell about the repetition of the experiment. He denied it, or at least claimed that many others had observed a bystander effect. He did not, however, produce any data, and it is not clear whether he was referring to a bystander effect of incorporated tritium, or other types of radiation.

The UMDNJ inquiry report did not consider the broader allegation that the bystander effect of tritium had been fabricated by Bishayee, because others in the laboratory could not reproduce it. DIO examined the paper published in 1999, which was a presentation of data showing the bystander effect on cell survival in cells exposed (in cell pellets) to cells containing incorporated tritiated thymidine. The paper also presented data supporting the Howell lab's explanation that the bystander effect (decrease in viability of cells not containing radioactive label) depended on the formation of gap junctions, since lindane, a chemical that blocks gap junction "communication," markedly reduced the effect (such as killing the bystander cells in the clusters).[96] The original data shown in the grant application was Figure 2, "Survival of V79 cells as a function of cluster activity of 3H-TdR" (tritiated thymidine). It displayed three curves: a curve showing a rapid decrease with time of the surviving fraction when 100 percent of the cells are labeled, a two-slope curve of greater survival (at equivalent dose of radioactivity) when only 50 percent of the cells are labeled (middle curve), and a curve with a shallow slope when only 10 percent of the cells are radiolabeled. In the accompanying text, it was stated that the two-component nature of the curve when 50 percent of the cells are labeled was "shown better in Attachment No. 1 (in the appendix) reference 66, the paper by Bishayee et al., 1999 (Attachment No. 2). In the text of the grant, it was stated, "the second component (of the curve) indicates that cells continue to be killed even though they are not significantly irradiated. This suggests that a bystander effect is responsible for killing of unlabeled cells." In the grant and in the paper, the data appears to be from only two experiments.

Similar data was presented in the second Bishayee et al. 2001 paper, where more evidence was shown for a bystander effect of tritiated thymidine and elimination of the effect by lindane and by dimethylsulfoxide, a hydroxyl radical scavenger. The authors (who included Hill) concluded that the bystander in clustered (hypoxic) cells may be initiated by free radicals and mediated through gap junctions. The number of experiments reported in this paper was between two and five for each condition where the bystander effects was claimed to exist (50 percent or 10 percent labeling of cells). It does not seem unreasonable for the inquiry committee, before dismissing the allegations, to have asked to see the data for the two or four experiments in the latter paper that were the basis for the claim of a bystander effect when 10 or 50 percent of the cells were labeled, respectively, and to examine Lenarczyk's experimental data, in which he did not observe a bystander effect. Instead, the Newark CCRI totally ignored Hill's claim, that both Lenarczyk and Howell could not reproduce the bystander effect claimed by the laboratory on the basis of experiments done by Bishayee, or else the committee was not aware of this allegation.

The allegation – that no one can repeat the basic observation of the bystander effect of tritium – called for verification of the data in this paper wherever a killing curve was claimed to have a second component (referred to as A2 in the Table 1). It would have been helpful, observed the DIO, to have the institution determine whether Howell attempted to repeat the bystander effect himself, and what result he obtained. The CCRI had recommended that Putterman ask Howell to take corrective actions to improve the conduct of research and the environment in his laboratory. Hill alleged to ORI that Howell's subsequent actions were to exclude her from the departmental laboratories, stop his collaboration with her, to not renew Lenarczyk's appointment, and to force Bishayee to leave the laboratory.

DIO asked Putterman to investigate the question of whether Howell demanded Bishayee's resignation immediately after the inquiry, because it seemed so much at odds with Howell's testimony regarding Bishayee's abilities and productivity. Putterman stated that Bishayee resigned, but he claimed it was because he was uncomfortable with Hill in the laboratory, and he went back to India (in February 2002). Bishayee's purported statement to Putterman proved untrue, and was documented in Hill's daily diary of events in the lab. Bishayee had returned to India to marry, and when he returned to the UMDNJ later that month, he brought his new wife to the division to introduce her; he and Howell were once again friends. It was not until the end of April 2002 that Bishayee moved out of Hill's son's apartment, just after he took a position with Debkumar Pain, Ph.D., a professor in the university's pharmacology, physiology and neuroscience department. Putterman informed the DIO that Howell had not written a letter of recommendation for Bishayee, but said it was not needed since his new job was also at UMDNJ. Putterman prevented Hill's exclusion from the departmental laboratories.

The DIO recommended that an investigation, independent of the former CCRI committee, be carried out, with input from persons engaged in cell biology, cell culture or related research on mammalian cells. It also recommended that the allegation that Howell knew that the data for the bystander effect could not be repeated in his own laboratory be evaluated by the second committee. It seemed very possible to the DIO that Howell failed to pay sufficient attention to the lack of verifiable primary data of Bishayee in 1999, included questionable data in a grant application, had failed to question preliminary experiments on mutation rate repeated, and that he retaliated in terms of their employment against Bishayee, Lenarczyk and Hill. A thorough examination of the data underlying the bystander effect, and a determination of the results (if any) of Howell's alleged attempt during the inquiry to repeat the key experiment should have been examined, the DIO report concluded, together with his actions as a supervisor of Bishayee and Lenarczyk. The ORI conclusion – and in parentheses "or just KLF" for Kay Fields – does/does not concur with the institution's determination that there was insufficient evidence to warrant an investigation. "DIO urges that the institution be encouraged to go forward with a more thorough investigation, and to include persons with more cell biology expertise on the committee." Fields, the ORI principal investigator, was in the "did not concur" column; her boss Alan Price overrode her and deleted the entire DIO conclusion. But Price's maneuver did not change the substance of Fields' and other DIO scientists' findings and most certainly did not absolve the UMDNJ of its responsibility to investigate as the DIO was originally set to recommend in August 2002.

After Hill called ORI, tensions ratcheted up in the radiology research laboratory. Funding for Dr. Edouard Azzam's three-year NIH grant began on March 4, 2002; he also received, later, an additional grant from the

U.S. Department of Energy. A little over a month later, on April 17, he informed Hill that she needed to move her equipment out of the large lab because Howell was bringing in a new postdoctoral fellow. Azzam was just the messenger. The day before, on April 16, Howell and Bishayee had been re-interviewed by Putterman for the ORI. In his interview, Howell stated that he was unaware of the date of the last performance evaluation he had written for Bishayee, and that he didn't keep track of when evaluations were due for his staff but that he did respond to notifications from human resources of the same. This was also the interview in which he told the aforementioned story to Putterman of how Bishayee came to leave the department. Howell further stated that he had not changed his opinion of Bishayee as "an excellent technician as a result of the misconduct pro-ceeding or for any other reason." Howell was aware, on this date, that Bishayee was working in the pharmacology, physiology and neuroscience department of the medical school, and admitted to running into him periodically. From Putterman's notes: "Dr. Bishayee has told him that he is applying for a position in radiation safety, and also that he just got married [two months prior]."

Regarding the bystander effect, Howell stuck to the story he had previously told the first Newark CCRI committee. But DIO wanted Putterman to ask for clarification, and she did. From her notes of the interview: "Dr. Bishayee performed the experiments that demonstrated the 'bystander' effect of radiation on neighboring cells which were the subject of Dr. Howell's first publication on this topic in 1999. Although no one else in Dr. Howell's lab has repeated these experiments," she wrote, "the bystander effect has been found by many other investigators in many types of cells both prior to and since Dr. Howell's 1999 publication. In fact, the bystander effect field has mushroomed, and there are at least 100 papers in the literature about it to date."

When Howell was asked about the purpose of the specific experiment in March 2001, which was the subject of the misconduct allegation, he declared to Putterman that it was to try a new experimental technique, cell sorting, to explore further the nature of the bystander effect; its purpose was not to discover new knowledge. "Cell sorting physically separates the irradiated cells from the neighboring cells," Putterman recorded his response, "while in the original bystander experiments both types of cells

were plated out together." Howell, she noted, had not repeated or continued these cell-sorting experiments in his lab since Bishayee left because
he did not have the staff to do so, but he attempted to assure Putterman
that he fully intended to take up the investigation "in the near future"
with the arrival of a new postdoctoral fellow. Howell did offer that it
would be "impossible to repeat *precisely* (his emphasis per Putterman) Dr.
Bishayee's experiment because the line of cells and the serum Dr. Bishayee
used are no longer available. Dr. Howell is unaware of anyone, such as Dr.
Lenarczyk, trying to reproduce the experiment in question. He supposed
that Dr. Lenarczyk could have done so without his knowledge, but it would
not have been easy. He also doubts that Dr. Lenarczyk has the technical
expertise to do so." Putterman noted that Howell had reviewed with her
Bishayee's Coulter counts for the experiment in question. "He criticized
them as being too low," wrote Putterman, "saying he would have turned
up the counter at least fivefold. But he did not think the variation in
the counts of each tube was unusually small. He pulled out one of his
own Coulter count experiments and showed that the variation he got was
very similar."

In a short interview with Bishayee for ORI, Putterman was fleshing out
why he had left the lab. Bishayee gave her another version of his departure,
each version having had a slightly different bit of information than the last.
Bishayee told her that he decided to leave Howell's lab at the end of the
CCRI inquiry for two reasons: one was the relationship with Hill (though
this differs from both her version and his actions toward her, both of which
are well documented) and the allegation she brought against him that
made him uncomfortable working together. The important reason, he told
Putterman, was his desire to "upgrade his technical skills in basic molecular
biology which he felt he could not accomplish by staying with Dr. Howell."
He wanted, she noted, to further his development and career with the goal
of becoming an independent researcher. At about this time, Debkumar
Pain, Ph.D., someone Bishayee had known from the past, had arrived at
the medical school and he took a job in his laboratory. Bishayee, contrary to
statements made by Howell and also Azzam's documented characterization
of Bishayee's departure and subsequent job search, informed Putterman
that he did not think Howell had "asked him to leave the lab." He stated
that he asked Howell to write a letter of recommendation for him for an

external position that he was exploring prior to Pain's arrival at UMDNJ, and that it was his understanding that Howell wrote the letter and sent it directly to the outside individual. Bishayee did not receive a copy of the purported letter. Pain did not require a letter of recommendation, he told Bishayee, because he had known him in the past. Bishayee confirmed to Putterman what Hill recorded in her timeline: that he and Howell had a good relationship, nothing had changed.

Putterman's line of questioning of Howell and Bishayee was an attempt to resolve the DIO's legitimate concern as to whether Howell demanded Bishayee's resignation immediately after the inquiry, "because it seemed so much at odds with Howell's testimony regarding Bishayee's abilities and productivity." The DIO's comments on Howell and Bishayee make it clear that federal investigators were not satisfied with much of what either testified to before the first Newark CCRI and it is clear from Putterman's interview of both that little useful information was gleaned.

Back in the lab, it had become known that Hill had called the ORI and from the re-interview of Howell, hostilities in the lab were palpable. Hill wrote in her timeline on April 19, 2002, that she'd passed Azzam's office and he offered to help her move out of the lab that morning. Then Azzam came by her office:

> *Azzam*: "I can read your mind – you are very angry. When is it going to end?
> *Hill*: "Don't know."

Azzam had "apparently learned that I contacted ORI," she noted. "He demands the data I have accumulated for him and Sonia [de Toledo], [and] storms out." Hill called Fields at ORI, and she confirmed Hill's suspicion that Fields had spoken to Putterman who, in turn, had talked to Howell and Bishayee. Fields encouraged Hill to speak to Putterman and to Azzam. "Azzam's behavior is retaliation," she continued, "because he is interfering with my ability to do my job. Azzam returned. I [asked] him to listen to me. He keeps talking: 'I am ruining lives.' 'Roger [Howell] is a sick man.' 'Roger is a young man with a young family, and I am destroying him and destroying his post-docs.' He (Azzam) had offers from Iowa and Sherbrooke. I tell him he should go. He was very agitated."

But agitation turned quickly to acting out. "[Azzam] slams [his] hands up and down. Says 'I am an Arab, I am an Arab!' 'I may be Christian, but I am an Arab!'" Hill took this as a threat. He wanted to know if Hill had copied Howell's notebook. "Yes." She wrote that Azzam thought it was despicable. "'I am responsible [he accused her] if Roger's children find out that Roger has done something bad.' 'I am responsible if Roger's post-docs are out of a job.' 'It is his [Azzam's] responsibility to take care of Roger's post-docs.'" Shortly after this confrontation, Azzam came back, "contrite" this time with the data he had taken away from her returned to her desk. "[He] says we can still work together. [But then] I realized that he is a party to the cover-up, and I can no longer work with him and Sonia. I will fund my own projects. I put the data back on Sonia's desk."

Hill's telephone call with Fields was revealing. While she had spoken to Putterman, it was clear that the latter was unaware that Howell's initial retaliation as described in the complaint had morphed into what Hill described to Fields as "shunning." "Did she, Putterman, know Anupam [Bishayee] had been forced to resign?" Hill asked Fields. "No." Fields offered to look into the matter. [This is an interesting response because Putterman had interviewed Howell the day before and had also just spoken to Fields regarding the interviews of Howell and Bishayee.] Fields also asked Putterman if she knew that Howell could not confirm the bystander experiment. "No," Putterman reportedly replied. "Fields said the original [CCRI] committee was stacked. Only Raveché [had] a decent publication record," Hill recorded in the timeline for April 19. "I [told her, Fields, that] I think Raveché is a good friend of [Thereza] Imanishi-Kari. [To which Fields] responded that Raveché was [an] inappropriate chair of the committee." Putterman had asked Fields why the ORI inquiry was taking so long. "I wonder about that as well," Hill continued. "Fields reiterated that ORI wants the universities to do more, that ORI does not investigate [because] they have much bigger fish to fry, [with a] caseload [that] is increasing and not enough staff or time." Fields informed Hill that the non-reproducibility of the results did not make it into the ORI report (the Newark CCRI never followed up on this). Fields knew as early as that spring that Price was going to override her on this point. "She was worried that [the] grant was based on phenomena that were not reproducible. Why was Bishayee fired after Roger defended him so strongly before the committee?" The latter point not only stood out for Fields but also for anyone reading evidentiary transcripts of the case. Fields informed Hill that a whistleblower cannot withdraw charges once that person has blown the whistle; the whistleblower is no longer in control, including no rights to any additional information and, further, no chance to rebut any statements made later. "Investigators like Roger ordinarily don't wind up losing when a post-doc has cheated," Fields iterated to Hill and Hill included in her timeline. "NIH does not like to take back grants thus his program officer would do nothing other than tell him to map out a new plan. NIH doesn't want principal investigators to think that if fraud is discovered, that is the end, otherwise no one will come clean. On the other hand, the university needs to convince the whistleblower that the job has been done fairly. [The] university should examine the data that went into all of Anupam's papers. Furthermore," she continued, "the members of the committee had no expertise in the areas of research that the complaint involved. Their publications were miniscule except for Raveché." From statements made by Howell and Bishayee to Putterman, both changed the direction of their stories and that Bishayee had left UMDNJ to "learn molecular biology."

Hill worked from the spring through the summer of 2002 making arrangements with the university's research office and foundation to fund her own research. That May, she made a notation in the timeline that Howell presented Bishayee's irreproducible results on Research Day in the radiology department. On July 1, Bishayee dropped by Hill's office to return the apartment key. She asked about his new job in the Office of Radiation Safety Services (ORSS), curious what happened with his job in Pain's lab. Bishayee told her that Pain had trouble with funding and the job was temporary. But Hill knew that Pain's NIH grant was good through September 2004. The ORSS job, she noted, was funded by the university – "hard money." She said to Bishayee, "'You and Roger seem to have made up.' He beats around the bush. [But] tells her he got the job through [Venkata] Lanka, [and adjunct professor and] the head of ORSS."

Having heard nothing from Fields since April, Hill called her for an update on August 6, 2002. The DIO report was written and was scheduled for presentation the following day; it was thirty pages, Fields told Hill, and it was likely that the ORI was going to recommend the case be referred to the university with conditions put on the Newark CCRI. She further conveyed that the ORI's caseload continued to be heavy and that there was a shortage of in-house attorneys. "[She] told me for the first time about Mosimann analysis. [This] refers to the experiments in 1999. [The] chance that Anupam's numbers are uniformly distributed is one in many billion [actually this is the aforementioned 1 in 10 trillion] but one of my sets is 1 in 40. [Fields] wanted to know if [the] Coulter counter could favor certain numbers. Later, I called Coulter to speak with David [of the company]. [The] serial number of our Coulter counter is 030788, model number is ZM." The Coulter company representative told her there was no chance that any number is favored. "Apparently, Putterman asked Roger about non-reproducibility of the bystander experiments, and he dissembled by saying that the bystander effect is very reproducible – which it is – and many have observed it. Catch is," wrote Hill on August 6, "he is the only one that has studied tritium; his observations are unique and have not been verified by anyone else." Fields asked Hill to do a literature search. "I searched on tritiated thymidine and bystander." She found four papers. "One [was] by Howell and Rao, and the other three weren't relevant." There were 119 related articles and none of them dealt with [3H]-dThd-incorporation. Another way that ORI detected falsified data was to compare standard deviations to square roots of the means. For normally distributed data that are obtained independently, the standard deviations should be equal to or greater than the square roots of the means. Apparently, Hill observed, the university sent along some other experiments that were done around the same time as the questioned experiments and the standard deviations were smaller than the square-root of the means (in her analyses of Bishayee's numbers, the standard deviations were about half of the square roots of the means). Putterman told Howell that the numbers were too small, but this seemed "to not bother him, a physicist."

Hill had spoken to Pain. "He moved to NJMS in October [2001], knew Anupam and that he was in the process of losing his [H-1B] visa," she continued. Bishayee had begun to work for Pain in January, went to India in February to marry and after about a month in his home country, returned to work for Pain for about four to five months. Prior to his employment with Pain, Bishayee had a protracted period without pay, of which Hill was acutely aware because she had been renting her son's apartment to him. When Hill asked Pain why Bishayee left his job in Pain's lab, he "couldn't say. [But he did tell her that] he treated him like a technician with constant supervision by a senior postdoctoral fellow. [Bishayee] never did anything on his own. [Then] Lanka offered him the job in ORSS. Pain wanted him to stay another month, but he left." Hill sent more data to Fields, but it was too late.

The ORI decision not to refer Hill's case for further investigation was not unexpected after her telephone conversations with Kay Fields. "Fields said she was so incensed [by that decision] that she walked out of the meeting and has been taken off the case [recall Price stated that she removed herself from the case per his September 5, 2002, letter to Hill]. It was argued," she wrote in her August 7 timeline entry, "that the Coulter counter data were not essential to the experiment which (a) is not true, and (b) Mosimann [protocol states] people are more likely to cheat when ancillary studies are involved." Fields conveyed to Hill that she was "furious. ORI [she explained again] wants to put more of the burden on the universities to do the investigating. NIH doesn't like ORI [its own watchdog], who they think are meddling around in the [world of real scientists]. She says it is normal that I would not have a chance to rebut. She [told Hill] that Anupam told the committee that Marek [Lenarczyk] and I were jealous and had a conspiracy against him because of the conflict between me and Roger." This was the first time she had heard of Bishayee's characterization of the Newark CCRI. Fields recommended Hill to get whatever she could via the Freedom of Information Act (FOIA). During this conversation, Fields informed Hill that she had a strong *qui tam* case and that if she won it, the university's penalty (up to that point) was $4.2 million. Fields went so far as to point Hill in the direction of lawyers who take *qui tam* cases.

A few days later, on August 12, Hill spoke again to Fields. Since there was no opportunity for an appeal to ORI, Fields provided more information about the *qui tam* option. "[Fields said] that Anupam told Putterman he left to learn molecular biology [at this point Hill had none of the evidentiary documents pertaining to the Newark CCRI, ORI or follow-up interviews Putterman had conducted on behalf of the ORI inquiry] – the same thing that Roger told her," Hill documented. "They are united in this lie. [Fields] can no longer talk to Putterman since she is no longer on the case. I have fewer rights," she opined just then. "I might have to sign a confidentiality agreement somewhere along the line. Any protection I might have would be gone after confidentiality was broken. In any case, I would only be protected [regarding] my employment. The only right I seem to have is to see the transcript of my testimony and to correct any mistakes in it. Regarding Roger," she continued, "I am asking for the world – that he retract the papers and talks with his grant administrator – besides, he is vigorously defending integrity in research."

When Hill suggested that her next move might be the Faculty Affairs Committee, Fields advised her to go Putterman first. When she suggested that she wanted to redress issues with Price, Fields reminded her that the ORI workload was heavy. "They [ORI] are triaging. [We] have three big-time falsifiers [at that time]." From those remarks, Hill wrote "I guess this [her case] is just little stuff."

Hill called Price two days later. "He gave me the 'cold shoulder,' the ORI director makes the decision [and although there] is no decision yet, [it] may be months, years. [It] will become a closed case, then it will not be available." Price told Hill that she was not the originator of the case but declined to tell her who it was; this was also untrue. Hill was the originator of the case to ORI. "What kind of nonsense was that?" she wrote in the timeline. "He says that I should send additional information regarding the numerical analysis that I have [to ORI]." She insisted she was the originator, "I had to be. He fudges. When I first talked with him last year, he was all sweetness and honey, now he is an ice cube."

Bothered by what Price said that she was not the originator, Hill called Fields that evening to confirm that there was no other complainant. There wasn't any. Fields recommended, again, that Hill go to Putterman, especially if Hill's ORI case information didn't make it to the university. She recommended, too, that if Hill went to Putterman, she take a statistician. ORI wasn't denying misconduct took place. But Fields didn't believe Price would change his mind and release the report in-full. She reiterated, too, that NIH "never takes grants away." Ultimately, September 5, 2002, Hill receives a letter from Price informing her, in part, that she would be receiving further correspondence from Chris Pascal, ORI chief, that ORI concurred with the Newark CCRI that "there is insufficient evidence to warrant an investigation." But this was just another falsehood of a string of setbacks in Hill's case.

The day after Hill received the news from ORI about her case, Lenarczyk e-mailed her eight Excel files – the eight V79 bystander experiments that showed, again, lack of reproducibility. On October 2, 2002, Lenarczyk sent Hill sixteen additional Excel files, all experiments with –A1, Howell's NIH grant. She also received, via mail, a Zip disk from Lenarczyk with additional files plus copies of those already sent via e-mail. With this information in hand, she prepared a statement for Putterman and set a meeting. When they met on October 24, Putterman refused to let Hill read her testimony from the first Newark CCRI, which Hill did have the right to do. When Hill told her she would consult an attorney, Putterman told her that no one else would see her testimony either. Further, no copies would be made. She'd take notes when she reviewed Hill's testimony, only if she had questions. Putterman promised to call Hill the following Monday; she didn't. On November 4, Hill met again with Putterman. She had read Hill's testimony and had questions. Putterman asked if Hill believed that Bishayee would confess. "The four [scientists] in [James] Mosimann's paper did once confronted with the numbers." Putterman countered that the original complaint Hill filed pertained to an experiment that never went beyond the notebook. But per the misconduct teleconference Hill participated in, once data is recorded in the notebook, misconduct has occurred. After this conversation, Putterman suggested Hill go back to the CCRI and make new allegations and provide all the new information Hill had gathered since the first university committee reviewed her charges and found them unsubstantiated. There was no mention of continuing retaliation against Hill in her division.

The second Newark CCRI committee was made up primarily of members who had served on the first committee; this was clearly against the advice of the ORI DIO. On November 11, 2002, Hill met with Anthony Forrester, chairman of the reconstituted committee. She left him with the Mosimann reprints, the graph she'd constructed from information sent to her by Lenarczyk demonstrating the lack of reproducibility (after eight experiments) of the bystander effect and the chi square analysis of the numbers. Lenarczyk would also later indicate in his deposition that he sent all of his experiments to the second committee, although they do not acknowledge this in the second CCRI list of articles examined. Further, there is no record in the minutes that they discussed these experiments during any of their recorded meetings.

Hill's meeting with Forrester didn't go particularly well. Forrester didn't want any details, just the bare bones. He wrote up what Hill told him about the Mosimann analysis and the graph but was clearly more interested in checking with the university's lawyer to find out whether Hill could have a copy of her previous testimony and anything written up by the second Newark CCRI. The university's legal counsel said "no." Hill then told Forrester about her August 1, 2001, meeting with Bishayee. Forrester didn't want to include it in the proceedings of the new committee; he was set to present just two of her allegations to the CCRI and on that basis the decision to pursue – or not – the case would be made. This was Forrester's first substantive case of scientific misconduct, Hill observed.

The second Newark CCRI met for the first time on November 25, 2002. Forrester introduced two articles, two statistical tables, and a copy of the ORI final report that pertained largely to the previous inquiry conducted by the committee involving the same complainant and respondent. Forrester further indicated that letters of notification to the parties had been prepared, if the committee decided to proceed to an initial inquiry. Importantly, Hill was making a new allegation based on new evidence, and that new evidence was the data analysis derived by Hill. The committee, after getting the basic lay of the complaint, developed three questions:

1. Who actually performed the data analysis?
2. What is the problem with the data produced by Bishayee?
3. Where is Dr. Bishayee's original data, and was it a printout or was it handwritten?

The committee debated whether the evidence in hand was sufficient to establish the possibility of scientific misconduct. Forrester reminded them that the only decision before the members was whether the evidence was sufficient to proceed with an initial inquiry. They decided to call Hill to speak to the committee.

When asked who performed the data analysis presented to the committee and whether the analysis was based on Bishayee's raw data, she told them that she had carried out the analysis, and that it was based on raw data of which she had copies. The previous committee had the originals of the raw data. Forrester asked her to explain how this analysis indicated scientific misconduct, as defined by the university policy. Hill then handed out additional material to the committee, consisting of a large comparative table of Coulter averages and three graphs labeled "Comparison of Mutants/Cell in Clusters – Bishayee versus Hill," "Comparison of Cluster Survivals – Bishayee versus Lenarczyk," and "Comparison of Cell Numbers on Day 3 – Bishayee versus Hill." Hill told the committee she had only five points to make.

In her first point, Hill observed that there were two experiments that she oversaw, the purpose of which was to see whether cells had hypoxia. She added that the experiments were very difficult to explain to non-biologists, an observation also made by Fields. Hill explained, too, that because she had been dissatisfied with the first committee's handling of her allegation against Bishayee, she contacted the ORI. Though she informed second CCRI members that the ORI very strongly supported her case, it chose not to proceed to an investigation. Further, she told them Fields, an ORI official, had faxed the ORI's preliminary findings to Hill. The preliminary findings (not the final report redacted by Price) were predicated on the statistical methodology used by ORI and based on the Mosimann protocol. The Mosimann material discussed the distribution of digits generated by a machine. The left-most digit is the most important. But as the user moved to the right, the digits become more random. Using Chi-squares, it is possible, she explained, to learn whether the digits are randomly distributed. Hill then referred to the large table of Coulter averages. According to her testimony, using the Mosimann method of analysis on the Coulter averages, there is a very high probability that Bishayee fabricated the data. In response to a question, Hill indicated that the Coulter counter does not produce a printout. The paper control would be hard copy and the way to do it, to clarify, is to mockup the cells and take a photograph of the resulting screen. Another control, which she used in this case, was to compare Bishayee's numbers with Lenarczyk's numbers.

In her second point, Hill continued that Fields had also suggested a second method using standard deviations. If independently obtained, the standard deviation should be the square root or greater. The highlighted average for each experiment shows Hill testified, the standard division and then the square root. According to Hill, Bishayee's figures are less than half the square root while Lenarczyk's around closer to half. Hill admitted to the committee that she had had difficulty using the Coulter counter and that her figures were "terrible." The committee seized on this description and asked Hill the difference between her figures which were "terrible" and Bishayee's that she alleged to be fabricated. She replied that if fabricated, the numbers, like Bishayee's, would be too close together and the square root would be less. The data would be "too good" suggesting that the data had been fabricated.

Hill's third point emphasized that Howell's grant application used data that showed that there wasn't much hypoxia in the clusters. In radiation biology, if cells are aerobic the conditions will be maximally sensitive, and the maximum number of mutants is developed. If cells are hypoxic there are decreased mutants. When Hill did the experiment, she got no increase in mutants but when Bishayee did the experiment he did get an increase. Hill produced a graph to illustrate her point. The grant, as submitted, indicated that there was an expectation for mutants to mirror survival. Bishayee's curves went down but Lenarczyk did the same experiment and got different results. The hypothesis was that there wasn't any hypoxia, which Hill told them was wrong. Two experiments, two weeks apart and the expectation is some variation but not "very different." The findings that clusters were hypoxic was inconsistent with the hypothesis. Hill told the committee she intended to skip the fourth point because "it wasn't very strong." She went on to tell them that she did the first mutagenesis experiment in September 1999. She could not participate in the replication but was present when it was completed. The experiment used fifty 100-millimeter dishes that had to be ordered and at the end there would be five replicates at each data point, ten stacks of five dishes each. At the end, the dishes needed to be fixed and stained.

Hill repeated the story: She asked Bishayee when he would fix and stain the experiment, he told her later that night. She iterated that she saw the dishes in the incubator and assumed that those were the dishes from that experiment. She looked at them closely for colonies and saw none. She went on to explain, in response to questions from the committee, that no stain is needed to see colonies on the plates. Even at the lowest dose, she emphasized there would be a lower number of cells. But she saw no cells on any plates. When Bishayee arrived at the lab, he told Hill he had fixed the dishes and that it "came out just fine." Asked about the dishes in the incubator, he told her that those dishes belonged to another experiment. Howell subsequently informed Hill that there was "no other experiment" and brushed off her concern. Later, Hill asked Bishayee for all the data from the experiment. In a story she would have repeated many times, Hill informed the members of the second CCRI, Bishayee told her he'd taken the data home. Bishayee brought the data to the lab the next day.

Hill reminded the committee that Raveché, during the first inquiry, advised Howell to replicate the experiments in the grant and the published papers. Lenarczyk did replications but could not reproduce the data. He kept trying because Howell was "frustrated," Hill informed the panel. The committee asked Hill for an explanation of how the statistical analysis was related to the graphs and she replied that the relationship was indirect. When asked why Bishayee would fabricate data, committee minutes reflected only that Hill informed them that he didn't do the experiment and the dishes were empty. But this is not the full story and certainly an incomplete transcription of what she tried to convey to the members of the second university inquiry.

After Hill was excused from the November 25 meeting, the committee discussed her comments and one the administrative staff referred the members to page twenty-one of the ORI report and pointed out that the ORI, in its review of Hill's first allegation against Bishayee, had raised new elements and questions about the previous Bishayee allegation, but then "closed the case." The committee decided to interview Putterman by telephone. In her opening remarks to the panel, Putterman went over the sequence of events that led to her contact with ORI investigator Fields and the letter she received from her that ORI wished to review the 2001 allegation against Bishayee. She told the panel that the ORI report concurred with the university that "there is insufficient evidence to warrant further investigation." She quotes the redacted Price report. Putterman further stated that Hill "is not aware that ORI did this analysis and has not received a copy of the ORI report." Hill had already seen the draft of the report – the Kay Fields version. Putterman continued: "The ORI report

(Price's final, striking and rewriting his investigator's findings) discussed the 1999 experiments on pages 14-17. ORI did not question the accuracy of the statements in Howell's grant application but only whether Dr. Bishayee's second experiment confirming Dr. Hill's first experiment was falsified. ORI analyzed Dr. Bishayee's data (page 16) in accordance with Mosimann's article and the methodology Dr. Hill herself subsequently applied in the current allegation; however, the ORI concluded that it was not possible to resolve whether these Coulter counts were fabricated or not, and that 'this evidence is not sufficient to warrant further investigation.'" The committee questioned Putterman about other statements in the ORI report, which criticized the UMDNJ committee for its apparent lack of expertise in radiation biology to which Putterman intimated that Fields had "developed a personal relationship with Hill, and that Fields' position was overruled by her superiors at the ORI." Putterman went on to say that ORI "is only interested in issues pertaining to Dr. Howell's federal grant and that ORI is not questioning the grant."

The second CCRI met again on December 2, 2002. Forrester made two recommendations, first that Bishayee be invited to speak to the committee in response to the allegation, and second that an expert opinion be sought as to the validity of Hill's data analysis. Neil S. Cherniak, M.D., responded that he believed the experts at ORI had already examined the evidence and closed the case and if that were, indeed, the case, there wouldn't be a need to pull Bishayee into another committee meeting to answer questions. Other members of the committee referred to specific comments in the ORI's final report (the Price version), quoting from page 17, that the ORI observed an unusual repetition of numbers and that it was questionable that this could be from a machine. Julie Kligerman, an attorney who served as the UMDNJ director of legal management, added that Hill believed that she had come forward with new evidence because this analysis was not completed during the first inquiry. Rita Turkall, Ph.D., professor and chair of clinical laboratory Sciences and associate professor in the UMDNJ Department of Pharmacology and Physiology, pointed out that the ORI cited a problem in its own analysis with the absence of proper controls and suggested that Hill herself was providing control data – again, the Price final ORI report. The ORI performed two analyses: one utilized data from March 26, 2001, compared to Hill's data from 1999; the other used Coulter counter data from September 24 to October 4, 1999, but without control.

The committee questioned whether Hill's data was appropriate to use as a control and pointed to the ORI concern for the lack of proper control in their own analysis; it was thus suggested that the committee get the expert opinion it had not up to that point, as well as an evaluation of the analysis itself. Daniel Fine, D.M.D., a member of both CCRIs, further suggested that, to be thorough, the committee should determine whether Hill's analysis utilized the same data as that in the ORI analysis. Cherniak made a motion to get the expert to advise the committee in the matter of Hill's data analysis and it was seconded and approved. But how to identify the "expert"? The CCRI opted to get names of prospective outside experts from contacts at the Cleveland Clinic and from Stanley von Hagen, Ph.D., UMDNJ adjunct assistant professor. The committee didn't intend to call back Bishayee until after it had the opinion of an outside expert regarding Hill's scientific misconduct allegation.

Turkall was stuck on the issue of which experimental data Hill used in her analysis, and whether the same data was, again, used by the ORI. Hill appeared to use additional control data, but if the data were the same as that used by ORI, and the ORI believed that their analysis had inadequate controls, then it is possible that Hill's analysis would be insufficient evidence to launch an investigation. Turkall believed that the committee, through the chair, should attempt first to get clarification from the ORI by speaking with the individual who did the analysis. What did the ORI mean by "proper controls"?

Having been made aware of the statistical anomalies in Bishayee's results, the second committee initiated a telephone conversation with members of the ORI. In her report, Fields had previously noted that she believed the first committee thought that Hill had been lying to them. In a transcript of this December 12, 2002 telephone conversation from the ORI, which Hill would later get in the discovery phase of her court case, Forrester, the second Newark CCRI chair, questioned the DIO's oversight report on what by then was a closed case. At the start of the conversation, Forrester and Price made the appropriate introductions, Price introducing John E. Dahlberg, Ph.D., the scientist who conducted the statistical analysis described in Price's final version of the ORI report dated September 5, 2002 (not Fields' preliminary report, which Hill had been previously sent).

In this previous confidential document, Forrester had two questions for Price and Dahlberg. He asked the meaning of the sentence (from the report, page 17): "However, given the absence of proper controls for this analysis, DIO does not find this evidence or the above inadequacies in the inquiry report sufficient to warrant further investigation in this case." Dahlberg replied that the statement referred to the DIO analysis of the right-hand digit in the Coulter counter data. The DIO lacked the comparable amount of data from an impartial party. Furthermore, while the analysis might be reason to believe that there was bias in the data, he did not know how the numbers were generated, that is, read from the Coulter counter or derived in some other way.

DIO director Price replied that when ORI wants to use digit frequency analysis as evidence to support a finding of falsification of data, DIO wants to have the unquestioned control data, by the same person or another reliable person, using the same instruments, that can be subjected to a parallel analysis. "DIO did not have such evidence in this case. Furthermore, ORI has not yet based a finding of scientific misconduct on statistical calculations alone, since there remains the argument that a rare event could have taken place," but he then made the insinuation, "Given the additional questions about the credibility of the complainants, DIO did not find sufficient evidence to warrant a finding in this case." Price had impugned Hill's integrity in one sentence and to her university colleague. But Price wasn't done.

Price also stressed that proper controls for the Coulter counter ("the instruments" to which he refers above) were not available and probably would never become available. Neither he nor the UMDNJ committee considered the possibility that others in the laboratory might have used the same Coulter counter and would therefore be able to provide controls. During discovery, Hill found numerous controls produced by others in the lab using the same instrument. The second committee came away from the telephone conversation with ORI with the message that statistics alone were insufficient indication of scientific misconduct. Relying on this, the second committee therefore backed up the ruling of the first committee. In doing so, the second committee completely ignored the evidence presented by Hill and Lenarczyk that Bishayee's results could not be reproduced. This evidence clearly demonstrated that there was more to the case than statistics.

Forrester's second question (also from page 17) asked the meaning of the sentence: "From the available evidence, DIO cannot resolve whether the Coulter counts were actually fabricated, and this issue for DIO remains unresolved." Price told Forrester that, as discussed previously, given the inability to prove that the respondent fabricated the questioned data, and the other shortcomings cited, DIO chose not to pursue the matter further and closed the case. He added that DIO had not asked the institution to do more investigation or analysis either, as DIO had in other cases where evidence was available to pursue. Forrester told him he'd read the last pages of the report and understood that statement. Dahlberg added that the underlying concern about relying on statistical evidence is that such analysis can result only in probabilities. Statistical analyses by themselves are "insufficient to prove misconduct."

Chapter 3 –
Qui tam

What happened next in Helene Hill's allegation of scientific misconduct would set the stage for her *qui tam* case. When Anthony Forrester, Ph.D., reconvened the second Newark Campus Committee on Research Integrity (Newark CCRI) after the new year, on January 14, 2003, he indicated that Anupam Bishayee, Ph.D., had been invited to appear at the meeting to respond to Hill's current allegation. Before inviting him into the room, Forrester described for the entire committee his December 12, 2002 telephone conversation with the Office of Research Integrity (ORI). Forrester reported that Drs. Alan Price and John Dahlberg stated that: (1) statistical analysis, in the absence of other valid empirical evidence, is insufficient justification to proceed with an investigation of scientific misconduct; (2) that in the case in question, there was no independent evidence of scientific misconduct, and (3) that control data were not possible to achieve under the particular circumstances of the case. At the end of the conversation, Price made clear that the ORI closed the case for these three reasons and did not expect UMDNJ to pursue the matter any further. Fine and other committee members noted for the record that the CCRI had conducted this preliminary inquiry with fairness and, he added, open-mindedness.

After much discussion among the committee members, the second Newark CCRI concluded that since, in the opinion of the ORI, statistics regarding random numbers did not produce sufficient evidence to warrant an investigation, there was no reason for the committee to consult an expert in the field. Further, the committee also concluded that the lack of independent control data with which to compare the experimental results generated by Bishayee meant that the questions raised in Hill's allegation might never be answered scientifically. Bishayee was then ushered into the room to answer questions.

Forrester started the Bishayee interview by reading the description of the new allegation against him, as described in the letter from the committee to Bishayee dated December 19, 2002. At the conclusion of introductory remarks, Forrester asked him, "Did you falsify experimental data?" Bishayee responded, "No, I did not."

Bishayee described, again, the experiments in question. This time he told the committee that the Coulter counter is not very accurate and that since running the counts multiple times leads to different results, any particular count is not of critical importance. He selected the counter results, he said that he wished to record. The protocols, he assured them, could be completed without the Coulter counts. Interrupted by a scheduled fire drill, the meeting was reconvened, but Bishayee wasn't immediately recalled to the room. Forrester asked the members whether they found Bishayee's explanation credible. Daniel Fine, D.M.D., and Rita Turkall, Ph.D., believed there could be bias in the choice of Coulter counts, but wondered how it was that Hill and Marek Lenarczyk were able to produce counts that were random in the right-hand digit. Fine speculated that Bishayee was not a seasoned investigator, and that Hill may have produced her data using different assumptions.

Bishayee was invited to return to the meeting. Forrester asked him whether he was solely responsible for the running of the Coulter counts or whether he had been supervised. Bishayee replied that "from time to time" Howell checked the counts to see whether cells were lost in the course of the experiment. Sometimes Howell, he offered, checked the counts by examining the cells under the microscope. Bishayee noted that the Coulter counter was difficult to use. There were many variables, he told them, that can cause changes in sensitivity. Clogging the machine could cause different results. He stressed that overall, the experimental outcomes were unrelated to the counts. Forrester then asked Bishayee whether selecting the preferred Coulter counts was an accepted standard within the experimental protocol. Bishayee responded that the protocol left it up to the investigator and that the second count is only a check carried out in the middle of the experiment. Forrester asked him how others could have gotten random numbers doing the same experiment. Bishayee answered that the counts can be very different depending on the people carrying them out; their selection of counts might differ.

Fine wanted to know why the number of cells mattered. Bishayee replied that the experiment was a study of the bystander effect, that is the damage to non-irradiated cells by their proximity to irradiated cells. The cell clusters in his study were a mixture of irradiated and non-irradiated cells. The experiment was supposed to be conducted with clusters that were "half and half." What was crucial to the experiment, he informed the panel, "is to start with the proper proportions of each type." Bishayee added that the Coulter counts were made on a sample of the cells, and the counts had been calculated based on the aliquot. The process, he explained, was very time-consuming because if the counts were not completed quickly, the cells could cluster again and jam the machine. Turkall wanted to know why out of three counts, one would sometimes be high and sometimes be lower. "What are the factors that influence the counts?" Bishayee told her there were no specific guidelines. Pointedly, he told the panel that Howell gave him no instructions as to what the guidelines were for the experiment. There is a difference between "there are no specific guidelines for Coulter counts" and "someone didn't give me the guidelines – the protocol and parameter – of the experiment."

When asked if he had engaged in any procedure that would lead to suspicion of falsification, Bishayee replied that he honestly did not. When Forrester handed Bishayee a graph titled "Bystander Effect" and asked for his comments, Bishayee told the committee that the graph showed an experiment repeated by someone who did not get the same results he did; there could many factors involved to explain the differences. The cells used could have been, he suggested, from different batches or already have mutations. The other investigator could have used different sera that could have yielded different results. The sera must be kept, he explained, at the same temperature, must be of the same age. There are many different variables that could impact on the findings he was seeing on the graph. Forrester asked if Bishayee could account for the eight repetitions of the experiment that were so different? He replied that they could be from different batches of cells with different sensitivity. Fine wanted to know whether anyone had been able to reproduce Bishayee's data. Bishayee answered the question, stating that the bystander effect had been demonstrated by different laboratories. Fine pressed and asked what was different about Howell's experiments. Bishayee told Fine that Howell was using a different model and different kinds of cells.

Fine asked why Hill's data was so different. Bishayee replied that the use of different serum and/or different timing could result in different results. Forrester followed up, asking if the line in the diagram was a fair representation of Bishayee's findings. He told him it was, but there were other contributing factors, such as vibration in the instrument that would make the experiment impossible to duplicate exactly but this answer didn't put the issue to bed. Forrester asked if Bishayee's findings had been replicated. "Yes," he replied, "but not exactly." Others, Bishayee suggested, had observed damage to bystander cells. Forrester asked who the others were who had made this observation, to which Bishayee offered the names of Tom K. Hei, Ph.D., of the Center for Radiological Research at Columbia University, and John B. Little, M.D., just then the James Stevens Simmons Professor of Radiology, in Department of Genetics and Complex Diseases at the Harvard T. H. Chan School of Public Health.

Little's work focused on the mutagenic and carcinogenic effects of ionizing radiation in cells directly targeted by radiation – his work was groundbreaking. He discovered, according to the public health school's living legacy page devoted to Little's work, what has become known as "non-targeted" biological effects of radiation including radiation-induced genomic instability, whereby genetic effects occur in the offspring of irradiated cells, and the "bystander" effect whereby genetic effects occur in unirradiated cells near irradiated ones. Today the bystander effect is a well-accepted phenomenon in radiation biology, but its biological implications – whether the effects are universally detrimental for us or possibly even beneficial under certain circumstances – "remain hotly debated."[97]

Hei had a grant with the NIH National Cancer Institute titled "Radiation Bystander Effects – Mechanisms," a project that was begun on February 1, 1997, and did not end until June 30, 2015. Hei's NIH abstract informs that the thrust of the study was radiation induced non-targeted/bystander effects have been demonstrated with a variety of biological endpoints using mammalian cell cultures, 3D human tissues, and more recently, in *Caenorhabditis elegans*, a free-living (not parasitic), transparent nematode (roundworm), and in mice; however, neither the mechanism nor the relevance of the bystander response to human health is clear, he wrote at the time of submission to NIH. Hei continued that "this program project brings together and links three highly integrated projects aimed at shedding new light on the mechanism and health relevance of the non-targeted/bystander phenomenon using both in vitro and in vivo approaches. The overall hypothesis of program project is that the radiation induced non-targeted (bystander) response can be initiated both in vitro and in vivo by oxidative stress and propagated by multifactorial signaling events involving cyclooxygenase-2 (COX-2) and junctional communication, and is modulated by the status of the Rad9 protein."[98] Hei crafted three projects, each highly interactive in goals and research approaches and further linked together by a technical core, which would provide specialized irradiation facilities, data analyses and state-of-the-art gene expression profile related technologies. Project 1 addressed the role of Rad9, a DNA damage response protein, in the radiation induced bystander effect in mouse embryo fibroblasts (MEF) and mice. The underlying hypothesis to be tested is that Rad9 controls the bystander process via regulation of COX-2, p21Waf1 and other downstream targets. Project 2 built on the preliminary findings that radiation induces non-targeted, out of field mutagenesis in lung tissues of gpt delta transgenic mice to test the hypothesis that COX-2 mediates radiation-induced bystander mutagenesis in vivo and that the bystander cells are genomically unstable in genetically susceptible populations. Project 3 addressed the central hypothesis that the in vivo cellular micro-environment modulates gap junction gating that determines the nature of signaling events propagated between directly irradiated and bystander cells. The projects were conceptually linked and technically interactive such that they complemented crosstalk and strengthened each other. The observation that the progeny of non-targeted cells showed an increase in genomic instability as evidenced by an increase in delayed mutations and chromosomal aberrations many generations post-irradiation indicated the need for a comprehensive assessment of the bystander issue, particularly among genetically susceptible population, to understand mechanism(s) and impact on human health.

Bishayee only briefly described the work Hei and Little had done with bystander effect. Fine brought the discussion back around to Howell's experiments, asking again what made them different. Bishayee responded that they were making cell clusters as a model of in vivo phenomena. Fine queried further, asking whether clusters made it harder to quantify. "Yes," Bishayee responded, in addition to "lots of other variables." Bishayee added that Hill was the coauthor on the data in question and at the time she had supported the data's inclusion. Then Forrester passed Bishayee a copy of the graph titled "Comparison of Mutants." Bishayee explained that the graph depicted an experiment he had done when he was new in the lab and unfamiliar with the science. The experiment was intended to confirm that he could perform mutation studies. Howell was ill, he continued, and was absent much of the time. The data had not been published because it wasn't yet funded. Later, Bishayee observed, Hill found different levels of mutations. "Was this preliminary work?" Forrester asked. Bishayee told him he didn't do another mutation experiment after the one shown on the graph. Someone else came, he deflected, and worked with Hill on mutation studies. Then Forrester handed Bishayee a graph titled "Comparison of Cluster Survivals." After examining it, he told the committee that Lenarczyk had used different cells after they had lost the cells and were forced to culture a new batch. The repetition of the experiment thus used the new batch store at a different temperature.

Forrester had one more graph to show Bishayee, this one titled "Comparison of Cell Numbers on Day 3." The objective, he testified, was to show that there were differences in the cell numbers for certain doses. He pointed out that at some points the cell numbers were close. The differences, Bishayee insisted, could be radiation effect or how the cells were handled. Furthermore, the cell numbers could be different. The differences, Bishayee continued, were not random so different factors could explain this. What differences?

Fine wanted to know if there had been discussions or debates in the lab when there were differences in findings. Bishayee answered that he "didn't recall." He remembered only an "argument" about counting. He also stated that he wasn't certain how Hill did her counting, perhaps by microscope. When asked whether the results in the last graph had been replicated, Bishayee replied that he did only a few experiments with the external gamma beam. But it was possible, he said, that others performed these experiments. He didn't seem to have any specifics.

Turkall commented that the methods used across the comparisons weren't identical. Bishayee concurred and stated that even within the same laboratory the same person couldn't repeat the experiment because of "the uncertainty of the biological model." It wouldn't be possible, he said, to anticipate all of the variables that could affect the results. Bishayee further offered that he was not an expert, just someone who was trying to establish whether the experimental methods could be used.

When it was clear, from the minutes of the meeting, that no one had any further questions for Bishayee, he was asked if he had anything else to say. Bishayee did. He told the committee that Hill had originally been the co-investigator on the grant and at that time had had no problem with the data or the protocols. It was only after she was taken off the grant, he suggested, "when she started all of this." This was a "new theory" for Bishayee, who had up to that time not stated this as the reason for Hill's insistence that misconduct had occurred. Before the second committee was convened, the excuses shape shifted to fit the tenor of the question asked. Bishayee told the committee that he had stopped doing scientific research and had a new job at the university in another field. He was then excused.

Fine wondered about Howell's role in what was happening in his laboratory but suggested that the committee put the matter to rest, given the lack of credible evidence of scientific misconduct. Turkall still had problems with the way Bishayee chose the numbers from the Coulter counter. But she also saw Bishayee as someone with little experience who was purportedly under supervision. She didn't realize how much he had been left to his own devices in the lab. Lack of supervision was a problem. Fine reiterated that Howell should have supervised Bishayee, and the problems that arose should have been rectified early in the process. At the end of the hearing, the second Newark CCRI determined that the data was not there – again – to support the charge of scientific misconduct. Fine added Hill only rejected the data once she was taken off the grant. Forrest asked for a motion from the committee of a "finding of no cause due to insufficient credible evidence of misconduct in science to warrant further investigation." The motion was made, seconded and unanimously approved. As a result of the successive findings of "no cause" by the

Newark CCRI, UMDNJ never disclosed the details of its findings to the NIH as its policies obligated it to do in the event there was substantial evidence of falsification and/or fabrication of data submitted in support of a grant application. Nor had the UMDNJ undertaken, according to Hill's subsequent *qui tam* filing, to withdraw the scientific literature that was generated as a result of this data.

Frustrated by her inability to convince two Newark Campus Committees on Research Integrity (Newark CCRIs) and the United States Public Health Service (PHS) Office of Research Integrity (ORI) that scientific misconduct had occurred, Helene Hill decided to file a case for *qui tam*, just as Kay Fields, Ph.D., had recommended she do during the course of Fields' investigation of Hill's case. After searching the Internet for legal counsel, she found three local lawyers and one large law firm in Philadelphia that purported to deal with *qui tam* cases. "Negotiating with a firm in Philadelphia was not attractive so I called the three locals. None had had any direct experience with *qui tam* work. The first sounded very old, the second was curt and impatient, but the third expressed interest," Hill recalled later. "I prepared a one-page summary of what had gone on so far and met with him on April 10, 2003, my seventy-fourth birthday and the two-year anniversary of my presentation to Howell, Baker and Raveché regarding [her investigation] of Bishayee." Hill had settled on Sheldon H. Pincus, of Bucceri and Pincus, a small, three-lawyer office, who took her case though he didn't take contingency cases. In going forward on Hill's case, he made exceptions, going so far as to agree to charge half his normal fee. She learned quickly that he liked to be called by his nickname – "Shelly."

Pincus was in his early fifties when Hill first met him. "Shelly was tall, slender and had a mop of curly hair set back from a wide forehead. His manner was brusque but friendly," Hill said. But she felt comfortable talking to him. "Our first interview was at the cost $250, and I could take as long as I liked. I briefly outlined the facts of the case, but we mostly talked about the process." That's when Pincus explained that he didn't work on contingency and she would have to pay his $4,000 retainer. "His hourly rate was $350, which I would have to start paying once the retainer was spent. He also said that he did not charge for 'the learning curve.' There was to be quite a lot of that as time went on," she remembered. "The money we were talking about did not seem too horribly much. I certainly did not envision the case dragging on for more than a year or so or costing much more than the original outlay. I left him with the one-page summary and the document that I had given to [Dr. Karen] Putterman before the second committee was convened. I expected that he would recommend that I drop the whole thing."

Pincus grew up in Brooklyn, moving to northern New Jersey as a teenager. After he received his Bachelor of Science degree with distinction from Pennsylvania State University in 1973, he attended law school at Rutgers School of Law at Newark, where he earned his law degree in 1977. Pincus and his family moved to Montclair to a street of elegant old houses set back from the roadway behind long sloping tree-shaded lawns. Though Hill was admittedly apprehensive about the small size of Pincus' firm, she came to appreciate the individual attention and personal consideration it might afford her. Pincus was best known, she discovered, for his work in labor relations and employment law, and had had a prominent role at that time in the New Jersey chapter of the National Employment Lawyers Association and was on the executive committee of the Labor and Employment Law Section of the New Jersey State Bar Association.

"The next morning, when I got to my office in Newark, there was a call from Shelly." Still convinced that he was going to tell her she had no chance of winning, Hill wasn't optimistic when she picked up the telephone. To her surprise he was interested and took the case. "He reduced his hourly fee to $175 per hour if the case, predicated on winning, were to garner a larger award. "At that time, I was not interested in the money." Hill was focused on getting the two Bishayee et al. papers retracted, Howell's NIH program director informed and, she hoped, to be reimbursed for her expenses. Pincus explained that if she went forward, she shouldn't expect expeditious relief. The complaint would be filed in the name of the United States government and sealed immediately. Thereafter, the government had sixty days to decide whether to take the case with the caveat that they could also ask the federal district court for extensions that might go on for years.

In the written submission, Pincus explained to Hill that everything had to be laid out – a difficult but not impossible challenge. Those involved in the fraud must have acted knowingly. In other words, an individual is deemed to have acted knowingly in regard to a material element of an offense when: in the event that such element involves the nature of his or her conduct or the circumstances attendant thereto, he or she is aware that the conduct is of such nature or that those circumstances exist; if the element relates to a result of the person's conduct, he or she is conscious of the fact that it is substantially certain that the conduct will precipitate such a result, and in Hill's case, that person or persons acted in reckless disregard. Hill would need to obtain convincing statistical analysis of Coulter counter controls. Pincus emphasized that the key to success was in the preparation of materials that accompanied the complaint, information that a lay person could read and understand. Pincus told her he would have to educate himself and that was the fun.

Pincus laid out all the rest. Notice of the complaint would go to the United States Attorney in the District of New Jersey, Christopher J. "Chris" Christie[99], and ultimately to the United States Attorney General in Washington, D.C., and it was the latter which would assess the merits of the case. The statute of limitations in Hill's case, he told her just then, was six years from the initial act, which had taken place in October 1999, and she was already more than three years past when federal investigators should have notified. With the days and months ticking by, the six-year statute of limitations only took Hill up to October 2005. The government should have been notified, without question, at the time Howell filed his first progress report in July 2001 and three years from that point was July 2004. While Hill, in consultation with Pincus, was still within the time limit, the two didn't discuss the fact that the government had been informed by Hill of the problems in Howell's laboratory with the filing of her inquiry to the ORI in the fall of 2002 and that they had rejected her findings.

Pincus explained that Howell, Bishayee and the UMDNJ were conspirators – this did not imply that they had met together to conspire, or even that each was necessarily working with the others or knew the end objective, intentional or not. They could be charged individually and severally. If the government decided to take the case, there would be no cost to Hill; however, if the government didn't take it, she would also not be able to just walk away without their permission. Conversely, if the defendants could prove that Hill's case was frivolous, the penalty to Hill could be staggering. Pincus believed that Hill was insulated from the latter circumstance, but they could still threaten it. The award would be triple damages – three times the amount of the fraud, which at that time was based on the amount of Howell's first grant – $1.4 million – minus plaintiff attorney's fees and expenses. Pincus reiterated that he was not interested in contingency, but he also expected the expense to Hill would not be inordinate, amounting to a figure between $5,000 to $10,000, a figure that must have been based, Hill later surmised, on the government taking the case and absorbing the greater expense of prosecution. The greater expense for Hill would be for discovery and trial, if the government did not take the case. Again, Pincus tried to assure her that he wasn't a neophyte in federal court and that he liked taking a federal case.

Still concerned about the small size of Pincus' firm, she questioned it could be as effective as a large, experienced one. Pincus explained that large law firms have small sections that focus on particular legal specialties, and he was not concerned. When she revealed that the ORI had turned down her case for further investigation, Pincus told Hill they'd have to "deal with it." When it came to Howell's laboratory data, they'd need access to all the pertinent notebooks. Hill looked worried. She thought Howell might, if he hadn't already, destroy some of the evidence. While Pincus told her that Howell had a legal obligation not to destroy anything, known as spoliation of evidence, and if he could be proven to have done so, Hill would win the case. Coincidently, on March 17, 2003, just three days after the memo from the second Newark CCRI's chairman Dr. Anthony Forrester, there had been a broadcast e-mail to UMDNJ personnel from Dr. Karen Putterman instructing faculty and staff that original research data and materials were to remain indefinitely on the campus. Hill wondered just then if there was any connection that might be drawn between Putterman's message and her case.

For the next few months, Pincus and Hill worked on the federal complaint and Hill continued, as she had been doing, to commute to the City College of the City University of New York (CUNY). The opportunity to work off the Newark campus came after Hill ran into an old friend, Karen Hubbard, Ph.D., then an associate professor in the biology department at City College. Hubbard and two other colleagues, one also at City College and the other at Sloan Kettering Institute – Memorial Sloan Kettering

Cancer Center, were involved in a study of molecular parameters in skin samples taken from patients with malignant melanoma. She invited Hill to join the project and to supervise the studies involving DNA damage in mitochondria – the sub cellular organelles that provide energy to cells. Hill was more than happy to do so and for about two to three years commuted to Upper Manhattan to work at the bench in Hubbard's laboratory.

During this period of Hill's work at City College, away from UMDNJ, she was informed that Howell was once again ill and when she was on the UMDNJ campus, "Azzam continued to give me a hard time about the small laboratory that we shared. He removed equipment that belonged to me and supplies disappeared. I was told that my things had been moved to a storage area but was never told where that was located and never given access to it," Hill recounted of that intensely uncomfortable period in Howell's division. "He made me feel very uncomfortable when I was there and eventually, I abandoned all hope of being able to accomplish anything there. He threw his weight around boasting about all the students he expected to have working for him and that a 'lady from New York' who, he implied was a physician of some kind, was going to need space in the lab." Hill encountered the unnamed female researcher from New York only once or twice, to her recollection. The woman wasn't talkative, and she lasted only a couple of weeks and Hill never saw her again. "Numerous students failed to materialize as well. Every few months, Ed [Azzam] would remove more of my stuff from the lab. When it was eventually restored to me several years later, what [remained] was returned in two small cardboard boxes containing parts of electrophoresis apparatuses – incomplete and unusable – and some pipette tips – useless without the holders. These were the remains of a one-thousand-square-foot laboratory and office suite that I had at one time called my own."

In early September 2003, Pincus had a draft of the complaint for Hill to review. The two discussed whether Edouard Azzam should be included as a defendant, but Pincus recommended that he not be named. While Azzam had been involved in what Hill characterized as retaliation, he had no direct knowledge of misconduct involving Howell's NIH grant. "He did have a role in the cover up," Hill iterated, but at the time, because the case had been closed by the university, he didn't have an obligation to report anything that he knew had gone on in lab.

Hill described the complaint Pincus had drawn as "a monster document" made up of three parts: the complaint, the written disclosure, and a description of Howell's NIH grant and the importance of the research to further understanding of the mechanism of radiation effects in diagnosis and radiation therapy. There were sections that followed describing Hill's interactions with the first and second Newark CCRI inquiries, and the ORI. Nothing had been left out. "It was a very imposing document." Pincus wrote the ninety-two paragraphs of the written disclosure, as is standard practice, in the voice of Hill as the relator. The last paragraph read that she believed "the data that was used for preliminary results and in the eight publications in which Bishayee participated must be closely analyzed. The Mosimann technique must be applied where applicable. Those publications that are found not to be true must then be withdrawn." She couldn't know it then but it would be seven years before Hill would finish analyzing all of the data produced by Bishayee and was able to show that, after his first few weeks in the Howell laboratory, the conclusion could be drawn, in her words, "that there was not one experiment after that that had not been fabricated."

Pincus filed charges in camera and under seal in the United States District Court of the District of New Jersey with Hill listed as the federal government's relator on October 14, 2003. Listed as the defendants were the University of Medicine and Dentistry of New Jersey, Dr. Roger W. Howell and Dr. Anupam Bishayee. The filing, as in prior *qui tams*, cited violations of the False Claims Act (the "Act"). "The violations of the Act involve the defendants' application for, and subsequent receipt of, federal grant monies (Grant No. R01CA83838) based upon the knowing submission of a grant application to the United States Department of Health and Human Services, National Institutes of Health. That application," it read, "as well as the findings of certain experiments that were subsequently undertaken, were supported with data, statements and records that were false or fraudulent."

Howell's grant proposal was promising at the beginning; it raised significant issues in diagnostic and therapeutic nuclear medicine. His proposed studies would have been of significance to patients, since the risk of radiation insult can be drastically underestimated and potentially lead to increased risk of inducing cancer. In contrast, some patients can be over- and under-treated in radionuclide therapy for cancer. Both scenarios could thus present adverse consequences in the outcome for the patient. It was, thusly, critical that patients not be misled about the results of Howell's research. The United States Public Health Service didn't know anything about Hill's *qui tam* filing until November 24, 2003, when it received a letter from the civil division of the United States Department of Justice delivered via the Department of Health and Human Services Office of the Inspector General. "I have no way of knowing what was done with this information," Hill opined, "but it must have sat in some file [or on someone's desk] while Howell's grant [renewal] was [being] reviewed, [and] while Howell and even Bishayee obtained new grants supported by the NIH."

Two months after the United States Attorney's Office started to investigate the charges, Hill and Pincus hoped the initial probe would lead to federal prosecution of the case. Evident from prior *qui tam* case law, some of them already discussed herein, there are three options once the federal investigators begin peeling back the layers evidence in discovery: (1) the United States attorney can take the case and if this happens, there is a ninety-five percent chance it will be won; (2) the United States attorney clears the relator to pursue the case on his/her own, with only a twenty percent chance of winning, and (3) the United States attorney can rule that the case has no merit, in which case it is over before it gets started. It took nearly four years for the United States attorney to decide not to take Hill's case instead allowing her to pursue it as the relator. In the interim, much of the material released to Hill and her attorney during discovery became available by subpoena, affording the opportunity to examine it in the United States Attorney's Office and, later, as .pdf files. In all, there were more than 30,000 .pdf files made available in this case. Out of all of it, there were roughly 400 experiences performed by Bishayee and others in the laboratory that were available for examination. There were plenty of controls, contrary to the characterization to the contrary by Price at the ORI) and there were plenty of additional experiments Bishayee had performed but failed to offer up when questioned by two Newark CCRIs and that the ORI's federal investigators didn't have in hand to consider when ruling whether to investigate Hill's allegation of scientific misconduct.

Hill knew that the attorney general's response was due on December 14, 2003. Pincus called her on nine days before the deadline to inform her that he'd had a conversation with assistant United States attorney Stuart Minkowitz, who at that time was the chief of the civil division in the United States Attorney's Office in Newark. Minkowitz was asking that Hill agree to enter an indefinite stay, which would mean her case was being removed from the active case list until such a time as it could be reinstated or dismissed. An indefinite stay would give Minkowitz's division time to investigate the validity of the claim and the viability and extent of any recovery. The defendants would be given a redacted copy of the complaint. Pincus conveyed that the average time of such a stay was two years; it would, in truth, take almost twice that long to investigate the case. At the end of the indefinite stay, the federal prosecutor had three options. Simply put, again, it boiled down to the federal government either taking the case and prosecuting it at the government's expense, allowing Hill to take the case as the relator and assume the financial burden of the prosecution or Minkowitz could determine that the case had no merit and dismiss it completely. If the government opted to take the case, the whistleblower, in general terms, typically got about fifteen percent of the triple damages, as determined by the monetary value of Howell's NIH grant. If she, as the whistleblower, pursued the case she would have to share up to thirty percent with the federal government. In the first two scenarios, she would be entitled to reimbursement of legal fees and expenses.

Hill had done her homework. She knew the odds of Minkowitz's office prosecuting the case were slim. The United States Attorney General's office rejects approximately seventy-five percent of the *qui tam* cases it reviews for potential prosecution. But if the government took the case, it won over eighty-five percent of the time. The odds went way down if the whistleblower pursued the case, winning only twenty percent of the time. The government can jump back at any time into a case on the merits. Many whistleblowers who lose have no regrets because they know that they have done the right thing, more often at devastating financial cost. "We would now sit and wait. Shelly would call Minkowitz about every three months to see how things were going." The federal government's pace was going *very* slowly.

The first glimmer of hope came on February 11, 2004, when Pincus called to tell Hill that Minkowitz wanted to meet with them on March 11 at 10 o'clock in the morning. Pincus and Hill met on March 9 to prepare for the meeting. She was palpably nervous. He cautioned her to answer questions directly without elaboration and that, no matter how friendly they were, to be guarded – "they were basically cops." Pincus also instructed Hill to bring nothing to the meeting and that her responses should be guided by the complaint and accompanying disclosure. There would be no going off message. But he also didn't want Hill to give the impression that she was holding back, and, importantly, avoid making un-favorable inferences about the defendants. Pincus needed Hill to emphasize the importance of the work from the point of view of patient diagnosis and care. They would be looking for what Hill had done, actions she had taken, and trying to trip her up.

It was a cold blustery but sunny day when Pincus and Hill arrived at the Newark federal building. Security screening necessitated they wait their turn for half hour to gain entry. They were met by Minkowitz, a youngish balding man in his thirties with a friendly demeanor and rapid-fire talk. He did not look like a cop. They were ushered to a small conference room with the usual table in the center, but walls lined with boxes and papers piled helter-skelter on top of each other. Minkowitz explained that his office had been temporarily quartered in the spaces Hill observed as they made their way to the meeting space, and that he expected to move to a larger space on another floor, hence the boxes piled high all around them. Hill was shown to a seat at the head of the table, what she later described as "the hot seat." To her left was the investigator from the United States Public Health Service, Office of the Inspector General, Susan Schlow[100], and next to her FBI special agent Suzanne Beck. Next to Beck was an assistant United States attorney from the criminal division. Minkowitz was seated at the opposite end of the table and to his left, a paralegal specialist. Pincus was seated to Hill's right.

Once the introductions were over, Minkowitz got serious and informed Hill that if she was found to have done anything wrong, everything she said could be used against her – she was essentially read her rights. He also explained that in her *qui tam* case, the federal government was suing the state government and while she could be compensated for my attorney's fees, she would receive no monetary award and the UMDNJ would not be penalized with triple damages. This was a terrible blow, she later recalled, and meant that this was not a federally prosecuted *qui tam* case. Later, on the way home from testifying, Hill told Pincus that if he wanted to, he could charge her his usual fee of $350 per hour rather than the half he had been charging, but he declined, saying that it did not matter because he really believed in her cause.

Minkowitz informed Hill that she should fully disclose everything she knew, including what "was in her head" as well as already on paper. She provided a brief synopsis of her life to date and proceeded to summarize as best she could her observations up to that time, the conversations she'd had including Edouard Azzam's outburst, Bishayee's revelations during their lunch at the church, her conversations with Kay Fields, and the Mosimann analysis. Minkowitz's office asked Hill to get into the experiments that she had observed in great detail and here, she erred. "I mixed up the experiment of 1999 with the experiment [she monitored] of 2001 and got all twisted in my explanations. Once I realized what I was doing, I backtracked," Hill opined, "[and] noticed that there was a flip chart that I could use and started over. The lawyer from the criminal division left in the middle of my presentation – an indication that this was going to be a civil not a criminal case in which the standard of proof would be much more stringent."

Hill's testimony continued until about 2 o'clock in the afternoon. She was exhausted and asked what the chances were that the case would go forward. Minkowitz acknowledged that this was not the strongest of cases, that the statistical analysis had never been put to the test and might not hold up in court. He iterated that the strongest element in the case was the non-reproducibility of the bystander effect in the spring and summer of 2001. Further, he informed her that an in-house probe would start by talking to Fields and getting the visa and unemployment compensation information for Bishayee. Minkowitz made it clear that she might have to provide information and testimony to other parties, and he wanted to know if there was anyone outside the UMDNJ his office might consult. Hill told them that Marek Lenarczyk was at the University of Colorado at Fort Collins, and they might want to start with him.

"As if he had some insight as to what was going on behind the scenes, Azzam now went after our shared lab with a vengeance. A very expensive electrophoresis apparatus that belonged to a colleague had disappeared, cabinets that had contained my effects were now cleared and my supplies were decimated," Hill journaled, "I had cells growing in one of the incubators and, without looking (or so he said), he turned off the CO_2 [carbon dioxide] input to the incubator which thereby killed my cells. We wound up in a nasty argument and he defended removing my things by saying that he did it for my own good." Hill continued to record everything that occurred in Howell's laboratory in detail.

A couple of weeks after their meeting at the federal building, Minkowitz informed Pincus that Hill would not be compensated for her legal expenses. When Pincus told her, he suggested to Hill that she could back out just then if she wanted to, but even not knowing what the future held, she said no. She was in it to the end, whatever that might be. As it turned out, Minkowitz was wrong about what he had told them about the federal government suing a state government. The statement he'd made about the UMDNJ not being penalized with triple charges was based on his belief that the UMDNJ had sovereign immunity from the claim. In some instances, the Eleventh Amendment precludes the federal government from seeking damages against a state. This was not one of them.

On March 22, 2004, Schlow and Beck called the ORI to discuss Hill's accusations of misconduct. The content of the conversation would not be known to Hill until she found it among thirty thousand .pdf files obtained by federal authorities during discovery. Fields spoke to them in the presence of Alan Price and an attorney. Schlow and Beck's report stated that John Dahlberg, Ph.D., analyzed the numbers that Hill had sent and felt the control data were not reliable. Fields wrote the original ORI report, and Price edited it since her opinion differed from that of other scientists. Fields believed that ORI should have asked UMDNJ to conduct an internal investigation based on Hill's expertise and lack of credible explanations proffered by Bishayee and Howell. "She believed that I was credible, and the statistical analysis needed further investigation. The other scientists – there were apparently eight of them on the committee – felt the evidence was weak and the statistical analysis unconvincing," Hill recounted. "Price felt the photos were not convincing, were not properly dated and an investigation was not warranted. He drafted the final letter." But, as Fields had previously forewarned Hill, none of the scientists present who reviewed her information at ORI were statisticians. Further, no one suggested that controls could be obtained by looking at Coulter counts generated by others in Howell's laboratory. Neither of these omissions were apparently mentioned to Schlow and Beck.

Based on the outcome of federal investigators' visit with the ORI, Pincus now prepared, with Hill's help, a supplemental disclosure that she signed on April 6, 2004. The supplement included additional information about experiments performed by Lenarczyk and Howell. Attached to the first exhibit was a revised graph of Howell's and Lenarczyk's bystander experiments, each of which, she observed, were counted by Lenarczyk except for the last one, which deviated sharply from the others, dropping to a lower plateau and that Hill believed had been counted by Howell after Lenarczyk had left UMDNJ and returned to Poland. "In my own mind," Hill later wrote, "I have referred to this experiment as the 'wishful thinking' experiment – Howell was trying his best to make the survival drop exponentially but was not quite dishonest enough to make up colony counts, which, as we [learned] later, was what Bishayee did. Colony counting," she continued, "is fairly subjective – it is done using a dissecting microscope, the colonies are very heterogeneous and decisions have to be made about overlapping colonies – are there 2 colonies or only one, about whether a given colony has enough cells in it to be counted (the rule of thumb, rarely heeded, is that, to be counted, the colony must contain at least twenty-five cells). Howell's method of counting," she iterated, "was prone to errors – the colonies were marked on the bottom of the dishes with a lab marker and the count was kept in his head. I had a counting marker that kept track of the numbers, removing at least some of the uncertainties. Marek [Lenarczyk], too, used my counting marker." The supplement hit the point that Howell contended that the cells had changed and that would explain why he and Lenarczyk were unable to show a bystander effect; however, since his experiment did show some indication of a bystander effect, his supposition of the cells changing would have to be patently wrong.

Additional exhibits in the supplemental contained a document previously omitted from the first disclosure; it analyzed the 1999 experiment and a copy of the document that Hill had provided to Putterman when they met in October 2002. Exhibit 5 was the timeline derived from Hill's chronology, the diary – a journal – she'd kept since the report was made to the first Newark CCRI. The last part of this newly prepared document contained two graphs, the first showing the marked difference between Bishayee's and Lenarczyk's results of irradiations in the Helena tubes – Bishayee's curve suggesting little or no hypoxia and Lenarczyk's indicating severe hypoxia, which coincided in the second graph to hypoxic results with V79, the same cell line used by Howell, Bishayee and Lenarczyk, taken from the literature. After Pincus filed the supplemental, he indicated to Hill that he believed he would be taking a backseat in the case from that point going forward. "How little did he know," Hill wrote, disclosing her disappointment.

About two weeks later, on April 21, Schlow interviewed Lenarczyk by telephone; he was, as Hill indicated, in Colorado. Hill wouldn't find out what was discussed in that interview until it was disclosed at discovery. Lenarczyk told Schlow that although he had used the same type of cells employed by Bishayee, he could not replicate his results. "He also indicated that he felt Bishayee was not very honest," Hill offered, "and that he advised me [while Lenarczyk was still in the lab] that something was wrong. [Lenarczyk indicated to Schlow that] we took photographs and that I reported Bishayee to the university. [Lenarczyk informed Schlow that he] believed that I had reacted too quickly and Howell was furious that he [Lenarczyk] had not advised him of the situation earlier. He also reported [to Schlow]," Hill continued, "that I was insulted when Howell had implied that I could have contaminated the 1999 experiment that Bishayee and I had done together. He was right about that."

Two days later, on April 23, Hill traveled to Saint Louis, Missouri, to attend the annual meeting of the Radiation Research Society. "We had lived in Saint Louis for three-and-a-half years from 1973 to 1976, and still had many good friends there," Hill recalled. Hill managed to avoid staying in the hotel where the meeting was being held by staying with a good friend in a nearby suburb. The friend even lent her a car to ease the commute. "When I did go into the meeting for the opening reception, I immediately ran into Howell and Azzam, much to my dismay. I was fated to run into them almost every time I turned around."

Despite the discomfort of bumping into Howell and Azzam with some regularity, Hill found one of the encounters of interest. She'd gone to the hotel dining room for the luncheon buffet and was shown to a table for four against a wall with a long bench running along it and lined with similar tables sandwiched close together. She took a seat in one of two chairs facing the wall. Four young men in town for a dance were seated to her left. The table to her right was empty except for a briefcase placed on the seat of one of the chairs. Hill assumed the owner of the briefcase and whoever might be seated at the table had gone to the buffet. She also assumed that whoever was sitting there was probably attending the Radiation Research Society meeting. She was right. "Of the several hundred attendees at the meeting, I had chanced to sit almost on top of Ed Azzam and a colleague of his from the University of Iowa [Carver College of Medicine], [Douglas R., Ph.D.] Doug Spitz. While I did not know him personally," she noted, "Doug was certainly known to me. He is a good scientist and quite well known in the [radiation oncology] field. Ed was probably as taken aback as I was, but he swallowed his pride and introduced me. Spitz asked if I was a scientist and Ed sang my praises as an expert on ultraviolet, melanoma and DNA repair."

Hill later documented that Spitz asked her what she did, and she told him about her interest in DNA damage in mitochondria. He was very interested, Hill acknowledged, the more so when he found out that Hill was doing polymerase chain reaction (PCR). "He immediately jumped the gun and said that I should be a 'specific aim' in the grant application that he was preparing to submit to the Department of Energy. I was floored but said I would be glad to participate. Spitz wondered why I did not work with Azzam and Azzam dissembled." Hill did become part of Spitz's grant application, although she it wasn't without, as she observed, "Azzam trying to block it, telling Spitz when he tried to get a hold of me that summer that I was not around. I learned of that by putting a call in to Spitz myself only to learn that Spitz had written me off. My proposal was added back in, and the grant was subsequently funded with Azzam in charge of the work in Newark," Hill observed. "For two years, Spitz's lab sent me DNA samples, and I did the studies that had been proposed, and sent him some rather nice data. [Spitz] wrote some papers without including my work and while I was glad that it kept me busy, I was sad that nothing ever came of it. Things eased up a bit between Azzam and me and I did some studies for him as well. Unfortunately," she continued, "the results I got for him conflicted with the results I got for Spitz using the same cells. Since this was one of the very issues being dealt with in the *qui tam* case, it would not have been good to make too much of that. I believe that there is a logical explanation for the discrepancies that could easily have been resolved in an experiment or two, but those experiments were never done and the whole thing has been chalked up to experience."

Pincus had given Hill strict instructions not to discuss the case with Lenarczyk, and she never did. When she saw Lenarczyk at the Saint Louis conference, he made a few comments about being interviewed by the FBI, but she didn't respond to what he'd said, heeding Pincus' counsel. But she did take Lenarczyk to her friend's house for dinner and drove him around Saint Louis to show him more of the city.

Hill regarded Azzam as the star of the Saint Louis conference. "His results were interesting and exciting. Roger's post doc from Italy [Massimo Pinto, Ph.D.] made a nice presentation of the bystander effect that had avoided using the Helena tubes. Roger by now was avoiding using them – he may finally have caught on to the fact that they were hypoxic. It was painful for me to listen," Hill wrote later, "especially to Ed. I had wanted so much to participate in that work – it was my field. My love, all my scientific career, has been radiation. And now I was the bystander – excluded from doing what I loved best and held back from retiring as I had resolved not to do so until the *qui tam* was over." Being a whistleblower was already proving a painful experience for Hill. "You have to do what you know is right but the shunning, the loss of friendships, the distain of others who don't know or understand the story is agonizing." Back at UMDNJ, Hill had pointedly tried to sidestep encounters with anyone in radiation research. She'd plan her lunches in the cafeteria to eat early to avoid running into division colleagues. She'd use the back stairs and seldom went to the faculty lounge to get around gatherings that were inevitably uncomfortable for Hill. She wondered what other faculty members thought of her.

Around campus, Howell and Azzam were "the stars," in Hill's words. She figured the UMDNJ faculty, those not as much "in the know" must have wondered, by then, why she was no longer associated with either of them, perhaps drawing the wrong conclusion that she was the one who'd done something wrong. "[The] *qui tam* has stolen my twilight years from me – years when I could have been retired, spent my time doing things that I love to do – volunteer work, docent at the science museum, embroidery, knitting, enjoying my grandchildren and my great-grandchildren," she could hold back no more. "Those are the things that have cost me the most. But I could not and would not quit and Shelly has even told me that he would carry on if something should happen to me."

FBI special agent Suzanne Beck filed a report on Bishayee's work history on April 28, 2004; he had, indeed, received unemployment benefits after he left Howell lab. Had Bishayee resigned, as Bishayee and Howell claimed, he would not have been eligible, and this was the conclusion reached at this point in the investigation. The machinations it took to circumvent New Jersey labor law to get Bishayee unemployment benefits led Hill to conclude that "the university and the [radiology] department, Howell included, must have colluded to allow this to [take place]." The Bishayee resignation versus firing issue would continue to ripple through Hill's *qui tam* case. That same day, Beck also reinterviewed Kay Fields again by telephone, an ORI lawyer listening in on the call. Fields acknowledged that she wrote the ORI report except for the DIO recommendation, which had been written by Price. Why didn't she write the DIO recommendation? Fields explained to Beck that she didn't agree with Price's decision that false claims had not been used to obtain the grant. Fields' support for Hill was predicated on several points that she laid out for Beck: Hill's testimony; the non-randomness of the Coulter counter numbers; the fact she did not find Howell's testimony believable, including that Howell excused the fact that there were no notes about the new cell line, and lastly, she found Howell's supervision of Bishayee at best lackadaisical. Fields expressed concern that Howell had not repeated the questioned experiment, that Bishayee did not have an explanation for the extra tubes that were seen in the incubator and that he did not comply with scientific standards associated with conducting experiments. The contents of this report would not be revealed to Hill until discovery was underway.

In July 2004, Susan Schlow set up a meeting with William "Bill" T. Wells in the grants administration branch of the NIH's National Cancer Institute to discuss the case, and on August 19, Minkowitz told Pincus that he and someone from the criminal division were going to Washington,

D.C., to talk with Fields. This information was encouraging to Hill and implied that the NIH might become a partner in the suit. There is no evidence, however, in the documents later obtained in discovery that either of these meetings ever took place.

Less than two months later, the clock on the case still ticking, on October 6, Pincus and Hill met in his office with Susan Schlow and a man named Frank Elmo, who turned out to be another federal agent. Pincus commented later that Elmo seemed to be the most interested in Howell's intent, but, for the most part, he sat quietly. Schlow asked most of the questions. The objective of the Schlow and Elmo meeting was preparation of a subpoena for the judge. Hill had two lists, one of the people who should be interviewed and the other of documents that would be required. Both lists were extensive. Hill's list of potential interviewees included Howell's NIH program director, Howell's current postdoctoral fellows, personnel in the FACS laboratory, Azzam, de Toledo, those involved in the UMDNJ inquiries to include Baker, Cook, Pain, Putterman, Saporito, and, of course, Howell. She also had Fields and Price on her list. Hill wanted to know if Howell's current postdoctoral fellows knew that the Helena tubes were hypoxic and, importantly, if they'd repeated Bishayee's 1999 experiment. Lenarczyk wasn't on her list at the October 6 meeting, perhaps because, she surmised later, he'd already been interviewed by the FBI.

The documents that Hill believed should be requested included laboratory notebooks starting before Bishayee's arrival and continuing forward; the radioisotope records; the FACS data; Howell's computer and the lab computer that Bishayee had used; reports and notes of the two campus committees and of the ORI; Bishayee's bank statements, work records, unemployment records, visa status, and job interviews he may have had after he left; annual progress reports for the grant, e-mails, especially Bishayee's "how's the weather" correspondence to Hill when he wanted to visit the laboratory without Howell present. Hill's record of the "how's the weather" e-mails had disappeared from her inbox in what she described later as an odd and mysterious way in about March 2002. "I was shutting down my computer one evening and chanced to hit a key, which, I don't know, and the screen went blank. The next morning when I turned my computer on again, all of the e-mails from my Inbox for the past six months – which included the messages I had received from Bishayee – were gone," she admitted. "I did not realize until *years later* that while I had lost the e-mails he had sent me, I still had my responses to them in my sent box, so I was able to retrieve a good many of them. I doubt that the loss was due to any nefarious behavior on the part of the university; it was just some kind of glitch in the operating system."

The greater question that remained was how to get controls for the Coulter counter. Hill suggested that it would be possible to attach the counter to a printer and obtain counts approximating the magnitude of counts made by Bishayee for analysis, or two or more unbiased observers could simultaneously record a series of counts. The Mosimann analysis could thus be applied to any numbers so obtained.

Pincus talked with Schlow about a week later. She told him they were preparing the subpoenas, and she was planning to talk with Bishayee the following week. Schlow asked if Hill knew Bishayee's working hours to which Hill asked if Schlow knew where he lived. Elmo replied that he was certain that they did – after all, they were law enforcement. Pincus speculated that Schlow might try to get Bishayee to turn the blame on Howell.

Just before Christmas, Pincus spoke with Minkowitz, who reported that the subpoenas had been served, and his office was expecting a response from each party served within a month; interviews were ongoing. "We would never know who they actually interviewed, if anyone," Hill reported. She received an e-mail on January 13, 2005, from Putterman that the UMDNJ had received a subpoena from the Department of Health and Human Services (HHS) for all notebooks relating to the bystander effect research between January 1996 and March 2002. Hill immediately submitted two of her notebooks and the Zip disks Lenarczyk had sent her. Everything was returned to Hill about three weeks later. A few days later, Azzam cleaned out the incubator in the laboratory shared with Hill. After that, Hill recalled she rarely went near the lab again, instead continuing her work on mitochondrial DNA for approximately four more years. But at that point, she later observed, this marked the end of forty years of having her own laboratory space.

In late April 2005, Minkowitz informed Pincus that his office was still trying to get the subpoenaed documents from UMDNJ and that Fields was on tap to review them. He intimated that there were hundreds of documents involved. A month later, in late May, Hill received a letter to her West Orange, New Jersey home address from the UMDNJ director of legal management reiterating that a subpoena *duces tecum*, a writ ordering a person to attend a court and bring relevant documents, had been issued in a more serious attempt to get the university to comply. Failure to respond could lead to criminal charges.

Baker called a departmental faculty meeting on July 6, which had been mandated to be held once a month but this rarely, Hill offered, ever happened just then in the radiology department. Hill ran into Anthony V. Boccabella, Ph.D., J.D., former chairman of the anatomy department who'd transferred to radiology when anatomy morphed into the department of cell biology and injury science. Hill admittedly rarely saw Boccabella but on that particular day, she bumped into him in the faculty lounge. "He told me Baker had instructed him to tell me to be sure to attend the department meeting. At this time, the university was in the news almost every day for questionable billing practices and in applying for reimbursements from Medicare and Medicaid," she explained. "Baker informed the department about the the other cases of wrongdoing and told the radiology faculty the FBI had taken documents from his office. He had given them all that they had asked for but didn't understand why they had been there. I wondered if there was a message in this for me," Hill speculated, "that the documents involving Bishayee's termination had at last been turned over to the United States Attorney by the FBI."

The Radiation Research Society met in Denver in the summer of 2005, after Baker's departmental faculty meeting. Lenarczyk was still working in Fort Collins, Colorado, but things were not going well, and he was looking for another job, according to Hill. "He told me about the call he had received from the FBI two years before and that in January, Putterman had asked him for his notebooks which were in Poland, but he got them and sent them to her about a month ago in roughly May of 2005." Hill cochaired a symposium on mitochondria at the conference but the buzz in the society just then was about Steven A. "Tony" Leadon, a former professor of radiation oncology at the University of North Carolina in Chapel Hill (UNC-Chapel Hill) who'd trained with Philip C. Hanawalt, an American biologist who discovered the process of repair replication of damaged DNA in 1963. Hanawalt is also considered the co-discoverer of the ubiquitous process of DNA excision repair along with his mentor, Richard B. Setlow, Ph.D. and Paul Howard-Flanders, Ph.D. He holds the Dr. Morris Herzstein Professorship in the department of biology at Stanford University, with a joint appointment in the dermatology department in Stanford University School of Medicine.

Back to Leadon, considered by his colleagues a leading authority in the field of transcription-coupled DNA repair (TCR). The UNC-Chapel determined that Leadon had fabricated and falsified data in his research on DNA repair in 2003. Three years later, in 2006, the ORI came to the same conclusion, finding that Leadon engaged in scientific misconduct by falsifying DNA samples and constructing falsified figures for experiments done in his laboratory to support claimed findings of defects in a DNA repair process that involved rapid repair of DNA damage in the transcribed strand of active genes, included in four grant applications and in eight publications and one published manuscript. In the wake of the investigations, Leadon's papers were retracted from several journals including *Science* and *Mutation Research*, while more articles were partially retracted from journals including *Proceedings of the National Academy of Sciences* and *Molecular and Cellular Biology*. Leadon was ordered to withhold from applying for federal funding for five years "The regular exclusion period is three years, but because Dr. Leadon's conduct extended over quite a few papers, quite a few years, we decided to take this exceptional action," John Dahlberg, then director of the ORI's division of investigative oversight, told *The Scientist*.[101] Since one of Leadon's doctoral students raised suspicions about his findings in 2001, the number of publications alleged to contain or rely on altered or falsified data had grown to four grant applications, eight studies, and an unnamed manuscript.

Leadon's misconduct had enormous ramifications. Researchers at Lawrence Berkeley National Laboratory, at Stanford and in Switzerland publicly retracted attention-grabbing research published in 1997 in the journal *Science* because the article contained data that, the lead scientist said were faked by Leadon, a close collaborator.[102] "I am devastated. It's been the worst experience of my professional career," said Priscilla K. Cooper, Ph.D., head of the molecular biology department at Lawrence Berkeley and lead author of the 1997 article. Leadon was a staff scientist and head of the DNA repair and radiation biology group at Lawrence Berkeley from

1986 to 1991, but he and Cooper did not collaborate during that period. Reporting indicates after he moved to the University of North Carolina in Chapel Hill, where he was director of radiobiology, Leadon and Cooper began collaborating in research long distance. They co-published the now-retracted 1997 paper along with Thierry Nouspikel and Stuart G. Clarkson of University Medical Center in Geneva. Nouspikel subsequently moved to the Stanford biological sciences department.

Cooper, in news reports, stated she was stunned to discover the fabrication of laboratory data by former colleague Leadon – a molecular biologist trained in radiation biology – in research that she, he and two other scientists published in 1997.[103] This 1997 paper detailed, according to media reporting, why, at the molecular level, cells in babies born with a rare and invariably fatal childhood disease called Cockayne syndrome can't efficiently repair DNA damage initiated by molecules called oxygen radicals, thus leaving these newborns vulnerable to their own metabolic processes. The implications of this retraction can be difficult to appreciate. The 1997 paper attracted a significant amount of attention from the scientific community and was subsequently referenced by more than 70 other papers and Cooper stated one authoritative source told her the number of citations had risen to about 250.

It was this media coverage about Leadon and the 1997 paper that had everyone at the 2005 Radiation Research Society talking. Hill observed that Hanawalt was just as stunned as Cooper and, most certainly, a laundry list of Leadon's other collaborators. Hanawalt had a grant application recently rejected and believed it had much to do with his long association with former student and colleague Leadon.

Barbara Fadem-Chenal told Hill in October 2005 that some of the women faculty in the medical school had gotten together to investigate the possibility of salary discrimination. A task force had been formed, spearheaded by two women in the neurosciences department, both of whom were professors who'd successfully applied for several federal grants between them. They'd discovered that several men in their department had been far less productive from the career and grant perspective but nonetheless commanded salaries considerably higher than their own. Hill began attending task force meetings and quickly learned that her own salary was also less than those of the men in her department. "Since I was *persona non grata* anyway, I was happy to join forces with these ladies," she observed of her participation in what eventually became a class action representing women faculty members in all of the colleges and schools within the university.

By early December 2005, the prosecutorial clock had been running on the United States Attorney's Office for two years. When Hill asked Pincus for an update, he called Minkowitz, who, he learned, was being replaced on the case by Susan Steele, a registered nurse-turned-lawyer who was set to take charge of the civil division in Newark. The scandals swirling around the UMDNJ had grown to mammoth proportions and forced Minkowitz's office to expand its health services expertise. A new FBI agent had been assigned to the case –Mary Beth Gardocki, who had a Ph.D. in biology from Wayne State University in Detroit. Pincus asked if Fields was still involved and learned that she was and that Fields and HHS were still supportive. But Minkowitz was still waiting for information requested via the subpoenas to be turned over. He also wanted to meet with Pincus and Hill again.

Later that month, UMDNJ dean Robert L. Johnson, M.D., F.A.A.P., called a town meeting to inform the university about circulating rumors about the rampant fraud in the hospital and the medical school. The United States Attorney for New Jersey, Chris Christie, had apparently appeared at a meeting of the university board of trustees, Hill noted, and it was thereafter that the dean called his meeting to assure everyone that issues were being addressed. He conveyed that nothing was wrong and while there may have been a few bad actors in the past, present faculty and staff were honest and it was a positive outcome that the University Physicians Associates (UPA), the private practice organization set up for the physicians, and the Faculty Practice Program had cleaned house. In truth, Hill observed, the dean, UPA and the hospital had been at odds for years, and Johnson had far underestimated the scope of wrongdoing on the UMDNJ's campuses. George F. Heinrich, M.D., the president of the UMDNJ foundation as well as the dean of admissions, reported that donors and parents were calling to express their concern about the way funds were being allocated, and one major donor was threatening to pull out.

When Christie threatened to prosecute the school for fraud – which would have disqualified it from receiving the federal funds that make up a vast amount of its income – administrators agreed to allow a federal monitor to investigate its operations. Johnson announced on January 3, 2006, that the board of trustees had accepted the federal government's deferred prosecution agreement. To have failed to do so could have closed down the entire university. The federal monitor, a former federal judge named Herbert Jay Stern, was a well-known public figure. Stern had served as assistant district attorney in Manhattan where he investigated the assassination of Malcolm X; had worked for the Department of Justice in the organized crime and racketeering section, and starting in 1969, had worked his way up through the ranks to become the United States Attorney for New Jersey. President Richard M. Nixon appointed Stern to the federal court in New Jersey in 1973, where he remained for thirteen years before entering private practice. Stern became better known in 1978 when he presided over the Berlin hijacking case – *U.S. v. Tiede* 86 FRD 227 U.S. Court of Berlin (1979) – by insisting East German (GDR) citizens Hans Detlef, Alexander Tiede et al., had the same rights guaranteed by the Constitution as any United States citizen, to include a trial by jury. He later wrote a book about it, *Judgment in Berlin*, that was subsequently made into a movie. He authored several other legal books on trial advocacy. His appointment as federal monitor by Christie provided a boon to Stern's law firm, said to be on the order of $8 million in legal fees.

Federal monitor Stern was empowered to look at the corporate structure, the effectiveness of compliance, especially Medicare and Medicaid, at UMDNJ. Stern further scrubbed the UMDNJ's contracts, security, salaries, bonuses, and conflicts of interest. Faculty and staff would be required, in the fallout of his investigations, to take an ethics course. The academic structure, including research, would not be looked at, presumably because there was no problem in that area. "But little did they know," Hill interjected. A whistleblower hot line was set up and the university would have to pay back the overpayments received from Medicare and Medicaid. There was much work to be done at UMDNJ, an inordinate amount of it yet to be uncovered. Stern's term as the federal monitor was two years.

A couple of weeks later, Pincus and Hill met with Susan Steele and Mary Gardocki in the United States Attorney's Office. Steele apologized for Hill's investigation taking so long, but it was now high priority, and they wanted to proceed quickly. Hill's enthusiasm after hearing this news was short-lived. The meeting had been what Hill called "a replay" of the meeting she and Pincus had two years earlier when Minkowitz was in charge. The discussion and subject matter were nearly identical – Hill went over the two experiments she and Lenarczyk witnessed, mixing them up again and having again to back track and start over. About this time, Pincus had run into a colleague who was knowledgeable about *qui tam* cases. He was told that the United States Attorney's offices in Philadelphia and Boston were the best for *qui tams* and that Newark had to coordinate with them, which inevitably led to numerous delays. While Pincus' colleague told him the Newark staff were nice enough to work with and Steele was good at her job, he advised that *qui tam* cases can drag on for many years, citing one that had taken sixteen years to adjudicate. Hill wasn't certain she'd last that long. In any case, Pincus' colleague told him that the next phase could take another two years. As Hill eventually found out, it would take considerably longer than what he'd told Pincus.

Pincus and Steele spoke occasionally during the months that followed. Steele informed him that she had asked the United States Public Health Service for more help and Pincus offered more from Hill and his law office. Federal monitor Stern was turning up more incriminating information of wrongdoing in the UMDNJ administration and two deans on other campuses would soon be in trouble along with a legislator who held a no-show job in exchange for his support of the university. One of the deans and the legislator would later do jail time. Later in the spring, Steele told Pincus that discovery had yielded an incredible volume of documentation. The UMDNJ had delivered eleven boxes of documents, and she wanted Hill to help prioritize the contents. She met Gardocki in the United States Attorney's Office and was confronted with a hodge podge of documents in no particular order. Each box was inventoried but the lists were cryptic and virtually unintelligible. Hill wasn't initially allowed to examine the contents, only to have a quick look at the contents and prioritize the inventories.

Fields came up to Newark in middle of July and went over some of the subpoenaed material leaving Post-its here and there apparently to draw Hill's attention when and if she got the opportunity to review the documents herself. Hill was away at a meeting and missed a first in-person meeting with Fields. Despite being initially told she was to stay hands-off the contents of the UMDNJ subpoenaed material, Hill was asked to come in to analyze them shortly after Fields made her visit to the federal prosecutor's office. "My summer with Box 6 was about to begin," Hill declared.

When Hill first began to go over the material in the boxes, she was watched closely by Gardocki and all her work with the documents was performed in the United States Attorney's Office. Once she arrived in the building, Gardocki would come down to the lobby to escort her up to a room where the boxes were laid out for Hill's examination. But she quickly zeroed in on Box 6, where she found a number of Bishayee's tritium experiments. Gardocki, she observed was restless having to sit and watch Hill go through the thousands of documents in the boxes, so she changed the venue to the FBI building at 11 Center Place in Newark, a place Hill once described as a forbidding building overlooking the Passaic River with no windows on the street side. There was a guard at the entrance to the roadway who had Hill's and Gardocki's names, and who directed Hill her parking space for the day – frequently under the shadow of a no parking sign.

Once inside the lobby, Hill was retrieved by Gardocki, who led her through a maze of dimly lit hallways to the elevators, and then up to the ninth floor, down a long, dark corridor of closed doors. Gardocki deposited Hill in a cubicle in an air conditioned but drab and windowless room. This would be her workspace for the time. Box 6 awaited on an insufficient tabletop and she got to work sifting through piles of Bishayee's experiments. Hill wasn't certain what to look for but decided to focus on pages that contained Coulter counts. Though she was not permitted to bring a computer into the building, Gardocki supplied her with a laptop that wasn't connected to their network. What they'd handed Hill contained no useful software except the necessary word processing capability she required.

Hill spent, in all, about eight lengthy sessions at the FBI, at all times monitored by Gardocki and if she had to leave the room, another FBI agent. Much of the time Hill could work in the cubicle without obvious observation. But Gardocki, she believed, was unhappy being Hill's minder and had Hill's work shifted back to the United States Attorney's Office. The atmosphere there was more pleasant. Hill had a conference room to work, one with adequate space to spread out the contents of Box 6, which now held her complete attention. Nobody watched Hill while she worked and the organization of the contents of box 6 went much faster though, she subsequently noted that there was much duplication of documents which confounded her organization. Every page was numbered – what she would later learn were Bates stamps – and every document presented during discovery would bear a Bates number. "I was unaware of their importance and failed to note the Bates numbers on the documents that I studied," she opined. "I was still not allowed to bring in a computer, but I was allowed to use a calculator. I was also told I must remove nothing from the room, but I was given permission to take notes and also take the notes home." Hill copied Coulter counter numbers, took the information home and analyzed it.

As Hill worked, it became obvious that she needed a timeline for the experiments. She discussed this with Gardocki, and she concurred, ultimately promising that the United States Attorney's Office would hire a professional to organize the material chronologically; this never happened, and Hill continued to plow through the material on her own. The chronology would become very important later on.

During Hill's analysis, she came across the six experiments done or supervised by Howell in 2001, four of which failed to demonstrate a bystander effect. These results confirmed and supported the bystander data that Lenarczyk had previously sent Hill and were considerably different from Bishayee's results that had been reported in the two papers. In Bishayee's experiments, in fact, there were more than 100-fold more killing of the cells than in Lenarczyk's and Howell's experiments. Bishayee's results showed a dramatic bystander effect, Lenarczyk's and Howell's experiments showed little or none.

Up to this time, Hill had had no reason to question the exponential decline in survival resulting from Bishayee's 100 percent experiments, so she had planned to focus only on the fifty percent or bystander experiments and leave the 100 percent experiments alone. Lenarczyk, too, had focused on the 100 percent experiments, those included on the Zip disk he had sent Hill in the fall of 2002, but she had never even looked at them. But now, for the sake of being thorough, Hill decided to look at Howell's two 100 percent experiments, expecting them to back up Bishayee's results. "I looked at the graph of Roger's first experiment and it seemed very similar to his bystander experiment graphs which is to say, even at high doses of radioactivity, only about half of the cells were killed, whereas, in Anupam's experiments, most of the cells were killed," she noticed. "I found this very puzzling. My first thought was that I had misread the protocol, and this was actually a 50 percent experiment. But, no, it definitely was a 100 percent experiment. I looked," she continued, "at the second experiment and the results were the same. I was dumbfounded. I turned to the experiments that Marek had sent me and found that all his 100 percent experiment results were similar to Roger's."

Later, it occurred to Hill that explanation for the discrepancies she documented between Bishayee's, Howell's and Lenarczyk's experiments had to do with the biological properties of thymidine, a precursor of DNA but also with the fact that the tritiated thymidine was radioactive and could destroy the cell's DNA. Cells in culture, like cells in the body, divide periodically to replace cells that have been lost or worn out or in order to increase the population in question – this latter would apply to cancer cells, too. The process of cell maturation and division is known as the cell cycle and is divided into four parts. The first is G1, G stands for gap reflecting the ignorance of what was going on many years ago when the cell cycle was first defined, Hill explained. After G1 comes S which stands for synthesis – that is, the period during which the cell's DNA replicates or doubles. After S comes G2 during which time the cell is preparing to divide and this is followed by mitosis or M, the act of dividing during which the chromosomes split and move away from each other so that two new cells can form. The V79 cells studied in Howell's lab and which were, in part, at the center of controversial experiments, had a cell cycle – the time between two mitoses – of about eleven hours. G1 takes about two-and-a-half hours, S about five-and-a-half hours, G2 about two hours and M about an hour.

Thymidine is a building block of DNA and will only be incorporated into DNA during the S-phase when DNA is being synthesized. But Howell's protocols – all of them – called for exposing the cells to tritiated thymidine for about twelve hours or a little bit more than one complete cell cycle. In the radioactive form of thymidine – tritiated thymidine – some of the hydrogen molecules have been replaced by tritium, a radioactive form of hydrogen. Tritiated thymidine, like thymidine, will incorporate in the cell's DNA during the S-phase and will remain there, where it will decay producing radioactive particles known as beta rays. The beta rays disrupt the DNA and kill the cells – and only the cells that contain tritiated thymidine in their DNA. If all the cells take it up – which would mean that all had passed through S-phase during the twelve hours, all or almost all the cells would be killed in an exponential manner. Hill pointed out that Howell's – and Lenarczyk's – results indicated that only about half of the cells had incorporated the tritiated thymidine and only half of the population were killed. Hill needed to resolve how this had happened.

The answer came to Hill as she continued to work through the material in the box, after she'd discovered Howell's 100 percent experiments. The explanation, she resolved, was this: under normal conditions – no tritiated thymidine – all the cells in Howell's experiments would pass through S-phase during the twelve hours; however, the conditions were not normal. Tritiated thymidine was present, and it blocks the cell cycle so that the cells exposed to it do not enter the S-phase and therefore are not killed by the decay of the tritium. The cells that were already in S-phase at the time the tritiated thymidine was added would be killed – those would be the 50 percent of the cells that died during Howells's and Lenarczyk's experiments. The remaining 50 percent never entered S-phase and thus were not killed by the decay of the tritiated thymidine. There were important ramifications of this finding. In the first place, it explained why Howell's and Lenarczyk's bystander (the 50 percent) experiments disagreed so dramatically with Bishayee's experiments. But even more than that, it meant that *every single experiment* (emphasis added) that Bishayee had performed using tritiated thymidine had been fabricated, according to Hill's analysis. The results of Bishayee's 100 percent experiments were impossible and the results of his 50 percent experiments were also impossible.

How could it be that Howell did not realize this, Hill wondered. Howell's training was in physics, not biology and certainly not radiation biology. He did not know any biology and he certainly did not know any radiation biology. Bishayee was trained in his native India with his doctorate in pharmacology. "He, too, was unfamiliar with the vagaries of radiation biology, but it didn't take him long, once he started working in Roger's lab, to figure out how to get the results that Roger wanted," Hill observed. This would matter going forward in the case, most especially by the time depositions started.

About this time, there was a sea change at the United States Attorney's Office. The scandals that first put federal monitor Stern in charge of the UMDNJ, had begun in 2003 when a former member of the New Jersey Medical School filed a whistleblower suit, triggering a federal investigation of the cardiology program. The cardiology fraud that involved the double-billing of Medicare and Medicaid, had now spun out of control. Stern had enough to charge the UMDNJ in December 2005 with defrauding Medicaid of nearly $5 million. But now, after investigators had taken a long, hard look at the books, the federal monitor called that number a mere fraction of the potential fraud, waste and financial abuse at UMDNJ. The new tally of overall mismanagement had risen to $243 million – and was expected to go higher. That figure came from a report released on July 20, 2006, by the monitor, former federal judge Stern. "It's very disheartening," said John Inglesino, counsel to the monitor in a statement released to the media, "but I become less and less surprised at what we find."[104]

In the UMDNJ Newark hospital's cardiology unit, the monitor had come upon a case of financial impropriety that was still vigorously underway, implicated high-ranking members of the medical faculty and that was directly related to the school's core mission of delivering health care. Stern's examination of the cardiology kickback scheme involved community cardiologists who'd been given the title of assistant professors at the university hospital – with salaries – and were supposed to work there part-time, performing specific teaching, research, and patient-care duties. But Stern concluded that they were expected to do little more than refer patients to the cardiac surgery program to ensure that it would not lose its state accreditation. He also accused the then-interim president of UMDNJ, Bruce C. Vladeck, Ph.D., and other university administrators of intentionally misleading investigators and attempting to bury previous concerns raised about the illegal referral program. In all, eighteen cardiologists took part in the scheme. Stern criticized hospital officials for failing to alert federal authorities to the kickback arrangement, even though it was the subject of a critical audit by the accounting firm J. H. Cohn. Moreover, school administrators recently paid a multimillion-dollar legal settlement to a former UMDNJ physician who had been dismissed after objecting to the referral plan.[105] He'd been the whistleblower.

All but one of the cardiologists involved in this scheme entered criminal guilty pleas or civil settlements with the government. The settlements typically required the cardiologists to pay the government twice the amount they had received under their contracts. The guilty pleas and settlements were highly publicized by the United States Attorney's Office, generating considerable negative publicity regarding the cardiologists, observed attorney Richard B. Robins, who represented the only cardiologist who refused to settle with the government. Robins' client had entered into an employment contract with the UMDNJ that included a non-exclusive list of services that the government considered no different than the other seventeen cardiologists charged in the case.

The government charged that the cardiologist had violated the federal Anti-Kickback Statute and the Anti-Self-Referral Act (also called "Stark Law"). The Anti-Kickback Statute, 42 U.S.C. §1320a-7b et seq., explained Robins, makes it illegal to knowingly and willfully offer, pay, solicit or receive something of value to induce business reimbursed under a federal health care program such as Medicare. The Stark Law, 42 U.S.C. §1395nn, et seq., makes it illegal for a physician to make referrals for services to an entity in which the physician or an immediate family member has a financial relationship (including an employment contract or ownership interest), unless the physician meets an exception such as having a "bona fide employment relationship" for fair market value. The government also charged the cardiologist with violating the federal False Claims Act, by causing UMDNJ to submit "false claims" – bills filed – to Medicare for the patients whom the cardiologist referred to UMDNJ. A finding of liability under the False Claims Act will result in the award of triple damages, civil penalties, and investigation and litigation costs against a physician. But the cardiologist argued that while he'd performed a substantial number of services listed in the contract, he would have done nearly all of them had he not been denied that opportunity by UMDNJ. Robins' client denied any knowledge of the kickback scheme. When the case went to trial in March 2011 – the only case to get to federal court of the eighteen originally charged – the jury found in favor of the cardiologist on all issues.[106]

Robins called his client's case a cautionary tale of lessons that all physicians need to learn, among them the fact that physicians should not assume their employment contracts with hospitals are lawful, and the fact a doctor rendered necessary, valuable and legitimate medical services will not lower the amount of damages awarded against a physician for a False Claims Act violation, if the court concludes, he noted, that the Anti-Kickback Statue or the Stark Law was violated and that "false claims" for payment were submitted to Medicare due to those violations.

In a July 21, 2006, *Philadelphia Inquirer* story about the UMDNJ fraud scandal, which by then was front page news across the country, university officials defended their position, noting that the quarter-billion-dollar figure included overspending, no-bid contracts, and other issues already addressed. But the monitor didn't concur. Already the amount of potential Medicaid fraud identified by the monitor from 2001 to 2005 has swelled to $35 million, most of the fraud originating from the school's billing Medicaid twice for the same services. But the potential fraud before 2001, according to Stern, might top $49 million. The UMDNJ, with a $1.6 billion budget, had five campuses.[107]

The monitor's oversight of the broken UMDNJ turned out more than the Medicare and Medicaid fraud. Probing the university's financial records, Stern's investigation turned up profoundly troubling fraud, waste and abuse that led to the resignation of Robert Michael Gallagher, dean of the university's School of Osteopathic Medicine in Stratford in South Jersey, effective that April 30, 2006, and Robert A. Saporito, D.D.S., UMDNJ's senior vice president for academic affairs, whose last day was March 31, 2006. Saporito's involvement in Hill's case was also under review by the United States Attorney's Office at that time. Gallagher's departure left all three of the UMDNJ's medical schools without a permanent dean.

Stern also investigated William Wallace, a dean at the Stratford campus who also served as a Gloucester County freeholder and chairman of the Delaware River and Bay Authority. The FBI raided his office in 2006 after receiving a tip that his secretary was shredding documents. Stern later accused Wallace of steering a no-bid catering contract to a friend and helping his daughter get admitted to the school, even though she did not take the required tests.

The United States Attorney for New Jersey, Chris Christie, had also opened a criminal probe into Camden Democratic state senator Wayne Bryant, who landed a $36,750-a-year no-show job at the osteopathic school in 2003. Among other things, investigators investigated whether the university had hired Bryant, chairman of the powerful senate budget and appropriations committee, to lobby himself for state aid. It said the university's government affairs office spent $3.8 million on "lobbying services" from 1999 to 2006 without proper authorization. A federal grand jury in Trenton indicted Bryant and Gallagher, a former UMDNJ dean, on corruption charges on March 29, 2007. Gallagher was convicted in 2008 on charges that he bribed former senator Wayne Bryant with a no-show position at the UMDNJ's School of Osteopathic Medicine. In exchange, the senator funneled the school $10.5 million in state funds as head of the budget committee. Gallagher was sentenced to eighteen months in federal prison and Bryant to four years.

Christie's office had more than it could handle coming out of UMDNJ. In early September 2006, his Newark office chief Susan Steele informed Hill that she and Mary Gardocki wanted to meet with Hill again. The meeting was set for September 13 at three o'clock in the afternoon at Steele's office. Pincus planned to attend, and Hill prepared a PowerPoint to summarize her findings from the evidentiary material gleaned by federal investigators.

The day of the meeting turned out to be dark, windy and generally unpleasant, a fitting backdrop to what was about to take place. Pincus met Hill at the building's entrance they went to the eighth floor together. Though Hill had her PowerPoint ready to put up, she couldn't get it to work and resorted to the three printed copies she'd brought along, just in case, and it was a good idea that she'd brought them. Steele didn't have Microsoft Office on her computer and couldn't have run Hill's PowerPoint. Steele sat at the far end of the conference table looking pained, Hill observed, and Gardocki sat to the right, chewing gum with her mouth open throughout Hill's presentation. Pincus sat hunched over and silent to Hill's left. She went through each slide with the increasingly sinking feeling that it was not going well. When she finished, Steele finally spoke.

In short, Steele told Hill that none of what she had just shown them was in her original charge and that there was no point to going after an organization that has no money anyway. Perhaps an oversimplification on Hill's part – or not – it begged the question, in sum, as to why a federal prosecutor with intimate knowledge of the federal monitor's work at UMDNJ had come to the conclusion that the university didn't have any money. Hill learned later that the UMDNJ's insurance against lawsuits covered it up to $21 million. Pincus was outraged by Steele's comment, pointing out that additional facts are intended to be uncovered during discovery[108] and that what Hill exposed was not a new charge but an extension of the original charge. Steele then made another statement that astounded Hill and Pincus when she informed them that she and Gardocki couldn't communicate with one another and neither one knew what the other was doing regarding Hill's case. This was not good news. Steele offered to "consider" turning Hill's case over to the criminal division. There were clearly no guarantees offered and, as it turned out, no intention to pass the case over for criminal investigation. Hill later characterized how she felt leaving the meeting as being "violated and used." She realized that they'd not wanted her to find anything, that her hard work with Box 6 had just been a charade. Pincus asked if they would like Hill to continue digging through the boxes and Steele said, without hesitation, "No."

As Hill and Pincus left, they stopped by the conference room where the contents of Box 6 lay spread out on the table like the innards of a great whale. Hill picked up her notes but left the Post-its on which her calculations and comments had been made. She and Pincus headed for the parking lot. By then it was drizzling and cold. Pincus gave her a big hug and tried to console her, but she already was fairly convinced that all had been lost.

Over the next few weeks, Hill conferred with Pincus to consider her options, keeping in mind that she was still under court order to remain silent about the federal prosecutor's investigation of her case. Hill had options. She could report to Stern, the federal monitor, perhaps ask ORI to reconsider in the light of the new information she found in Box 6; take her complaint to the faculty affairs committee of the medical school, or she could ask to have her name taken off the paper in *Radiation Research* on which she was named as a coauthor.

Pincus spoke again with his colleague who takes *qui tam* cases who advised it wasn't time to throw in the towel just yet. Pincus had already filed an extension of her claim and indicated to Steele that it was not a new filing, although it strengthened the original, we had already filed an extension of the claim and Shelly had indicated to Susan that this was not a new claim, it strengthened the original case. Pincus' colleague wondered, too, if Hill had been used.

Just as Hill was losing hope, on June 15, 2006, *the New York Times* reported that a Columbia University chemistry professor had, in March of that year, retracted two papers and part of a third published in a leading journal and was just then retracting four additional scientific papers. The professor in question, Dalibor Sames, Ph.D., was the senior author of all the papers in question, in which experimental findings included in his published work could not be reproduced by other researchers in the same laboratory. Bengu Sezen, Ph.D., one of Sames' former graduate students, performed most of the experiments described in the papers.[109] To do this was a brave action on the part of the professor, Hill observed. "I would not do anything like this lightly. To retract a paper is very difficult for any scientist,"[110] said Sames, when asked about the latest retractions. Absent taking Sames action and retracting the Bishayee et al. papers, Howell had missed the opportunity to minimize the fallout.

Hill continued to push Pincus to let her take on of the actions she'd already spelled out, topping the list, taking her case to Stern, going back to the ORI and getting her name withdrawn from the paper she coauthored. She expressed her willingness to withdraw the charges to be released from the confidentiality imposed by the seal. Pincus told her he'd already drafted a letter to Steele dated September 25, 2006, regarding the action taken by Sames at Columbia in which he further emphasized that Hill was not trying to bring new charges. Shortly after this, of course, another case made headlines: the Elizabeth Goodwin, Ph.D., case at the University of Wisconsin, required reading in the introduction to this narrative. Cases of scientific misconduct had begun to crop up with uncomfortable regularity.

Hill's e-mail exchanges with Pincus reflected her frustration that the case was dragging on and going nowhere. She'd begun to believe that Steele was incompetent, and it'd be better for all concerned if she forgot the entire case. Pincus told her to wait and see what might happen, cautioning her to abide by the confidentiality implied by the court seal. He was worried she'd open herself to slander and libel lawsuits if she pushed too hard and broke her silence. But her failure with two Newark CCRIs, and the ORI would follow her until 2010, when the presiding judge, Dennis Cavanaugh ruled against her.

In her frustration, Hill put in a call to Kay Fields. The two spoke for about an hour, as Hill recalled the conversation, both worried that if anyone found out they'd talked, it could spell trouble. But they continued to talk anyway. Fields told Hill she still supported her, revealing what happened on her summer trip to Newark. While Fields had been impressed with Gardocki, she believed the government had already lost interest in Hill's case. During this conversation, Hill learned from Fields that Steele was supposed to have subpoenaed the minutes of second Newark CCRI, she didn't think they'd done it. The ORI, she told Hill, disliked working with the UMDNJ because they were generally uncooperative. Pincus had already informed Hill that Steele had told him they had sent Hill's PowerPoint presentation to the ORI, but Fields denied having seen it, although she would claim to have seen it later. Hill clung to the hope that the ORI would come to support her case and help me out. But that never happened. Fields admitted that she was taken off her case because she persisted on providing Hill advice.

The newspapers were now full of the cardiology scandal confirming Hill's belief that this was, indeed, keeping the United States attorneys busy and had, in the process, preempted her case. With Pincus' help and permission, Hill sent two follow-up letters to Steele explaining the science involved. While the letters were scholarly, concise and clear to any radiation biologist, they probably weren't any of that to Steele, Hill thought later. Certainly, another news item that put the medical school on edge was the favorable report of a commission charged to look into a merger of UMDNJ with Rutgers University. But with Stern also looking closely into powerful New Jersey lawmakers pushing the merger, it would take several more years for this to take place.

Pincus and Steele had several telephone conversations over the next few weeks. Steele did send Hill's PowerPoint to the ORI and was awaiting their response to it. Pincus promised he would urge her to release Hill from the confidentiality requirement, but warned her that once the seal was broken, the contents of Box 6 would no longer be available to her. He was under the impression that Steele's office had interviewed more people on the medical school campus but neither he nor Hill found out who they were or what was learned. There were notes in the documents that Pincus eventually obtained from Schlow that showed that Schlow talked with Lenarczyk, Fields and Hill. "I will never see any notes from Mary [Gardocki]," Hill opined. "She played her cards very close to her chest and Steele told Shelly that I had no believer in her [Gardocki]. So much for scientific help from that quarter."

In early December 2006, Hill had another long conversation with Fields. She had, by then, seen Hill's PowerPoint but not the two additional letters she had sent Steele. This led Hill to assume that they'd not been included in the mailing to ORI. Fields and her colleagues understood and accepted the PowerPoint, but she wasn't certain that the Department of Justice intended to act on it. They do, Fields explained, triage and she repeated that her office disliked working with the UMDNJ, as they had done sloppy work in other cases of scientific misconduct. This indictment of UMDNJ did not appear in any of Fields' written reporting on Hill's case. What she suggested next, if Justice backed down, was that Hill should ask the ORI to formally re-review the case, including the work of the second Newark CCRI and the contents of Box 6. She further recommended that Hill send copies of pertinent documentation to her congressman and to John D. Dingell Jr., the Democratic congressman from Michigan who had been concerned for over a decade with the issue of scientific integrity. For Dingell, it was more than just passing political folly – it was a mission that Hill could understand.

Dingell's knowledge of misconduct in medical research was clearly laid out in the paper he wrote[111] for the *New England Journal of Medicine*, published on June 3, 1993, more than a decade before Hill's *qui tam* was filed in federal court. In it, he spoke to "our sometimes-misplaced sense of optimism" in the public's confidence in science and the scientific method. "We believe that honest intellectual inquiry can lead us to theories or laws that, after testing and prodding, will yield truths that will light our path to the future." He addressed the public trust that elected representatives knew the full story when federal money funded science and research. "The foundation of public support for science," he wrote, "or for any public endeavor, is trust – in this case, trust that scientists and research institutions are engaged in the dispassionate search for truth. We are willing," he propositioned, to spend great sums in the service of a higher value. And no value is believed to be more dear to a scientist than the truth." But were scientists capable of staying in the "truth lane."

Certainly, Dingell was concerned that in the modern era, where cutting-edge scientific discoveries moved rapidly from the laboratory to the legislature to the regulatory agencies, and scientists, hospitals and academic institutions convened press conferences to publicize their findings and satisfy a public "both eager for scientific advance and troubled by every new report of a threat to health and the environment," mistakes could take place. "Americans," he wrote, "are not a patient people. The public is rarely content to five, ten or fifteen years for further research and study." The consequences of this, he believed, could be serious. He was largely concerned with scientific misconduct. "[T]here is a different between honest error and various forms of misconduct, such as plagiarism and fraud. Blatant forms of misconduct attracted the interest of the Subcommittee on Oversight and Investigations of the United States House of Representatives, and the subcommittee's interest has, in turn, caused the scientific community to pay attention to the issue of misconduct." It was just then that Dingell and his subcommittee tried to define misconduct, though it was never sought to explore the outer reaches of this concept, he noted.

Relevant to Hill's case, Dingell's subcommittee investigators had pursued cases in which published papers claimed to present data that, in fact, never existed or described the results of experiments that had never been performed. Investigators had pursued cases in which researchers abused the peer-review process to obtain advanced copies of papers and appropriated their insights or accomplishments. Further, he observed, they'd followed cases in which evidence of misconduct was covered up or whitewashed by institutions apparently more interested "in the appearance of integrity than in the reality of it."

A chagrined Dingell importantly noted that the nature and the purpose of the subcommittee's inquiries had been misrepresented, quite purposefully, by those who had everything to lose from the truth coming out. Some had accused the subcommittee of unfairly putting science on trial. "That is false," he wrote. "Others have said that we seek to punish people for honest errors. That is false. It has been suggested that we do not understand that mistakes are inevitable in scientific inquiry. False again." There is a corollary, Dingell explained, to that inevitability of errors, and it is that when errors are recognized, scientists have a duty to make them known. But that latter proposition had, in Dingell's and his subcommittee's experience, been turned on its head: "it was said that since errors are inevitable, there is no need to acknowledge or correct them."

Beyond the policy implications, Dingell's subcommittee owed a greater accountability to the American public, to the taxpayer inevitably footing the bill. "As chairman of an oversight subcommittee with responsibility for the National Institutes of Health (NIH), I have a duty to see that taxpayer dollars appropriated for scientific research are actually used for research," he iterated. "Most biomedical research performed in this country is federally funded." He also understood that every time a researcher takes taxpayer money and publishes fabricated, falsified, or plagiarized findings, the taxpayer has, in effect, been swindled. Further, and this certainly applied to Hill's motives for filing her *qui tam* case, for each dollar wasted on a dishonest researcher is a dollar that might have gone to another, more worthy candidate who might have made a real contribution. "In short," he concluded, "there is an opportunity cost for each grant that is abused."

Dingell exposed another truth that played into the underpinning of Hill's case: that the nature of scientists and the scientific method is to build on the interesting results obtained by others. "Every paper published with fabricated or falsified data will spur scientists using still other federal grants to try to replicate or extend the results, wasting even more money and time." So how widespread was misconduct? Dingell did not know just then. But one indicator he'd seen, a study performed by the Acadia Institute, found that approximately forty percent of the deans of the nation's major graduate schools knew of confirmed cases of scientific misconduct occurring in their own institutions within the previous five years. Another indicator, he noted, was a survey sponsored before his article was published by the American Association for the Advancement of Science that found twenty-seven percent of a group of scientists surveyed stated they'd personally encountered, during the prior decade, research that they suspected was falsified, fabricated, or plagiarized. Those twenty-seven percent reported, on average, witnessing at least two such incidents. Further, close to half of them stated that the incidence of fraud was on the rise, whereas only two percent thought it was declining. More than half of those in the study characterized university investigations of misconduct as slipshod, a point Hill could, no doubt, could appreciate.

Perhaps even more telling, Dingell observed, were the scientists' responses when asked what they'd done about the misconduct once it had been recognized. The vast majority, he learned, had done little to nothing. The few who took action generally confined themselves to discussion of the incident privately with a few people, and only two percent brought the matter to the public's attention.[112] The irony of this was clear. Although large numbers had seen misconduct, virtually none who had seen it had acted, and although the majority viewed university investigations as ineffectual, nearly all claimed that scientists should monitor themselves and that outsiders should not become involved.[113] Ethically bound to do the right thing, Hill was in the extreme minority at UMDNJ who was willing to go the distance to report the misconduct in science that happened in her radiology research division's laboratory.

Dingell's subcommittee had also found what he described as "disturbing attitudes" when investigators contacted some twenty leading scientists to solicit their views on misconduct. "In private interviews, these scientists cited examples of misconduct they had witnessed, whistleblowers they had seen harassed, or other matters engendering concern," he documented. "Yet none were willing to testify, write open letters, or even have their names used publicly, for fear of retaliation." Hill, of course, knew about all the above firsthand.

The oversight subcommittee further found that there was a cordon of academia who believed that misconduct should only be taken seriously if it is widespread. This would be akin to suggesting, he continued, that a single bank robbery shouldn't be investigated unless and until it could be proved that that particular type of bank robbery is rampant – it's a ridiculous argument.[114] He had a revelation for them: that the field of science had duty to look at each allegation of misconduct fairly – and that, until his committee began giving the issue a hard scrub, had not yet happened. He regrettably reported that the case studies pursued by the subcommittee had only turned up more evidence of widespread and unchecked scientific misconduct.

Dingell cited the cases, in short, that had caught the attention of his subcommittee, all of them merely prequel to the many that would follow between the early-1980s and publication of Dingell's article. Certainly, some of these cases were known to Hill when she persisted in prosecuting her *qui tam*, despite repeated losses in university hearings, the ORI and courts of law. The first one that got the subcommittee's undivided attention involved John Darsee, M.D., of Harvard Medical School, whose federally supported research in cardiology and his long list of publications marked him as a rising star until it was discovered, almost by chance, that he had committed an act of fraud," Dingell wrote. "Two investigations were conducted at the medical school – the first by Darsee's immediate supervisor and the department chairman, and the second by a committee of faculty members from Harvard and elsewhere that the dean had appointed. Both reported," Dingell continued, "no misconduct in Darsee's published research." A third committee, appointed by the NIH, discovered a massive fraud – the data for a number of Darsee's published experiments did not exist, forcing Harvard to retract thirty of Darsee's papers and abstracts in February 1983. Subsequent review of Darsee's earlier work while he was at Emory University resulted in the retraction of an additional fifty-two papers and abstracts published during his tenure Review of Darsee's earlier work at Emory University led to the retraction of an additional fifty-two papers and abstracts published during his tenure there. But that wasn't the end of it.

The Darsee investigation went back even further, to when he was an undergraduate from 1966 to 1970 at the University of Notre Dame, where it was soon discovered that he'd also falsified data in his research. Exposed and with little recourse, Darsee claimed he couldn't remember any wrongdoing. He apologized, left Harvard and took up a clinical fellowship at Ellis Hospital in Schenectady, New York, that is, until June 1983. The following year the New York State Board of Regents pulled his license to practice medicine. Of interest, all of this was doubly awkward for Harvard, which had maintained up to the point of the NIH investigation, that it had reviewed the data fully – the same data that later turned out not to exist at all.[115]

Then, of course, Dingell pointed to another major case involving Stephen E. Breuning, Ph.D., a professor at the University of Pittsburgh and one of the United States' most influential researchers in the field of mental retardation treatment. Breuning's research purportedly showed that the condition of severely retarded children improved markedly when they were taken off certain tranquilizers. From 1979 to 1983, he reportedly published a third of all the articles in his field, Dingell noted. Some states even changed their treatment protocols predicated on his findings. But in 1983, in a letter to the National Institute of Mental Health (NIMH), Breuning's mentor, Robert L. Sprague, Ph.D., blew the whistle on his protégé, raising serious concerns about the integrity of thirty-four-year-old Breuning's research. The University of Pittsburgh investigated and reported to the government that it found no problem, a phrase that was beginning to ring all-too familiar to Dingell. The NIMH sat on the entire affair for *years*, and the whistleblower, Sprague, suffered considerable reprisals. Breuning, in the meantime, continued to travel the country, Dingell wrote, promoting his theories further to an unwitting public.[116] As it turned out, among his many transgressions, Breuning's data, in large part, never existed, and subjects had never been tested; he'd published deliberately deceptive scientific papers.

Whistleblower Sprague's allegations were eventually substantiated in almost every detail, Dingell iterated. In a May 24, 1987, *New York Times* article[117] about the case, it was revealed that the Breuning case "appears to be one of the most significant in a growing trickle of cases of fraudulent scientific studies that have come to public attention in recent years. Unlike many cases in which fraudulent research has affected only the research work of other scientists who have wasted time following false leads," it read, "Dr. Breuning's work has had great influence on the treatment of patients and on public policy. Based on the research reports that have now been judged fraudulent, Dr. Breuning, after only a few short years of work, achieved the status of a major worker in the field of mental retardation, the federal report said. His deceptive reports not only had 'a significant impact' on the entire field of study but also strongly influenced social policies for treating the mentally retarded, according to the study." The *Times*, pulling from the federal findings in the case, reported that some states, notably Connecticut, amended policies governing treatment practices to be consistent with Breuning's supposed findings. As for Breuning, he was found guilty of two felonies and served time in a halfway house. Amazingly, even after Breuning was "unmasked," as Dingell called it, several scientific journals fought the efforts of collaborators to retract articles that he had coauthored.

Then, of course, there was the previously discussed case involving faculty members from Tufts University and the Massachusetts Institute of Technology (MIT). Dingell had paid attention to that one, too. Whistleblower Margot O'Toole, Ph.D., was a junior postdoctoral fellow in molecular biology at MIT in 1986 when she raised uncomfortable questions about a senior researcher's work. According to a March 22, 1991, *New York Times* article[118] on the fallout of the case, David A. Baltimore, Ph.D., a Nobel laureate who was a coauthor of a research paper in *Cell* that used the disputed work, described her as a "disgruntled postdoctoral fellow." O'Toole lost her job, her house and feared that her husband, Peter Brodeur, Ph.D., who also worked in Imanishi-Kari's laboratory, would lose his job. Vilified and effectively driven from her profession after revealing that the article coauthored by Baltimore and O'Toole's supervisor, Thereza Imanishi-Kari, Ph.D., had relied on large part on data that were falsified. Senior scientists at Tufts and MIT contended that O'Toole's allegations were unfounded and reported that the paper was virtually error-free. These senior scientists, Dingell wrote, informed O'Toole that the article could not be corrected because any correction would damage Imanishi-Kari's career and because "the scientific literature was so full of error anyway that one more error would not matter."

Baltimore, for his part in the scandal, which was considerable, crafted a letter to Herman Eisen, Ph.D., a professor of biology and the founding member of the MIT Center for Cancer Research, that he didn't favor retraction of the paper because of possible adverse consequences for another scientist who had been listed as an author, even though he, Baltimore, admitted that the paper was grossly defective and that he would never fully trust Imanishi-Kari's work again.[119] Even after forensic analysis by the United States Secret Service had established that the data Imanishi-Kari had used was largely fabricated, Dingell opined, she and her coauthors continued to portray themselves as innocent victims. The fallout of the case was far reaching. At the end of a protracted investigation by the NIH's Office of Scientific Integrity (precursor of the ORI), Baltimore retracted the *Cell* article and later resigned from the presidency of Rockefeller University. But there was a hitch. After some time had gone by, in a maneuver designed to circumvent public attention now drawn to the scandal swirling around Imanishi-Kari, he attempted to retract his retraction of the article. Imanishi-Kari, as mentioned in the introduction to this narrative, faced intense investigation by a grand jury.[120] But the United States attorney in Baltimore, Robert Bennett, formally declined to criminally prosecute her in July 1992. In statements to the media, as well as in correspondence with Dingell's subcommittee and the Secret Service, Bennett stressed that declination was based on the difficulty of presenting the complex scientific facts to a lay jury, which would have to understand them fully to find guilt beyond a reasonable doubt in a criminal trial. Bennett emphasized that the declination wasn't an exoneration of David Baltimore or Imanishi-Kari, nor did it reflect doubt on the part of the prosecutor's office that the data had been falsified. But according to Charles W. McCutchen, Ph.D., a retired physicist formerly of the Laboratory of Cell Biology and Genetics, National Institute of Diabetes and Digestive and Kidney Diseases, National Institutes of Health, the Imanishi-Kari case quashed substantive efforts "to do something about the crookedness of science. Dingell," he noted, "was trying to get fair treatment for the ordinary scientists, but he could not reach them. His message was drowned out by the blare of 'Dingell is anti-science' from establishment organs."[121] As for MIT, where the work in question took place, the university decided it was time to revamp its procedures for investigating misconduct.

So, what happened to O'Toole, the whistleblower in this case? At the conclusion of its investigation, the NIH called her a hero. "Dr. O'Toole suffered substantially for the simple act of raising questions about a scientific paper," it concluded in a report about the Imanishi-Kari case (also called the David Baltimore affair). "Notwithstanding the losses and costs she incurred, Dr. O'Toole maintained her commitment to scientific integrity."[122] The Office of Scientific Integrity's draft report stated, in effect, that O'Toole had been right all along – crucial data in the paper based on the work of Imanishi-Kari, had been faked. The scientific paper described findings suggesting that transplanted genes could stimulate a recipient's immune system. The findings haven't been confirmed by other researchers. "One of the most surprising things to me is the way so many members of the scientific community and the scientific press were ready to denigrate Dr. O'Toole,"[123] said Mark S. Ptashne, Ph.D., who was a researcher at Harvard University Genetics Institute when he was interviewed by the *New York Times* about O'Toole's predicament. O'Toole was hired in 1990 to work in Ptashne's institute in the immunology of breast cancer after years of being unable to find work in science. "They [Tufts, MIT, Baltimore and Imanishi-Kari] were willing to go to battle with absolute certainty, without bothering to read the paper and think about the likelihood that the paper was wrong," Ptashne concluded.

In described her congressional testimony to the *Times*, O'Toole explained that Republican congressman Norman Lent, of Nassau, New York, dismissed her, saying "She had the attention and consideration of almost a score of eminent scientists from very high up in the ladder who have reviewed her complaint and found it wanting." Lent's statement that her complaint was determined to be unfounded by multiple reviews – a refrain Hill had also heard of repeated – equated to suborning eminent scientists behaving badly. Describing her examination of laboratory records, Lent remarked, "If I had someone in my office who did that sort of thing with my notes, they'd be out of here in a flash, and they wouldn't be rehired by anybody I could call up."[124] But despite all of this, O'Toole weathered the storm, supported by Linus C. Pauling, Ph.D., Dingell, and Ptashne. Like O'Toole, Hill would do the same.

While Dingell reported that the cases examined by his subcommittee, of which there were many, invariably differed in their particulars, common threads ran through all of them. First, in most cases, the dishonest scientist engaged in misconduct for *years*. Commensurately, those in a position to know ignored the warning signs or turned a blind eye. Second, he observed, in nearly *every* case, the whistleblower was treated badly, often far worse than the offender, for having put truth ahead of personal convenience or career advantage. Third, in virtually every case the subcommittee examined, the internal procedures that were supposed to allow universities and other research institutions to police themselves failed – miserably. At best, Dingell wrote, internal investigations appeared careless and inept; at worst, they appeared to be conducted in bad faith. Fourth, the offices within the NIH that are responsible for monitoring issues of scientific integrity showed a marked disinclination to perform their job, a reluctance, he iterated, to pursue scientific misconduct cases aggressively, a dilatoriness in developing evidence, and an overall lack of dedication to the goal of confronting scientific misconduct. Fifth, many prominent scientific misconduct cases were – and are – "accompanied by a barrage of erroneous and misleading reports in the media in which the long-suffering whistleblower is labeled as vindictive and irresponsible, those investigating the allegations are lambasted as ignorant troublemakers, and blatant acts of dishonesty – such as the wholesale theft of another researcher's article or the fabrication of data – are trivialized as minor mistakes, misrepresented as differences of interpretation, or dismissed as mere communication problems"[125] when nothing could be further from the truth.

Back to Fields December 2006 conversation with Hill. Field told Hill that Howell had gotten a good score in the study section on his NIH grant renewal but one of the reviewers was knowledgeable about the effects of tritiated thymidine on the cell cycle and had suggested that Howell use tritiated deoxycytidine instead. Howell had again cited what Hill contested as falsified material and included it in one of the figures in the grant renewal. She recommended that Hill request a copy of Howell's grant renewal through a Freedom of Information Act (FOIA) request to NIH. Lastly, Fields informed Hill that she'd been on probation for a year but it had everything to do with her challenge of another ORI investigator and nothing to do with Hill's case. As she reviewed her notes from this conversation four years later, Hill strongly suspected that Fields had had a run-in with John Dahlberg, who'd moved into Price's job when Price left the ORI.

Hill pursued Howell's grant renewal application via NIH FOIA. She was informed by NIH's FOIA coordinator that to obtain a copy of Howell's paperwork, he would have to be informed about who was making the request and that Howell would then have the opportunity to redact the copy sent to the requester but only the parts that might be marked proprietary information. Hill made her request in mid-December 2006, and it would take more than two months for her to receive a copy. All the figures had been removed from Hill's copy, and it was so severely redacted that there was not much left of any use. But she was able to determine that Howell had cited, again, both Bishayee et al. papers published in *Radiation Research* that contained what she leveled was false information in over twenty instances each in the little text that remained after Howell's heavy redaction.

By the end of February 2007, Pincus was beginning to suspect that things were definitely not going Hill's way. "The ORI had had mixed feelings about my PowerPoint," she opined. "It was too much hypothesis. [United States Attorney Susan J.] Steele, in trying to justify her worming out of the affair was even wondering whether the results I was questioning were key to Roger's getting the grant. In other words," she continued, "it would have been all right to present false information in a grant application if that information was not all that important to the purpose of the grant. Then there was a question whether Roger even knew what was going on. We would find out more about that during discovery."

Pincus' feeling that the case wasn't going Hill's way proved true and by April 4, the die was cast. The federal government was not going to take Hill's *qui tam*. "Steele felt in her heart of hearts that there was something there, but she worried they could not prove it," Hill wrote later. "She would write us a letter lifting the seal, allowing me to go back to the ORI and/or report to the monitor and otherwise explaining what my options would be." This United States Attorney's Office's back and forth took nearly three and a half years to come to the decision *not* to prosecute the case. In her letter to Pincus, Steele noted that the decision to decline intervention was not a statement regarding the merits of the case, that the United States retained the right to intervene later and would be entitled to the majority of damages and penalties. "Even so, the government was not necessarily on our side when it came to releasing documents and information during discovery and Shelly [Pincus] would be required to submit formal requests which might not necessarily be granted," Hill subsequently recalled. "That is to say, that, even though we were pursuing the case on behalf of the government, the government was not necessarily going to help us but would take the largest share of the pot if we should win." Steele's letter also remarked that "if we would file an amended complaint, that would cause everything to start over again with a new seal and, it would go without saying, a new opportunity for the government to extend and delay the proceedings," she continued. "Furthermore, we would only be allowed to settle the case if the government agreed to the settlement. The settlement would, at least, include payment to me by the defense of the expenses I had incurred plus between twenty-five and thirty percent of the award. At its close, the letter gave me permission to consult the monitor but said nothing about appealing to the ORI for reconsideration." The formal "notice of election to decline intervention" was signed on April 6, 2007, by Susan Steele. The unsealing order was signed shortly thereafter by federal district court judge Dennis M. Cavanaugh, who was also the presiding judge in the case. Hill would later write that drawing Cavanaugh, for reasons that would become apparent during the case, was just bad luck.

Cavanaugh, who retired in from the federal court in January 2014, was appointed as a judge of the United States District Court, District of New Jersey by President William J. Clinton in September 2000; he had previously served as a United States magistrate judge from January of 1993 to 2000. During his tenure on the federal bench, Cavanaugh had the opportunity to manage thousands of cases and conferences, assist in settlement and preside over trials of every type of civil and criminal matter that comes before the federal court.[126] In a twist that came less than three years after Hill's last appeal of her case, Cavanaugh retired from the bench and went to work as an attorney for the defense counsel in Hill's case – McElroy, Deutsch, Mulvaney and Carpenter.

On Hill's seventy-eighth birthday, Pincus laid out how the case was going to go after Steele declined to prosecute. "He would be called before magistrate [judge Mark] Falk who would set up the rules for discovery and specify the time limits," Hill explained. "The pretrial costs would be a minimum of $100,000, which would come out of my pocket." Hill would not only have to find and pay for expert witnesses but also stenographers, duplication fees, transportation costs and many other incidentals, too many to mention. The defense would – and will – drag their feet as much as possible, requesting extensions as often as allowed and even missing a deadline but it will not hurt them, Pincus told her. The magistrate would try to force both parties into mediation. Hill would be lucky if the magistrate opted to fashion his own settlement, but that was not to happen. This was going to be an uphill battle – the odds went down when the government pulled out. Pincus further advised that they not ask for and would, in fact, not want a jury trial. Jurors would never understand the science involved. Further, and far more personal, Hill had concluded that what he really meant was that the jury "could easily be swayed and charmed by the handsome young Roger [Howell] being pursued unmercifully by the aged, ugly and vindictive witch [Hill]." Was there any good news? "We could still plan on going to the monitor after the seal was removed, so I continued to be ever hopeful," she continued. "But we have to proceed carefully in order not to tread on Steele's toes."

Once the seal was removed, Hill's complaint would be served on Roger Howell, Anupam Bishayee and the UMDNJ. "I thrilled a bit as I imagined the impact on Roger and hoped it would give him pause," Hill quietly admitted. After the case was ordered unsealed and all parties served, Hill demurred. Though she was no longer under court order to say nothing of the proceeding, she'd gotten accustomed to being silent about it. "[I]n any case, I felt terribly uncomfortable accusing a colleague of fraud," she confessed, "so I remained quiet, speaking only in general terms, never mentioning names. I resolved," she explained, "to remain in the employ of the university until the case was settled, but this was going to be hard. As time passed, I longed more and more to be able to retire, to spend time as a volunteer and to savor the cultural opportunities in and around New York City. I speculated that we would stay in the New York area for a year or so once the case was over and then move to Kalamazoo [Michigan] where I could enjoy the two little granddaughters who lived there. I longed to have the time to be a grandmother for those two and a great-grandmother for the three little ones that lived not that far away in Parkersburg, West Virginia. She observed that "our income was more than adequate but I certainly never anticipated how much the cost of *qui tam* would cut in to our cash flow and limit our activities," she opined. "It was like spending time in prison and much to my dismay, [the] *qui tam* hovered over me like an evil cloud and came to define me. I would mull things over as I went to sleep. On sleepless nights I would toss and turn dreaming of winning, terrified of losing. My children and their significant others expressed their contempt for my persistence." The two lawyers in Hill's family, a son and a daughter's partner, "looked on my foolishness with their foreknowledge of almost certain failure. The only support I had would come from my husband, George, and my lawyer, Shelly [Pincus]."

Hill's best friend at work, Barbara H. Fadem-Chenal, Ph.D., a professor in the department of psychiatry, met with her for lunch several times a week. "She never wanted to learn the details of the case and I think secretly hoped that I would fail," Hill journaled later. "She [Fadem] had been the defendant in a *qui tam* case in the early 1980s. She is an animal behaviorist and got her Ph.D. at Rutgers University at Newark," Hill explained. "[There, she studied] the effects of sex hormones on opossums and was well funded for many years by the National Science Foundation. She obtained her original breeders from a colleague at Rutgers who expected her to include him as a coauthor on her papers, which she did for a time, but then, as he was contributing nothing and she could do the breeding herself, she began to leave him off. He failed to get tenure, blamed it on her and pursued her for many years in the courts. It sounded as though his case had little merit, but he made it his life's work. So Barbara's sympathies lay with the accused, even though I was her friend," she concluded. "In her defense, I must say that she had little respect and much contempt for Howell and [Edouard] Azzam." But at the end of the day, while Hill hoped her case would become public, as time wore on, what became most important to her she later confided, was to outlast the grim reaper and to journal what was happening in her case. "If nothing else, I was determined that the world should know how this one-woman scientist had been treated, how respected scientists could behave and how a university could subvert all knowledge of fraud and deception," she wrote.

In early July 2007, the defense, on advice of counsel, waived service of the summons and complaint so that these documents could be delivered without the necessity – and additional expense – of a process server. This was a bit disappointing, Hill remembered later. "I had conjured in my mind a picture of Howell being accosted by such a delivery person and having to sign for the papers, but he was spared that embarrassment."

Now that the case was really getting off the ground, or so Hill thought, her expectation of what might happen next was quickly deflated and much of that disappointment was driven by the fact that though Howell, Bishayee and the UMDNJ were served the complaint and disclosures, Howell, most especially wouldn't acknowledge the evidence in front of him that Bishayee had committed misconduct in science. "He would see my diary kept during the shadowing affair with Marek and the photographs, although he had probably seen those when he went before the first [Newark CCRI] committee. He would see the analysis of Anupam's numbers in the first suspicious experiment in 1999 and would have an opportunity to put that together with my report that there were no colonies on the mutation dishes," she hoped. "How easy it would have been at this time to admit his error and to retract the two tritium papers. His first grant had run its course and his renewal was underway. So why did he not come clean?" she wondered. "I think there are several reasons: he might lose face with the radiation biology community if he had to back down from his position that he had demonstrated a bystander effect for tritium and that it was accomplished through gap junctions. He would have to face me with his admission that he had been wrong. I think this was [a] very compelling [reason]. He had such a deep hatred of me," Hill journaled later, "that it was impossible for him to eat crow in front of me. He may still have feared for his grants, unlikely though it was that his [grants] would have been in jeopardy at this late date. Or [perhaps] the university lawyers may have told him to hang in there, he still might win. So if he had any inclination to do the honorable thing, he could have rationalized that now was not the time. [But] perhaps the most compelling reason for Roger to refuse to back down was that when the story came out, he would look the fool. Anupam had done an amazing job of leading him down the garden path," she explained, "and now he would have to retract the papers." Howell had published eight papers with Bishayee, two of them tritiated thymidine but the other six involved other radioactive isotopes. "None of us, except, of course, Anupam, knew at the time that virtually all of his work had been fabricated."

The first step in the legal process was for the defense to file an answer and to make a counterclaim. The defendants were represented by McElroy, Deutsch, Mulvaney and Carpenter, the third largest law firm in New Jersey with more than 280 lawyers and offices in six states. The firm was formed by merging two prestigious firms in May 2004. The attorneys assigned to the case were John P. Leonard and Scott S. Flynn. The first document the defense presented to the court was their answer to the charge made in October 2003, followed by their counterclaim, which was filed on July 30, 2007. Each item in the original complaint was either denied, admitted or dismissed for lack of knowledge. Many of the items admitted to were obvious, including mundane information like job titles and home addresses. The defense denied, among other things, that Howell terminated Bishayee and Lenarczyk, locked Hill out of the laboratory and created a hostile work environment. Howell would later admit under oath in his deposition that he did compel Bishayee to sign the resignation Howell had written up for him, and it was also true that he'd retaliated against Hill and locked her out of the lab. The legal volleying that went on was only a sampling of the tough road ahead. One example that struck Hill was language of the original charge and the defense counsel's answer and counterclaim on item nineteen. The charge from Hill's October 14, 2003, filing states:

> [...] on two occasions preceding the submission of the revised grant application, Hill observed Bishayee engaged in preliminary experiments. Hill's observations led her to believe that Bishayee was falsifying the data underlying the experiment and the conclusions reached by Howell from those experiments.

The defendants countered on July 30, 2007:

> Defendants lack knowledge and information sufficient to form a belief as to the truth of the allegations set forth in Paragraph 19 of the Complaint and leave Plaintiffs to their proofs thereon. To the extent that a response to Paragraph 19 is required, Defendants deny the allegations set forth therein.

"One curious thing that I may have failed to note originally was that there is no mention in the complaint [of October 14, 2003] of my interactions with the ORI," Hill recalled. "However, these interactions are evident in exhibits 36 through 42 that accompany the written disclosure filed alongside the complaint." On reading the July 30, 2007, answer and counterclaim, Hill had to deal with what she described as her "feelings of rejection and irritation at the put down that was insinuated throughout the document. It was stated in various ways that [my] actions are clearly frivolous and vexatious. 'It is apparent based on Dr. Hill's continued pursuit of the allegations in spite of the conclusions of the Committee and ORI that the Complaint was brought primarily for purposes of harassing

Defendants.'" Nowhere was it stated, to her great disappointment, that I was just pursuing the truth. The defendants conclude by demanding judgment against me and asking as relief that the complaint be dismissed, that the findings be found frivolous and that the defendants be awarded reasonable attorneys' fee and expenses," she continued. "This latter really made me shiver in my boots. Whatever would happen if I would have to pay not only Shelly [Pincus] but Leonard and Flynn?"

On August 8, Hill sent Marek Lenarczyk a document she had written titled "I am a Whistleblower," as well as the analysis of the tritiated thymidine experiments and the PowerPoint she'd presented to [FBI special agent Mary Beth] Gardocki and Steele. "I did it because now I could speak out and he was one of the few people who knew that Anupam was fabricating data and who, himself, had expressed outrage at Anupam, whom he persisted in calling 'Mr. Maharajah.' Marek, after all, had started the whole thing," Hill documented of this critical piece of the case. She thought just then that if anyone would back her up, it would be Lenarczyk. "How wrong I was. He wrote back that he had received my package and that it was 'fascinating story.' That was my last contact with him."

Hill also hoped she had another ally in ORI investigator Kay Fields. But she, too, backed off. "My other supposed ally was Kay Fields. I put in a call to her the next day [August 9]. She was picking her words more carefully now. She recommended that I call Alan Price, her former boss, who had by now retired and was running his own consulting business. He was, she said, very fair but I would call him only to be rapidly dismissed because he had signed an agreement with the government that kept him from having anything to do with his former cases," Hill opined. "It is hard to imagine that Fields did not know about that – and she would run into the same thing herself in a couple of years. Fields let it slip that the person who had opposed her when she sided with me was John Dahlberg [Ph.D., chief of the ORI investigative division] who had by now replaced Price as her boss. I would have to deal with Dahlberg again in three years. [Fields] thought she would be able to testify after she retired, but this was wrong. She suggested that I reapply to the ORI based on the new grant and she warned me that the ORI had a seven-year [statute of] limitation[s]. We discussed," she continued, "the statistics and the fact that the analysis of my terminal digits in my 1999 experiment did not look too good: the

Chi-square p-value for one set of my data was 0.025, but overall, my results were not significantly different from uniform. In the end, she let it be known that she was having her problems at the ORI and their new [tactic] was to bypass her."

With information gleaned from discovery, it was up to Hill to locate expert witnesses to support her case. She thought she would need at least three, to include a statistician, a radiobiologist knowledgeable of cell cycle and hypoxia, and an ethicist. "This was really hard for me," Hill conceded. "I did not want to ask colleagues in radiobiology, [and] I didn't know any statisticians except the ones at the medical school – or [any] ethicists." She started the search, which Hill began on August 18, 2007, she sent four letters requesting assistance Hill included her "scientific uncle" Richard "Dick" Setlow, Ph.D., a biophysicist recognized internationally for his research on DNA damage and repair and a senior scientist emeritus at the United States Department of Energy's Brookhaven National Laboratory. She also inquired with W. Gillies McKenna, M.D., Ph.D., and a Scotsman, was in the Department of Radiation Oncology at the University of Pennsylvania School of Medicine, where he'd become chairman and the Henry K. Pancoast Professor of Radiation Oncology before he moved the United Kingdom to head of the Department of Oncology at the University of Oxford and became director of the university's Gray Institute for Radiation Oncology and Biology, which is jointly funded by Cancer Research UK, the Medical Research Council and the University. McKenna was also the former president of the Radiation Research Society. Working with McKenna at the University of Pennsylvania was his wife, Ruth J. Muschel, M.D., Ph.D., an assistant professor of pathology and laboratory medicine, and she, too, moved to the University of Oxford in 2005 as a professor of molecular pathology and of radiation oncology and biology. There, Muschel's research continued to focus on the role of signaling pathways in the response of cancer cells to radiation therapy. Hill made her pitch for help to both. She also asked Sara Rockwell, Ph.D., a professor of therapeutic radiology and associate dean of scientific affairs at Yale University School of Medicine. Rockwell had notably served on advisory panels for the NIH, Department of Defense (DoD), National Aeronautics and Space Administration (NASA), Office of Science and Technology Policy (OSTP) and other governmental agencies.

When McKenna and Muschel were at the University of Pennsylvania School of Medicine, Hill commuted three days a week for about a year in the late 1990s to work in McKenna's laboratory trying to purify a protein that rescued mouse melanoma cells from radiation and chemotherapy. "My efforts were, unfortunately, unsuccessful. Ruth and Gillies were wonderful, honorable people for whom I had the highest respect," Hill recalled warmly. "Both are modest, shy and unassuming." By the time Hill was looking for expert witnesses, they had left the Pennsylvania medical school for Oxford, England. and had moved to Oxford, England, and that is where she contacted him and Muschel.

In addition to her positions at Yale University School of Medicine, Sara Rockwell was the editor-in-chief of *Radiation Research*, the journal in which the two articles on tritiated thymidine had been published. Hill had known Rockwell for many years. "She was knowledgeable in radiation biology, but I also wanted to talk with her about policies for retractions when experimental results could not be verified. She also taught ethics at Yale," Hill observed.

Setlow was quick to respond, saying that he did not feel competent in the area she needed him but recommended that Hill call or write several others whom he listed off. "I received a long e-mail from Ruth Muschel urging me to give up," Hill said. "I was quite surprised. I thought she and Gillies would be as outraged as I was, would offer help and would urge me on. This may have been good advice – perhaps she knew more than she was saying. It was obviously given with concern for my welfare in mind." Then there was Rockwell. "I spoke on the phone with Sara Rockwell who told me that *Radiation Research*, to her knowledge, had never had to deal with such a problem, that if they would have to, they would turn to the ORI for guidance. So much for that."

Hill's request for help from faculty at Columbia University's Center for Radiological Research was the most remarkable on several levels to include just how incredibly difficult it was going to be to find expert witnesses for her case. In June 2007, Hill called Hei and asked him to meet with her about a research project of hers. She also asked about the availability of two of Hei's colleagues at Columbia, Eric J. Hall[127], Ph.D., and Hongning Zhou, M.D., M.P.H., a research scientist in Hei's laboratory; Zhou had just arrived that April to work at the university. In response, Hei agreed to meet with Hill, and arranged a meeting at Columbia. At the time, Hei believed that Hill was interested in proposing to the Columbia professors a collaborative research project in their common field of radiation biology/oncology.

On August 23, 2007, Hei, Hall and Zhou met with Hill at Columbia University as previously agreed. "When I arrived, he greeted me warmly and I apologized for involving him in the sordid affair," she recalled. But Hei just looked at her, staring blankly. She asked if he'd had an opportunity to read the documents she'd sent over. He looked at her again, puzzled by her comment. At the meeting, Hill told them that she had a "whistleblower" lawsuit against the UMDNJ, her own university, Howell and Bishayee. Prior to this meeting, Hei was unaware of her *qui tam*, and didn't know that Hill wanted to confer with him, Hall and Zhou about it.

How was it that Hei didn't receive the documents Hill sent over? The buildings at the Columbia Medical Center are a complex intertwining of hospital and research buildings. Hei's and Hall's facilities were buried inside another group of buildings and FedEx had gotten lost in the maze and failed to deliver her package. "This set things off immediately on a bad footing," Hill noted. "We went into the conference room and were joined by Eric [Hall] and Hongning Zhou, Hei's right arm for whom I had written a letter some years back supporting his application for a green card. I started off," she continued, "by trying to explain my observations and what a *qui tam* was all about. Eric was obviously put out." There was a reason for Hall's discomfort. He'd been previously involved in a case of scientific fraud case years before. Then he turned to Hill and asked: "[A]nd do you know where that fellow is now? He is on the Radiation Study Section at the NIH." With that, Hill observed, he stormed out of the room. "I was left with Tom and Hongning."

Hei would later make it clear that he had no personal knowledge of, and did not participate in, the research, experiments, or grant applications referred to in Hill's case. During their August 23 meeting, Hill asked Hei to assist with her case by agreeing to serve as an expert witness. At the end of the meeting, Hei told Hill that he was not interested in becoming involved in her lawsuit and would not agree to serve as an expert witness.

What Hill was asking was out of the question for Hall. He was, just then, the director of the Center for Radiological Research at Columbia

University and professor of radiation oncology and radiology. He was born and raised in England but had lived in the United States for many years at that time. He, too, had been president of the Radiation Research Society, was the author of arguably the best textbook in radiation biology entitled *Radiobiology for the Radiologist*, now in its seventh edition. "I ha[d] known Eric for years and [found] him charming, engaging, humorous and somewhat unprincipled. For several years, in the 1990s, I wrote questions for the radiology board exams, as did Eric," Hill explained. "At one time, he confided in me that when his residents took the exam, he would give them each one hundred dollars if they would bring him one question from the exam. This was, of course, totally unethical, but he found it amusing and did not mind boasting about it."

"Though I had known [many of them] well, both on a personal and scientific [level], [in fact], I thought they were my friends," Hill observed later of Hei and Hall. "These researchers had published a paper [about four years earlier] in *Cancer Research* that [was similar] to two of the papers published by Bishayee and Howell [Hill had been the coauthor on one of those Bishayee et al. papers]." They used Chinese hamster ovary cells that were closely related to the cells that Bishayee had used, and they had incubated tritiated thymidine labeled cells with unlabeled cells in the cold – they were looking for and found a bystander effect. The major difference from Howell's protocols was the tubes. The Columbia group had used small tubes with a wider bore that allowed for a sizable cushion of air above the cells; in fact, the conditions would not have been hypoxic. Howell's conditions were, Hill was and continues to be convinced, very hypoxic. There was no room for air in the Helena tubes.

Hill's meeting with Hei was anticlimactic. She later journaled that Hei had "questioned Bishayee's results and that their postdoctoral fellow who had done research would have more information about that as he had spoken directly to Howell. We were unable to locate this man [Hei's postdoctoral fellow Rudranath Persaud, Ph.D.]." Sometime after their late August meeting, Hei received by mail a letter and package of materials from Hill – her second attempt to get the information to Hei and several to implore Hei's help. In her letter, Hill requested Hei and Hall's "input and assistance" with her lawsuit. "I believe that your lab could help to resolve the question regarding hypoxia in the Helena tubes and I expect," she wrote to Hei, "that you might have more information regarding the plateau for survival in the 100 percent experiments and the effects of ^{3}HdThd on the cell cycle." Hill's letter concluded that she would "understand" if the professors could not, or did not want to, assist with her case. Though Hei and Hall had told Hill "no," this was not the end of the story with the Columbia researchers.

"By now I abandoned the idea of finding an expert witness in the Columbia group, but I still wanted to know what they knew about Roger's and Anupam's experiments," Hill explained. She'd gotten the impression from Hei that he had had some questions about Howell's results, but that Persaud had spoken to Howell and "straightened things out. I did not want them as expert witnesses, I wanted them as witnesses. I then tried to locate Persaud, but that was not easy. The Avon factory complex where he would be working as a biochemist is in Suffern, New York, just on the New Jersey-New York border. Finding a phone number for that facility was no easy task but I managed to get through to an operator only to be informed that there was no such person there. I called in several times, hoping to get more information and eventually learned that he had been in the toxicology division, but was no longer there. Had Hei known that and just happened to forget to tell me?" she wondered. "Finally, I did talk with someone who had known him and told me that he was now working at L'Oréal, located in Clark, New Jersey. I was able to determine that he did work there but was never able to reach him to have a conversation. Shelly and I talked it over and decided that he would have to be subpoenaed in order to depose him," she continued. "Shortly thereafter, a subpoena was issued to him at L'Oréal, only to be returned as undeliverable. Shelly then determined that we would have to subpoena Tom Hei. This made me very uncomfortable, but I acquiesced."

Attached to the four letters that Hill sent to McKenna and Muschel, Rockwell and Hei, were two documents that been vetted by Pincus and Hill's husband George. The first was a six-page exposé providing them with a synopsis of her story up to that time. She described the two experiments she eye witnessed, the proceedings of the two Newark CCRI inquiries, and the ORI, the filing of the complaint, and inquiries conducted by the Newark CCRI and ORI, the filing of the complaint, the monitor, her interactions with the FBI and the United States Attorney's Office, her discoveries in Box 6 and finally the rejection of the case by the federal prosecutor. "I also related the forced resignation of Bishayee, locking me out of the lab, the subsequent shunning, and the demeaning treatment by Azzam; in fact, I told them more than they would have wanted to know and probably more than they should have known about the case," Hill recorded later. The second item was the 50-slide PowerPoint she'd prepared. "The first eight slides outlined the cell cycle – probably known to everyone already – and Roger's protocols – a couple of slides had even been taken from a slide show of Roger's that he had posted on the web," she continued. "Then there were eight slides describing the first suspicious experiment, comparing Bishayee's mutant curve – big increase in mutants with dose – to my mutant curve – no increase at all. There followed five slides of the second suspicious experiment including photos of contaminated plates and two sets of Helena tubes in the incubator after one set should have been processed. Next came analysis of Bishayee's 100 percent experiments found in Box 6 at the United States Attorney's Office compared to Howell's and Lenarczyk's experiments. Bishayee's results showed an exponential (straight line) drop in survival while Howell's and Lenarczyk's survivals stopped dropping – reached a plateau – at about 50 percent survival. The 50 percent experiments were similar – Bishayee's survivals showing an exponential drop while Howell's and Lenarczyk's survivals again plateaued but this time the leveling off was at about 65 percent," Hill observed. "I argued that that was the expected level based on the knowledge of the survival of the uniformly exposed cells that made up half the mix."

"I made one mistake in the PowerPoint presentation that I later felt obligated to correct," Hill noted. "This sort of thing drove Shelly [Pincus] crazy. What is good science makes very bad law. Lawyers," she observed, "like to have everything written in stone and Shelly was constantly exhorting me not to change things. But I had to. During this period, I was frantically searching the literature to find papers that would back up my interpretations of the data and further my knowledge of the effects of thymidine and tritium on the cell cycle. In my wanderings," she explained, "I came across a web-based organization known as the Science Advisory

Board. It consists of posts of protocols, e-mail exchanges, forums on a variety of topics, product reviews and lots of other useful information. A year or so earlier, I had found there a protocol for synchronizing[128] mammalian cells in tissue culture using the double thymidine block." But she couldn't confirm the information she'd located that stated the block would not work on cells that had wild type (not mutant) p53. The p53 is a famous protein – it won the molecule of the year award by *Time* magazine some years before and its fame came from the fact that it is the master regulator of cell division and when it mutates, cells can go crazy. Hill explained that p53 has gone awry in a number of cancers. She believed that the report she'd found demonstrated promise but when she couldn't locate it again, Hill wasn't comfortable including it in the materials she previously found. This was disappointing because it was encouraging to have read a report that showed that Chinese hamster cells hamster cells in culture had mutant p53 and thus would be sensitive to cell cycle blocking by thymidine and, by extrapolation, tritiated thymidine, too. Later, Hill corresponded with a scientist who told her that he was perfectly able to synchronize Hela cells, which have wild type p53, with thymidine. "Mutant p53 or not, Chinese hamster cells can be synchronized – hence arrested – by the thymidine block. But I had emphasized the point in PowerPoint, it was unnecessary" and she didn't want the presentation to continue to include it.

Back to Hill's bid to involve Tom Hei as an expert witness. Despite Hei's prior insistence that he did not want to be involved, on June 18, 2008, with no prior notice to Hei, Hill had him served with a subpoena at his Columbia University office, directing him to testify in her case on July 9, 2008. Hei's deposition was adjourned without a date so that he could make his motion to quash. Hei was represented in his opposition by Paul J. Fishman who replaced Christopher J. "Chris" Christie as the United States Attorney in Hill's district when Christie was elected governor of New Jersey.

The fallout of Hill's meeting with Hei turned adversarial with Hei's motion to quash was filed on July 25, 2008. Hill's subpoena also didn't identify the subject matter of Hei's testimony. In her responses to defendants' initial interrogatories in her *qui tam*, however, Hill had stated that

Hei "has performed research and writing on the bystander effect and/ or tritiated thymidine." She also stated that Hei had knowledge of "discussions with [the] plaintiff [Hill]" in which Hei purportedly "expressed suspicion with Dr. Howell's findings regarding the bystander data reported in Dr. Howell's papers." Hill also claimed, according to Hei's motion, that "the results of [Dr. Hei's study are] at variance with the results of Dr[s]. Howell and Bishayee."

Though she would subsequently write that Hei seemed supportive enough when they had first met, when she subpoenaed his testimony, Hei claimed that Hill was attempting to use him as an unpaid expert. "The court should quash the subpoena served on Dr. Hei," per the motion. "Rule 45 expressly authorizes the court to quash a subpoena that requires 'disclosing an unretained [unpaid] expert's opinion or information that does not describe specific occurrences in dispute and results from an expert's study that was not requested by a party. This rule was designed," it was argued, "precisely to address the 'growing problem' of litigants using subpoenas 'to compel the giving of evidence and information by unretained experts." Hill vigorously denied this charge.

Hei just as vigorously claimed no personal knowledge of any facts relevant to Hill's case. "Because of Dr. Hei's expertise in the field of oncology, however, Dr. Hill asked Dr. Hei to serve as her expert witness in this case. Dr. Hei respectfully declined," according to the memorandum to the court. "Apart from calling Dr. Hei to recount the substance of their irrelevant and inadmissible [hearsay] conversation, it appears that [the] plaintiff is seeking Dr. Hei's testimony not about [the] defendants' research, which is the subject of this lawsuit, but about Dr. Hei's research, which is not. Apparently," according to Hei's motion, "[the] plaintiff intends to ask Dr. Hei about the results of Dr. Hei's own 'research and writing' on the 'bystander effect and/or tritiated thymidine' to contrast Dr. Hei's methods and findings with those of Drs. Howell and Bishayee. But testimony by an expert, even if elicited solely for such comparative purposes, is still expert testimony." Hei's counsel also suggested that Hill's attorney, Sheldon Pincus, had also made it clear that Hei would not be asked for his opinions, but rather he would be asked only fact questions about Hei's own methods and findings, arguing to the court that simply limiting the deposition to "fact" questions does not transform an expert like Dr. Hei into a "fact" witness; nor does it change the Rule 45 analysis. Rule 45(c)(3)(B)(ii) applies, by its plain terms, not only to "opinion" testimony, but also to "information that does not describe specific occurrences in dispute and results from the expert's study that was not requested by a party."

Hill, it was further argued, could not show that Hei's testimony would have any bearing on her lawsuit. According to Hill's complaint, her False Claims Act case turned on whether certain grant applications submitted to the federal government were false because they relied on falsified data derived from an experiment performed by Bishayee in 1999. Hei had personal knowledge of that experiment, nor did he participate in the research, the experiments or grant applications referred to in the case. While he admittedly performed his own experiments and published his own study on the bystander effect (also the subject of Howell and Bishayee's research), Hei's argument continued to be that he had nothing to offer of any relevance to Hill's case.

> In support of Hei's motion to quash, counsel for the defendants addressed a pointed letter summarizing Hill's case before Mark Falk, United States magistrate judge in the United States District Court, District of New Jersey, dated August 7, 2008. In it, the defendants' counsel wrote:

> On its face, Dr. Hill's identification of certain Columbia University scientists in her Answers to Defendants' Interrogatories, including Dr. Eric Hall, Dr. [Tom] Hei, Dr. Hongning Zhou and Dr. Rudranath Persaud, and her current efforts to depose certain of those scientists suffers from two primary and overriding defects. Either Dr. Hill is improperly attempting to force these scientists to provide unretained expert testimony in violation of Fed. R. Civ. P. 45 (c)(3)(B)(ii) [Rule 45] or she is merely seeking to elicit only factual information from these individuals that is wholly irrelevant to the claims alleged in this matter and thus not properly discoverable.

The defendant's protest of Hill's attempt to subpoena Hei concluded:

> It goes without saying that no one at Columbia University, and certainly not the scientists named in Plaintiff's Answers to Interrogatories and recent subpoenas, has any relevant firsthand knowledge of the facts and circumstances of these prior investigations [referring to the two Newark CCRIs and the ORI investigation] or the facts at the heart of both those investigations and Plaintiff's Complaint. Unless Plaintiff is in fact attempting to elicit expert testimony from these scientists, which they have already told her they are refusing to voluntarily provide her with, then any testimony they may provide will be irrelevant to this case and only result in Defendants and Columbia University bearing unnecessary litigation costs.

Undeterred, Hill replied through counsel to Hei's motion to quash, arguing that she was not seeking Hei's testimony as an expert because she merely intended to ask him factual questions about his "recollections of experiments done in his lab." Hei's response to this, filed with the court on August 11, 2008, reiterated that Hill had not attempted to explain how or why Hei's own experiments, done in his own laboratory, were remotely relevant to her lawsuit, which pertained to experiments performed by Bishayee. On that point, Hill suggested that Hei's experiments and studies were relevant because the *defendants* [emphasis added by Hei's counsel] had placed those experiments at issue. Specifically, according to Hill, "Defendants have asserted that, because other researchers have been able to replicate the bystander effect using tritiated thymidine, my claim should necessarily fail."[129] Recall that it was Bishayee who raised Hei's and Little's work in his testimony before the second Newark CCRI; it was Bishayee, not Hill, who thus involved the Columbia University scientists in this case by virtue of his citing their work. But as a threshold matter, Hei's counsel argued that Hill failed to document when and where the defendants supposedly took this position [involved Hei's body of work], or how such a position, even if it was asserted by the defendants, would constitute a defense that would defeat her claim. "In any event," they wrote on Hei's behalf, "even if [the] defendants had taken such a position [citing Hei's work] at some point in the past, they are plainly not taking that same position today."

Neither Howell nor Bishayee were relying on Hei's study or his experiments to defeat Hill's claims. To tie Hei's work to the experiments performed by Bishayee, Hill suggested in her reply to the motion to quash, that Hei "has intimate knowledge of the procedures and protocols that were followed by the Howell lab," per an August 4, 2008 letter brief crafted by Sheldon Pincus. But in the same breath, according to Hei's counsel, "Hill admits that these same protocols 'were clearly defined in papers and published well before the Columbia studies took place.'" This is an undisputed fact. Hei's motion carried with the federal judge on the case, Dennis M. Cavanaugh, who ruled in Hei's favor and quashed Hill's subpoena for Hei to testify in her *qui tam* case.

Several months after that, Hill later wrote, "I received a curious phone call from [my] old friend and mentor, a highly regarded photo- and radiation biologist [at the Brookhaven National Laboratory], Dick Setlow. He told me," she continued, "that he had received a phone call from Dr. Hei asking him [Setlow] to tell me to forget the whole thing [the *qui tam* lawsuit]." What Hei had told Setlow more precisely was "tell Lanie to forget the whole thing." Hill wondered if this was an attempt to obstruct justice. What reason could he have for doing this? Hei remains an influential and groundbreaking researcher in the radiation biology community. During the period in question, per his Columbia University biography, in addition to his Columbia professorship, Hei is listed as an adjunct professor and Ph.D. mentor, Chinese Academy of Sciences, China (1995-); adjunct professor and Ph.D. mentor, School of Public Health, Soochow University (2006-); Chinese Academy of Sciences special appointment professor, Hefei Institute Physical Science (2011-); distinguished visiting scientist, National Institute of Radiological Sciences, Chiba, Japan (2008-2014); director of cancer research core, National Institute of Environmental Health Sciences (NIEHS) Center for Environmental Health in Northern Manhattan (2005-2009); editor, *Advances in Space Research*, Radiation Biology Section (2008-2014), and also associate editor, *Journal of Radiation Research and Journal of Translational Medicine* (n.d.). He was also a member, Cancer Etiology Study Section, NIH 1995-2001, a post he returned to in 2014, and

Hei remains editor in chief, *Life Sciences in Space Research* (2014-). He was president of the Radiological Research Society from 2012 to 2013. Hei has maintained a significant connection to the NIH, serving as a member and chair of the Special Emphasis Panels since 2001, and a member of the NIH Intramural Review Board from 2001 to 2003; he has also significantly furthered his research via NIH grants. From Hill's perspective, Setlow's call from Hei was the first of several indications that the radiobiological community didn't want her to vet her findings in a court of law.

Hill ultimately located Persaud. He had found his way to Columbia University Mailman School of Public Health as an adjunct assistant professor of environmental health sciences. When she told Pincus, he told her to forget it. "If we subpoenaed him the Columbia lawyers would just file a new motion to quash."

Hill's list of possible expert witnesses, contacted but not retained, had grown to a list with fifteen names on it. Setlow had recommended Antone L. Brooks, Ph.D., a professor in the Environmental Science Department at Washington State University Tri-cities. His career included positions as laboratory senior scientist and section manager at Battelle, Pacific Northwest National Laboratory, manager of the Cellular and Molecular Toxicology Group at Lovelace Inhalation Toxicology Research Institute, Albuquerque, and technical representative in Washington D.C. for the United States Department of Energy, Office of Health and Environmental Research. Brooks' professional affiliations included the Health Physics Society, the Mutation Research Society, and the Radiation Research Society. Brooks, nearing retirement in 2008, sent Hill a warm reply that he did not have the time to invest in her case. But he did recommend Edwin H. Goodwin, Ph.D., at Los Alamos National Laboratory, and Hill's UMDNJ colleague Edouard Azzam. Hill's letter to Goodwin was returned, addressee unknown. Goodwin had moved from Los Alamos to Fort Collins, Colorado, as the scientific founder of KromaTiD™, Inc.[130] Azzam was out of the question for obvious reasons, Hill offered later, "although he would certainly [have been] one of the best qualified to serve as an expert for me."

Kay Fields had recommended her former ORI boss, Alan Price, now retired and head of his own consulting company, but he couldn't help Hill because he wasn't permitted by the ORI to input on cases he'd adjudicated for the government. "Fields should have known that" Hill noted. "Price does have a very interesting website where he gives a long history of his positions and accomplishments. He also posted pictures of his retirement party at ORI, so I was able to narrow down Fields to one of the two women in the pictures. I would learn later," she continued, "when I had the opportunity to read notes from the ORI that Price had a very low opinion of me and [had] insinuated in a telephone conversation he had with [Karen] Putterman that I was dishonest."

Hill had also contacted Marian Passannante[131], Ph.D., then an associate professor in the UMDNJ Department of Preventive Medicine and Community Health. She told Hill since she wasn't a statistician she'd turned over Hill's documentation to Bart K. Holland[132], Ph.D., M.P.H., the UMDNJ director of Educational Evaluation and Research, who politely informed Hill that he was not allowed to represent anyone in opposition to the UMDNJ. The latter message also turned out to be the case for Donald R. Hoover, Ph.D., M.P.H., a professor in the Department of Statistics at the Rutgers University Institute of Health, Health Care Policy and Aging Research (IHHCPAR), as he, too, was in the same position since Rutgers, like UMDNJ, is a state university.

"I wrote to Vilhelm [A.] Bohr, M.D., Ph.D., [D.Sc.], a senior investigator at the National Institute on Aging, on Setlow's recommendation. Wendy Pogozelski[133], [Ph.D.] and I had taken Vilhelm to lunch at the Radiation Research meeting in Denver after he gave a talk at a symposium on DNA damage in mitochondria that she and I had co-chaired. Vilhelm," she explained, "is the grandson of the great physicist Niels [Henrik David] Bohr, who [made foundational contributions to understanding atomic structure and quantum theory, for which he] received the Nobel Prize for physics in 1922, and the son of Aage [Niels] Bohr, a Danish nuclear physicist who shared the Nobel Prize for physics in 1975." Vilhelm Bohr, Hill observed, would have made an excellent expert witness because of his reputation in the field of DNA repair. "I never received an answer from Vilhelm. Scientific fraud may have been too much for him."

Hill did have a long talk with Samuel H. Wilson, M.D., who was, at that time, acting director of the National Institute for Environmental Health Sciences (NIEHS). "I had shared a very small lab bench with Sam at the Harvard Medical School where I was a postdoctoral fellow and he was a medical student who had taken off from the University of Colorado after his second year of medical school. Sam later transferred to and graduated from Harvard Medical School," Hill recalled, "and became a well-known and well-respected researcher in DNA synthesis. I prize him as one of the finest people I have ever known in science. Of course, there was nothing that he could do for me, but he recommended that I speak with David B. Resnick, J.D., Ph.D., a bioethicist at NIEHS, which I did. But he, too, had little that he could offer and it became clear to me that I only needed two experts – one statistician and the other a radiation biologist."

Hill connected with other old friends but was unable to get anyone to help her out. "One was an old friend I had come to know at meetings. She had taken a tenure track job at a prestigious medical school in the South, and it looked as though she was doing well from her postings on her website. Unfortunately, this was not the case. She had published several papers while there but had failed to get an R01 grant (the kind of NIH grant that Howell had) and consequently [she] had been denied tenure. There was no longer any support for her position, so she was in the process of leaving," Hill later documented. "I felt devastated for her. I may have had my troubles, but I did have tenure, and it allowed me to pursue [the] *qui tam* without having to worry about being let go. I also had a very good salary that was a great help in paying the massive bills that were accruing." Hill was afraid if she asked her to be an expert witness, defense counsel would break her down in the deposition phase. "I did use her as a consultant, but I could not use her as an expert."

Beyond the obvious disappointments, Hill's witness hunt turned up some unexpected contacts. She'd posted a request for an expert in radiation biology on the Science Advisory Board website, and this turned up a response from a user who called himself jethrod1, who asked Hill more about the case. This should have warranted a cautionary, if not a red, flag. But she replied with a lengthy e-mail laying out everything about her case and naming the UMDNJ as the university. Jethrod1 replied that he had hired an Armenian radiobiologist as a postdoctoral fellow in the prior fall semester named Bogdan Ivanovich Gerashchenko, M.D., Ph.D. From the information that Hill sent him, it didn't take much to put two and two together that Gerashchenko had worked for Howell, the principal investigator referenced in Hill's correspondence. He was correct. Gerash-chenko was Bishayee's replacement in Howell's laboratory. Hill didn't have much contact with him because Howell had, in Hill's words, excluded her from staff activities, most especially division meetings. Howell put Gerashchenko to work "on a rather useless project that avoided the use of the Helena tubes so he would not have had any personal information about their hypoxia," Hill iterated. Jethrod1 didn't know Howell but was acquainted with Azzam and made it clear he considered Azzam "a man of integrity." "So much for that," Hill opined. "He was worried that [Gerashchenko] may have been involved in the fraud, but I reassured him about that. Unfortunately, since it turned out that he was so close to the problem [and those potentially involved], it was not possible for him to act as an expert for me; however, [Jethrod1] did give me some additional names, none of whom were of much use to me." Gerashchenko had, despite no obvious implication in misconduct, published with Howell, who again put his postdoctoral fellow's name first on several articles, one of which was titled "Proliferative response of bystander cells adjacent to cells with incorporated radioactivity, published in *Cytometry Part A*[134] in July 2004 (this was Gerashchenko's third article with Howell as the only coauthor). "In a recent study, we showed that cells irradiated with gamma-rays stimulate cell growth of unirradiated (bystander) cells, when the two populations are co-cultured as a mixture. Direct cell-to-cell contact appears to be a prerequisite for the proliferative response of the bystander cells," read the summary of the article. "The aim of the current work is to investigate the possible proliferative bystander effects caused by intracellular irradiation with incorporated radionuclides, specifically the short-range beta particle emitter, tritium ((3)H)." This was one of seven papers published between 2003 and 2007 in which Gerashchenko is listed as a coauthor with Howell while he worked as Howell's postdoctoral fellow, and one of four on which he is listed as first and as the only coauthor with Howell.[135] From the UMDNJ, Gerashchenko went to work with John Dynlacht, Ph.D., at Indiana University School of Medicine, Department of Radiation Oncology.

Hill sent an e-mail request to Dr Christopher S. Lange[136], D.Phil., pro-fessor and associate chair of radiation oncology at the SUNY Downstate Medical Center in Brooklyn. Lange's medical field focus, which made him a logical sounding board and potential expert witness for Hill, concentrates on biophysical and biological research relating to human medical problems, particularly cancer and its treatment. He has studied stem cells, cancer stem cells, aging, embryology and regeneration, the molecular basis of cellular radio sensitivity, DNA double-strand breaks (DSBs) and their repair in mammalian cells, and mammalian chromosome structure. Aside from this expertise, Lange was, just then, and continues to do so in the present, running the research division of the radiation oncology department, as well as the radiation biology division, for teaching residents. Lange has continued to conduct studies of cancer stem cells and their role in determining patient treatment outcomes. Lange changed the fields of embryology and regeneration when he explained the mechanism of anterior-posterior (AP) polarity control. He showed that the AP morphogenetic field, as described in the textbooks, incorrectly predicted the results of transplantation and regeneration experiments. This is yet another model of his that changed paradigms and textbooks and remains unchallenged. Additionally, Lange was the first to show that DNA double-strand breaks in mammalian cells are reparable. Along with Joseph Y. Ostashevsky, Ph.D., he also created a quantitative basis for the molecular understanding of cell survival (the DSB model). He made several significant changes in this field. Also, with Ostashevsky, he invented the clustered loop model of higher eukaryotic (mammalian) interphase chromosome structure.[137]

"Chris is another old friend from meetings, is very intellectual and erudite," Hill observed. "I have great respect for his integrity. I sent him the whistleblower statement and the PowerPoint on the evidence for the fraud. He responded with a very thoughtful critique of the latter, but no mention of the former – too much for any reasonable scientist to deal with," she surmised. "He felt he was not familiar enough with the effects of tritiated thymidine on the cell cycle, remarking shrewdly that 'the nature of legal fact and scientific fact are not quite the same.' He complimented me regarding the PowerPoint, stating that it was a nice review, suggesting some changes that would make the scientific statements stronger and [recommending] that I write a paper on the material." Hill drafted a paper she titled "Tritiated thymidine and hypoxia," and sent him a copy. "By then, I guess, he had had enough because that was the last I heard from him. I saw him occasionally after that, but he made no mention of the case.

This," Hill believed, "is the general attitude of scientists when confronted with things that they do not want to hear."

Lange had suggested that Hill contact John T. Lett, Ph.D., by then retired from the Department of Radiology and Radiation Biology, later the Department of Environmental and Radiological Health Sciences, at Colorado State University (CSU), Fort Collins, Colorado, where he worked in the same department with Charles Waldren, Ph.D. (Waldren had worked closely with Akiko Ueno, Ph.D.). Lett coauthored papers Hill cited in her PowerPoint. "I did talk with Lett who told me that he had left CSU under a cloud and was unable to get involved with anything without his lawyer's permission. He also said that he was having trouble remembering things, intimating that he [might] be suffering from Alzheimer's disease." Lett, who grew up in London, England, and took his doctoral degree in physical organic chemistry from the University of London, immigrated to the United States in 1968 to take the position of professor of radiation biology at CSU, where he taught classes and carried out research until his retirement. Lett's research focused on the ability of cells to repair their DNA after having been damaged by radiation, and he performed some of the definitive experiments that describe cellular repair after damage by accelerated heavy ions such as would accost astronauts in flight. In particular, he studied damage to the retina and the formation of cataracts after irradiation. His research with heavy ions and with radiation-sensitive cells has impacted the theories of cellular and tissue radiosensitivity and he published more than 100 articles in scientific journals.[138] Hill never found out what he meant by "under a cloud." Lett died on October 6, 2009, in Fort Collins at the age of seventy-five, not too many months after Hill contacted him.

Though he was unable to help Hill himself, Lett suggested that she contact one of Lett's former students, which Hill did. After a lengthy conversation, she promised to put Hill in contact with a colleague who had been a statistician for the United States Air Force. "But I never heard from her again."

Going back to the Internet, Hill found the Round Table Group[139], Inc., a company founded by Robert K. Hull, J.D., Russ W. Rosenzweig, M.B.A., and Chris Crone that offered expert witness search and referral services for litigators, intellectual property attorneys, in-house counsel, and legal professionals but also expert witnesses, including academics, corporate executives, and industry subject matter experts. The company was founded in 1994 in Rockville, Maryland, with additional offices across the United States. "In desperation, I contacted them, and they put me in contact with two potential radiation biology experts. One was a professor at Princeton University who wasn't really a radiation biologist per se but a nucleic acid biochemist and molecular biologist." Hill read his resume and realized quickly that he had little experience with cell cycle, although he had impressive credentials, including a long list of publications in prominent journals. He'd coauthored several papers with equally renowned scientists in his field; his fee was $390 per hour. The second Round Table Group recommendation was Michael E. Robbins[140], Ph.D., a radiation biologist and associate director, Brain Tumor Center of Excellence, section head, radiation oncology-radiation biology and professor of radiation oncology-radiation biology at Wake Forest School of Medicine in Winston Salem, North Carolina. Robbins was a transplant from Great Britain, much like Lange and Lett, and he had a remarkable curriculum vitae, but more importantly, he oversaw teaching radiation biology to medical residents – doctors in training – in radiology, radiation oncology and nuclear medicine in the medical school. This meant to Hill that he had to know the effects of radiation on the cell cycle and, further, he would be familiar with the principles of nuclear medicine, even though, unfortunately, his research was not in these areas. Robbins' work was focused largely on the effects of radiation on the nervous system and the brain. Of greater concern to her, he hadn't used tritiated thymidine in any of his research studies. His price was high, more than $500 per hour – "and I decided to pass him over."

After several false starts and delays, largely on the part of the defense, depositions started. The original scheduling order indicated that discovery had to be completed by October 1, 2008, a schedule, by practical circumstance, would not be finished until January 7, 2010. Hill's attorney told her the depositions were a waste of time and that they were required to permit the trial to proceed more efficiently. She didn't know, just then, that the case would end without much of what was vetted in discovery getting a fair shake in front of the court. The depositions weren't such a waste of time after all.

Chapter 4 –
The Lenarczyk Deposition

Marek Lenarczyk, Howell's former postdoctoral fellow and the one who Hill attributes with having begun the entire allegation of misconduct after he reported his suspicions about the March 26-30, 2001, experiment to Hill, had moved to Milwaukee, Wisconsin, in May 2005.[141] The first deposition was thus conducted there on October 28, 2008. At Hill's expense, she and Pincus flew to Milwaukee. The defense sent attorneys John Leonard and Scott Flynn, who would depose Lenarczyk, a non-United States citizen working on an H1-B visa; he had never participated in a pretrial deposition, adding to his anxiety. He also didn't have his own legal counsel at the proceedings, an expense that he could not afford. Lenarczyk's 260-page deposition equated to hours of questioning.

Lenarczyk was palpably uneasy during questioning, reluctantly acknowledging that he left the UMDNJ because of the intensely unpleasant environment. He acknowledged, too, that sixteen experiments he performed had failed to replicate Bishayee's results. Howell was never able to produce five of those experiments, which pertained to the bystander effect, although Hill's counsel requested them several times. But Lenarczyk kept copies of the data on spreadsheets and provided those to Pincus. The originals would have been more useful had the case gone to trial. During his deposition, Lenarczyk struggled with his words and the way in which he had been characterized in Hill's original filing. But important passages proved revealing. The dark side of the moon – the dysfunction and chaos of the laboratory – that the public rarely, if ever, sees – is never more apparent than in the depositions taken in the Hill case and most especially Lenarczyk's questioning. The depositions further demonstrate just how difficult it is for the whistleblower to get anyone, even a colleague, to testify truthfully and fully about what he may have witnessed – what he may know – when the "big science club" is on standby to blackball the first one who talks outside the lab and airs the dirty laundry.

Leonard: "Can I direct your attention to paragraph 24. Again, this is Dr. Hill's statement that she has filed with the court. 'As a result of their [Bishayee and Howell] actions, Hill and Lenarczyk concluded that Bishayee had, in fact, fabricated the experiment's data and engaged in scientific fraud.' Is that what you concluded that he engaged in scientific fraud?"

Lenarczyk: "I don't think so. I don't recall those words, 'engaged in scientific fraud.'"

Leonard: "Okay."

Lenarczyk: "What I found I just told you. We found that experiment, which was conducted by Dr. Bishayee, was a little bit suspicious to us what he found out. It was not clearly connected to the data which he showed us."

Leonard: "Understood."

Lenarczyk: "But I never said 'scientific fraud.' I mean – ."

Leonard: "Okay. So as it relates to you, that first sentence in paragraph 24 is not true?"

Lenarczyk: "To whom I supposed to conclude that, that Dr. Bishayee was – ."

Leonard: "Apparently Dr. Hill put it in there, that the two of you – I mean, these are her words – that you concluded – ."

Lenarczyk: "Yeah, but this is not my report, so I didn't – ."

Leonard: "I understand."

Lenarczyk: "I never saw this report. So, you are asking me now about something that was not prepared by myself in that report."

Leonard: "That's true."

Lenarczyk: "So, that's why this is hard to me, because something is related to me because my name is here, but – ."

Leonard: "I understand."

Lenarczyk: " – but none of this paragraph, none of even one word I prepared in that report."

Leonard: "Please, you have to understand, I didn't put your name there."

Lenarczyk: "So, I'm a little bit afraid what I'm saying, because maybe you are thinking that I'm trying to cover or something. No, I'm not, because I'm trying to figure out what this was and trying to do my best to explain what I can understand after this, because is the first time when I can that document."

Leonard: "I see."

Leonard: "In fairness to you, there are certain statements that are made in here by Dr. Hill about what you concluded, about what you think. And what I'm saying to you is, do you agree with the statement she's making about what you concluded or not? I mean, I didn't write this either, Dr. Hill did. So, when Dr. Hill says, 'As a result of their actions, Hill and Lenarczyk concluded that Bishayee had, in fact, fabricated the experiment's data and engaged in scientific fraud,' is that true or not true as it related to you?"

Lenarczyk: "No, I didn't conclude that."

Leonard: "Did you conclude he fabricated data?"

Lenarczyk: "I don't remember that. I mean, what I said. I told you I found what was did by Dr. Bishayee at that time was not clearly understood by me, because his observation was not related work for his data, what he was showing."

Leonard: "Okay."

Lenarczyk: "That's why we have this suspicion about that. Plus at the same time I was running an experiment very similar to his experiment."

Leonard: "Understood."

Lenarczyk: "And then my data from my experiment were not clearly confirming the data which was made by them even before I came to the lab."

Leonard: "Understood."

Lenarczyk would repeat many times that he knew nothing of the documents he was being shown. As the deposition progressed, his discomfort with the process only increased.

Lenarczyk: "I was only involved in that situation because Dr. Hill, I think, tried to do something more when that case was closed at the University of Medicine. So, that's why I was involved in the process."

Leonard: "Dr. Lenarczyk, I completely understand. But you have to understand our position [as the defense counsel]. I understand exactly what you're saying, but Dr. Hill has gone on and created a lot of documents, and your name is all over them as somebody who supports her in these allegations. Okay? She's prepared documents and given them to staff at Columbia University – and we'll go through all of them – and your name is all over them, because what she is saying you agree with her there's been scientific misconduct, there's been fabrication of data. We know you didn't write the documents. I'm here to just say is that your opinion or not. And I think what I'm hearing – and tell me if I'm wrong – is, no, that's not your opinion."

Pincus: "Again, objection to the form of the question. You may answer."

Leonard: "You can answer."

Lenarczyk: "That's everything what I want to say. I never prepared any kind of documents. I was sharing my observation with Dr. Hill at that time what was happening in the lab, because, you know, that was like a normal situation. Every time in every lab when the people are doing something they are observing, they are sharing their own ideas what is going on. Particular to me was something difficult to understand, because I was running essentially the same kind of experiment as they did before and I couldn't confirm that data. So, I was asking myself maybe I'm doing something wrong, or I was suspicious for myself, because I would like to do the best job at that time. So, when I was trying to share my observation with Dr. Hill then everything was – everything went so quickly. I mean, she took everything and she tried to make that case. And she went somewhere to university to report the case, and that's why everything was happening at that time."

Leonard: "I understand. We'll work our way through it. Paragraph 26 says, 'After the committee's report' – By the way, the initial committee report came back and said no evidence of scientific misconduct."

Lenarczyk: Yes, I know that, because I have a letter from them at that time.

Leonard: "So then paragraph 26 says, 'After the committee's report, Howell proceeded to terminate the employment of Bishayee and Lenarczyk.' Were you terminated by Dr. Howell?"

Lenarczyk: "No, I didn't."

Leonard: "Okay."

Lenarczyk: "I talked to Dr. Howell that I would like to – I would like to terminate my employment."

Leonard: "Okay."

Lenarczyk: "He didn't fire me. Okay?"

Leonard: "Okay."

Lenarczyk: "I was hired for one year, and I was terminated even probably a month before formal term of my first year working for him."

Leonard: "Okay."

Lenarczyk: "I just explained to him that I'm not feeling comfortable to working in that kind of situation when something is not going, maybe not right, but very difficult to explain. So, I said I don't like to be involved in that kind of situation to work in the lab when we don't know what we are doing essentially.

Leonard: "I understand. And you have to understand this is one of the things we're going – ."

Lenarczyk: "I was not terminated by him."

Leonard: "You were not terminated by him?"

Lenarczyk: "No. No. I asked him to agree that I would like to leave that position on that particular time when I left."

Leonard: "So that sentence in paragraph 26 is not true?"

Lenarczyk: "In terms that I was terminated, probably, yes, not."

Lenarczyk was repeatedly asked if he had any part in the drafting of Hill's complaint filed in the federal court in the District of New Jersey. Each time, he replied "no." But Leonard kept probing Lenarczyk's involvement.

Leonard: "Okay. Did Dr. Hill reach out to you in or about October 2003 when the complaint was filed to ask you for your assistance in drafting the complaint?"

Lenarczyk: "No."

Leonard: Did her attorney reach out to you to confirm your firsthand knowledge of the facts in the complaint?

Lenarczyk: "No."

Leonard: "Okay. Do you have any firsthand knowledge of fraud committed by Dr. Howell?"

Lenarczyk: "What do you mean 'first'?"

Leonard: "In other words, firsthand knowledge is not what somebody's told you. [It's] what you think, what you've observed, you personally."

Pincus: "Objection to the form of the question in that it calls for a legal conclusion.

Leonard: "Alright."

Pincus: "You may answer."

Leonard: "Alright. Let me give it to you this way. If you walked outside and you saw a car accident, you would have firsthand knowledge that there was a car accident."

Lenarczyk: "Okay."

Leonard: "If I told you I just came up from the lobby and I saw a car accident, you would have heard about it from me, but you personally wouldn't have any firsthand knowledge of it."

Lenarczyk: "Okay. I see."

Lenarczyk eventually understood Leonard's line of questioning.

Leonard: "Know something yourself, observe it, experience it, yes. Do you have any firsthand knowledge of fraud committed by Dr. Howell?"

Lenarczyk: "By Dr. – ."

Leonard: "Howell."

Lenarczyk: "Fraud?"

Leonard: "Yes."

Lenarczyk: "No."

Leonard: "Okay. Do you have any firsthand knowledge of fraud committed by Dr. Bishayee?"

Lenarczyk: "No."

Leonard: "Do you understand that Dr. Hill, through her complaint, is seeking to impose penalties and fines in excess of $4 million against the university and Dr. Howell and Dr. Bishayee individually?"

Lenarczyk: "Do I know about that?"

Leonard: "Yes."

Lenarczyk: "No."

Leonard: "Okay. Do you understand, do you have any understanding that Dr. Hill would be entitled to a fee of $1 million if she is successful in this lawsuit?"

Lenarczyk: "No."

Leonard: "So, I take it Dr. Hill hasn't offered you a portion of the $1 million fee?"

Lenarczyk: "No."

Pincus: "Objection to the form of the question. We'll call that a rhetorical question with a rhetorical objection."

As questioning resumed, Leonard pressed Lenarczyk for an answer. Did Lenarczyk have any firsthand knowledge of misconduct committed by Howell? Lenarczyk answered "no." Leonard asked if Lenarczyk had firsthand knowledge of any misconduct by Bishayee. Pincus reiterated his objection. Lenarczyk replied "no."

Leonard moved on to a line of questioning focused on Lenarczyk's employment in Howell's lab (he had previously been employed by Howell between 1995 and 1996). When Lenarczyk returned to work in Howell's lab in 2000, he explained that he was supposed to work with Bishayee because he was running similar experiments using the same biological model, the same protocol.

Leonard: "So you're using the same protocol, but you're looking for different things?"

Lenarczyk: "Right. And, I mean, for some extent we were using the same protocol, because we were supposed to treat the cells exactly in the same way, and then proceed then to get some data out of that experiment. So, I was working on, essentially on something which you may call mutagenicity theory. Like looking for the mutagenic effect of something which was treated by something. In that case cells were treated by radionuclide."

Pincus: "Radionuclide. Right?"

Hill: "Radionuclide."

Lenarczyk: "I'm sorry about the pronunciation." So, essentially, we were working on the same cells. We had the same kind of cells, and he was proceeding his experiment, I conducted my experiment."

When Howell's NIH grant started, Lenarczyk was working on it exclusively.

Leonard: "And beginning in 2001 is it fair to say that you were primarily working on experiments with Dr. Hill?"

Lenarczyk: "What do you mean – I mean, we were working in the same lab. Okay? We were sharing the same lab. So, what we were doing, we were working – Yes, we may say it like that. I mean, I was running through March 30, 2001. Okay?"

Leonard: "Okay."

Lenarczyk: "Okay."

Leonard: "What did you observe that you found troubling?

Lenarczyk: "We observed nonconsistency [*sic*] between the data which was coming out of the experiment which was running by Dr. Bishayee at that time."

Leonard: "Okay."

Lenarczyk: "And that was very related to what I used to work on at that time."

Leonard: "Okay."

Lenarczyk: "Because part of my experiment was to confirm some data which they were found before, I believe, by Dr. Bishayee who was working earlier at the same lab before I came. So, the scientific design of my experiment, which I was asked to do, was that I had to go through the same kind of experiment which was done before by Dr. Bishayee, because that was just a normal situation in that kind of assay which we were using."

Pincus: "If I may – ."

Lenarczyk: "What I was doing, I was working on the experiment, which technically part of my experiment was to confirm, or to at least verify the data which was prepared before by experiment in the small lab, which I think that that was given to Dr. Hill at that time. So, she had at that time small room with that lab, and I was working in that room. Because that was only the one place when I had an opportunity to do that experiment, because of the technical reason."

Leonard: "Okay. Did there come a point in time where Dr. Hill terminated the project – excuse me – Dr. Howell terminated the project that you and Dr. Hill were working on?"

Lenarczyk: "If I what? Sorry."

Leonard: "Did there come a point in time when Dr. Howell terminated an experiment that you and Dr. Hill were working on?"

Lenarczyk: "No. I left from Dr. Howell's lab when the grant was still running, I believe. So, if you asking me about experiment, I'm thinking about whole grant which was dedicated to run experiments."

Leonard: "Not the whole grant. I mean, there were different experiments that were going on within the grant, correct?"

Lenarczyk: "Yes, because we were working – it's not like one experiment is for one grant."

Leonard: "No, I understand."

Lenarczyk: "You can run an experiment, even a couple of them at the same time if you can handle that, but you have to start the experiment and you can finish the experiment."

Leonard: "Right. But wasn't there one experiment that you and Dr. Hill were working on, and at some point Dr. Howell said he wasn't seeing results and wanted to terminate that."

Lenarczyk: "I don't remember that. I'm sorry.

Leonard: "That's okay. If you don't remember, you don't remember."

Lenarczyk: "Sometimes you are trying to be very precise, but I don't recollect everything that happened seven years ago. So, if I can say I don't remember, this is my best knowledge at this point, which is 2008."

Leonard pressed on what was going on in the lab at the time Hill claimed Bishayee committed the misconduct.

Leonard: [...] "So you're in the lab. And at some point you observed something in the lab with regard to Dr. Bishayee that you find troubling?"

Lenarczyk: "Right."

Leonard: "Okay. And we're talking. I guess, March 26 through March 30, 2001. Okay?"

Lenarczyk: "Okay."

Leonard: "What did you observe that you found troubling?"

Lenarczyk: "We observed nonconsistency [*sic*] between the data which was coming out of the experiment which was running by Dr. Bishayee at that time."

Leonard: "Okay."

Lenarczyk: "And that was very related to what I used to work on at that time."

Leonard: "Okay."

Lenarczyk: "Because part of my experiment was to confirm some data which they were found before, I believe, by Dr. Bishayee who was working earlier at the same lab before I came. So, the scientific design of my experiment, which I was asked to do, was that I had to go through the same kind of experiment which was done before by Dr. Bishayee, because that was just a normal situation in that kind of assay which we were using."

Pincus: "If I may – ."

Lenarczyk: "What I was doing, I was working on the experiment, which technically part of my experiment was to confirm, or to at least verify the data which was prepared before by Dr. Bishayee. It's like, okay, in simple words okay, experiment consists of two parts, A and B. I was carrying part A and B, Dr. Bishayee was carrying part A. So, when I was doing my experiment, which is A and B, I had to go through A. So, practically I had to confirm something which was discovered or which was reported before I came by him. And that was difficult for me, because I was working for a couple of months in that lab and I didn't make any similar data which was found before by Dr. Bishayee – probably Dr. Bishayee, because nobody worked on this project except for him, I think. So, I had a hard time, because I was doing my experiment and I couldn't get the same data. And then – ."

Leonard: "Did you ask Dr. Bishayee why that was?"

Lenarczyk: "I asked him couple of times to help me with something, because part of the experiment was to treat the cells with radionuclide. And that was a very precise treatment, because we had the record for that. But the treatment by itself is very influential by the hand who is doing that treatment.

"So, I was afraid that maybe my data are coming because I did something differently than he did before. So, when I asked him to show me, technically he didn't show me. I mean, he couldn't try to be helpful at that time.

"So, then I was running the experiment by myself, and we had the lab notebook of everything what I did. And later on when I had the difficulty on confirm the data and nobody – I mean nobody – he couldn't help me to try to find out if I'm doing something wrong maybe, that's why I had different data.

"But what happened is that we were talking, of course, with Dr. Hill, we talked about this. And then one day when I found that something is very difficult to explain I was trying – I started to be a little bit suspicious what's going on, because if you are showing on the paper the data, that data is supposed to come from real experiment which you are carrying on, and there was no consistency between this. And then we were talking about that with Dr. Hill basically. And then from that point everything was happened. I mean, Dr. Howell was later on tried to be very suspicious what's going on. So, he asked couple of people – ."

Leonard: "Could we back up before we get too far down the road? You said you talked to people. Did you ever go to Dr. Howell and talk to him about the situation?"

Lenarczyk: "I don't remember."

Leonard: "It was his grant, correct?"

Lenarczyk: "Yes, it was his grant."

Lenarczyk didn't take his concerns to Howell – at least he didn't remember it. When Leonard asked him to get more specific about the discrepancies he observed between his work and Bishayee's experiment, he focused on the protocol.

> *Leonard*: "So, you see some differences in your results and the results Dr. Bishayee did using a similar protocol?"
>
> *Lenarczyk*: "Right."
>
> *Leonard*: "And there's a concern by you that maybe you're doing something wrong initially?"
>
> *Lenarczyk*: "Right, because, you know, this is like the first thing which you can concern. Okay? Because if you are doing something you can say, 'I cannot do the same as somebody did before, why is that?' So, I asked myself maybe I'm doing something wrong with my hands, using my hands. So, then I tried to ask Dr. Bishayee to help me, or at least to observe what I was doing."
>
> *Leonard*: "And he was of no help?"
>
> *Lenarczyk*: "No."
>
> *Leonard*: "Okay. And you have no recollection of speaking to Dr. Howell about it?"
>
> *Lenarczyk*: "About not giving me any help?"
>
> *Leonard*: "Not giving you any help, or the fact that your data wasn't in line with Dr. Bishayee's data.
>
> *Lenarczyk*: "No, I don't have that recollection."
>
> *Leonard*: "Okay. You did speak to Dr. Hill?"
>
> *Lenarczyk*: "Yes, because, see, we were working at the same time on the same project. So, it was like an automatic situation that when she got something out of his experiment, her experiment, and when I got something out of my experiment, the first communication was between us, because we were working in the same lab area. And we were, in fact, involved in the same kind of experiments. Okay? So, we were talking, like sharing our data, what we did. Okay?"
>
> *Leonard*: "I'm sorry. And what did she say when you brought this to her attention?"
>
> *Lenarczyk*: "She was – I don't remember, but – I think that she found that something is very suspicious about that, because we found something which was not really clear to us. And then she started

to think that something is making up, I mean, you know, like you can make data not exactly out of the experiment you run or – ."

Leonard: "She did this or you did this? Who made these sort of conclusions?"

Lenarczyk: "I don't know if that was a conclusion at all. But, you know, it's like when you observed what was happening in that lab in particular with that experiment which Dr. Bishayee was running at that time, we started to be suspicious that something is not right, because – ."

Leonard: "I'm sorry, you keep going."

Lenarczyk: "Because he was showing – I think that he was showing the data which we found out that that data cannot come through that experiment he was running at the same time."

Leonard: "How do you know that? How do you have personal knowledge of that fact?"

Lenarczyk: "First of all, some of the tubes which he – 'tubes' means plastic tubes – he supposed to use them to run the experiment, to finish the experiment, to collect the data – they were still in the fridge at the time, because which was the environment where we kept the small samples."

Leonard: "Did you ask anybody if they were the same tubes or if there was an explanation for that?"

Lenarczyk: "I don't remember that happening, if I asked somebody."

Leonard: "Well, don't you think that would be an important piece of information to know?"

Lenarczyk: "That would be an important piece of information, but – I don't remember. I mean – I'm sorry. I don't remember if I asked anybody."

Leonard: "When you spoke to Dr. Hill about the situation, did she tell you that she had problems with Bishayee before?"

Lenarczyk: "I don't remember. I don't remember that."

Leonard: "Whose idea was it to engage in this covert operation to watch Dr. Bishayee and take pictures of his work and document what he was doing?"

Pincus: "Objection to the form of the question. You may answer."

Lenarczyk: "Whose idea to do that? I mean – ."

Leonard: "Yes."

Lenarczyk: "Technically, we did that together. Okay?"

Leonard: "Why?"

Lenarczyk: "I don't know why was it, what might be the explanation of that. We just want to collect the data based on which we may be able to explain the nonconsistency [*sic*] between the experiment and the data."

Leonard: "Well, you took about 17 photographs, correct?"

Lenarczyk: "I took a photograph. What number, I don't recollect."

Leonard: "So, you're familiar with the photographs with your hands. You copied pages out of notebooks, correct?"

Lenarczyk: "My notebook? I have my notebooks from – ."

Leonard: "From Bishayee's notebook, copied relevant pages of the notebook."

Pincus: "Well, objection to the form of the question. No foundation. You may answer."

Lenarczyk: "I don't recollect that."

Leonard: "Searched through trash cans?"

Lenarczyk: "I don't remember that."

Leonard: "Okay. Tested tubes for radioactivity?"

Lenarczyk: "I don't think so. I don't remember that."

Leonard: "Okay."

Lenarczyk: "I don't think so that I tested the tube for radioactivity."

Leonard: "In one of her papers Dr. Hill refers to it as shadowing Dr. Bishayee. Do you know what she meant by that?"

Lenarczyk: "No."

Leonard: "Did either of you speak to Dr. Bishayee about your concerns or what you were doing?"

Lenarczyk: "No, I don't think so."

Leonard: "Okay. What was to be gained from engaging in these activities without disclosing what you were doing to Dr. Bishayee?"

Lenarczyk: "What was the reason for that?"

Leonard: "Mm-Mm."

Lenarczyk: "I don't know. I mean, we just did that. I mean, no reason."

Leonard: "I'm just trying to understand."

Lenarczyk: "I tried to collect this data because, first of all, I couldn't get any help out of him to explain something. I was working on the experiment, which was even published before, which I couldn't confirm. And when we tried to discuss this one it was almost like no discussion about the data.

"So, when I was facing the trouble with my experiments, and I couldn't confirm anything what was happening, then I tried, started to be suspicious that something is not right, maybe the experiment which was done by him was not correct, or maybe something – I have no good explanation why I couldn't confirm something out of my experiment with that experiment which was done before."

> *Leonard*: "Yes, I understand that completely. The problem I'm having is understanding why, as a person of science, you wouldn't take the two sets of data and sit down with Dr. Howell and Dr. Bishayee and Dr. Hill, or whoever else was in the lab, saying, 'This is what I'm finding.' To me a person of science or not, electing not to do that, and digging through trash cans, and taking photographs, it just – I don't know – I find it hard to understand why somebody would engage in that behavior."
>
> *Pincus*: Objected to the form of the question again.
>
> *Lenarczyk*: "I don't have an explanation for that."
>
> *Leonard*: "Okay. Was the goal to get Dr. Bishayee? Was the goal to get Dr. Bishayee fired?"
>
> *Lenarczyk*: "No."
>
> *Leonard*: "Okay. So, what conclusions did all this bring you to, you personally"
>
> *Lenarczyk*: "All of this what was happening at that time?"
>
> *Leonard*: "Yeah, your investigation."
>
> *Lenarczyk*: "See, the very important part which was happened we never discussed yet. Okay? Which maybe that would be the conclusion, not necessarily what I saw.

"But when we found that my data are not confirming his data, I mean Dr. Bishayee, and when everything was starting, I mean, that whole process about when Dr. Hill went to university authorities to report everything, then I think that we were obligated to not talk to each other, because that was like were investigating. And up to that point when we were asked to go there we were supposed to not talk to each other about that.

"So, anyway, by the time when Dr. Howell figured out that something is not right he asked a couple of people in that lab, including himself, me, Dr. Bishayee, and another person who was working for Dr. Azzam at that time, tried to do the same experiment by three of us, because that was like finding the proof of confirm that something is really either my fault, or maybe something magic is going on which we cannot confirm the previous data.

"So, one day Dr. Howell locked the lab when we were working that radionuclide stuff, and then we were asked to come over, three of us at the same time, because Dr. Howell, as the boss, he was able to run the experiment. So, we are running [parallel] at the same time, three of us, the same experiment from the same cells, just to avoid any influence that somebody is preparing by himself, somebody is preparing by – ."

Leonard: "When you say 'the same cells,' you mean each of the three of you had the same cells, not necessarily the same cells that were used in the prior experiment?"

Lenarczyk: "Right. Right. Because the prior experiment was probably done a year before."

Leonard: "Exactly. I'm sorry. Go ahead."

Lenarczyk: "This was not exactly the same cells in terms of the piece of the cell."

Leonard: "Right."

Lenarczyk: "That was the same strain of cells. We were working on that. So, we prepared the cells. Dr. Howell asked us to run the experiment, three of us, just to find out what's going on. Because I think that he started to be suspicious too at that time. That's my guess. Maybe I'm wrong."

Leonard: "He never said that to you?"

Lenarczyk: "No, he never said to me."

Leonard: "Okay."

Lenarczyk: "But based on his reaction, which was happened, I was really scared about that, because he really, really have to do something, which has never happened to me before.

"So, he asked us to do that experiment. So, we run experiment, three of us, Dr. Bishayee, Dr. Howell, and me at the same time. We processed the cells. And the experiment took like one week from the time when you are processing to see the data, to see the results. So, by that time the cells were in incubator, which is lab equipment. And that incubator was in that piece of lab. When we were done with the first part of the experiment, Dr. Howell locked the lab, nobody can get in just to not interfere with something. And then after one week when we supposed to see the results of the treatment we went to that lab, three of us, and we checked that piece of work done by each of us, and everything was done like I did. So, they didn't confirm either. Dr. Bishayee or Dr. Howell, anything what was before – ."

Leonard: "Right."

Lenarczyk: " – discovered by them before I came."

Leonard: "Right."

Lenarczyk: "So, that was like confirmation that what I did was, what I got from this experiment was right. Because they worked on that experiment before, even before I came to work with them."

Leonard: "Right."

Lenarczyk: "But when I came and found out that I couldn't find the same data out of my experiments as they did before, after this experiment which was conducted by three of us, everything was, looked like I did."

Leonard: "Right."

Lenarczyk: "Which means that they couldn't confirm anything what they found before. So, and that was not a very pleasant situation, because I was feeling that something is not right. I couldn't – we couldn't find a good explanation for that. They couldn't provide a good explanation why this was going like that."

Leonard: "Who is 'they'?"

Lenarczyk: "Dr. Howell and Dr. Bishayee. Nobody working in that lab because, obviously, everybody knows what's going on, because we were working in that lab. So, I was just running the experiment what they asked me to do to verify kind of hypothesis, but my data was completely different from their data – not completely – yeah, completely different from what they found before."

Leonard: "Right."

Lenarczyk: "So, later on – ."

Leonard: And just so we're clear, you're not suggesting that just because your data is different that that means the other data was falsified or fraudulent?"

Lenarczyk: "No, no, it's not like that, because different means like – okay. You can mix two colors, and then one person has a different color, and another person has a different color."

Leonard: "Right."

Lenarczyk: "Why this is happening, it's supposed to not, because you are using the same product of the bigger one, okay, and you are getting different results, which that is something which supposed to not be seen, because otherwise means that something is not comparable."

Leonard: "Right. Okay. Can you explain to me what you believe to be the goal of Dr. Bishayee's experiment?"

Lenarczyk: "What was the goal?"

Leonard: "Mm-Mm. What was the focus of his experiment?"

Lenarczyk: "I think that this goal was to find out what is the response of the cells after treatment with radionuclide at that time was, I think, three years in the lab, using the end point which we are calling survival of the cells."

Leonard: "So the focus of his experiment, would it be fair to say, was on the radioactive cells?"

Lenarczyk: "Right. I mean, that was like a little bit complicated experiment, because I think that that was like that. You have cells which they were irradiated with radionuclide, irradiated or treated with radionuclide. And these cells after treatment are supposed to die, because was killing them, that treatment. But they were not dying immediately like that. Okay?

"When you have a parallel, the population of the cells, which they are not treated, they will repopulate, and you can control that repopulation by different kind of end points; like survival, for example.

"Now, if you would mix the cells in known parts, each one, treated versus untreated, then you will predict what will be the survival if you will mix cells which they were treated with the cells which they were not treated. Because you will assume that all of the cells which they were treated will die. All of the cells which were not treated will survive. However, the hypothesis was that by the treatment of these cells with radionuclide, when you would mix them with the population of untreated cells, then by the time when the treated cells are dying they will influence the survival of the cells which were not treated – ."

Leonard: "The bystander effect."

Lenarczyk: " – which we call bystander effect."

Leonard: "Understood."

Lenarczyk: "And that was measured by him and by me at the same time."

Leonard: "Understood."

Lenarczyk: "So, I couldn't find any evidence that the survival is showing the bystander effect, which means that untreated cells, when they were mixed with treated cells, they're supposed to show different survival based upon – ."

Leonard: "Because they've been influenced."

Lenarczyk: "They supposed to show lower survival, because the treated cells will release some, that factor, okay, to influence that survival."

Leonard: "Okay."

Lenarczyk: "I couldn't find that working for a couple of months."

Leonard: "Okay."

Lenarczyk: "They found that before."

Leonard: "Right."

Lenarczyk: "Even before I came."

Leonard: "Understood."

Lenarczyk: "And that was essentially everything what was different. Okay?"

Leonard: "Okay."

Lenarczyk: "And then when we were running, after three or maybe four months, I don't remember now how long I was working on my experiments, to find out that I couldn't confirm the data. But when Dr. Howell, they asked us, okay, that's it, we have to run experiments, the three of us, because something is wrong. Either

Marek [Lenarczyk] is wrong, or something is wrong with the cells, or maybe – we don't know what is wrong."

Leonard: "Okay."

Lenarczyk: "When we run this experiment, three of us, and then it was evident that everything is like I did."

Leonard: "Right."

Lenarczyk: "So, there was not that kind of bystander effect they reported before."

Leonard: "Right."

Lenarczyk: "Then I found that this is not a good situation. Okay? I didn't like to be involved in that kind of experiment when the people they have no idea what is going on. And there was no reason to work on that, because I couldn't find – I mean, even if I would go father to run experiments later on, if I would find some evidence with mutagenicity, which is the part of that – ."

Leonard: "You did see bystander in that population?"

Lenarczyk: "I don't remember, but you can check the notebook. Okay? There is probably tons of data which I couldn't tell you what is exactly. But you can track the data.

"By hypothetically if you would find out some effects of the mutagenicity, when part of this experiment is showing survival, which they published before, only that part, then if I want to discuss the results, I have to discuss the results referring to the results which they found before, which was contradictory, because we have a different result. And for many, many weeks I was afraid that something was wrong with my subject, because maybe I'm doing something wrong. But when we did the experiment, the three of us, I was rather convinced that what I was doing was right."

Leonard: "Okay."

Lenarczyk: "Because we cannot confirm any data which they published or they showed before."

Leonard: "Okay. Contradictory results, but you have no personal knowledge of any fraud or fabrication of that?"

Lenarczyk: "No, I didn't know about that, because most of this data done before I came, I believe."

Leonard: "Understood. So, you have no personal knowledge of – ."

Lenarczyk: "No.

Leonard: " – the protocol, how it was done, or what was in the lab at that time?"

Lenarczyk: "No."

Leonard: "Okay."

Lenarczyk: "And I was having a very nice time working with Dr. Howell before, so why I have to think about that even."

Leonard: "Did Dr. Hill ever tell you that she wanted to have Bishayee fired?"

Lenarczyk: "She wanted to fire Bishayee?"

Leonard: "She wanted to convince Howell to fire Bishayee."

Lenarczyk: "I don't remember that. Okay? And this is what I'm saying now. When I'm saying I don't remember that, it's really I don't remember that. Okay? Some questions are related to, which I mentioned before, seven years before."

Leonard: "I understand."

Lenarczyk: "So, when you can show me some kind of paper then I can at least read this and I can say, okay, this is right, this is wrong, but something is – ."

Leonard: "You'll agree with me, won't you, that what you and Dr. Hill did with Bishayee's experiment contaminated that experiment?"

Lenarczyk: "Mm-Mm."

Pincus: Objected to the form of the question – again. "You may answer."

Leonard: "I think he already did."

Lenarczyk: Can you ask again? That Dr. Bishayee did experiment with contaminated cells? [Lenarczyk hadn't understood Leonard's question.]

Leonard: "No. That what you and Dr. Hill did in removing tubes and taking pictures, that in and of itself contaminated his experiment."

Lenarczyk: "By doing that you may contaminate his experiment?"

Leonard: "Yes, by removing tubes, by taking photographs, by – ."

Lenarczyk: "No, we cannot contaminate the cells by taking them. I mean – No, you cannot do that. I mean, you have to contaminate the cells usually before you can put in the tube. Because when they are in the tube they are locked, so we'll have no physical possibility to touch the cells. By taking the tube, by keeping the tube, because tube is just a tube with a cap, and the tube are locked by the cap. So when you are taking the tube in your finger you are not touching the cells, which are inside."

Leonard: "How about when you're testing it for radioactivity?"

Lenarczyk: "Technically you may contaminate, because you have to take them out. But I think that usually – I don't remember exactly now this protocol, but – I don't remember how the protocol was at that time. But I believe that you can take the cells to measure their radioactivity in the incubator when the environment is aseptic."

Leonard: "Is that what you did with Dr. Bishayee?"

Lenarczyk: "I didn't do that. I mean, I don't recollect that I did that. I mean, if you can show me the protocol I may be a little bit more precise, because I don't remember now the protocol, how they measured the radioactivity."

Lenarczyk denied testing the Helena tubes for radioactivity. When Leonard tried to ask if he had checked for radioactivity and would doing so contaminate the experiment, Pincus objected. The question, as posed, called for speculation. Contamination could occur if the cells were removed from the tubes. Lenarczyk denied knowing how Bishayee's March 26-30, 2001 experiment pertained to Howell's grant goals. Further, Lenarczyk denied any collaboration with Hill in filing the allegation of misconduct against Bishayee with the Newark CCRI. In a string of questions, all trying to get at his involvement in Hill's university inquiry, Lenarczyk replied "I don't remember."

Leonard: "Do you know you're down as a co-complainant on that complaint [at the university]?"

Lenarczyk: "If I was, than I was."

Leonard: "Well, do you have any recollection of that?"

Lenarczyk: "No, I don't have recollection."

Leonard: "Did Dr. Hill ask you if it was okay to include you as a co-complainant?"

Lenarczyk: "I don't remember that."

Leonard: "Okay. Prior to her filing the complaint with the committee, are you aware of any efforts either you or Dr. Hill made to raise issues internally within your department?"

Lenarczyk: "Say again?"

Leonard: "Sure. Prior to filing that complaint that you're listed as a co-complainant, did you try to resolve the issues within your department; that is to say, did you go to Dr. Howell?"

Lenarczyk: "I don't remember."

Leonard: "Do you agree that it would be an unusual step to bypass going to the person whose grant you're working on to file a complaint directly with the committee?"

Pincus: Objected to the question. "You may answer."

Lenarczyk: "I don't have response for that."

Leonard: "I'm sorry, what was your answer?"

Lenarczyk: "I don't have response for that. I don't know." Because you asked me if that would be the normal way to?"

Leonard: "Well, let me rephrase. Why didn't you do it that way?"

Lenarczyk: "Which way?"

Leonard: "To try to resolve it within the department. Why wouldn't you go to Dr. Howell before taking it to a committee?"

Pincus: Objected to the form of the question. No foundation. "You may answer."

Lenarczyk: I don't know. I don't have an answer for that."

Leonard: "In retrospect do you regret not taking it to Dr. Howell first?"

Lenarczyk: "Probably, yes."

Leonard: "Okay. You were interviewed by the committee?"

Lenarczyk: "At UMDNJ?"

Leonard: "Yes."

Lenarczyk: "Yes, I was."

Leonard: "Did Dr. Hill ever tell you not to discuss this situation with Dr. Howell?"

Lenarczyk: "I don't remember that."

Leonard: "You don't remember her telling you that it would be better for her to tell him rather than you?"

Lenarczyk: "For not telling him?"

Leonard: "No. Do you recall her telling you that she should tell Dr. Howell, not you, about what's going on?"

Lenarczyk: "I don't remember that."

Leonard: "Okay. You were interviewed by the committee?"

Lenarczyk: "Yes."

Leonard: "You testified honestly?"

Lenarczyk: "I think so."

Leonard: "Any reason to recant or change your testimony?"

Lenarczyk: "The reason to change at that time?"

Leonard: "Now, as you sit here today, do you have any reason to change – Is there anything that you know now that would cause you to change your testimony which you previously gave?"

Lenarczyk: "No, I don't think so."

Leonard: "Okay."

Lenarczyk: "So, essentially what I'm saying now is exactly what I said before, except for that which I don't remember. Is that right? I mean –."

Leonard: "I don't know if it's exactly the same except for you don't remember, but – ."

Lenarczyk: "If I don't remember something I said I don't remember."

Leonard: "Okay."

Lenarczyk: "If I remember something, I believe I said the same at the time when I was investigated by the committee."

Leonard: "Okay. Are there any feelings of guilt that Dr. Bishayee's reputation at the university was being called into question before the committee?"

Lenarczyk: "Any recollection that Dr. Bishayee was – ."

Leonard: "No. Did you ever feel any feelings of guilt that his character was being called into question by the committee?"

Lenarczyk: "No, I don't remember."

Leonard: "You don't remember or you didn't feel guilty?"

Lenarczyk: "If I'm feeling guilty that – I couldn't understand your question. Can you – ."

Leonard: "Okay. Yeah. Did you ever feel guilty that Bishayee's reputation, okay, was being called into question by the committee based on what you and Dr. Hill – ?"

Lenarczyk: "Before or – ."

Leonard: " – after you and Dr. Hill went to the committee?"

Lenarczyk: "I don't know that."

Leonard: "You don't know if you felt guilty about it afterward?"

Lenarczyk: "If I feel guilty about him or if he was guilty?"

Leonard: "No. Do you feel guilty about having reported him, because of his reputation and people were looking at him?"

Lenarczyk: "I don't think so. I mean, why are you supposed to be guilty?"

Leonard: "No."

Lenarczyk: "Am I supposed to feel – I mean – ."

Leonard: "Do you and Bishayee have a good working relationship?"

Lenarczyk: "Up to the point when I asked him and have no question, I mean, yes, we were kind of friends, normal, like the normal lab environment, but he was not very helpful every time when I asked him something."

Leonard: "Did you ever have any arguments?"

Lenarczyk: "No."

Leonard: "Okay. Eventually you learned that the committee concluded that there was insufficient credible evidence of misconduct in science to warrant any further investigation?"

Lenarczyk: "I think that I got similar letter like that kind of statement."

Leonard: "Okay. Did you agree or disagree with that finding?"

Lenarczyk: "Nobody asked me about this."

Leonard: "I'm asking you. I should say do you support a finding that Dr. Bishayee committed scientific misconduct? I think that was the actual finding."

Lenarczyk: "I don't like to discuss this comment. I don't know. This is the comment by the committee, not by myself. So, you are asking me to judge the committee, what the committee – that's not my – I think that's not my responsibility."

Leonard: "Were you upset with the committee's finding?"

Lenarczyk: No, I'm not. I mean, what committee found, that was found."

Leonard: "What was your relationship with Bishayee after that investigation?"

Lenarczyk: "My relationship with – ."

Leonard: "Bishayee after the investigation."

Lenarczyk: "What was my relationship with him?"

Leonard: "Mm-Mm."

Lenarczyk: "Like before. I mean, normal. We were working in the same lab."

Leonard: "Were things tense between you?"

Lenarczyk: "No. We even go for the conference. We shared the same room at that time, so it was no problem. I mean, we were not fighting each other, we were trying to work in the same lab. Okay?"

Leonard: "Okay. When was that conference?"

Lenarczyk: "It was in Puerto Rico."

Leonard: "Do you remember the year, by any chance?"

Lenarczyk: "I think it was supposed to be 2001, I believe. The conference was in March, I think, the beginning of the springtime. So, it must be 2001, because when I came it was 2000 and that was May. So, I think it was 2001."

Leonard: "Okay. Did Bishayee, while you were sharing a room, ever say to you, 'Marek, good work. What are doing?' or 'What are you trying to do to me? Why did you do this?'"

Lenarczyk: "No, we were not talking about this. We were trying to talk about that. I think it was not very good subject to talk about that between us. I mean, he – No, he didn't ask me. We were trying to keep a normal relationship when we were working together."

Leonard: "Did you like Bishayee?"

Lenarczyk: "Like everybody else, I like people, so why I don't like him. I mean – ."

Leonard: "But you don't dislike him, let's put it that way."

Lenarczyk: "No, I don't dislike him."

Leonard: "How was your relationship with Howell after the investigation?"

Lenarczyk: "I couldn't see very much difference from my side. Okay? I mean, I was still working in the lab doing the research. Maybe he was a little bit upset what was happening. I don't know. You would have to ask him about that."

Leonard: "Well, did he ever tell you he was upset?"

Lenarczyk: "I don't remember."

Leonard: "What was the atmosphere like in the lab after the investigation?"

Lenarczyk: "I think that not like before."

Leonard: "How do you mean?"

Lenarczyk: "Everybody knows that something is going around, which is not probably nice, because we are trying to do something which is not nice to see."

Leonard: "Okay. After the investigation was Dr. Howell ever hostile towards you?"

Lenarczyk: "What do you mean?"

Leonard: "Aggressive. Angry."

Lenarczyk: "No."

Leonard: "Okay. Did you ever witness him acting that way towards Dr. Hill?"

Lenarczyk: "No."

Leonard: "Did you ever witness Dr. Howell taking any retaliatory acts against Dr. Hill?"

Lenarczyk: "What is 'retaliatory'?"

Leonard: "Treat her poorly. Did you ever see them treat her differently or poorly after the investigation?"

Lenarczyk: "I would say no."

Leonard: "Okay. How about Dr. Baker or Dr. Azzam? Did they ever treat her differently or poorly after the investigation?

Lenarczyk: "Me?"

Leonard: "No."

Lenarczyk: "Her?"

Leonard: "Her."

Lenarczyk: "I barely see Dr. Baker. I mean, I have no contact with Dr. Baker, because I think that he was the chairman at that time of the department, and he was not working with us."

Leonard: "Okay."

Lenarczyk: "So, I really don't know him. Dr. Azzam? No, it was like before."

Leonard: "Okay. How was your relationship with Dr. Hill after the investigation?"

Lenarczyk: "I think that we had a good relationship, I mean, like before. There was nothing different."

Leonard: "Did Dr. Hill ever express any dissatisfaction with the committee's findings to you?"

Lenarczyk: "Dr. Hill express – Yes, she expressed that, because I think that that's the real reason why she went later on for the next step."

Leonard: "What did she say?"

Lenarczyk: "I don't remember exactly, but definitely she was not convinced or happy what was found."

Leonard: "And you're aware that Dr. Hill subsequently pursued claims with the Office of Research Integrity?"

Lenarczyk denied knowledge of ORI investigation and had no comment for defense counsel. He was never asked to participate as co-complainant in the ORI phase of Hill's allegation of misconduct against Bishayee. But Leonard handed him a document and it was case information pertaining to the ORI investigation.

Leonard: "I want to direct your attention to – from the top – "Inquiries and Investigations" – 1, 2, 3, 4, 5 – where it says "Complainant(s)."

Lenarczyk: "Okay."

Leonard: "You're listed as a complainant."

Lenarczyk: "Right."

Leonard: "Did you know at the time that you were listed as a complainant?"

Lenarczyk: "That is document which was prepared for that investigation which was done by the university or later on?"

Leonard: "Later."

Lenarczyk: "No, I don't think so that I saw that document."

Leonard: "This was dated August 27, 2002. Did Dr. Hill ask you if it was okay to list you as a complainant?"

Lenarczyk: "I don't remember that. But definitely I didn't prepare this document."

Leonard: "Did you see it before it was filed?"

Lenarczyk: "I don't think so."

Leonard: "We covered this previously that you were, in fact, not fired by Dr. Howell. Is that correct?"

Lenarczyk: "Yes."

Leonard: "Do you have any animosity toward Dr. Howell?"

Lenarczyk: "Animosity?"

Leonard: "Any bad feelings or anger?"

Lenarczyk: "To Dr. Howell?"

Leonard: "Yeah."

Lenarczyk: "No, I spoke to him a few months ago at the conference."

Leonard: "Did you ever tell Dr. Howell that you weren't happy with Dr. Hill's actions with regard to complaints?"

Lenarczyk: "I don't remember."

Leonard: "Do you remember sending him an e-mail to that effect?"

Lenarczyk: "No, I don't remember, sorry. I don't know."

Leonard: "We can come back to that. We'll go back to that when he comes back."

Lenarczyk: "Which doesn't mean that I sent it, it's simply I don't remember."

Leonard: "No. And we have an e-mail we'll show you. I understand."

Lenarczyk: "If you have an e-mail, that's probably what I did."

Leonard: "Dr. Lenarczyk, please understand nobody is questioning – [your honesty]."

After and exchange about Hill notifying Lenarczyk about the ORI inquiry and findings, Leonard brought him back to the document with his name at the top, listing him as the co-complainant.

Leonard: [...] "We talked about this previously where it says you're a co-complainant. Do you see that up on top?"

Lenarczyk: "Yes, I see that."

Leonard: "You understand this is for ORI now, this is not for the committee at UMDNJ."

Lenarczyk: "Okay. I understand that."

Leonard: "Okay. So, I just want to make sure you're clear on that."

Lenarczyk: "Yes, I'm clear about that."

Leonard: "Okay. So, did you assist with the filing of the complaint with ORI?"

Lenarczyk: "'Filing' means preparing the paper to send it to somewhere?"

Leonard: "Yes."

Lenarczyk: "I don't think so."

Leonard: "Okay. Did you assist Dr. Hill in pursuing the matter with ORI?"

Lenarczyk: "Means to encourage her to do that?"

Leonard: "No, to help her."

Lenarczyk: "I don't remember."

Leonard: "Do you remember anything about the ORI investigation?"

Lenarczyk: "Definitely the call from the FBI agent."

Leonard: "That got your attention?"

Lenarczyk: "Of course. I mean, you know, if somebody is calling you from the FBI – ."

Leonard: "Just so you understand, that has nothing to do with the ORI investigation. You got a call from the FBI because this complaint was filed in federal court."

Lenarczyk: "Oh, okay."

Leonard: "And the United States Attorney was trying to decide whether or not they were going to prosecute the case."

Lenarczyk: "Oh, I see."

Leonard spelled out the sequence of investigations for Lenarczyk, who continued to respond by repeating much of what Leonard had just told him. But fuzzy recollection melted once Leonard produced the e-mail that Lenarczyk wrote to Howell dated October 10, 2001, in which Lenarczyk informs Howell that he did not like what Hill was doing with the ORI complaint.

Leonard: "Dr. Lenarczyk, I'm going to hand you what has been marked Lenarczyk Exhibit 4. It's an e-mail from you to Dr. Howell [...]. Would you take a look at that, please."

[Discussion off the record.]

Leonard: "Do you recognize this e-mail?"

Lenarczyk: "Sure."

Leonard: Okay. Can you tell me the paragraph that starts with 'Lanie [Helene Hill] Story. Lanie story is really not very easy to understand.' What do you mean by that?"

Lenarczyk: "I think this is referring to any additional steps which was done after that case was investigated at UMDNJ, I think."

Leonard: "So by this sentence you're referring to the ORI and the additional complaints that were filed?"

Pincus: "Objected to the form of the question. I mean, just the timing of his e-mail just makes it an improper question. You may answer."

Leonard: "This is one you wrote, so you have to tell us what you meant."

Lenarczyk: "Yeah, but, you know, once again, few years ago. I don't remember now, really."

Leonard: "So, you don't know what you meant when you wrote 'Lanie story is really not very easy to understand?"

Lenarczyk: "No, I don't understand – I don't remember now what I was meaning by that."

Leonard: "How about the next sentence, 'I am not happy because all work has happened, too?'"

Lenarczyk: "That was because of that whole investigation, because – I was thinking that everything will be discussed first in the lab and then try to find out the solution for that and not to go – solution for that in the lab – I mean, inside, like you said, in the department of – ."

Leonard: "And not what? I'm sorry."

Lenarczyk: "And not to go so far with that kind of investigation."

Leonard: "Okay."

Lenarczyk: "Because I didn't feel that this is necessary to do. I mean, I think that if you have that kind of situation usually you're supposed to deal in the lab first. That was my – probably this referring to that."

Leonard: "Okay."

Lenarczyk: "But this is what I can make comments of now. Okay?"

Leonard: "Okay. How about the last sentence? 'I know that I am not an angel, too, but perhaps we should shake our hands and try to build the peace at least.' Peace between whom?"

Lenarczyk: "Just, you know, form of word, but it's not like – I was feeling that something is not right between us when I left when everything was found at that time. Definitely probably Roger was not very happy about that. Because who would be happy? I mean, it's not a nice situation for him. So, it's maybe kind of apology or something like that, you know. What has happened has happened, and I think that now we have to move on, and that's it. I mean, that's what my intention, I mean, to keep our relationship on a good, in a good way. I mean, not fight, not to – There is no reason why we have to. I mean, he might have that kind of feeling, but I don't know."

Leonard: "No, in the discussing the last two sentences does that refresh your recollection about the first sentence when you said, 'Lanie story is really not very easy to understand'? Do you now have a better idea of what you meant by that sentence?"

Lenarczyk: "I think that it might be like – I mean, maybe I'm off the time frame, because, you know, this is so far. But for somehow I probably felt that what was found at the committee by university supposed to be end of that story. That's probably referring to that. I'm not sure. Okay? I was rather surprised that everything is going so far again, and again, and again, because I was not having that kind of situation before, and I was not feeling that – okay, if you complain something, like committee can say, okay, there is no reason to find something which you complain about that, that's it. I mean, that's –

"For me it was not a very big problem. Okay? I mean, working in science you can get sometimes different data. Okay? This is the science. That's why we call it research, research, research. So, we are facing all the time we troubleshoot. Okay? Which means that we have to try to confirm something, try to disprove something, because this is just science. So, when something has happened like that – it's difficult to explain at the moment when you can see that, but if you have no reasonable – a reason how to explain that, which at least I had that situation before, because I tried to do my best. I mean, I tried to – then I decided, okay, it happened. Okay? I don't know what happened before I came to that lab. I know what is happening now when I'm working, and I have difficulties to confirm something which was published before, which we couldn't. Because that was not only my finding, that was also the finding all of us at the time. So, I said, 'Okay, fine.' That's not my project, in fact. I mean, I'm not really responsible for all the hypotheses which are in that project. I'm just a person who is trying to verify the hypothesis which was prepared by, at that time, Roger Howell.

"So, when I went back to call on them I tried to – I couldn't forget about everything, of course, because that was kind of a very, very rare situation which happened in the lab. So, maybe I was feeling that Roger is not so happy about that, which probably he was not so happy about that.

"As we discussed before, I was not fired by him. He didn't throw me away because of that investigation. I mean, I'd rather try not to work on that anymore, because it was very difficult to me. I mean, I don't like to argue with something which cannot become verified in a logic way, okay, at least for me. So, I quit the job. I went back, and then we were having that kind of conversation."

> *Leonard*: "Okay. Are you aware that after ORI determined that no further investigation should be conducted that Dr. Hill filed yet another complaint with the UMDNJ committee?"

Leonard explained the timing of the third complaint, the second to the UMDNJ committee. Lenarczyk was in Poland at the time Hill filed the second Newark CCRI complaint but told Leonard he didn't believe that he and Hill were talking at that time, for obvious reasons, "because of the distance. Maybe we shared some e-mails. I don't know," he told Leonard. "It's really too difficult to recollect everything for me now." Leonard pushed back. "By 'conversation' I mean communication, e-mails." Leonard wanted to know about a package of documents Hill mailed to Lenarczyk. The line of questioning was intended to get Lenarczyk on the record as assisting Hill in preparation of the complaints, most especially the second Newark CCRI inquiry.

> *Leonard*: "Do you remember any conversations or any communications between the two of you regarding her filing the second complaint with the UMNDJ committee?"

Lenarczyk: "I don't remember now."

Leonard: "How did you come to learn that she did, in fact, file a second complaint with the committee?"

Lenarczyk: "Maybe we had that kind of e-mail conversation. I don't remember now. I mean, I – ."

Leonard: "But you remember knowing at or about the time?"

Lenarczyk: "No, I don't remember. I mean – ."

Leonard: "Do you have any understanding of the basis for the new complaint?"

Lenarczyk: "If I don't remember – No. I don't remember that. I mean – ."

Leonard: "Would it refresh your recollection if I tell you that the second complaint is based on allegations that a statistical review of Dr. Bishayee's data suggest a lack of uniformity or randomness? Does that mean anything to you?"

Lenarczyk: "Sounds like kind of statistical evaluation of some data."

Leonard: "Right."

Lenarczyk: "But I don't recollect that I did that statistical evaluation."

Leonard: "Are you a statistician?"

Lenarczyk: "No, I'm not."

Leonard: "Do you have any statistic training?"

Lenarczyk: "Formally, no, but I use statistics for evaluating the data."

Leonard: "And you've never looked or attempted to do any statistical analysis of any data relating to this case?"

Lenarczyk: "To which case? To – ."

Leonard: "To the case that – to the complaint of Dr. Hill to Dr. Bishayee's work."

Pincus: Objection to the form [of the question]. Are you referring to the federal complaint? In other words, are you referring to – ."

Leonard: "Dr. Bishayee's data."

Pincus: "Yeah, but you used the words 'the complaint.' That's where we're asking for clarifications."

Leonard: "Have you ever employed any statistical analysis for use on Dr. Bishayee's data?"

Lenarczyk: "Did I do a statistical analysis by myself of his data?"

Leonard: "Yes."

Lenarczyk: "I don't think so."

Leonard: "Okay. Did Dr. Hill ever discuss any statistical findings that she performed with you?"

Lenarczyk: "It might be in those documents which I gave you. I don't remember exactly, but there might be something in that letter which I gave you, because I believe that there is something about the evaluation of data."

Leonard: "Which you didn't read that?"

Lenarczyk: "I mean, I read that, but if you will ask me what is precisely in particular document, I don't remember now, because I read like, you know – ."

Leonard: "Well, if I show you that document now would you understand what that document purports to show?"

Lenarczyk: "No necessarily everything. I mean, if there is some, let's say, statistical analysis then I'm not responsible for judging this one, because I have not that kind of knowledge of statistics. Okay?"

Leonard: "Good enough. Did there come a point in time where you learned that UMDNJ, once again, found inconsistent credible evidence of misconduct in science to warrant further investigation of Dr. Hill's claims against Dr. Bishayee?"

Lenarczyk: "I don't remember that."

Leonard: "Do you remember speaking to Dr. Hill about the fact that, once again, the committee found no reason to go forward with the investigation?"

Lenarczyk: "It may have happened that she informed me about that, but, you know, I really don't remember that."

Leonard: "You have no specific recollection of her telling you that?"

Lenarczyk: "I don't have very specific – I mean, we were not talking like, you know, each day or each week. So, when we were sharing couple e-mails, or sometimes we talk to each other, that was very, very rare. So, I really don't remember everything, you know."

Leonard: "Do you recall her expressing any anger or being upset about the fact that now three different bodies have come back and found insufficient evidence to proceed?"

Lenarczyk: "I don't remember."

Leonard: "Do you remember that complaint that I showed you earlier today?"

Lenarczyk: "Which one? This one?"

Leonard: "No, the big one underneath it."

Lenarczyk: "Yes, I remember."

Leonard: "Did you know at or about the time that Dr. Hill filed that in October of 2003?"

Lenarczyk: "Did I know about that?"

Leonard: "Mm-Mm."

Lenarczyk: "I don't remember. But it might be possible that she informed me about that just to say that she's going for something. I don't remember."

Leonard: "You know, I don't want to be flip with you, but I just want to be sure. You don't remember or you're not sure that she told you about the second ORI filing [the second inquiry of the three prior to the federal complaint] – ."

Lenarczyk: "I don't remember. Okay? It would be – ."

Leonard: "Just let me finish the question. The second ORI filing, the third committee filing, and now the complaint filing. You have some understanding that all these things took place, but you don't have any specific recollection of her telling you about them?"

Lenarczyk: "I don't remember exactly what was happened, but I think I know what she was doing."

Leonard: "How?"

Lenarczyk: "Maybe she send me some e-mails."

Leonard: "Did you ever speak by phone?"

Lenarczyk: "I don't remember. Sorry."

Leonard: "You don't remember how you talked to her?"

Lenarczyk: "No, I don't remember."

Leonard: "Did you ever call her or she call you while you were in Poland?"

Lenarczyk: "In Poland?"

Leonard: "Mm-Mm."

Lenarczyk: "No, I don't remember. I mean, I would be surprised if I would call her, because of the distance."

Leonard: "Because of what? I'm sorry."

Lenarczyk: "Because of the distance and time and everything."

Leonard: "You spoke earlier about speaking to the FBI; is that correct?"

Lenarczyk: "Yes."

Leonard: "Do you remember telling the person who interviewed you on April 21, 2004 – ."

Lenarczyk: "I don't remember the name. That was she, I think."

Leonard: "Yeah, Susan Schlow."

Lenarczyk: "I don't remember the name, but definitely that was she."

Leonard: "I will represent to you that that's who it was. Okay."

Lenarczyk: "Okay."

Leonard: "Do you remember telling that special agent that you thought Dr. Hill may have acted too quickly?"

Lenarczyk: "That Dr. Hill what?"

Leonard: "Acted too quickly. Do you remember saying you were not sure why Dr. Hill reported Dr. Bishayee to the university so quickly?"

Lenarczyk: "I don't remember that, but maybe I said that. I mean – ."

Leonard: "Okay. I'm going to show you some documents."

[These were marked as Lenarczyk Exhibit 5.]

Leonard: "Dr. Lenarczyk, I'm going to hand you a document that's entitled Lenarczyk, that's been marked Lenarczyk Exhibit 5. the document is entitled 'I Am A Whistle Blower' by Dr. Hill. The documents are Bates numbered beginning 000730 through 734 consecutively. I say that because it's not stapled together. Would you take a look at that document, please. Just take a minute and read that opening paragraph. Or actually read the whole thing if you'd like. Take a few minutes. It's not that long."

Lenarczyk: "Okay."

Leonard: "Have you ever seen this document before?"

Lenarczyk: "I don't know if something is about the whistle blower is that letter."

Leonard: "I'm sorry, you have to keep your voice up."

Lenarczyk: "Something might be – because I really don't know what does it mean 'whistle blower,' but I think that something like that was in this letter. You may check those documents which I gave you."

Pincus, Hill' attorney, interjected that Lenarczyk had identified the document by Hill, an eight-page statement that was sent to Lenarczyk as part of a packet of case material that he brought to the deposition in response to the defense's subpoena. All documents in Lenarczyk's packet, received by mail from Hill, were made part of the official deposition record. But they weren't mailed to him in Poland; they had been sent by Hill to Lenarczyk to his Allis, Wisconsin home address, an apartment on South 102nd Street. There was no cover sheet among the documents Lenarczyk shared with Leonard and Flynn but a photocopy of the outer envelope was included in the material Lenarczyk laid out on the table and it, too, was entered as evidence. Among the documents Leonard sorted out on the table and labeled as exhibits were the e-mail exchange between Hill and Lenarczyk and a document she had sent him titled "Summary of *Qui Tam* Case Against the UMDNJ Proposed by Helene Z. Hill, Ph.D." Also included was a document that was eight, nine pages long, front and back, titled "An Analysis of the Findings in Box Number 6," a slide show presentation. Leonard produced a series of additional exhibits, to include a two-page document titled "The Effect of Tritiated Thymidine on the Cell Cycle of Chinese Hamster V79 Cells and its Implications in Experiments Done in the Laboratory of Roger Howell, Ph.D., 1998-2001"; a letter from Dr. Hill to Susan Steele of the United States Attorney's Office dated November 16, 2006; a two-page document, front and back, titled "The Impossibility of an Exponential Decline in Survival of Chinese Hamster Cells in the Presence of Tritiated Thymidine"; Hill's whistleblower statement, and "Analysis of Coulter Counter Counts by Bishayee and Hill." With the documents properly labeled with sequential exhibit markings, Leonard proceeded to ask Lenarczyk pointed questions.

> *Leonard*: "We just want the record to reflect that we have now marked all the documents that Dr. Lenarczyk brought with him [most of which he claimed to have no recollection of] in compliance with the subpoena, and I've returned to him the originals of those documents."
>
> *Leonard*: Dr. Lenarczyk, previously I gave you a document entitled "I Am A Whistle Blower." Have you had a chance to look through that document?"

Lenarczyk: "I have not read that document. Do you like me to read this document now? That's essentially probably the same document which is – ."

Leonard: "It's not."

Lenarczyk: "It's not?"

Leonard: "Why don't you just glance through this one."

Lenarczyk: "Okay."

Leonard: "Now, if you would turn over the second page where it says 'Fabrication of Results in 2001.'"

Lenarczyk: "Right."

Leonard: "Okay. It says, 'In the fall of 2000, Howell hired Dr. Marek Lenarczyk, a postdoctoral fellow, to work on his research that was supported by an NIH grant. Howell told Lenarczyk to work with Bishayee. Bishayee was very elusive and Lenarczyk had great difficulty pinning him down. Lenarczyk became suspicious of Bishayee and frequently voice his concerns to me [to Dr. Hill.]' She's the author of this. Is that true, you frequently voiced concerns to Dr. Hill?"

Lenarczyk: "I mean, we were talking about this one, because I was feeling that something is not right, because I was finding completely different results from the one that was published by Dr. Bishayee before I came [the 1999 experiment]. So, obviously, I would like to share something and try to find out what is the reason why now I'm having different results, when before was made by another person having the same biological model that I was using. It was just a clearly scientific discussion about the results."

Leonard: "The next sentence says, 'When he' – meaning you – 'complained to Dr. Howell, Howell brushed him off.' Is that true?"

Lenarczyk: "I don't remember that. Okay?"

Leonard: "I thought your testimony previously was you didn't have any discussion with Dr. Howell."

Lenarczyk: "And I believe that this is rather true, because if I would have that discussion with Dr. Howell I probably would not go for any other further steps. So, I believe everything would be decided in the lab."

Leonard: "Okay."

Lenarczyk: "I mean, that's my assumption, but I'm not sure about that."

Leonard: "Do you have any recollection of Dr. Howell ever brushing you off?"

Lenarczyk: "No, he didn't brush me off."

Leonard: "Do you have any recollection of ever bringing any complaints to Dr. Howell?"

Lenarczyk: "Complaints about the results which we are having or – ."

Leonard: "About Bishayee. It says, 'When he complained to Howell, Howell brushed him off.'"

Lenarczyk: "I don't remember that. I mean, I definitely didn't feel that I was brushed off by him. Okay? We were fully talking to each other. I mean, it was not a problem to talk to Roger about anything that was related to the science or the results. I don't remember that he brushed me off."

Leonard: "So, this statement is not an accurate statement?"

Pincus: "Objection to the form of the question. You may answer."

Lenarczyk: "This is not my statement. Okay?"

Leonard: "No. No."

Lenarczyk: "I don't feel particularly that I was brushed off."

Leonard: "Okay. You have to understand I know you feel like sometimes I'm – Nobody is accusing you of anything. We know you didn't write this, but I've got to know. This is what somebody else is writing about something that you supposedly told them. If that's not the case, then we need to know that."

Lenarczyk: "I don't remember that I was brushed off."

Leonard: "Do you remember ever complaining to Dr. Howell?"

Lenarczyk: "No, I don't remember."

Leonard: "Okay."

Lenarczyk: "I'm sorry."

Leonard: "That's okay. Later down in that paragraph it says, 'We agreed not to inform Howell' – 'we' being you and Dr. Hill – ' to inform Howell, because we both were concerned that Howell would protect Bishayee and terminate the experiment, thus destroying any evidence of malfeasance.' Do you believe that to be true?"

Lenarczyk: "That they will destroy the experiment to – ."

Leonard: "That Dr. Howell would destroy any evidence. If you went to Dr. Howell, that he would destroy the evidence of wrongdoing, of any wrongdoing."

Lenarczyk: "I have no reason to believe that. Okay? Because I never had that situation with Roger Howell before. So, I don't know if he would do that. I mean, I don't – I don't feel particularly that he would terminate the experiment."

Leonard: "Okay."

Lenarczyk: "Especially as everybody knows at the time that something is going on. I mean – ."

Leonard: "This is somebody writing something that they say you felt or you told them, and I just want to be sure, because that's a very – that's a very strong statement to say somebody would destroy evidence. And if you believe that, fine, say that, but if that's not what you believed you need to say that as well."

Lenarczyk: "I have no reason to believe that. I mean – But this is statement which I can make now. Okay? I mean, I don't believe that Roger will – If I will work with him, okay, I don't think so that he will ask me to discontinue the experiment if something is not really right. And I wouldn't like to do that anyway, because experiment is just experiment. If you can get the data, you have data. You have to try to analyze the data, even if the data are not as you expect. Okay?"

Leonard: "Okay. Understood. The last sentence says, 'Lenarczyk's and my observations support, strongly support our belief that Bishayee fabricated the results here of a second experiment.' And I think we spoke earlier, about you have no – Do you believe he fabricated the results? Do you have any personal knowledge?"

Lenarczyk: "He fabricated?"

Leonard: "Yes. Do you have any personal knowledge that he did that?"

Lenarczyk: "I mean, I couldn't caught him. I mean, I didn't see that he fabricated. I mean, I just had difficulty to find out a logical explanation why we cannot – why I'm doing the same and I cannot confirm something which is already done."

Leonard: "Okay. Could you turn to the last page where it says, 'On April 6.'"

[Leonard has to point out the passage he wants Lenarczyk to see.]

Lenarczyk: "Okay. Which part? Sorry."

Leonard: "The second paragraph from the bottom it says, 'In this *qui tam* case, the stakes are very high. If the defendants are found guilty' – that being Roger Howell, Dr. Bishayee and the university – 'the penalty is three times the amount of the fraud, in this case, $4.2 million. The whistleblower – that would be Dr. Hill – 'generally receives thirty percent of that, in this case more than $1 million.' Do you understand that in this litigation she's seeking to get a judgment against Dr. Howell, Dr. Bishayee and then university for over $4 million?"

Lenarczyk: "Did I understand?"

Leonard: "Do you understand that?"

Lenarczyk: "Now?"

Leonard: "Yes."

Lenarczyk: "I understand, because it says that here. So, it's clear to me now."

Leonard: "Do you agree with her position that they should be held liable for in excess of $4 million?"

Pincus: "Objection. Calls for a legal conclusion. You may answer."

Leonard: "You can answer."

Lenarczyk: "I don't know why university supposed to be losing that $4.2 million by this case."

Leonard: "How about Dr. Howell? Do you think he should have to pay $4.2 million?"

Pincus: "Same objection."

Lenarczyk: "I don't understand that law. But, you know, I don't feel that this is what will happen. I mean, I don't know. Really, I don't know. I don't like to make a comment of that, because this is something which is coming out of that as a consequence of the law. Okay? So, I don't know. I definitely didn't know at the time that might be the consequence. Okay?"

Leonard: "The paragraph above says, 'In order to present the strongest possible case, I will need expert witnesses in the following areas: statistics, cell cycle, hypoxic effects and radiation studies.' Are you an expert in any of those areas?"

Leonard, again, has to direct Lenarczyk to the paragraph's last sentence "On April 6..."

> *Lenarczyk*: "Okay. Not really. I'm definitely not an expert in statistics."
>
> *Leonard*: "Cell cycle?"
>
> *Lenarczyk*: "I didn't work with hypoxic effects."
>
> *Leonard*: "Okay."
>
> *Lenarczyk*: "Cell cycle for some extent, yes, but not forever. I mean, this is broader meaning of the science. Okay. I don't know what that means, cell cycle. I mean, I understand the cell cycle, but – ."
>
> *Leonard*: "What area in cell cycle to you feel that you have expertise?"
>
> *Lenarczyk*: "I would say I'm not expert in cell cycle. Okay? That's a better conclusion."
>
> *Leonard*: "Okay. And you've never served as an expert in any case or legal proceeding?"
>
> *Lenarczyk*: "No, I never."
>
> *Leonard*: "Okay. Alright. I would like to get this material marked as Lenarczyk 7. Before I get to this, going back to the document 'I Am A Whistle Blower,' and the fact that your name appears throughout this document, would it surprise you to know that this document was given to Dr. Hall in the Radiology Department at Columbia University?"
>
> *Lenarczyk*: "Would that surprise me that was given to Dr. Howell?"
>
> *Leonard*: "Yeah – No, not Howell, Hall and folks at Columbia University."
>
> *Lenarczyk*: "I didn't know that. I mean, how can I know that?"
>
> *Leonard*: "No, I'm saying it was. I'm representing to you that it was. Did you know that she intended, Dr. Hill intended to give this out to other experts in the field at other universities?"
>
> *Lenarczyk*: "Maybe that was the logical way, because maybe she was looking for some expertise there. Columbia is not a bad place to look for that kind of expertise. They are people working in radiobiology."
>
> *Leonard*: "Do you have any feelings about the fact that all this information about you is in there?"
>
> *Lenarczyk*: "I don't like my name anywhere in this document."

Leonard: "That's what I'm asking you."

Lenarczyk: "But, you know – ."

Leonard: "I mean, there's some pretty strong accusations in there."

Lenarczyk: "I mean, it's written by Dr. Hill. So, what can I say?"

Leonard: "Do you think she should have verified this information with you before she circulated this document?"

Lenarczyk: "At least I would do something like that. I mean – ."

Leonard: "Okay."

Lenarczyk: "I will not write down on behalf of somebody not showing him what I'm writing down, because it's like – The consequence of this document is that this is the document. Okay? This, I would say, made by blah, blah, blah, or something like that. I would expect that some people who are in the document will know at least at the time the document is prepared and what is the intention."

Leonard: "I'm going to show you, Dr. Lenarczyk, Exhibit 7. It's a document entitled "Attachment 1b." It's been marked Lenarczyk 7. Would you take a look at that. Oh, I'm sorry. The Bates range – No, no, I apologize – are 000065 through 66. Have you ever seen this document?"

Lenarczyk: "I don't remember now."

Leonard: "Can you turn to the second page."

Lenarczyk: "Yes."

Leonard: "And about four sentences down where it says, 'It is my hope.' Do you see that?"

Lenarczyk: "Down from the top or – ."

Leonard: "No, I apologize, in the third paragraph it starts, 'On Monday.'"

Lenarczyk: "Oh, third. One, two, three, four down."

Leonard: "To the right it says, 'It was my hope.'"

Lenarczyk: "Okay. Yeah, I see that."

Leonard: "It says, 'It was my hope that by careful documentation I would be able to present a convincing case to Dr. Howell regarding Dr. Bishayee's incompetence and I could then persuade Dr. Howell to terminate his appointment.' Do you see that?"

Lenarczyk: "Yes, I see that."

Leonard: "Did you know at the time that that was Dr. Hill's intention to get Dr. Bishayee fired?"

Lenarczyk: "I don't remember that. That's a very strong argument. I don't remember that."

Leonard: "She wrote this. This is her statement. This is her position. The only question is, did she share her feeling on this with you?"

Lenarczyk: "I don't remember that. Okay?"

Leonard: "Okay."

Leonard had another document, titled "Scientific Misconduct at the New Jersey Medical School," marked as Lenarczyk Exhibit 8. He asked him to take a look at it. Had he seen it before? The document had Lenarczyk's name all over it and it had been circulated. Leonard called Lenarczyk's attention to the page that started "The committee's report and its aftermath."

Leonard: "The second paragraph it talks about how Dr. Howell told Dr. Hill that both Dr. Bishayee and yourself 'would be leaving at the end of the month.' Dr. Hill states she knew that you wanted very much to stay, and that Dr. Howell didn't permit that."

Lenarczyk: "I never asked him to stay. I asked him to agree to leave, terminate my position. I don't remember that he asked me to leave."

Leonard: "Okay. So, at the end of the paragraph, the sentence that says, 'However, Dr. Howell dug in his heels and Dr. Lenarczyk returned to Poland on July 26.' That's not your recollection of how your departure transpired?"

Lenarczyk: "I don't think so, because I wrote the letter to Dr. Howell to ask him to agree to terminate my position."

Leonard: "Okay."

Lenarczyk: "Even one month before I have that contract at the time. And I explained to him why. One of the reasons was because I had a ticket already, a return ticket. I never heard from him, as I recollect now, that he would like to fire me. I mean, he didn't tell me that."

Leonard: "The last paragraph on that page, the next-to-the-last-sentence says, 'Dr. Howell likes to play around with fitting data into mathematical models.'"

Lenarczyk: "Which is that? Last paragraph of the same page?"

Leonard had some difficulty getting Lenarczyk to the proper passage but even harder was getting him to answer the question. "I know what was the mathematical model at the time, probably describing that reason, that experiment," Lenarczyk answered, "but I don't know if Roger likes to play around. I don't know." Then Leonard produced Lenarczyk Exhibit 9, Hill's document "Answers to Defendants' Initial Interrogatories." Under Interrogatory Number 2, "Indentify any and all persons who have knowledge of any facts that are relevant to this matter, and summarize the facts of which each such person has knowledge," Lenarczyk's name is listed. Leonard directed him to the last two sentences.

> *Leonard*: "'Dr. Lenarczyk observed defendant Bishayee in March of 2001 when cultures were found to be contaminated; and complied with Bishayee's request to provide clean cultures.' Okay? 'Dr. Lenarczyk observed that Dr. Bishayee left the original samples in the incubator, and thus must have substituted other cells for the contaminated ones.' Do you have any personal knowledge that Dr. Bishayee substituted other cells for contaminated cells?"
>
> *Lenarczyk*: "No, I don't have that knowledge."
>
> *Leonard*: "Okay. With respect to that response, did either Dr. Hill or her attorney call you to confirm your knowledge before sending that out, do you know?"
>
> *Lenarczyk*: "This document?"
>
> *Leonard*: "Mm-Mm."
>
> *Lenarczyk*: "No, I don't think so."

Leonard then had Flynn mark Lenarczyk Exhibit 10, Hill's PowerPoint presentation titled "Evidence Supporting Allegations of Fraud at the New Jersey Medical School." He directed Lenarczyk to the section pertaining to what Hill had labeled "The Second Suspicious Experiment: Spring, 2001." The last dotted line read that Hill and Lenarczyk shadowed Bishayee until the experiment was completed. "Can you tell me," asked Leonard, "your understanding of what that means?"

> *Lenarczyk*: "I don't know what does it mean. Maybe that means that we were waiting for the time when the experiment is finished.

Is that the meaning of 'shadow'? I remember what 'shadow' means when my kids were shadowing somebody, as to follow somebody, to observe what these people are doing, but that was part of the educational process of some class. But I don't know what 'shadow' – I mean 'shadow', I know what is shadow by the meaning of 'shadow.'"

Leonard: "Well, is it possible she meant that the two of you were following and observing Dr. Bishayee?"

Lenarczyk: "Not like that. I mean, I was not following him."

Leonard asked Lenarczyk about "Hill and Lenarczyk's Observations," part of the presentation. He wanted Lenarczyk to read those and tell him which of them he had observed. Lenarczyk replied, again, that he didn't remember, claiming that it was "too long a time ago." Leonard pressed. "Do you remember doing any of these things?" He admitted only to the possibility he provided Bishayee with fresh cells. Leonard took Lenarczyk to another statement from the same document in which it read "Hill and Lenarczyk conclude that Bishayee mocked up the FACS separation using new cells he received from Lenarczyk." He wanted to know if Lenarczyk had any firsthand knowledge that Bishayee had used the cells Lenarczyk had given him to do that – to mock up the FACS separation. "I don't know what he did with the cells," Lenarczyk told Leonard, "even if I gave him that cell, because we were not talking about that." Turning to the next page of "The Second Suspicious Experiment," Leonard asked Lenarczyk to read down to the line "Hill and Lenarczyk conclude." "I want you to tell me which, if any, of these conclusions did you personally draw?" Lenarczyk denied drawing any part of the conclusions spelled out on the page because "that's not my document." When asked by Leonard if any of Lenarczyk's conclusions about Bishayee's experiment, he answered "I don't remember." After Leonard finished, Hill's counsel, Sheldon Pincus began his questioning of Lenarczyk. Shown documentation to the contrary, many of Lenarczyk's statements to Leonard fell apart.

Lenarczyk's first exchange with Pincus was directed at a nickname Lenarczyk had for Bishayee. He called him "Mr. Maharaja."

Pincus: "Do you ever recall referring to Dr. Bishayee as Mr. Maharaja?"

Lenarczyk: "I referred him to as a maharaja? I don't know. Maybe I did something like that. I mean, I don't remember, I mean – ."

Pincus: Can you recall ever making such a statement or a reference to Dr. Bishayee to Dr. Hill?"

Lenarczyk: "It might be possible. I don't know. It's like that was my recollection because – ."

Pincus: "What was the basis for you referring to him as Mr. Maharaja?"

Lenarczyk: "I think that maharaja is something very related to Indian people, like the history of."

Pincus: "To the extent that you utilized the phrase Mr. Maharaja in reference to Dr. Bishayee, what is your recollection in terms of what you were referring to Dr. Bishayee about?"

Lenarczyk: "Nothing spectacular. I mean, just, you know, different name. People they have kind of nickname or something like – it's like that. I mean, somebody can call me gray-haired guy or something like that. I don't remember that there was any particular meaning of that."

Pincus: "If Dr. Hill were to testify that you had made such a statement to her on occasion, or more than one occasion, do you have any reason to doubt the fact that you had done so?"

Lenarczyk: "Statement about what?"

Pincus: "Referring to Dr. Bishayee as Mr. Maharaja."

Lenarczyk: "If I have any reason to do that?"

Pincus: "Would you have any reason to doubt the accuracy of Dr. Hill's recollection that you had referred to Dr. Bishayee as Mr. Maharaja?"

Lenarczyk: "I don't understand that."

Pincus: "You're doing exactly what I wanted you to do. If Dr. Hill were to testify under oath that, in fact, you had on an occasion or more than one occasion referred to Dr. Bishayee as, quote, 'Mr. Maharaja,' unquote, would you have any basis to dispute that?"

Lenarczyk: "No, I don't think so."

Pincus asked Lenarczyk about his return to work in Howell's laboratory in 2000 and showed him a document labeled "Lenarczyk 11." He asked if Lenarczyk was familiar with it. "I think this is the paper where they published this one before I came to the lab," Lenarczyk responded. "This is a paper that was published in a periodical called *Radiation Research*, Volume 152, pages 88 to 97 in 1999, and it's by Dr. Bishayee, a Dr. Rao, and Dr. Howard," Pincus told him, "and you were familiar with the contents of this paper at the time you commenced – [Leonard objected to the form of the question and Pincus replied] Let me finish the question, please." Lenarczyk replied, "With the contents of that when I came to the university?" Pincus: "Yes." Lenarczyk answered that he was not familiar with the paper at that time.

Pincus: "Did you have occasion to review the paper when you did come to the university?"

Lenarczyk: "I read that paper."

Pincus: "You did read it?"

Lenarczyk: "Yes. I mean, that was part of my job to find out what this was."

Pincus: "So you were familiar with the contents and the results which Dr. Bishayee had reported. Is that correct?"

Lenarczyk: "Yes."

Pincus: "Okay. Thank you. I'll take that. You can just lay it on the side for the moment in case we have to go back to it." Pincus then had another exhibit – number 12 – marked.

Pincus: "Dr. Lenarczyk, I show you what I've marked as Lenarczyk 12. This is another article that was published in *Radiation Research*, Volume 155, in 2001, I believe. Are you familiar with this document?"

Lenarczyk: "I was familiar more at the time when I was there probably, because I read it."

Pincus: "Do you recognize it?"

Lenarczyk: "Yes. Yes."

Pincus: "And, to your knowledge, and to your recollection, had this document been accepted for publishing at the time you arrived back in Dr. Howell's lab in 2000, or was it subsequent, do you know?"

Lenarczyk: "I don't remember that."

Pincus: "Fair enough."

Lenarczyk: "This is not my paper, so I didn't know about it. I don't remember that."

Pincus: "But you do recall reviewing it as part of your job?"

Lenarczyk: "Yeah, because as part of my job I'm supposed to read some papers, especially this one which came from the lab when I was working, and they are related for whatever I was doing at the time."

Pincus: "So, is it fair for me to say that you were familiar with its contents and the results which the authors, including Dr. Bishayee, had reported?"

Lenarczyk: "Yes, I will say that I was familiar."

Pincus: "Okay. You can lay that one on the side for a moment." Pincus then had the recorder mark an exhibit number 13.

Pincus: "I show you Lenarczyk 13. Are you familiar with this document?"

Lenarczyk: "I don't remember, but probably not, because this looks like this was – ."

Pincus: "Was it your understanding that it was this grant that served to pay for your fellowship commencing in about 2000?

Lenarczyk: "I understand that my salary was paid out of this grant." [A clear answer.]

Pincus: "So your understanding is that your salary was paid out of this grant?"

Lenarczyk: "I understand that Dr. Howell has to find the money to pay my salary anyway, and I believe that that was out of this grant, because this is the grant which is referring to."

Pincus: "Were you familiar with the, or did Dr. Howell describe to you, the preliminary results that had been presented in support of this grant at the time you started reworking in his lab?"

Lenarczyk: "I don't remember that."

Pincus: "So, sitting here today do you have any recollection of reviewing this grant, or any of the specific aims that would guide your research while working in Dr. Howell's lab?"

Lenarczyk: "I don't remember that. I mean – ."

Pincus: "Did you have any oral discussions with Dr. Howell in regards to the specific aims that would guide your research?"

Lenarczyk: "I believe that we were talking about what I was supposed to do, okay, because that was part of our job. I mean, I couldn't believe that I was asked to do something without discussing with Dr. Howell. But I don't remember now exactly if that was happened, and if so, when. Okay?"

Pincus: "In general, however – ."

Lenarczyk: "In general usually the boss when you are hired is showing you or explaining to you what you will be involved in working on the project. So, I believe that we were talking about that. I mean – ."

Pincus: "Based on the custom and practice that existed in – Let me ask the whole question. Based on the custom and practice that existed in the laboratory at that time – [Leonard continued to object to the question] is it fair to say that Dr. Howell explained to you what the nature of your research was about?" [Leonard objected again.]

Pincus: "You may answer."

Lenarczyk: "I will assume that."

Pincus: "Okay."

Lenarczyk: "If I will be in the same position as Roger Howell, if I would be the boss, and I would hire somebody to work on my grant, definitely I would do that. Because otherwise how is the person who is hired going to be know what's supposed to be done." Pincus then marked exhibit number 14.

Pincus: "Going to show you what's been marked as Lenarczyk 14. This appears to be an experiment that was performed – ."

Lenarczyk: "By me."

Pincus: " – on October 2, 2000. Are you telling me that this was an experiment performed by you?"

Lenarczyk: "I believe, because it's like my hand notes on the side of this document, I think. I mean – ."

Pincus: "Let's just look at the documents themselves."

Lenarczyk: "Probably nobody will mark 13:40 as a time here, except somebody who is coming from Europe."

Pincus: "Alright. But if you would look at the documents which comprise this packet that has been marked as Lenarczyk 14, to your knowledge, does this represent the documents representing the protocol for the experiment and the results that you obtained as a result of performing the experiment?"

Lenarczyk: "Yes."

Pincus: "Would you be kind enough just to – [Leonard objected]. Would you be kind enough to just walk me through the various sections. So, for instance, if we look at the pages, the first page that you have before you that have the Bates stamps of B019439 and B019440, can you identify what – ."

Lenarczyk: "What is what?"

Pincus: " – that is, please. Specifically, is that what we refer to as the protocol?"

Lenarczyk: "Yes, I think that this is on the first page on this, whatever 19439, is exactly the protocol you have to follow to do that experiment."

Pincus: "Okay."

Lenarczyk: "This is the table showing here what is the radioactivity you can prepare."

Pincus: "Let's continue on. Bear with me, please. The second page of the exhibit, which has Bates stamps B019441 and B019442, what is the second page? Is this also the protocol?"

Lenarczyk: "Yes, I think that the first part is protocol up to the -42, and then there is my additional handwritten notes about what I did for mutagenicity to follow all this protocol."

Pincus: "The part that's B019442, what is that?"

Lenarczyk: "This one? I think that this comes from Coulter counter, I believe. It's an instrument which is counting the radioactivity. That's my guess. Okay?"

Pincus: "Okay. But to your knowledge, is that part of the experiment that you commenced on the second of October?"

Lenarczyk: "I believe so."

Pincus: "Continue on to the third page of the exhibit, the pages that are marked B019443 and B019444. Do you see where I'm referring to at the bottom of the page? No, no, no."

Lenarczyk: "Okay."

Pincus: "I'm giving you a Bates stamp reference. So, if I say third page, but I'm just calling your attention – ."

Lenarczyk: "This is like two pages in one?"

Pincus: "That's correct. So what are these two pages, if you would?"

Lenarczyk: "This page is showing here the result you can expect from probably experiment with mutagenicity for HPRT gene, which is showing here probably the number of colony – no, first one is – first one is showing you what is the activity of medium you are measuring in the small tubes. And the second page, which is Table 3, this is also about the radioactivity."

Pincus: "The Coulter counts?"

Lenarczyk: "Right."

Pincus: "Now, if you go to the fourth page. And, again, I'm referring you to Bates stamps B019445 and B019446. What does the right-hand portion of this page represent, please?"

Lenarczyk: "Right page?"

Pincus: "No. I'm sorry, the left."

Lenarczyk: "The left one?"

Pincus: "The left one."

Lenarczyk: "The left one is a readout from the Coulter counter, which is the instrument to measure the radioactivity, I believe." […]

Pincus: "And the right-hand, Table 2, what is that, please?"

Lenarczyk: "This, again, something which if referring to the radio-activity of probably cell suspension, because is this marked on the table, which is essentially measuring the radioactivity. It says, 'Suspension,' I think."

Pincus: "Are you familiar with what a scintillation counter was?"

Lenarczyk: "Yes, we were using this one for measuring the radioactivity."

Pincus: "If you would go to the next page, the Table 4 – ."

Lenarczyk: "Okay."

Pincus: " – on B019447. What was that, please?"

Lenarczyk: "It's probably data out of survival experiment, which is the number of colonies these particular cells, we were using to flag them to check what are the survival of the cells.

Pincus: "And the page to the right, B019448?"

Lenarczyk: "I think that this page of my notebook exactly the same which is on the left."

Pincus: "Okay."

Lenarczyk: "That's my guess. Yes, that's probably like that."

Pincus: "I want you to turn the page, please, and you'll see that it has a different Bates stamp, 00934. Are you familiar with – this indicates that, or appears to relate to the experiment that we're talking about, October 2, 2000. It indicates that you're the investigator. Can you identify this?"

Lenarczyk: "That was the program which we designed, I think, under Excel to – I don't remember what was done. This is the record of something which is related to the radionuclide measurement, or the activity measurement, I think."

Pincus: "And the following pages, is that all part of the same program that you were just referring to?"

Lenarczyk: "I believe so. I mean, I think that this is something which is showing some radioactivity measured by the Coulter counter for this particular experiment."

Pincus: "And if you would go to the pages 938, 939, and these charts. What is that, if you would, please."

Lenarczyk: "This is a graphical explanation of the results which you can get from this experiment in terms of what is your survival of the cells versus radioactivity you measure for these particular cells."

Pincus: "Is it your recollection that this was the first experiment that you did for Dr. Howell that involved these V79 cells 100 percent labeled with tritiated thymidine?" [Leonard objected to the question.]

Lenarczyk: "I don't remember if that was my first experiment or the next one."

Pincus: "Or the next one, is that what you said?"

Lenarczyk: "Of the next one. I don't know."

Pincus: "But you do recall – ."

Lenarczyk: "It's definitely my experiment."

Pincus: "And the documents that I've shared and reviewed with you, to your knowledge, all comprise the documentation that was associated with this experiment, is that correct?"

Lenarczyk: "By 'this,' which are you referring? To that paper?"

Pincus: "No, what we're just looking through now, which would be Lenarczyk 14."

Lenarczyk: "Yeah, that's – ."

Pincus: "That's the documentation – ."

Lenarczyk: "This is kind of standard documentation for the experiment which we were doing at that time in the lab. So, it was probably full sets from the beginning how to go for the experiment, what you have to do, how to measure, and then how to express your data on the graphical – or on the pictures."

Pincus: "If you go to the first page where its says 'Experiment Name.'"

Lenarczyk: "Yes."

Pincus: "Where it's the '3H-TdR toxicity (cluster 100 percent labeling).' What does that mean?"

Lenarczyk: "I think that this is, I believe, if I'm correct now, that this is referring to the situation when you will assay all of cells which they were irradiated."

Pincus: "That's what the 100 percent reference – ."

Lenarczyk: "That's 100 percent labeling. That means that what you did with this experiment, you labeled all of them and then you tried to find out what is there, the result of that."

Pincus: "When you look at this experiment and the results, what conclusion or results did you determine here by performing this experiment?"

Lenarczyk: "Let me check."

Pincus: "Okay. I understand."

Lenarczyk: "You're asking me about something – ."

Pincus: "I appreciate that. So, you check. Or let me rephrase."

Lenarczyk: "If you will look at the page 98, which is survival versus uptake."

Pincus: "That would be this chart here?"

Lenarczyk: "Right."

Pincus: "Okay."

Lenarczyk: "I don't remember what was corrected versus uncorrected now. But the points which are above the line, which is 1, it's showing you essentially that you have no effect. Because the survival is 1, which means that nothing has happened."

Pincus: "When you say 'no effect,' you mean no bystander effect?"

Lenarczyk: "That was not even the bystander effect, because that situation was referring to non-bystander situation. That situation was referring to something which is everything labeled. So, we are not looking for bystander, what are looking for – ."

Pincus: "So, what are [you] looking for here?"

Lenarczyk: "You looking for survival, what the treatment can do for the cells in terms of survival. So, if you would have a situation like here when all of the points are above the line, that means that you have no effect. Now, if some points are below, that means that this is the effect, which means that these cells for some extent they were killed by the treatment, because the survival is lower than 1. Because 1 is assuming that this is the controlled survival, and you can normalize everything for the cells which they were not treated."

Pincus: "How would you compare your results to the results that Dr. Bishayee represented in the two earlier papers from *Radiation Research* that I showed you?" [Leonard objected to the question.]

Lenarczyk: "This not easy now. I cannot make any comment, because I have to read this paper again, okay, because I don't remember exactly what was in the paper."

Pincus: "Can you take a moment to see if you can do that."

Leonard: "Without any backup data you're asking him to compare research experiments?"

Pincus: "We're going to go through it."

Leonard wasn't going to let it go. "I just want to be very clear, Shelly [Sheldon Pincus], that is what you're going to do this afternoon is sit and try and have Dr. Lenarczyk to act as an expert by comparing and making comparisons to Bishayee's work and what that means, that's against court rules." He continued, "We've been doing this road with the Columbia [University] people. This man is not given any real data. He's sitting here with photocopies, and you're asking him to make determinations or give an expert opinion, when he isn't even qualified, under his own testimony, to be an expert about that." Pincus retorted that Leonard had already walked Lenarczyk down the path of this line of questioning. "You asked him questions," Pincus observed, "about what experiments he performed. He identified the fact that his results differed markedly from that which Dr. Bishayee had been doing, the same kind of experiments, and which they had done in the lab preceding his arrival. I'm going," he iterated, "I'm pursuing that somewhat, and I want to ask him questions about that. So, it's not a question…" Leonard cut him off "…which it appears you elicited expert testimony in this latest desperate attempt…"

Pincus was undeterred. "Don't go there because you have no basis, you know, for making that assertion," Pincus argued. "This gentleman worked in the lab, as did Dr. Bishayee and Dr. Howell. He performed experiments. I'm asking him questions about his protocols, and asking him in terms of when he identifies the fact that his results were different from those which Dr. Bishayee performed, I'm asking if, in fact, this was one of those experiments. So, let me pursue that, please." Leonard admonished Pincus and told him to go get an expert if he wanted answers to the question he'd posed to Lenarczyk but he allowed it, as long as Pincus didn't ask Lenarczyk to start opining on experiments performed by other scientists in the lab before his arrival in 2000.

Pincus turned back to Lenarczyk to continue questioning him about details of his experiments and results, most especially differences in his outcomes and Bishayee's, lab notes and whether Lenarczyk recognized more documents associated with his tenure in Howell's laboratory. Lenarczyk had no recollection when asked about those differences in experiment results and whether he'd discussed them with Howell. "Do you have any recollection of any explanation for these differences being provided to you by either Dr. Bishayee or Dr. Howell?" asked Pincus. "Objection. Asked and answered," Leonard interjected. Pincus pressed. "No. Do you have any recollection of receiving an explanation from either of those gentlemen?" Lenarczyk had already answered this question for Leonard and after a lengthy exchange with Pincus, he answered it again, just the same: "I don't remember." Pincus, in total, asked Lenarczyk about a series of 100 percent experiments he performed as the investigator in Howell's lab on December 14, 2000, and May 3, May 21, and June 21, 2001, and whether he had discussed his results, especially any differences in results, with Howell or Bishayee. "I don't remember," Lenarczyk replied each time. Then Pincus moved on to a series of A1 [Chinese hamster ovary cell line] and V79 cell line experiments Lenarczyk performed on November 10 and 28, 2000, and February 19, 2001.

"When you said earlier in your testimony today that when you were doing these kinds of experiments you found that your results were very much different than the ones that Dr. Bishayee was performing, and/or which had been done in the lab prior to your arrival, you were referring to these experiments, were you not?" Pincus asked. Leonard objected to the form of the question. Lenarczyk had gone over all of the same material when Leonard deposed him. "Answer the question, please," Pincus insisted. "I had to refer for this experiment," Lenarczyk replied, "because, as I explained before when John [Leonard] asked me, when I came to this lab to do experiments, my experiments, when I was doing my experiments I had to go through the same phase of experiment which was done before." Pincus kept pressing for a different answer. "Now, again, I want to be clear having reviewed these groups of, these various experiments here. Does it refresh your recollection in any fashion as to whether, given the differences in your results as compared to the results that were reported, you know, by Dr. Bishayee, do you have any recollection of discussing those differences with him?" Leonard objected again, "Asked and answered twelve times." The question didn't change Lenarczyk's answer: "I don't remember that." When Pincus asked him if he'd discussed the same – again – with Howell, Lenarczyk replied the same: "I don't remember that." Leonard renewed his objection.

Pincus produced a document marked exhibit number 22 [Lenarczyk 22], a 100 percent experiment performed by Howell on July 16, 2001. Lenarczyk seemed startled by the document and at first told Pincus he had left the lab by the time the experiment was done.

> *Pincus*: "To your knowledge, were you gone from the lab by that point in time?"
>
> *Lenarczyk*: "I was – I think I left in June earlier than that experiment was done."
>
> *Pincus*: "I would like you to just take a moment – go ahead. I don't mean to interrupt you, sir.
>
> *Lenarczyk*: "No, no, no, wait a minute. There is something in the paper, in the materials which seems to me that – ."

Pincus: "Were you still there?"

Lenarczyk: "I think that I left one month before that."

Pincus: "In June you think you left or was it July of 2001?"

Lenarczyk: "I don't remember, but looks like – I can check this one. I don't remember now."

Pincus: "It's not that critical. I'm just, you know, curious as to your recollection. But, in any event, this appears to be an experiment that Dr. Howell performed. Do you recognize his writing?"

Lenarczyk: "But it looks like there – it's probably also my writing."

Pincus: "Really! What are you looking – what page are you looking at, if you'll give me a Bates stamp reference in the lower right-hand portion of the document, leading you to believe that you participated in this experiment?"

Lenarczyk: "See, like in page 7438."

Pincus: "7438."

Lenarczyk: "Dr. Howell is using number 7 without crossing. If you can see on the title of this."

Pincus: "Yes, I see. He's not using what I call the European 7, right?"

Lenarczyk: "Right. And this is definitely his writing, his handwriting. But in the table of numbers – ."

Pincus: "I see."

Lenarczyk: "…and I think that this is my handwriting."

Pincus: "So, you may have worked on this experiment [a 100 percent experiment]?"

Lenarczyk: "I may have worked on this experiment."

Lenarczyk was still confused about his role in the experiment. "Do you recall," asked Pincus, "just for purposes of refreshing your recollection, do you recall ever having, or Dr. Howell ever sending any experiments to you in Poland?" Lenarczyk didn't remember that ever happening. Pincus pressed Lenarczyk for an interpretation of the cell survival in Howell's experiment. Clearly, Lenarczyk's handwriting was in the document. "So, it might be that we started the experiment together," Lenarczyk speculated, "or – I don't know, because it looks like later on because the experiment was started on 16th [of July] as it's marked on the first page on 434." He continued, "And then when you go down then you will see that, for example, in the table, which is on 437 is July 20, which is three days or four days after, and this is probably my handwriting. And then the next one is…" Pincus cut him off, "You're judging that on the European 7s again, correct?" "Even not like that," Lenarczyk replied. "See, when you would go to page 438 there is some text, which is on the top, which is just a title of that one. And if you can compare that lettering with that one which is down are completely different. This is probably done in my writing, and this is Roger Howell's writing."

Lenarczyk told Pincus that the results of the experiment, shown in Table 3 on page 440 of the document, "looks like they were put in the table by Roger. This is definitely Roger's hand. That's my guess. Okay? I will make a 7 with a cross. And also that text is very similar in terms of characters that he was using in the previous page." Pincus returned again to the cell survival data in the experiments. "…[I]f you were both working on that experiment," Pincus asked, "and you saw this biphasic[142], you know, result, did he mention to you that this appeared in any way, shape or form to be different than the results that Dr. Bishayee had reported?" Leonard objected to the question. But it didn't matter. Lenarczyk replied: "I don't remember that. Sorry." Pincus moved to another Howell experiment, dated September 27, 2001. "I was in Poland," Lenarczyk told him. "You were in Poland. You're certain of that?" Pincus asked him. "Right, I'm certain about that," Lenarczyk iterated. "Do you have a recollection of Dr. Howell sending you a copy of this experiment or in any manner communicating to you the results of this experiment?" Pincus wanted to know. "I don't remember," said Lenarczyk. He wanted to know if Lenarczyk had any comment on Howell's survival data. Leonard objected again.

Pincus directed Lenarczyk to a page that provided the results of the experiment. "Again, I just want to put an objection on the record that at this point you're having him look at Howell's experiments," said Leonard, "which he has no firsthand knowledge at all, and asking him to interpret results of, you know, experiments that he didn't participate in, nor apparently had any knowledge of, and so being handed twenty-some pages of documents…" Pincus acknowledged the objection and asked Lenarczyk to take a look at a chart that pertained to survival. "…[B]ased on your understanding of these types of experiments, as you've described during the course of your testimony, what were the results of the survivals that

Dr. Howell represents in this chart associated with the September 27, 2001 experiment?" Leonard objected to the question. But Lenarczyk replied, "If you will compare this chart with the picture which is in the published paper there is not very much consistency between the data." But in what way was there a lack of consistency, Pincus wanted to know. "Because you have a dose [of irradiation] from five to twenty, and then you have here the dose from zero to ten. Now, here you have a survival going down by three log phase."[143] Lenarczyk was referring to the Bishayee paper. Confused, Pincus clarified Lenarczyk's statements. "The data that is represented in the chart [he had shown Lenarczyk] in Dr. Howell's September 27, 2001 experiment you identified as being representative of one log phase, correct?" Lenarczyk replied, "Yes." To understand, Pincus suggest that if the results from Howell's paper was transferred into the chart in the Bishayee paper, a much steeper decline would be observed. "Yes, I would expect that," said Lenarczyk, "because, see, he [Howell] applied the doses like ten, whatever units is here, and he still has survival in one log phase. Once in the paper for ten he's down to third log phase, which is a big, big difference in the magnitude, because this is the log phase." The Howell experiment demonstrated a biphasic response compared to the Bishayee paper, which was clearly not biphasic and more like "a normal linear regression decline."

To sum up his questioning, Pincus asked Lenarczyk – to Leonard's objection – "So, if I do the math correctly, it seems that we've reviewed some nine 100 percent experiments using Chinese hamster cells that were performed by either yourself or Dr. Howell between October 2000 and July 2001, as well as this most recent, or tenth experiment that was performed by Dr. Howell alone in which the survival is biphasic, or tending to flatten out in the first decade, correct?" He continued, "Is what I've described to you consistent with what you've testified to here? To what we reviewed?" Lenarczyk answered, "I think that you are correct." Lenarczyk didn't have enough information to determine how the discrepancies in Bishayee's papers, which Lenarczyk had described as being in exponential decline, had occurred. Asked if Howell had explained the discrepancies to him, Lenarczyk didn't remember. Leonard objected to the question, noting that the question had been asked and answered "about 47 times at this point." Yet Lenarczyk, when questioned by the Newark CCRI in

April 2001, knew and testified to the same that the results of Bishayee's experiments couldn't be replicated. Lenarczyk wasn't interviewed by the second Newark CCRI because he was no longer at the university. But he did send them his notebooks, at their request, even though Howell had a copy of it in his office. The UMDNJ wanted the original.

Pincus showed Lenarczyk documentation from the first Newark CCRI, a log of activities with which Hill was familiar. "...[D]o you recall whether the committee shared this document with you or gave you the opportunity to review it?" he asked Lenarczyk. "I don't remember," he replied. "There's a lot I don't remember, because it was so, so far away from today." Then he asked if Hill had sent him the document. "I don't remember that. Maybe yes, maybe no. I don't remember." Lenarczyk had performed an experiment on December 26, 2000, using V79 cells – a 50 percent experiment. "Yes, it's showing here by 'fraction of cells labeled' .5 on the first page." There's a difference between a 50 percent experiment and a 100 percent experiment. The 100 percent experiment, in this context, indicated that all of the cells were irradiated or labeled with radionuclide. In the 50 percent experiment, only half the cells were irradiated and then mixed with similar half cells that were not irradiated, to thus put them together to get 100 percent. Bystander effect was designed to be a 50 percent experiment. Lenarczyk's experiment didn't show a bystander effect and he couldn't confirm Bishayee's results regarding the same and which had been reported in the *Radiation Research* paper. "You might say like that," said Lenarczyk.

Asked by Pincus if he discussed the results of his December 26 experiment with Howell, in which he couldn't confirm the bystander effect, Leonard objected as asked and answered but Lenarczyk replied that he hadn't been asked about that particular experiment. "It was a completely different line of experiments," he continued, than those about which Leonard had previously questioned him. "He said he's never, during my examination, he said he never talked to Howell about any of this," Leonard opined. "I don't remember that," Lenarczyk offered. "Did you discuss the results of this experiment with Dr. Bishayee?" asked Pincus. "Again, I don't remember that," Lenarczyk answered. He showed Lenarczyk another 50 percent experiment he'd performed on January 15, 2001. This experiment, like the previous one, couldn't confirm the bystander effect. Pincus shows him an experiment he did on February 5, 2001. Again, he could not see bystander effect in his results. Then Pincus rolled out Lenarczyk's June 14, 2001 experiment – same result.

"Shelly, does it move it along at all – I mean, I think we all agree that it [the bystander effect] hasn't been replicated in that lab. I don't know if we need to sit and go through every single…" Leonard reiterated. "I'm almost – I'm going to go through these. I'm sorry. It's very boring to us. Okay," Pincus retorted. "You're coming to the same conclusion; it's not replicated," Leonard continued, "I mean, we're willing to [stipulate] to that." Despite the defense stipulation, Pincus showed Lenarczyk his July 5, 2001 experiment to have him confirm whether he did or did not show bystander effect. He hadn't. Then came a 50 percent experiment Lenarczyk had done on November 20, 2000, using A1 again – no bystander effect – followed by another performed on November 28, same year. Pincus would go on to enter 50 percent experiments for February 15 and April 12, 2001, and still no replication of bystander effect. Lenarczyk, no matter how many of his experiments had been entered into the record, could testify as an expert witness. Pincus asked Lenarczyk if he had, at any time, provided Hill with a copy of the April 12, 2001 experiment. "I don't remember," he replied.

During Leonard's questioning, he asked Lenarczyk about an experiment he had performed with Howell and Bishayee. "Do you recall your testimony?" asked Pincus. "I recall you indicating that there was a point in time where you were, so called to use your words, 'locked.'" Edouard Azzam observed the experiment. Lenarczyk didn't remember much about it. "But I seem to recall it was your testimony that during the course of that experiment that you, Dr. Howell and Dr. Bishayee performed, in the presence of Dr. Azzam, it was a 50 percent experiment, correct?" Pincus prompted. "Correct?" Lenarczyk said, "Yes." Then Pincus produced the 50 percent experiment performed by Howell on April 19, 2001, to which Lenarczyk had no firsthand knowledge. Lenarczyk could not know for certain whether Howell was or was not the investigator. "He's just reading a line on a document," Leonard objected. "Based on your review of the document," Pincus said, "am I correct this appears to be an experiment that was performed by Dr. Howell?" Leonard continued to object. "He's not identified here as the investigator by the name of Howell," Lenarczyk interjected – at least not for another five sheets into the document. Asked if he had any involvement in the experiment, Lenarczyk replied that he didn't remember. Asked if the experiment showed bystander effect, he replied that it appeared that there was "no bystander effect."

Pincus showed Lenarczyk another Howell 50 percent experiment dated May 3, 2001. "Object. Shelly, he has no firsthand knowledge in the world whether that's correct. He's reading the same thing you're reading. The fact that somebody types it into a document doesn't mean anything," Leonard continued. "You should ask Dr. Howell about this." Howell was identified as the investigator on the May 3 experiment to which Lenarczyk had also had no involvement. "Do you recall sending in the results of this experiment to Dr. Hill?" Pincus asked him. "I don't remember that," he replied. Asked if there was bystander effect in this experiment, the answer was again no. "Are the rest of these all Howell's experiments?" Leonard asked Pincus. "I think I have one more here," Pincus replied. "Before you start, I'll get this out," Leonard put on the record. "He has no firsthand knowledge of any of these experiments conducted by Roger Howell. He is not qualified as an expert to render an opinion as to what they say, nor does he have any specific knowledge of any of the data that went into these." Pincus brushed off Leonard's objection. "I mean, we can go through this exercise of handing him 20 pages at a clip and having him say 'yes, yes, yes, yes,' but it's of no evidentiary value. I don't know why we're wasting our time," Leonard continued. Pincus pushed on, showing Lenarczyk Howell's June 28, 2001 experiment. Lenarczyk didn't recall that he was involved in that one either. "Do you recall whether you sent the results of this experiment to Dr. Hill?" asked Pincus. "I don't remember." Undeterred, Pincus asked what the experiment showed. "Like before," replied Lenarczyk, "it looks like there's no evidence of bystander effect."

Pincus showed Lenarczyk his own April 2, 2001 experiment. He then showed him a memorandum dated April 6, 2001, from Howell to Stephen R. Baker, Ph.D., regarding "integrity of data." Asked if he had seen it before, Lenarczyk replied that he didn't remember it. "Whether you did or didn't," said Pincus, "I'd like to call your attention to certain statements that Howell makes in the body of this. If you'll look in the first paragraph, approximately a little more than halfway down, there's a statement by Dr. Howell to Dr. Baker. 'As I told you this morning, I have requested my postdoctoral fellow, Marek Lenarczyk, repeat some of Dr. Bishayee's experiments as a check of the validity of the data.'" But did Lenarczyk ever "get the memo." "I don't remember," he told Pincus, "but it might be possible that he asked me to do something." Pincus asked again. "Sitting here today, do you have a recollection of Dr. Howell requesting you to repeat some of Dr. Bishayee's experiments as a check of the validity of Dr. Bishayee's data?" Asking another way didn't get a different answer. "I don't remember."

Lenarczyk was never able to replicate any of Bishayee's 100 percent experiments using V79 cells either. "And you hadn't been able to replicate Dr. Bishayee's results in any of the 100 percent experiments in which you used the A1 cells, correct?" Pincus asked. "Yes," he replied. "You hadn't been able to replicate any of Dr. Bishayee's results in the 50 percent experiments which we reviewed in which you used A1 cells, correct?" The answer was he hadn't done so. "To your recollection, aside from Dr. Howell or Dr. Bishayee, did you report your results to Dr. Azzam?" Pincus asked and Leonard objected. "I don't remember that," Lenarczyk offered. "What about Dr. Toledo, Dr. Azzam's spouse?" But again, he didn't remember that either. "To Dr. Baker?" Pincus pressed again. "This one probably not, because I barely met Dr. Baker," he replied. "To anyone else inside or outside the lab?" Pincus asked. "We were talking with Lanie [Hill] about results," Lenarczyk answered. "So, you let Dr. Hill know?" Pincus continued. "The results. So, definitely she knows," he answered.

Pincus introduced the April 27, 2001 minutes of the first Newark CCRI, calling Lenarczyk's attention to page 13 and asking him if he'd ever seen the document. "Probably not. I don't remember," he replied. "Did Dr. Howell ever indicate to you that he felt that you had been nonproductive in your nine months as a postdoc in his lab?" Pincus wanted to know. "No," Lenarczyk answered. "Did he indicate to you – if you look at the paragraph down below – the very next paragraph, that he felt that you had produced no reasonable data and that your position was only guaranteed for one year?" Pincus continued. "No, I never heard that from him," Lenarczyk replied. "Was it your understanding that Dr. Howell was satisfied with your performance as a postdoc in his laboratory during this time period that we've been reviewing?" Pincus asked again. "He never mentioned this, too. I just was working, and that's it." The exchange continued.

Pincus: "Did he express to you in any form, orally or in writing, dissatisfaction with your performance?

Lenarczyk: "I don't remember that."

Pincus: "Did he at any time, that is Dr. Howell, express displeasure with you because you had been unable to replicate the results that Dr. Bishayee had produced in regards to the bystander effect?"

Leonard: "Objection. Asked and answered."

Lenarczyk: "No."

Pincus: [Produced another document, this one statements by Howell. Lenarczyk noted that he had never seen it.] "I want to call your attention to certain statements that were made by Dr. Howell and see if you have any knowledge concerning them, please."

Leonard: "Before we do that, Shelly, you know, and this is the same thing with this other document. These aren't statements made by Howell; these are notes taken of a conversation."

Pincus: "Okay."

Leonard: "Even in this, these aren't statements made by Howell; it's somebody writing down after the fact what they perceived to be a conversation. Same with this. In this case it was…"

Pincus: "You don't believe there's any basis to rely on the accuracy of the committee's minutes in any way, shape or form. John, is that what you're telling us?"

Flynn: "We've had to represent that on the record in questioning him."

Leonard: "The point we're making is that these are minutes, in this case by an unknown individual [recorder]. I have no idea what the minutes are. And it's not whatsoever, Shelly, it's – you know, these are not [exact transcription of] quotes. This is not a statement signed by Howell where he says, 'I see what you say and I agree with that. I sign it.'" [Pincus rephrased, acknowledging the objection. But he still wanted to know if Lenarczyk knew anything about what had been written.]

Lenarczyk: "No."

Pincus: [He produces notes apparently Putterman may have taken on April 16, 2002.] "Dr. Putterman reports that Dr. Howell had had discussions with Dr. Bishayee 'concerning ill will in the lab and 'uncomfortable working conditions' due to a dispute between Dr.

Hill and Dr. Bishayee.' Did Dr. Howell ever have any discussions regarding 'uncomfortable working conditions' in the laboratory with you?"

Lenarczyk: "No."

Pincus: "Did he have any discussions with you regarding the subject matter of ill will in the lab?"

Lenarczyk: "What do you mean 'ill will'?"

Pincus: "Well, it says 'ill will.' Any disagreements. Did he have any discussions with you regarding disagreements that may have existed between Dr. Hill and Dr. Bishayee?"

Lenarczyk: "I don't remember."

Pincus: "Did he ever indicate to you that – Dr. Howell, that is – that he had observed any aggressive behavior of Dr. Hill to Dr. Bishayee? [He rephrased after Leonard's objection.] Did you ever observe Dr. Hill behave aggressively toward Dr. Bishayee?"

Lenarczyk: "No."

Pincus: "In the fifth bulleted item there is a reference that says, 'No one else in Dr. Howell's lab has repeated these experiments,' referring to the bystander effect." He continued, "In fact, based on what we reviewed here today, prior to April 16, 2002, you had attempted to, and had, in fact, repeated the bystander experiments, had you not?" [Leonard objected to the phrasing of the question.]

Pincus: "Prior to April 2002 you had repeated the bystander experiments, correct?"

Lenarczyk: "I repeated something which we can call bystander effect experiment, but my finding was that I couldn't find bystander effect."

Pincus: "In other words, you repeated the experiments, you could not replicate the results?"

Lenarczyk: "Right."

Pincus redirected to the bottom of the first page to a comment attributed to Howell that stated: "'He also doubts that Dr. Lenarczyk has the technical expertise to repeat the experiments.' Do you believe you had the technical expertise to repeat the bystander experiments?"

Lenarczyk: "If I wouldn't, I wouldn't be hired."

Pincus: "So, you do believe you had the technical expertise?"

Lenarczyk: "Yes, I believe I have."

Pincus introduced a graphic of Lenarczyk's work, whose name appears first on the [PowerPoint].

Lenarczyk: "Which was accepted [the work] by the Radiation Research Society at the meeting which, I believe, was happened in Reno, Nevada. And that was probably 2002."

Pincus: "Now, would I be correct that – these PowerPoint slides are somewhat small – but do you recall whether in this presentation you present mutagenesis data from both yours and Dr. Howell's bystander experiments?"

Lenarczyk: "I don't remember that. It's not clear on that." But it was mutagenesis data. Responding to a further question from Pincus asking if there was a reason – or was it unusual to present mutations without showing the concomitant survival data, he continued his answer. "I wouldn't present that. Okay. I mean, it's supposed – if you would publish the data for mutagenicity, you have to show survival, because everything is related to the fraction of survived cells. So, without having knowledge what is the survival of the population out of which you are going to measure mutations, then you have no denominator by which you can explain the ratio or whatever number of the mutation." With nothing more of importance said, Lenarczyk's deposition ended a little over eight hours after it had begun.

Chapter 5 –
A Forgotten Summer Conversation

"The next to be deposed was Dr. Azzam," Hill later wrote of his December 12, 2008 oral examination. "He had little help to offer. He did not remember a conversation we had had in the summer of 2001 in which he had told me that he had persuaded Howell to get rid of Bishayee. Howell had come up with a list of possible explanations for the inability to replicate Bishayee's results," she continued, "and Azzam's hand was evident in the generation of that highly speculative list. An amusing thing occurred after his deposition had been declared closed. Azzam wanted it reopened so that he could go on record that he had not been a party to any cover-up. Curious. No one had made any mention of cover-up."

Azzam's deposition was all about fact gathering. He'd come to the university on March 1, 2000, as an assistant professor on the tenure track and had since become a tenured full professor as of July 1, 2008. But when Azzam first arrived at the UMDNJ from Boston, where he'd just completed his postdoctoral studies at the Harvard School of Public Health, he lived for a period of time in Hill's home. "This must have been," he recalled, "I was in with Dr. Hill from Monday to Friday, from March to July [2000] when my family came from Massachusetts, and we purchased a home in Livingston, so that was July 15 [when they took possession of the house]." Pincus asked him if during that period did he recall Hill or himself being engaged in any discussions regarding Bishayee. "Lanie was always – Dr. Hill was always – concerned about tissue culture habits of – but at that period of time, no concern about Bishayee, that I recall," he replied. "You say that you recall during that time period, March to July of 2000, Dr. Hill expressed some concerns about tissue culture habits?" Pincus asked again. "But not of Dr. Bishayee," Azzam reiterated. "My question was specifically relating to Dr. Bishayee," Pincus continued, "and your recollection, if I understand you correctly, is that she did not express to you any concerns?" Azzam did not recall anything.

Pincus asked if Azzam shared lab space with Hill when he first arrived at the UMDNJ. "Me and Dr. Hill? We were all in the same laboratory space because my own space was not available yet, so we shared lab space in Room 451 of the F-level at the medical sciences building," he replied. "Who else shared lab space?" Pincus asked. "Dr. Howell, and Dr. Hill was working on a project with Dr. Howell." Azzam wasn't working on that project, he told Pincus. But he was still in the space. "And was your wife using the space, too?" Pincus wanted to clarify. "Yes," Azzam answered. "And what about Dr. Bishayee? Was he using that space?" Pincus continued. Yes, he was, Azzam said. "Was Dr. Lenarczyk there at that point in time, to your knowledge?" Azzam responded no, not just then. "I don't think he was there. I do not remember, but I'm definite he wasn't there between March and July [2000].

> *Pincus*: "During that time, do you recall socializing with Dr. Hill?"
> *Azzam*: "Oh, yes. Certainly."
> *Pincus*: "And, again, during the time that you, and I assume your wife, also socialized with Dr. Hill?"
> *Azzam*: "Between March and July?"
> *Pincus*: "Yes."
> *Azzam*: "No, Sonia was in Boston."
> *Pincus*: "But subsequent to that point in time, did you remain social once you moved to Livingston?"
> *Azzam*: "Oh, yes. Yes. We shared meals and, certainly, yeah."
> *Pincus*: "When your wife was working in the lab, did she express any concerns about Dr. Bishayee?"
> *Azzam*: "No. To answer your question is no, unless you will ask me the question of when did I hear about the concern, then I will answer."
> *Pincus*: "Did she ever tell you that she didn't want Dr. Bishayee using something called a laminar flow hood?"

Azzam: "No."

Pincus: "Did she ever express to you any concerns about his lab technique, or his cleanliness habits insofar as doing experimentation?"

Azzam: "She expressed a concern about tissue culture in general in the Howell lab."

Pincus: "You say that you recall that Dr. Hill expressed a concern about tissue culture?"

Azzam: "Yes."

Pincus: "In general?"

Azzam: "A general concern."

Pincus: "When was this, please?

Azzam: "That was when we used to walk in the morning, Dr. Hill, and Dr. George Hill [Hill's husband], and myself, when I was staying in Dr. Hill – in the Hills' house – we used to go for walks every morning, and I recall very clearly, Dr. Hill, Lanie, expressing the concern, general concern about tissue culture."

Pincus: "What was the concern she expressed?"

Azzam: "The concern, we have to teach Roger and his people proper tissue culture habits."

Pincus wanted more specifics of the discussion and what Azzam understood Hill to mean when she spoke about the tissue cultures.

Azzam: "In general, about tissue culture, I – really, tissue culture is a – I have my own habits in tissue culture, so I – so, everybody has his own habits. Tissue culture is – has so many variables, so I assume you have to follow many variables, but my variables might be different from Dr. Hill's, from Dr. Howell's, from many other people. But," he continued, "I didn't share my tissue culture hoods at the time. I had my own to observe with Dr. Bishayee, or Dr. Hill, or Dr. Howell were doing. I know Dr. Hill and Dr. Bishayee were doing tissue culture, but I had my own hood that I cleaned. It was separate from the area that Dr. Bishayee – it was at the same lab, but he was in an anteroom, so I don't see him. I'm focusing on my own work, and I did tissue culture myself at the time, and Dr. Hill was in a totally separate room, and did her own tissue culture. I never cultured cells," he explained, "for Dr. Bishayee. I did take over from Dr. Hill a couple of times she went to New Hampshire, and she said, if you wouldn't mind to passage my cells for me. That I did. But I never cultured cells for Dr. Howell or Dr. Bishayee for me to observe any of their work. So, for me to pass a judgment, I don't like to pass a judgment on anyone." Pincus pressed for additional clarification of the March to July 2000 time frame. "I never did tissue culture with Dr. Bishayee or Dr. Howell ever from 2000 until as of yesterday," Azzam responded. "Working on a tissue culture that I see what his people do, I never taught them to subculture cells, or to [do] culture cells." But Pincus wanted to know, notwithstanding having never done tissue cultures with Howell or Bishayee, if Azzam interacted with Howell. "Yes, definitely," Azzam said. "Did Dr. Howell," asked Pincus, "in or about this time period – I'm going to, by way of parameter, from March of 2000 until approximately in, you know, 2001, did he ever express to you that there had been problems with Dr. Bishayee?" Azzam replied, "No."

The questioning turned to what Howell may have said to Azzam about Bishayee. "Did [Howell] ever indicate to you in or about this time period that Dr. Hill had related suspicions?" No, he replied, before Pincus could finish his question. "...that Dr. Hill had spoken with him regarding suspicions that she had of Dr. Bishayee making up data [from March 2000 through 2001]?" Azzam told Pincus that he'd gotten to know Hill had a concern about Bishayee about the time that she submitted a formal complaint, indicating not before April 2001. "Prior to that, as best as my knowledge would come, tells me I had absolutely no knowledge that they had serious issues among themselves," he said. "I got to know, I think – and I got to know about this matter, actually probably most likely from [Marek] Lenarczyk, when he came to my office, that there has been a formal complaint [an encounter Lenarczyk did not want to discuss]. As much as my recollection will come, that I did not hear from Dr. Howell, no." Azzam remembered that the time frame he learned about the complaint was between March and June 2001. "I'm going to represent to you for the purposes of moving matters along, and trying to pinpoint certain facts, that that occurred in early April of 2001. So, within the time frame," Pincus prompted. "But I'm going to make that representation to you, so when I refer to the complaint, we're going to be focused on that point in time." He went back to the comment Azzam made about Lenarczyk coming to his office and telling him about the complaint. "That must have been April 2001," Azzam explained, "so subsequent of Dr. Hill, I believe, submitting her complaint to whoever it was, I really do not know, within the university, but there has been a – the concern of Dr. Hill. That's when I got to know that there is an issue in the lab." To be clear, Pincus asked if it was only after Azzam learned that Hill submitted a formal complaint that Lenarczyk came to speak to him. "That is correct," he replied. "Was there anyone else present besides yourself and Dr. Lenarczyk at the time he spoke to you?" Pincus asked. "I'm not sure if Dr. Bishayee was there."

Azzam did observe something going on between Lenarczyk and Bishayee prior to April 2001. "About 9:30 at night, both of them – and I couldn't understand what it was between them, about some cells and things like that, and – but I have no comprehension that they had any issues among each other," he explained. "I used to give them rides home to take – to take them at that time. I don't even recall where Dr. Lenarczyk was, but I know where Dr. Bishayee was. Neither one had a car. I would drive them, but the little issue among each other – and I did not understand what it was. I was busy with my own work, but they had – they were unhappy with each other a little bit, although they continued to talk to each other." If Lenarczyk and Bishayee were "unhappy" with one another, what was it about? "Dr. Bishayee was – I think I got to know about [the complaint] from Dr. Lenarczyk, but who was there when he told me – I can picture him coming to my little office in 466. Who was there present," Azzam searched for the words, "either him and I, or someone else, I would picture Dr. Bishayee being the person – in there." Pincus wanted to know what Lenarczyk told him. Azzam described the conversation about the complaint as brief, explaining that Lenarczyk told him Hill had submitted it against work performed by Bishayee. Azzam is on record as telling Pincus that he had no specifics about Hill's complaint. "So, if he didn't relate facts or details," Pincus asked, "yet you recall him having other discussions with you about the complaint, what was it that he was discussing with you?" Azzam still claimed that Lenarczyk hadn't been specific about the complaint.

Pincus wanted to know how many times Azzam might have encountered Lenarczyk. "Five, I had – I had many encounters with Dr. Lenarczyk. I gave him rides, but they [the five times] were general discussions of a jovial nature where we exchanged jokes."

Beyond the initial contact between Azzam and Lenarczyk, beyond the lack of specific information being exchanged, Pincus asked if Lenarczyk expressed concerns regarding Bishayee's work. "Yeah. He expressed a concern, yeah." But he didn't elaborate, claimed Azzam. "Did you ever come to learn after the complaint was filed in early April 2001 by Dr. Hill, that Dr. Hill and Dr. Lenarczyk had followed or shadowed Dr. Bishayee's work in the lab for a period of time?" Pincus pressed Azzam. "Yes, I did," he replied. He'd learned about it from Howell – and from Hill. "Let's first deal with Dr. Howell. How soon after the complaint by Dr. Hill was filed," Pincus asked, "was it that Dr. Howell first spoke to you about this?" Like Lenarczyk, Azzam wasn't certain he remembered exactly. "But probably a week or two weeks after, and very briefly at that time, also." But what did Howell say? "I do not remember that." Pincus asked if Azzam had learned anything about the complaint from Bishayee. "No, he related, I understood, that there was something, but no specifics. No specific."

What about Azzam's interaction with Hill? Azzam admitted that it was Hill who probably mentioned to him that Bishayee had "fudged" data. "Again, an exact time [of that conversation], I am not really sure. I'm trying to put in the time frame to make me remember is at what time – at that same – whether it was in the same period that we were – Dr. Hill, Dr. Howell, myself, were in contact with Brookhaven National Lab, and we were putting the Department of Energy grant jointly, all of us together. And so it would have been during that period of time. I would say that this may have been in May?" he explained. "We used to meet together. We went to Brookhaven. But exactly the timing may have been May, June. At that time. That – probably that's – of 2001? No, it should be subsequent to that, because in 2001, then, that must be subsequent, way subsequent. Not in 2001, because I was in Dr. Hill's house until July, sometime early July. I must have taken the week off, and at that time, I do not recall that we had these conversations of the allegation, although – yeah – oh, no. That was 2000. 2000 that I was in Dr. Hill's house. So these events – Dr. Hill filed the complaint in April. Yeah, so – April 2001. So maybe Dr. Hill spoke to me. We exchanged more in May and June. May and June, I would say." Azzam told Pincus he never took any notes of the conversations.

Azzam wasn't on Howell's original NIH grant, which started before Azzam arrived on campus. But he was on it as co-investigator when it came up for renewal in 2005. Azzam told Pincus that he had provided Howell with the system to culture cells in three dimensions, "three-dimensional culture. I put a lot of input for his postdoc Massimo Pinto. So, it came to using the three-dimensional tissue culture system. I developed that system for normal cells, and his postdoc, Massimo Pinto, I did not participate in his work, but intellectually, I gave him the techniques of how to work. We showed him how to use that system," Azzam continued, "and Dr. Howell wanted to expand his work into novel avenues using human cells, and that's where comes my expertise, is in the human cells culture in a three-dimensional architecture. So, I suggested using the human cells, the tumor cells, cancer cells. I suggested to him an in vivo model using the testis model that he has expertise in. So, I said, why don't you use the testis model which you have and label cells and study what we call the bystander effect in vivo. And the system that I provided him, and in which I have quite a bit of expertise, generated it not only to Dr. Howell, but to many other people in the United States, and around the world, to culture cells under very controlled conditions in three dimensions." That was his input to the process. But had he reviewed the original grant to know where the project stood before continuing forward? He'd read it; he'd not "reviewed" it. The sections he'd first "reviewed" had been e-mailed to him in Boston, in 1999, before he arrived at UMDNJ. When Pincus asked if Azzam, after being named co-investigator in 2005, had reviewed the original grant for the purpose of understanding what had been previously done, he denied having done an assessment. "I happen not to have reviewed," Azzam told Pincus. "I never read Dr. Howell's grant in its entirety. I don't think I ever read it. I may have – I did read sections of it, particularly the introduction, as I was preparing my own grant."

Whether Azzam familiarized himself with the original grant, with the experiments that were performed prior to his coming aboard as co-investigator, was a fair question. "We hold meetings, journal clubs, where – really not from – when Dr. Bishayee would present some of his data, which was not extensive at that time, I participated with Dr. Hill, Dr. Howell,

Sonia [Azzam's wife], Dr. Bishayee and myself. We sat," he explained of the period Bishayee was still in the lab, "in Dr. Howell's office, and – but at that time, probably we discussed more survival curves that I obtained with incorporated tritium – tritiated thymidine – than what Dr. Howell's findings. I know about Dr. Howell's results," he admitted. "I – through presentations, and absolutely and only through presentations in the journal club by Dr. [Sergiy] Gerashchenko, Massimo Pinto, [Venkata S. V.] Prasad Neti [Ph.D., who arrived in the UMDNJ radiation research laboratory in 2002]." Pincus was still unclear. "You were aware – were you aware of the fact that before your becoming co-investigator, that Dr. Bishayee had done a series of experiments in the lab under the auspices of this grant?" Azzam answered "yes." He was also aware of Lenarczyk's experiments associated with the grant. But he claimed to have never reviewed any of the prior experiments in preparation for joining the grant as co-investigator. "No, I did not review the experiments, in the sense that I don't understand what you mean by 'review.'" Pincus' question had devolved into a game of semantics. Had he looked at Bishayee and Lenarczyk's lab notebooks? "No, absolutely not. Never, ever did I look, and I never looked at anybody's lab book." The denials kept coming before Pincus could get out a question. "No." "I did not do that." "Not anybody's lab book." Pincus wasn't suggesting that anything unprofessional or unethical had taken place. "It seems to me," he told Azzam, "when somebody agrees to serve as a co-investigator of a grant, you have to, I would think, and you correct me if I'm wrong, come to some conclusion, or some opinion that you want to be associated with that [the grant]."

"I agreed to serve as a co-investigator," Azzam replied to Pincus, "because the system – and I was very crucial in helping Dr. Pinto in his postdoctoral project because, essentially, when Dr. Pinto joined Dr. Howell, Dr. Howell was in the hospital with serious ailment, that we did not know, I got a phone call at midnight, whether he will make it or not make it that evening. And he was in the hospital," he continued, "when Dr. Pinto arrived in the laboratory. So, I welcomed Dr. Pinto, and I think I initiated that particular project to help Dr. Howell to work on his system. I thought for Dr. Pinto to look at the bystander effect, the system that we had developed to study another item in the radiation field, would be a suitable thing for Dr. Pinto to do a postdoctoral project." Pinto spent three years working in Howell's laboratory. Azzam claimed to also have never looked at his notebook of the project he'd set in motion. "…[B]ut he presented the data to me, and mainly – so my serving as a co-investigator into that grant comes from that area that – so my intellectual contribution to the work that Dr. Howell was doing." The bottom line, Pincus tried to clarify, was that Azzam had just testified he came onto a project relying on data that Howell provided for the purpose of Azzam making his decision to become a co-investigator. Yes or no. "No, because I have an interest in that field, really. I'm – I have been – the doctor, he did not give me the data. He presented the data." Again, semantics. What's the difference, Pincus asked, between giving and presenting? "No, giving me data, I receive a notebook, and look at that. I see a graph presenting a graph. So I don't see the fine, fine details of what Dr. Pinto had done, but I have no question about anything that Dr. Pinto – I think Dr. Pinto had done a – his – a careful investigator has a deep insight and understanding of the field, and I have no reason to believe that Dr. Pinto – I have – ." "Forget about Dr. Pinto," Pincus interrupted him. "What about the concerns that had been raised about Dr. Bishayee's data? Did that enter into your determination as to whether to serve as co-investigator on the renewal grant?" Azzam didn't concern himself with Bishayee, he told Pincus. "I don't understand that" Pincus retorted, "but notwithstanding that you say you don't know, you're now agreeing to associate yourself with the grant which Dr. Hill made a complaint about." Defense counsel Flynn objected. Azzam acknowledged Hill's concerns. "As far as the second grant from 2001 up to Dr. Howell submitted the second grant, a renewal of the first grant, I know, again, of the work done by Dr. Gerashchenko, again through the journal club presentations – not only the journal club presentations, again I help Dr. Gerashchenko in the tissue cultures of the cells." He plowed on, getting further from answering Pincus' original query. "[I] did not think of why you are here [his participation], and there are serious allegations of – of what I have seen, myself, again, because I have seen a lot of Gerashchenko and Pinto, the work done by those two." "We're not concerned with them," Pincus interrupted, again, trying to redirect Azzam's response.

Pincus showed Azzam an experiment done in April 2001. He wanted to know if he recognized it and whether Azzam was aware that the first

Newark CCRI had requested Howell repeat certain experiments after Hill reported Bishayee's misconduct. Azzam was aware of the request and the document in front of him. "I am familiar with my initial in here that I have initialed these. I sat on the chair in the Room 451A or B. I do not know the name of the room, exact number of the room. Dr. Howell was doing experiments, and I sat there on the chair that he was actually doing the experiments. I did not participate in the performance of the experiments," he explained. "That's as much as I did, and this happened at an extremely busy time for me, that was – I was annoyed that I was wasting my time." Pincus was curious how he came to be asked to "sit on the chair and waste his time." Who asked him to sit there? "Dr. Howell." Was it a choice to be there or not, or was it a professional courtesy? "As a professional courtesy." No one else had asked him to be there? Baker? Raveché? "No." Did Dr. Howell perform the experiments himself? "Yes." Was Dr. Bishayee present? "My recollection, and as clear as it can be, no one but Dr. Howell performed the experiment." Some of the writing in the report in front of Azzam, Pincus pointed out, was Bishayee's. He asked if Azzam recognized Bishayee's writing. "No." He then offered that he would recognize Howell's writing but not Bishayee's. Pincus had him look at the preceding page [Bates stamp 7367 to 7366] and asked him if he recognized Howell's handwriting. "Yeah."

> *Pincus*: "Okay. Would you agree with me that, at least from a lay person's perspective, that the writing on 7367 is different than that on 7366?"
>
> *Azzam*: "Looks different."
>
> *Pincus*: "So, does this refresh your recollection as to whether, in the course of this experiment being done – also, if you look at 7371, this page, in fact, has your initials in the upper right-hand portion."
>
> *Azzam*: "Uh-huh."
>
> *Pincus*: "Do you recall whose work you were initialing? Was this Dr. Bishayee's or was this Dr. Howell's?"
>
> *Azzam*: "I really cannot recognize the handwriting [the narrative begins to shift, and the recollection starts to dim]. It doesn't look like Dr. Howell."
>
> *Pincus*: "It looks like somebody else's though?"

Azzam: "But now as we're going into this in here, Dr. Howell was working on the laminar flow hood. Did Dr. Bishayee make the record of these counts here? It is quite possible."

Pincus: "I realize time has passed, and I'm only asking your best recollection – ."

Azzam: "It is quite possible, and things flash in my mind, that it is quite possible that he recorded these numbers and, if anything, now that I look at these, and you ask me the question, and probably the Coulter counter was right beside the chair where I was sitting. I was sitting in – to look at Dr. Howell that is doing the experiment. So that's the hood, this is the desk, and the Coulter counter was here. It is quite possible that Dr. Bishayee recorded that he put the sample on the Coulter counter, and he recorded these Coulter counter – these counts. So if it is, is it Dr. Lenarczyk or Dr. Bishayee that recorded the numbers, it might very well be Dr. Bishayee, and one of them Dr. Lenarczyk."

Pincus: "This experiment, if you go back to 7362, this is an experiment associated with the bystander effect. Isn't that correct?"

Azzam looked over the document. Yes, it was associated with the bystander effect. Pincus got Azzam on record as having observed the protocol in which 50 percent of the cells were irradiated, and half were not.

Pincus: " – did you have any responsibilities whatsoever insofar as verification or supervision as this experiment progressed?"

Azzam: The experiment was getting done. I didn't – did not go into details of the experiment."

Pincus: "You just sat and observed?"

Azzam: "That's correct."

Pincus: "Why were you initialing? I don't understand this."

Azzam: "The work was getting done. The work was getting done."

Pincus: "How did you determine whether the work was getting done or not? What led you to – let me finish, please. The fact that you initialed carries at least reasonable – it is reasonable for me, or any person, to surmise or to believe, that by virtue of your placing your initials – ."

Azzam: "Right."

Pincus: " – it was done for the purposes of establishing that somebody witnessed what was going on?"

Azzam: "Right. And I did witness that they were using Coulter counting. They were culturing the cells. If they were mixing the culture with each other, I was there. I was there. And certainly, I did see, but did I take a – obviously, to know if a culture is labeled with irradiated thymidine, a Coulter counter, a Geiger counter, cannot tell you much about it, right? But, yes, I know that the experiment was being done, certainly."

Pincus: "The protocol, if you look at number one, said that the cells that were being used were to be 80 to 90 percent confluent at the start of the experiment. Is that correct?"

Azzam: "If that's what it says, I guess so."

Pincus: "What does that mean to you?"

Azzam: "Eighty to ninety percent confluent? That would mean that almost 90 percent of the surface area on which the cells are growing are filled with cells."

Pincus: "Okay. And if there's – not 100 percent of the plate is filled, are you familiar with what's known as contact inhibition?"

Azzam: "Yes, I am, certainly. That does not apply to the V79, by the way."

Pincus: "But in order for there to be contact inhibition, it would have had to be – ."

Azzam: "These cells are never contact inhibited, sir."

Pincus: "I'm sorry?"

Azzam: "These cells will never get contact inhibited."

Pincus: "So, if they were not ever contact inhibited, would that prevent these cells from going into what's known as S-phase in the cell cycle?"

Azzam: "If they are not contact inhibited, yes, certainly, they will go into S-phase. Certain cells will go into S-phase of the cell cycle."

Pincus: "Certain cells?"

Azzam: "Certain, not all. In the V79 cells, they can fill the flask by lapping on top of each other; one layer, second layer, third layer, and so you will always have cells in S-phase. At least, I culture V79 cells, maybe millions of them, and billions of them for a period of – from 1978 until 1986, so I know them – at least, they come into different strains, so I'm very familiar with V79, yes."

Pincus: "In fact, when we go back to your curriculum vitae, which I had a few moments ago, this suggested that, at least looking at your writings and your experience, you've had extensive experience with the bystander effect?"

Azzam: "Yes."

Pincus: "At least from what I can see, you've had numerous talks on it?"

Azzam: "I could be, maybe one of the world experts on the bystander effect."

Pincus: "You've had articles and grants?"

Azzam: "Yes. Yes."

Pincus: "Insofar as this 80 to 90 percent confluence, would I be correct that that would mean when you talked about some cells going into S-cycle – ."

Azzam: "S-phase."

Pincus: " – that that would mean that anywhere from approximately 20 to 40 percent?"

Azzam: "It really all depends on – I am answering you a scientific – the scientific answer. It all depends on how the culture, when they were re-fed, the last time they were re-fed, how long before you came to use them. Actively growing cells – you would have actively growing cells. So, if they are 80 percent confluent being V79 cells, I would say – and you did flow cytometry analysis, you may get 30 – 30 percent being in S-phase, but you really have to do it. It all depends how."

Pincus: "I know you really have to do it, but I'm also focusing you on the particular experiment that you witnessed."

Azzam: "Yes."

Pincus: "So when you're giving me these numbers – ."

Azzam: "If I put my initial again in here again, from the time of 2001, I did put my initial on something that I did see getting done. So I didn't initial in the blind. So, that they – whatever number they recorded, or something that counts, I was there. I did see the counts that they have recorded."

Pincus: "Were you done? I didn't mean to interrupt you. The medium that was used to culture these cells is referred to as MEMB[144]. Is that correct?"

Azzam: "Yes."

Pincus: "What's that stand for?"

Azzam: "It's, again, A and B – MEM is the typical – is the name of the growth medium that's minimal essential – minimal. Okay. Now, I cannot define it exactly. [He searches for an answer – the wrong one.] The A and B, I think, is the nomenclature that Dr. Howell and Dr. Hill would use in there. So MEM is the typical name – is the typical name of the product, of the growth medium. A and B comes from Dr. Hill and Dr. Howell, and I do not remember what they mean by A and B exactly." [MEMB is the acronym for Modified Eosin Methylene Blue Agar (cell culture medium).]

Pincus: "Okay. In general, is it your recollection that one had calcium in it, and one did not?"

Azzam: "No, I don't remember."

Pincus: "You don't remember? Okay."

Azzam: "What I really wanted to pay attention, the V7 – so I didn't have an interest to know the details of what they were doing s – because I had my own work to take care of."

Pincus: "Did you, at any time, see something know as deoxycytidine added to the medium? Deoxycytidine?"

Azzam: "Deoxycytidine to this particular medium of this particular experiment?"

Pincus: "Yes."

Azzam: "What I know, if the added is tritiated thymidine was added."

Pincus: "I'm asking whether deoxycytidine was added to the medium?"

Azzam: "I do not recall."

Pincus: "Do you see anything in this protocol that would suggest that it had been?"

Azzam: "Deoxycytidine in here? I don't see event he deoxycytidine being mentioned here on page 362."

Pincus: "Okay."

Azzam: "I don't see any deoxycytidine mentioned here."

Pincus: "I want to turn your attention to 7365, and these were some Coulter counts, I believe."

Azzam: "These are not Coulter counts, these are colony counts."

Pincus: "I apologize. This appears to suggest, insofar as the experiment that you witnessed, that there was little or no killing of the bystanders?"

Defense counsel objected, noting for the record that Pincus' line of questioning bordered on Azzam being used as an expert witness, a violation of the rules, especially on a page where his initials didn't appear. Pincus wanted to know if Howell had discussed with him the results of the experiment Azzam observed. "He mentioned the results to me," he replied. "[…] He was telling me, and they were talking and discussing – he was discussing the data, and honestly, and I was listening, but my mind, I was not dwelling. I know that they did not find imitation response, and they prepared the poster. He prepared the poster and everything, and as he was preparing the poster – ." Pincus wanted to know "what poster" and what was on it. "A poster on mutation work that was submitted to one of the meetings," Azzam told him. "Probably, the meeting in Reno, Nevada." [The same meeting to which Lenarczyk referred in his deposition.] "I'm not talking about that right now," Pincus cautioned. "[N]ow I'm focusing on your being present while Dr. Howell and, possibly, Dr. Bishayee are conducting this experiment on or about April 12 [2001], and some days thereafter." He wanted more information about how the experiment ended – the results – and if Howell discussed them with Azzam. "Discuss the results? Is that I – again, what I recall, that they did not observe a bystander effect, that – yeah." Howell did not show bystander effect, just as Lenarczyk previously indicated and Azzam confirmed.

Pincus: "Do you remember, you know, looking at the colony counts, or the Coulter counts, or the FACS results?"

Azzam: "No."

Pincus: "You didn't look at that?"

Azzam: "No, I did not myself."

Pincus: "Do you know what Coulter counter was used for the purposes of this experiment?"

Azzam: "Yeah, a Coulter counter that Dr. Howell had. Yes, I know."

Pincus: "Do you know how long that Coulter counter had been used in the lab?"

Azzam: "That same Coulter counter, when I arrived, that Coulter counter was there."

Pincus: "So that Coulter counter was there in 2000?"

Azzam: "Right."

Pincus: "Is it still there, to your knowledge? Have you seen it?"

Azzam: "I really do not know. I really do not know. I have not – I know that Dr. Howell purchased a new Coulter counter."

Pincus: "When was that?"

Azzam: "That was a couple years ago. Two, three years ago."

Pincus: "Have you seen the counter that you recall seeing from the period of about 2000 through a couple of years ago? Is it still around? Have you seen it in the lab?"

Azzam: "I never really looked for it."

Pincus: "Okay."

Azzam: "And – yeah."

Pincus: "To your knowledge, do Coulter counters contain certain types of – I'll call it, hazardous materials? Is that correct?"

Azzam: "They contain – that particular Coulter counter contains mercury."

Pincus: "Would I be correct that mercury is a compound for which records have to be maintained when it gets disposed?"

Azzam: "When it gets disposed, certainly, it's a hazardous material, yeah."

Pincus: "So to the extent that that Coulter counter no long exists in the laboratory, based on – ."

Azzam: "I don't know if it doesn't exist. I have no idea."

Pincus: "Let's assume for the purposes of our discussion that it no longer exists within the laboratory. Based on your experience, your rank, you're being a man of science, there would be records that would have had to be prepared reflecting the disposal of somehow – ."

Azzam: "Certainly. EOHSS [UMDNJ Environmental and Occupational Health and Safety Services] also have to be the record. The environmental health – the safety committee at the university – because you would give them the mercury, and they will dispose of it. It's a regulated substance."

Pincus learned from Azzam that the EOHSS would have come to collect the Coulter counter to properly dispose of the mercury. Defense counsel had requested a record of that disposal. Pincus renewed the request to get the information on the Coulter counter's disposal. "Do you remember," Pincus asked Azzam, "in any of your experiments, you used the particular Coulter counter that was used in this experiment?" Azzam answered that he used the hemocytometer in most of his experiments. "So, the answer is no?" Pincus asked again. "The answer, we used the Coulter counter, but we did use but – what Sonia does in the lab, the students and until today, we strictly use the hemocytometer," Azzam continued. Pincus wanted to know if the lab maintained a record of the Coulter counters. "Coulter counters, as I mentioned to you, that in all our survival experiments, at least, I would say, 99 percent of them, it was a hemocytometer, and if anybody used the Coulter counter for my own work, it would be me, actually, for an experiment that I did. And – that I did, and I would have the records of it, certainly." Previously, when he was in Canada and Boston, Azzam had used only the Coulter counter.

Pincus produced documentation of an experiment performed on April 19, 2001, another 50 percent experiment involving tritiated thymidine. He asked if Azzam had ever seen it. "I remember that – I'm not sure." The reason Pincus asked was because the investigator on the experiment wasn't named on the document, and he thought that Azzam might recollect or not if he had participated in it. "As I mentioned to you," Azzam replied, "I did not participate other than observe. There's a big difference between both." Pincus asked if had observed the April 19 experiment. "I do not remember. Unless I see my initial, I don't remember." Whether there are initials or not, Pincus pressed, did Azzam recall whether Howell requested he observe more than one experiment. "I was there for several days that I sat," Azzam answered. "I wasn't there only for a single day, multiple days that I observed. I sat there for the full length of the day that he was performing the experiments so, the number of experiments, I don't recall. I may have at the end, I said, I cannot do that anymore. I got many things on the go. I may have said that. I do not want," he continued, "because I felt my time was entirely wasted, and it's against me – my own attitude. What am I? I felt very awkward to be sitting on that chair. I respect Dr. Howell, I respect all my colleagues in the – so, it was a very awkward situation for me, plus my time was precious." Pincus was curious. Why was it awkward? "Awkward to tell me to sit – if I sit there, it means I know there is something wrong. I hold the highest respect for my colleagues, you know." With that statement, Pincus asked what he meant by "something was wrong." Flynn objected. "The wrong – I knew by that time there is allegations from Dr. Hill against work done in Dr. Howell's laboratory. I knew that." Azzam recalled that he sat for at least a number of days on the April 12 experiment and his initials appeared for at least seven days, ending on April 19. But he couldn't remember if he sat in on an experiment that started on the day the first ended – April 19 – and though Pincus pushed, Azzam's initials do not appear on any of the latter experiment's documents. He also had no recollection of ever speaking to Howell about the results of the April 19 experiment.

"[I]n science, all along, if you discuss with me," Azzam told Pincus, "if you discuss with me, unless I am given the opportunity, or I am willing to go and dig in the data to see that there is something improper, you tell me whether a bystander effect exist or not, my attitude to it, you've got – if you don't observe it, it doesn't mean that it does not exist," he continued. "These experiments have time course. I maintain that the bystander effect is dose-dependent, and linear energy transfer dependent. I did not publish on it yet, but I am the first person to say it into discussions. Many people objected. They said the bystander effect is dose independent, linear energy transfer independent." Azzam testified that he had data that had been repeated three times that showed he was correct. "[B]ut the time that I looked at the effect, and the end point that I chose to look at the effect shows that I am perfectly correct. And I showed it in Argentina, I showed it in multiple places that the people objected to my initial statement that it is dose dependent, now are very much convinced that I should hurry and publish that work." Azzam's work – and ultimately his confidence – were predicated, Pincus observed, on his ability to show anyone doubting the data. He could repeat his experiment and get the same result. "Yeah. I repeated not using Dr. Howell's system, but I have very firm data, that if you mix radiated thymidine labeled cells, human cells with unlabeled cells, there is a transfer of the stress response from those tritiated thymidine labeled cells to the unlabeled cells." Pincus reiterated that Azzam's competence was based on that ability to repeat the result. "That I find a bystander effect, yes, it is very significant. It's my own grant application."

Pincus established that Azzam was not only co-investigator on Howell's 2000 NIH grant – the one in question – but also other grants with Howell in which one or both of them were either principal or co-investigator, among them grants for the National Aeronautics and Space Administration (NASA) and the United States Department of Energy (DOE). "I am the principal investigator on these grants and Dr. Howell is a co-investigator." Pincus asked about other grants. "Yes. Those two and University Hospital Cancer Center. I am a PI [principal investigator] on a grant in which he is an investigator, Dr. Howell, and I am co-investigator, on a grant with Dr. Howell and Dr. [Ronaldo P.] Ferraris." Azzam was working just then on four additional grants with Howell. The NASA grant was for $1.2 million; the DOE grant for $730,000, and the one for the hospital was small, by comparison, at $80,000. The grant with Howell and Ferraris was for about $600,000. All in all, Azzam was working with Howell on just over $2.6 million in grants.

Nearing the end of Azzam's deposition, Pincus returned to the subject of entries dated March 29, 2001, in which there was a reference to, around four o'clock in the afternoon, Bishayee being in a great rush when he asked Lenarczyk to give him some V79 cells, which he did. The notation also mentions that this was "observed by Ed" [Azzam]. Azzam didn't recall this ever happening. "I would say not only that I don't have a memory, I want – I don't recollect. I really do – nobody in the lab. We are all busy. We're certainly busy. Ask Marek [Lenarczyk] gave cells to Anupam Bishayee," he tried to explain, searching for the words. "For me to see that cells were given to Anupam Bishayee, that would mean – that would mean that I had to be present there, which is very unlikely, because if I recall, Marek Lenarczyk worked in the same room with Lanie Hill, which was not in the same place where I would work." Bishayee worked, he told Pincus, in another part of the lab from where he had space. "[S]o if I would witness that, it seems to me that I had nothing to do. I would be in their rooms, when I was busier than them, busier than Dr. Hill, and busier than Dr. Howell". He insisted under no circumstance would he have been sitting around observing Lenarczyk giving Bishayee V79 cells. What Azzam was telling Pincus was he had no firm recollection. But he also opined that it was before Hill made the complaint with the first Newark CCRI and he had no visibility on what was going on between Hill, Bishayee, Lenarczyk and Howell at that time, sticking to his story that it wasn't until the allegation went to the committee that he knew anything. "Whether or not there was a complaint or not at or about this time has nothing to do with whether you observed [Bishayee and Lenarczyk] engaged in a conversation [on March 29]," Pincus reminded him.

Certainly, one of the issues that stood out in the Hill case was Bishayee's departure from Howell's lab. Pincus wanted to know when Azzam became aware that Bishayee had resigned. "I – how did I learn of it? That Anupam is leaving the lab? That is, I may have learned of it from Roger [Howell] or from Anupam. I have no recollection. By either one I may have learned about it," he explained. But had Howell or Bishayee, Pincus asked, ever told him that a letter was given to Bishayee – a much different scenario than a resignation letter being received from an employee about to depart his position. "No," Azzam replied. "You didn't let me ask the question. Did Dr. Bishayee ever tell you this letter had been given to him by Dr. Howell with instructions to sign it?" Pincus pressed. "No," he answered again. Pincus asked if Azzam had had any other discussions with either Howell or Bishayee regarding his resignation that Azzam hadn't already identified. To clarify, he also included "any conversation." "Conversation? Certainly. I may have advised – advised or – to me, that allegations of misconduct in science, obviously, is – whether it's real or not real, is the end of a person's career," Azzam added. "Any hint of an allegation of misconduct in science, it raises eyebrows. Scientists – the scientist community – is a very small community. We all know each other. An allegation of someone – and I may have advised – I may have told Dr. Bishayee, why don't you go and work in radiation safety, one of the things that I did. I said, that's something related to your knowledge; it's an area of growth. You don't have," he continued, "to submit grants. You don't – it's a secure area where you can have a very solid position. I most probably would have said that to him." Pincus wanted to know when he might have provided this advice to Bishayee.

"No. The timing, no." Bishayee did move over to radiation safety. But did that lead Azzam to believe he'd heeded his advice? "I have no idea," he told Pincus. "I have no idea. But I – when I advised that, I have seen very prominent scientists from the radiation community – Clyde Greenstock is one of them, who published several papers – ended up into this area of the Radiation Safety Atomic Energy of Canada, so I had grounds to advise that [to Bishayee]. I had seen," he continued, "many prominent radiation researchers, not that – I say prominent. Joseph Borsa had left basic research to go into food irradiation. That's a – really made several discoveries. Preceded David Baltimore, he received the Nobel Prize. It was done by Joe Borsa before him at [the University of Pennsylvania] and he abandoned research to go into [he nervously repeats himself] food irradiation and I certainly, in some of the talks that I had with Dr. Bishayee, and I had – I continued on talking terms with Dr. Hill, Dr. Bishayee, Dr. Howell, Dr. Lenarczyk. All of them. I never stopped talking to any of them." Azzam remained in contact with Bishayee, including the day that Bishayee left to go to his new job.

Pincus shifted the discussion of Bishayee's departure to an August 2001 conversation Azzam purportedly had with Hill regarding Bishayee and Howell. "I had many discussions with Dr. Hill. Many. Whatever it was, whether it was in August. I always – I mean, until probably we stopped talking about her case. I had discussions probably with Dr. Hill exchange of – sitting in her office, meeting her in 2001, 2002." Pincus was specifically asking about August 2001. "Maybe I did. In August, more than likely I did, because we were in – I was trying to see what I can – if these two individuals can talk to each other rather than going – that was my wish for them, come, you folks talk to each other." By "you folks" Azzam was referring to Hill and Howell. Pincus reminded him that what was on his mind as "good intentions" and what he actually did were not the same. "Isn't it a fact that you told Dr. Hill that Dr. Howell had handed a letter of resignation to Dr. Bishayee? Told him to sign it, told him not to come on campus, took away his keys and identification, and told him not to apply for research positions on the campus or in radiation biology?" Azzam replied that he didn't remember that. "When you say you don't remember it, are you denying it or – making such a statement, or is it that you just don't remember making such a statement? There's a difference,"

Pincus reminded him again. "The statement that you just made, I do not remember saying that, but if I did say something, if I did, because in my mind I wanted to be a peacemaker. I know that Dr. Hill was very upset with Bishayee, that I may have – I may, I may, because as I say," Azzam continued, "the spirit is to – that, look, Dr. Howell, Bishayee will not be in the group, because she mentioned for many times, all what he has to do, Dr. Hill mentioned he has to get rid of Anupam." Pincus wanted to know when Hill said "get rid of Anupam" to him. "Many times," he told Pincus. He asked again for a time frame. "I don't remember but she said that at one time, probably we were going to a faculty meeting in radiology. That all what – he's got to get rid of Anupam." But was it before or after the complaint? "After the complaint." He continued, "Obviously, after the complaint. Before the complaint, Dr. Hill wanted to – I never could dream of anything that she had against Dr. Howell because they were – on the first – the second Gordon conference, which is the most prestigious conference in radiation biology, radiation oncology, my mentor, Jack Little, from Harvard School of Public Health, was organizing that conference, and – yeah, and Dr. Little wanted to invite Roger to come and give a talk at the Gordon conference and Roger was ill at the time. And we can tell that from my [curriculum vitae], when the Gordon conference [was] and what Dr. Hill said [was] let us maybe record Dr. Howell's talk, and so then it would be shown via video at the conference." So, what he was telling Pincus was that Hill's relationship with Howell was extremely cordial prior to the Bishayee allegation. There were no problems evident to anyone in the lab.

But didn't Azzam try to persuade Howell to get rid of Bishayee? Pincus wanted an answer. Azzam again told Pincus no, that he hadn't tried to get Bishayee to leave. "Did you ever try to persuade Dr. Howell to retract the papers that contain the data that couldn't be replicated or confirmed? Pincus asked. "No." Had he ever suggested the same? "No, because I don't know if anything is wrong. I – I – you know." Had he ever had any conversations or reported to Baker any suspicions he'd had about Bishayee? "No. I went to Dr. Baker and I – I – went to him to tell him one thing. I said [to Baker], can you do something to bring Dr. Hill and Dr. Howell to some understanding." Pincus asked when that conversation took place. Azzam testified that it must have been after Hill made the initial allegation. "I went specifically to his office." Pincus asked what Baker said to him. "He said [there was] nothing [he] could do. It's out of my hands."

Azzam opined that the entire Hill allegation negatively affected him. "Because this thing did cost me – for me – obviously, we are a very small group. I'm a new faculty member, that this was undermining my performance, my productivity."

Pincus: "Your performance?"

Azzam: "Of course."

Pincus: "Why is that?"

Azzam: "Talking to Dr. Hill, talking with Dr. Howell, it poisoned an atmosphere. Two of my colleagues that I respect very much have a major misunderstanding. I am trying to submit my grant, establish a lab, and everything was very, very disturbing, and the mind – it took a lot of my productivity. I'm sitting here today. I could be – I have eight, nine people in my lab. I could be there."

Pincus: "I understand you."

Azzam: "It's, you know – ."

Pincus: "Now, to your recollection, did Dr. Howell ever relate to you a discussion he had with – do you know who Dr. Putterman is?"

Azzam: "Yes, I know."

Pincus: "Who is Dr. Putterman?"

Azzam: "Her exact title, I am not sure, but she is a vice president at the university."

Pincus: "Did he ever relate to you a discussion he had with Dr. Putterman regarding these matters, these complaints of Dr. Hill?"

Azzam: "No specifics."

Pincus: "Did he tell you he ever met with her concerning this?"

Azzam: "Yes. He said [he had] to go and meet with Dr. Putterman regarding. He mentioned to me that, yes. And when I received your – the subpoena, the first thing I did, I called Dr. Putterman."

Pincus: "You did?"

Azzam: "I did."

Pincus: "Why did you do that?"

Azzam: "I had no idea that I am party to this matter."

Pincus: "Okay."

Azzam: "Because Dr. Hill really never mentioned anything to me, that I am somehow in here."

Pincus: "What did she say to you?"

Azzam: "Who?"

Pincus: "Dr. Putterman."

Azzam: "She – what she mentioned to me is that this is – it's a subpoena, then you have to appear."

Pincus: "Okay. Anything else?"

Azzam: "No."

Pincus: "Did you discuss your testimony at all?

Azzam: "Absolutely not. I wouldn't even know what Dr. Putterman looks like, because only that telephone conversation."

Pincus: "You never met her face to face?"

Azzam: "No, I did not."

Pincus: "She never spoke to you at all during the course – ."

Azzam: "No."

Pincus: " – of this – ."

Azzam: "No."

Pincus: "Let me again finish. During the course of all the investigations that went on?"

Azzam: "No, absolutely not."

Pincus: "Have you ever seen any documentation that Dr. Putterman prepared regarding Dr. Howell?"

Azzam: "No."

Pincus: "Before Dr. Bishayee resigned, did you ever – did Dr. Howell ever relate to your discussions that he had had with him about any kind of uncomfortable working conditions in the lab?"

Azzam: "Repeat your question, please."

Pincus: "I'll rephrase it. Prior to the time Dr. Bishayee resigned, did Dr. Howell tell you he had spoken to him about these uncomfortable working conditions in the lab that you described?"

Azzam: "No. No."

Pincus: "Did he ever say to you that he felt Dr. Hill wouldn't leave Dr. Bishayee alone, and that was why he felt it was best for him to leave the lab?"

Azzam: "No. No."

Pincus: "Dr. Howell never said that to you?"

Azzam: "Nothing that I remember, no."

Pincus: "Did Dr. Howell ever tell you that he had encouraged Dr. Bishayee to find a position outside the university entirely?"

Azzam: "No."

Pincus: "No such discussions?"

Azzam: "No."

Pincus: "When you recommended to Dr. Bishayee that – that he get out of research, and got to radiation safety, did Dr. Howell object to you, or voice any dissatisfaction with your having –."

Azzam: "No."

Pincus: " – made such a suggestion to him?"

Azzam: "No, not to me."

Pincus: "But you have no recollection of Howell saying that he wanted Bishayee out of the university completely?"

Azzam: "No."

Pincus: "Did Dr. Howell ever tell you that he wouldn't give any letters of recommendation to Bishayee?"

Azzam: "No."

Pincus: "Did you write any letters of recommendation to Dr. Bishayee?"

Azzam: "I did write letters of recommendation for Dr. Bishayee when he started to apply. That must be in the year before he went to his current position, which, I presume, is in Ohio, teaching positions in pharmacy. I did write letters of recommendation for him."

Pincus: "Did you make any comments regarding his research?"

Azzam: "His research? I made comments that I am – I – in his presentations, obviously, there were teaching positions that he gives clear presentations, which he did."

Pincus: "What about the quality of his research? Did you have any – did you have any occasion – ."

Azzam: "I may have."

Pincus: " – to comment?"

Azzam: "I have the letter."

Pincus: "You do?"

Azzam: "Yeah, I have the letter. It must be in the – wherever he was, that the observation of a bystander effect in a three-dimensional system is a very important observation, and it's no secret that that paper is quoted. [...]"

Pincus: "I'd like you, when you get back, I'll do a follow-up, to collect – was there just one letter of recommendation that you wrote for him or more than one?"

Azzam: "No, no. I must have written two or three. It would be the same letter that I sent everywhere."

Pincus turned the deposition to the relationship between Hill and Bishayee. He wanted to know if Azzam had observed any conduct on Hill's part that led Azzam to believe she wouldn't leave Bishayee alone after she made the complaint against him. "No. Dr. Bishayee, what I found it very weird, continued to live in Dr. Hill's, you know, accommodation, that Dr. Hill had an apartment that [continued to rent]. There was nothing," he continued, "that I had seen from Dr. Hill against Dr. Bishayee. In fact – and Dr. Bishayee got married, Dr. Hill came to his wedding reception. I was there. Dr. Hill came in. I don't think Dr. Howell was there but myself and my wife were there. Dr. Hill walked in, gave a present, and she left." He observed that Bishayee lived in Hill's son David's apartment. "That was the only time I entered the apartment. The day I – Anupam needed me to help him move. I went and I helped him move. And I entered the apartment. I did help him. He came newly married with his wife, and – [his words trailed off]." Pincus had more questions about this time. "At the time Dr. Bishayee submitted his resignation, did Dr. Howell ever express to you an opinion, or opinions, regarding whether he believed him to be an excellent technician?" Azzam appeared to question the characterization "excellent technician." No, he told Pincus. "I have no – whether he was good or he was bad." But had Azzam ever seen any performance evaluations Howell prepared for Bishayee? "No," he replied. "What about Dr. Lenarczyk? At the time he left the lab, did Dr. Howell comment to you in regards to the quality of his work?" Again, the answer was no. Asked if Howell had ever, at any time, expressed dissatisfaction with Lenarczyk, Azzam again said no.

There were other aspects of the Hill and Bishayee relationship that piqued Pincus' interest. He came back to the fact that despite the seriousness of Hill's allegation against him, Bishayee continued to rent an apartment from her in Newark's Ivy Hill section. Then Hill attended Bishayee's wedding reception at a restaurant in Montclair. Azzam had seen her there but didn't recall sitting with Hill or seeing her talk to Bishayee. "Did you see anything at that reception that led you to believe there was hostility between them? Pincus asked. "No. No." Azzam testified further that he'd never seen any hostility between Hill and Bishayee. "But I knew," Azzam offered, "that Dr. Hill, and he [Bishayee] knew himself, had very, very serious allegations against him. That is not a secret," he continued, "unless he was totally out of his mind not to understand what he was up [against], an allegation of misconduct, falsification of data." Pincus returned to the word Azzam had used to describe these things: "weird." "Yes. You know, if it was me, and somebody tells me – well, I have something called self-respect. I would – if it was me, I would have move out immediately, not two seconds after," Azzam said. "And probably wouldn't invite that person to a wedding reception?" Pincus observed. "Probably, by courtesy, not to hurt the person. This is – that's – each one is entitled to his own opinion in these matters, but we are all forgiving, and we are all – I am also a Christian, and a very devout Christian, that I probably would have a different attitude from what I just spoken now. But with," he continued, "the spur of the moment, my reaction would be negative, but later on, it will be try to be as positive as possible."

As Bishayee was leaving Howell's lab in July 2001, Pincus asked if Bishayee ever indicated to Azzam that rather than radiation safety he might want to go into molecular biology. Azzam answered no. He'd had no such discussion with Bishayee. Had Bishayee ever told Azzam that one of the reasons he was leaving was to upgrade his technical skills in technical biology? No, not that Azzam had ever discussed with him. "Did he ever indicate to you that he was leaving to further his [career] development becoming an independent researcher?" Azzam replied no again. Bishayee had, according to Azzam, applied for another postdoctoral fellowship at the University of Iowa. "And he got invited for an interview because Dr. Bishayee, myself, met Doug Spitz [Douglas R. Spitz, Ph.D. of the University of Iowa Carver College of Medicine] in Dublin, Ireland, and Dr. Bishayee's performance was very good. Actually, his presentation was beyond my expectation," he continued. "I must say that he did such a good job, and I presumed that when he – maybe when he contacted the people in

Iowa, Dr. Spitz invited him to come and give a seminar there, but that's as far as I know about this. Dr. Spitz never asked me for any recommendation letter, or anything in that regard, and I know Dr. Spitz very well."

Pincus brought the deposition line of questioning back to the subject of the investigation and Howell's list of possible explanations for the inability to replicate Bishayee's results, that list that Hill observed had Azzam's hand all over what amounted to a highly speculative document. Pincus showed it to Azzam. Had Howell ever indicated to Azzam that he was preparing such a document for the purposes of the investigation, he asked. "No," Azzam replied, "but I know that Dr. Howell mentioned to me, I have to – I am involved, I have to respond to things related to the matter, but I have not – he did not discuss any of that. I've never seen this document, no. I had obviously many discussions with Dr. Howell about why was this bystander effect not being observed, and it was – it was me, I would say, it could be the cells, could be the serum, could be any sum of these. So discussing these items, certainly, I did a few discussions on what could be the reason of the lack of repeatability." Pincus suggested that he was hypothesizing. "Yes," Azzam said. "Because if I understand you correctly, you know, for you to render any kind of opinion, as to what had been the reason why these experiments had not done – you would have, just as you said here today, sit and review everything that had gone on. Correct?" Azzam told Pincus he was correct. "And I wouldn't know if anything is right or wrong in those experiments. In my opinion, always when we do an experiment, [it] has to be to our best of our ability. Has to be an accurate, correct experiment. That's what science is all about." But what about Azzam's hypothesizing about what the reasons might have been that Howell couldn't get the bystander effect?

> Pincus: " – you say, it might have been the cells, might have been the serum. What else?"
> Azzam: "Certainly. In these things, many of the cells and the serum, the culture matters, the incubators. Tissue culture is very, very sensitive, has many variables to control. V79, I've seen them change."
> Pincus: "What did you tell him about the cells?"
> Azzam: "The cells? What I say? My understanding, that they had lost the liquid nitrogen Dewar. I know that they lost it. Because we

were – maybe Sonia was filling the liquid nitrogen. We would fill the – ."

Pincus: "Did you say viewer or fill?"

Azzam: "Dewar. To fill the liquid nitrogen, the Dewar." [He spelled it for Pincus and then explained that he was present when it was discovered that there was no nitrogen in the storage tank.]

Pincus: "When was this?"

Azzam: "It was subsequent to – I don't recall exactly when this had happened, and I don't even recall whether Anupam was in the lab or not."

Pincus: "But this happened after the complaint?"

Azzam: "I believe so."

Pincus: "After Dr. Hill's complaint?"

Azzam: "I believe after the complaint it has happened. I believe so, yes. I believe so."

Pincus: "So the – hold on a second."

Azzam: "After – also, I'm not 100 percent sure just to state myself – ."

Pincus: "Hold on. Let me interrupt you, please, because I want to make sure I understand what you're saying. There is something in which nitrogen is kept?"

Azzam: "No, you keep cells in a tank."

Pincus: "In a tank filled with nitrogen."

Azzam: "Filled with nitrogen."

Pincus: "And that tank that contained the cells which they used for this experimentation [Howell's lab] – ."

Azzam: "Yes."

Pincus: " – disappeared?"

Azzam: "No. No. It didn't disappear at all. That tank was there, but when we opened it, it had no nitrogen because that tank – my tank, you have to fill it every two weeks. Dr. Hill's tank, which I continued to fill until a couple of years ago, probably, you had – it maintained nitrogen pretty good. Every month, probably, you had to fill it. Roger's tank you had to fill – maintain – was a smaller tank, that it would maintain – the nitrogen will not evaporate. It would take up to three months before you would need to fill it, and it was a very well insulated tank, as opposed to mine. Mine is probably the biggest, the most expensive, but you have to really – it loses nitrogen, evaporates, so we have to keep on filling it – ."

Pincus: "Okay."

Azzam: " – continuously. And when we – because nitrogen – my use of nitrogen is, you order a big tank, we always have excess nitrogen. So, we share the nitrogen, in fact, among multiple labs. [...] These tanks can fail because if you bump them, if you come to my lab, you see my door, I have foam galore so nobody – because we're short of space – to bang the tank, and if you break the seal on these tanks, very easy – you easily can lose them. So someone who doesn't know what he's doing, they can break. If a janitor comes, he's waxing the floor, picking up the garbage – ."

Pincus: "So put a time frame on this. When did this happen?"

Azzam: "I do not really recall, Mr. Pincus, whether it was before the allegation or after the allegation. I do not recall."

Pincus: "So you don't remember whether at the time you actually observed –."

Azzam: "I don't recall."

Pincus: " – Dr. Howell try to repeat one of the bystander experiments, whether, when he went to get the cells, there was a problem associated with it?"

Azzam: "It was – it was before the allegations happened that he lost his tank. Maybe Dr. Hill would remember that better than me."

Pincus: "If that happened, would there be any report or anything that would be prepared about that?"

Azzam: "No. No. You wouldn't prepare."

Pincus: "So what else about the cells, if anything, did you hypothesize might be – and, by the way, let's assume there wasn't nitrogen in that tank."

Azzam: "The cells would die."

Pincus: "Okay."

Azzam: "So they are good for nothing."

Pincus: "Don't you oftentimes develop new generations of cells?"

Azzam: "You ask other people to give you cells, or if you have them stored elsewhere, it can – but in the case of the cells of Dr. Howell, these V79 cells existed in culture for many, many years, probably

30, 40 years they have been in culture. These are not – you know, beyond the point that I hypothesize, could be the cells – these are transformed cells. They are not normal cells. The cells that I have worked with have contact inhibition. Cells that lose contact inhibition on the way to transformation. Cancer cells don't have contact inhibition. They pile on top of each other, form a cancer. So, a cell can change from one subculture to the other. In fact, they do. Because they lose – their chromosomes will change, so they will change. I use," he continued, "I buy cells from the Coriell Institute [of Medical Research]. I freeze 30, 40 vials of them. I conduct my experiments only between passage 9 and passage 12, period. So, when I repeat, I go back to passage 9 to 12. The reason passage 9, because they sell them to us passage 7. V79 cells don't benefit from this luxury. Normally, Dr. Howell acquires these V79 cells, I presume – and Dr. Hill would know better than me – I think from [...] Harvard [T. H. Chan] School of Public Health [...] so these cells have been in culture for a long period of time. I'm not saying," he explained, "that they did something wrong, or something not wrong, but as a scientist, you tell me I work with V79. That's why I tell you, I don't work with V79. I want to distinguish and effect normal cells [...] hopefully in vivo, in a rodent, or – so, their cells, the V79, change continuously, so if you tell me the results, I'm not replicating, it may very well be that the V79 cells have changed. I've seen them change."

Pincus: "None of this did you attempt to relate back to the actual experiment?"

Azzam: "No. V79 cells will change, so to me, it could be a reason. Serum another reason. When I buy serum – I spent a lot of time, and we spend a lot of money to screen the serum. And Sonia [Azzam's wife], just this past year, she screened serum at least two times for the vendor to have sold a batch of serum that we see lacked, so if he worked with another batch of serum and now he went to a new batch of serum, serum is very critical in these observations."

Pincus: "And you suggested culture medium and incubators, too, to him?"

Azzam: "Definitely."

Pincus: "Anything else that you suggested to him [Howell]?"

Azzam: "That – I guess, these [are] the main things."

Pincus: "That's all I need to know. Thank you. And you've never seen this document?"

Azzam: "No."

Pincus: "He never shared that with you or discussed it with you?"

Azzam: "No, I have never seen that document."

Pincus: "Did he ever indicate to you that in trying to explain to the committee or to Dr. Putterman or any other administrator, that these things that you have been describing to me – ."

Azzam: "No."

Pincus: "Let me ask the question. – may have been the cause of his inability to replicate results?"

Azzam: "I didn't get it. Sorry." [Pincus had the question read back to Azzam.]

Azzam: "I don't remember."

Pincus: "Fair enough."

Azzam: "I don't remember. Whether at that time I mentioned the cells, the serum, I may have very well have said that to him. I may have said that to Dr. Hill as well because I was in, really, a lot of communication with Dr. Howell and Dr. Hill and on a daily basis. So, I may have mentioned that to both of them."

Chapter 6 –
Trouble in the Lab

oger Howell's December 18, 2008 deposition followed closely on the heels of Azzam's. "He did not appear very happy about it," Hill wrote later, "but anyone who is interested [in what he had to say] can judge for themselves." Perhaps more memorable than his testimony was the argument that ensued between plaintiff and defense counsel. Hill recalled that it was so heated, the lawyers "nearly came to blows." She had little else to say about it except that "I was quite upset when [Howell] implied that I was prejudiced against foreigners, Indians in particular, and that Azzam and [Pranela] Rameshwar [Ph.D.] concurred." She observed just then that such derogatory remarks hardly jibed with her Smith Medal citation, awarded by her undergraduate alma mater Smith College in 1997, which read, in part:

> It is said that you think with precision and act with passion. When you suffered the cuts of discrimination, you responded by clearing the path for other women to excel in science. A prized mentor, a gifted teacher and a fierce advocate for equal rights, your enthusiasm and optimism have led scores of women to scholarship. [...] Selfless in your industry, your work has made the world better for the sick, the learned and for all who yearn for equal opportunity.

Howell's deposition was taken in two parts to accommodate his schedule.

Pincus started Howell's deposition with the whereabouts of the Coulter counter that Bishayee used to conduct the various experiments in question and which he had also asked about during Azzam's deposition. He had an inventory tag for it; it was identified as SN030788. "I have been led to understand that that was discarded at some point in time. Is that correct?" he asked Howell. "That is correct." The Coulter counter was discarded, according to Howell's testimony, sometime in 2007. "Why did you discard it?" Pincus followed up. "It was not functional," Howell replied. "It was missing its mercury manometer removed by the technician who installed the new Coulter counter back in 2006 and it was no longer functional." Pincus wanted to know the name of the technician who removed the mercury. "He is from Beckman Coulter. His name is...Lanie [Hill was present at the deposition], you know who he is, the Beckman Coulter guy? He probably worked in your liquid scintillation counter. He's been there for years. It will come to me." Pincus prompted that Howell must have had to fill out some paperwork to discard university property. He had. "When a disposal is to be made, you send a request with the university identification numbers to – what are they called?" Howell explained, "The people who keep track of all the equipment. I forget what they're called, inventory control, or something like that." Since the Coulter counter contained mercury, the lab had to follow the protocol for hazardous waste removal. The mercury was disposed of through the EOHSS as Azzam previously testified. Asked how long the records of disposal [of the mercury, in this case] were kept, while Howell didn't know, he also believed the records of that kind would still be available.

"Are you aware of the fact that approximately in April 2008, I made a request to be provided with all of the documentation associated with not only this particular Coulter counter but an inventory of other Coulter counters in your offices?" Pincus asked Howell. "Yes, and I produced those documents," Howell replied. "When did you do that?" Pincus came back at him. "I don't know, April or whenever it was. I don't remember."

Pincus: "You collected all of those materials – ."

Howell: "Some of them. I don't know exactly what got turned over to you but I located – I requested from Bernie Sarrel, I believe, the list of – that disposal list."

Pincus: "And who is he?"

Howell: "He is the facilities manager, I think, for planning. I don't know what his exact title is, but he basically keeps track of all the research space. And I assume his responsibility also is to monitor the inventory, but he doesn't maintain the inventory, I think he just passes the papers from one place to the next."

Pincus: "And did you also locate and obtain the environmentally related documents?"

Howell: "No, I did not."

Pincus: "Why did you not do that if I may ask?"

Howell: "Why would I do it?"

Pincus: "Because it was requested."

Howell: "Nobody requested me for Environmental Health and Services about mercury so not that I know of."

Pincus: "Well, you said that in regard to this particular Coulter counter there would have been such documentation. Correct?"

Howell: "Yes, I did."

Pincus: "So sitting here as of today, you have never been requested to obtain that information?"

Howell: "No."

Pincus: Scott [Flynn] can you explain this please? What is going on?

Flynn: He doesn't work in Environmental Health and Services.

Pincus: Yeah. But we made these requests – ."

Flynn: "And I have made them to the appropriate people, and I have pushed since our last meeting so…"

Pincus: "Why haven't I gotten the documents that he said he turned over in April?"

Flynn: "That [he] and I talked about. You are going to get one document today that he did fill out. It is being Bates'd right now."

Pincus to Howell: "Were there any other documents that you recall were generated insofar as maintaining inventories of any Coulter counters used in your lab and/or the disposal of them that we haven't identified in the course of this questioning?"

Howell: "Not that I know of. That is the only Coulter counter that was actually purchased and that was purchased when I arrived in 1987 when I first came to the school. All Coulter counters that arrived since then, which – none of which I have used, were donated by pharmaceutical companies."

Pincus: "When they're donated is it correct for me to assume that they become university property?"

Howell: "In the past, no. More recently, yes, at least as far as I know. In the past, no."

Pincus: "So when you say in the past can you be more specific by way of a time frame as to when the practice changed?"

Howell: "That I don't know. But I used to insist that they not be designated as university property in the event that somebody wanted to move them elsewhere and nobody ever argued with me."

Pincus: "So to the extent that your wishes were respected, they would not have been tagged and inventoried?"

Howell: "Correct."

Pincus: "What about if you needed to discard them what would you have done?"

Howell: "Then we would have requested – in fact, we still have times that – small items that aren't tagged, you know. They don't tag every single – every single item, they go through and look 'do you want to tag that, this and that'? So, anything that you want to discard, you have got to fill out a form and ask them to discard and you will tell them it is not – it doesn't have a university identification number."

Pincus began to roll out the first of thirty-one exhibits, most of them copies of experiments performed in Howell's lab and certainly those in dispute. From the first exhibit, the exchange between Pincus and Howell was contentious. Pincus had asked Howell to make assumptions about the experiments he was shown to which defense counsel John Leonard objected. The back and forth between Pincus and Leonard got heated.

Leonard: "Shelly, you can't sit and interpret these documents. I mean if she [Hill] has questions then maybe she needs to write them out. But he is reading line by line trying to interpret – make assumptions."

Pincus: "You don't have to testify for him, John. If he can't answer the question – ."

Leonard: "He already said he can't."

Pincus: "He has been instructed not to. And I asked him to assume something, and I am simply asking whether he can do it. If he can't, we will move on."

Leonard: "Well then, I am going to tell him not to answer any questions based on assumptions. I mean this – if you're asking him to do – make scientific decisions or interpret scientific data – ."

Pincus: "How can you instruct him not to respond? Are you objecting as to form? Are you objecting as to privilege?"

Leonard: "I am objecting as to form because you're asking him questions based on numerous assumptions we are going through. He is helping you by finding letters in the protocol. She is writing, you know, until she is getting blisters on her fingers. If you can get the questions right – ."

Pincus: "Cut the commentary because that is not going to do anything."

Of the exhibits Pincus rolled out, those marked Exhibit Howell-5, Dr. Lenarczyk's V79 experiment dated October 2, 2000; Exhibit Howell-6, Lenarczyk's experiment dated December 14, 2000; Exhibit Howell-7, Lenarczyk's experiment dated May 3, 2001; Exhibit Howell-8, Lenarczyk's experiment dated May 21, 2001, and Exhibit Howell-9, Lenarczyk's experiment dated June 21, 2001, were shown to Howell.

"Alright. Let's just make sure that you're in agreement that Howell-5 appears to be an experiment that he [Lenarczyk] did with V79 cells, I understand to be known as a 100 percent experiment on or about October 2, 2000. Do you recognize this document as the documentation associated with Dr. Lenarczyk's experiment?" Pincus asked Howell. "No, I don't recognize it because I don't think that I have seen this document," Howell responded. "You have never seen – you were his supervisor, and you have never seen this? Pincus asked, surprised. "I do not recall seeing this. I see, I think, what is Dr. Hill's writing on here." Pincus showed him Howell-6, an experiment Howell did along with Lenarczyk in December 2000.

"Have you seen this?" Howell was still looking at Howell-5. Pincus tried to move Howell along. "No. I am going to finish – [looking at Howell-5]," he retorted. Howell still didn't recall ever seeing the document.

"Now looking at this document and this experiment [Howell-5], which was apparently performed in your lab, if you would look at the document that is Bates stamped 091453 and 019454 would you agree, looking at the results that are graphed, that insofar as this 100 percent experiment insofar as survival was concerned, Dr. Lenarczyk's results were biphasic?" Pincus prompted. "Biphasic – which one?" Howell asked. "What does this tell you in terms of the results about the amount of killing [of cells] that occurred during this experiment? Can you glean that from the data that is set forth?" Pincus asked. "I would have to go through and analyze – reanalyze the whole thing. That is a plot in front of me. I have no idea whether or not it represents the data within. I have no idea," opined Howell. "You can't answer my question?" Pincus came back at him. "Not without analyzing all the data." Turning that response around, Pincus asked what Howell would have to do for the purpose of doing that – to answer his questions. "I would have to determine the uptake of the cells. Okay? So that would require cell numbers, activity from a liquid scintillation counter, [and] I would have to look at the colonies and their dilution [and then] I would have to go through and analyze the whole thing. It would take a long time," Howell explained.

"Let's look at Howell-6 please, the experiment that you did along with Dr. Lenarczyk on or about December 14, 2000," Pincus moved on. "Does the documentation associated – does the documentation comprising Howell-6 appear to be all of the documentation associated with this experiment?" Howell had no idea. "You have no idea?" Pincus asked. "How can I tell you if it's all the documentation of the experiment? How could I possibly tell you that?" Howell responded. "Because you participated in it and they apparently come from documentation provide by your counsel, which previously had been supplied to the United States Attorney's Office in response to a subpoena *duces tecum*. And you're telling me," said Pincus, "you don't have any recollection as to whether there are any documents that are missing – ." "That is not what I told you," Howell replied. Leonard jumped into the exchange.

Leonard: "Hang on. Objection. Objection to form. (A) First of all, we represent all the defendants. And insofar as you did not serve each individual defendant with discovery, you don't know which documents are coming from whom. (B) The fact that documents are produced in no way shape or form represents that it is all the documents related to any particular experiment. (C) This experiment is dated 2000. To ask this man in 2000, a year – eight years later whether or not what is stapled here represents all the documents in connection to this experiment is an absolute joke."

Pincus: "Okay."

Leonard: "Any other questions?"

Pincus: "Yeah, I have plenty."

Leonard: "Let's get to them."

Pincus: "Oh, we will regardless of how long it takes."

Leonard: "Well, you got a day."

Pincus: "I may have more than a day. We had an agreement that we were not confining ourselves to the one-day rule. Is that correct?"

Leonard: "Let's go. Let's just move."

Pincus: "Am I correct?"

Leonard: "We did unless we are going to do this all day."

Pincus: "No, we had an agreement. Okay?"

Pincus turned back to Howell: "Am I correct that the results of this experiment show that the survival curves were biphasic?" Howell responded that, again, he couldn't answer the questions without going through and analyzing [the data]. Before he could finish his sentence, Pincus interrupted him. "Because I'm asking you to look at the graphs -. Let me ask the question. You asked me to give you the courtesy of asking a question and I will then give you the courtesy of giving a response." Leonard told Pincus to move on. The exchange became uncomfortably contentious between Pincus, Howell and Leonard.

Pincus: "Are you ready?"

Howell: "I am ready."

Pincus: "Okay. May I ask my question now without being interrupted?"

Howell: "Yes, you may."

Pincus: "Thank you. If you look at B019492 – ."

Howell: "Yes."

Pincus: "- the graph that appears in the upper right-hand portion of that document. Does that indicate to you that survival was biphasic?"

Howell: "The points on that graph indicate that whatever is graphed is biphasic."

Pincus: "Thank you. And is there any question in your mind that you participated in this experiment with Dr. Lenarczyk?.

Howell: "No."

Pincus: "Thank you. Let us look at Exhibit 7 please, Howell Exhibit 7. Do you recall Dr. Lenarczyk performing a 100 percent cluster experiment with tritiated thymidine on or about May 3, 2001?"

Howell: "No, I don't recall him performing an experiment."

Pincus: "If you would look at the graphs that appear on page 19472, the upper right-hand corner – ."

Howell: "Excuse me, I've got another call. [A recess was taken.]

Pincus directed Howell back to graph at the top of page 19472. "Am I correct," he asked Howell, "that the points of that graph suggest that survival was biphasic? Howell answered that he thought they were biphasic. Pincus showed him Howell-8, Lenarczyk's May 21, 2001, experiment. But Howell didn't recall having ever seen it. Pincus pressed on, asking again whether the experiment's results indicated it was biphasic. The exchange grew tense when Howell asked Pincus if he actually understood what "biphasic" meant. "Yes, I do," Pincus told him. "Then how can you ask the question?" Howell retorted. "I just did," Pincus replied. "Then it is a ridiculous question to ask," Howell rejoined. "You can't answer it?" Pincus asked, pushing back. "I can answer it," Howell replied. "It shows to me a single component. That means exponential, not biphasic."

Pincus, undeterred, produced Howell-9, Lenarczyk's June 21, 2001, 100 percent experiment involving tritiated thymidine. He asked if Howell had seen it, as he would continue to do with all documents he produced during the deposition. Howell responded that he hadn't read it. He asked again about a biphasic result. Yes, Howell said, it showed a biphasic outcome. "And so by biphasic what we are referring to is that insofar as survival there is a decline followed by a plateauing of survival. Isn't that correct?" Pincus asked. "There is an initial decline, which has a steeper slope than the second component of the survival curve," Howell replied. "Do you have any reason to deny the fact that Dr. Lenarczyk conducted the various experiments that I have just shown you, which we have marked as Howell-5 through -9, in your laboratory?" asked Pincus. Leonard objected to the phrasing of the question. "Do I have a basis?" Howell asked Pincus. "Any factual basis to deny that these experiments were performed in your laboratory by Dr. Lenarczyk?" Howell answered: "I would need to go and see the original notebooks. And assuming it is his writing then, I guess, I would assume that he did the experiment." Pincus came back at him, barring seeing the notebooks, did Howell have any reason to doubt that Lenarczyk had conducted the experiments in Howell's lab. Leonard objected again, telling Pincus "I don't even know how you'd answer that." Howell interjected, "I am puzzled as to what you're trying to get at."

Pincus: "In other words if they're in his lab notebook, you're – ."

Howell: "Are you asking if Marek did experiments in the lab?"

Pincus: "I am asking if he performed – do you have any reason to deny that he performed these various experiments that I have just shown you?"

Leonard: "Objection to form. I think the problem is, if I can, Shelly, I think the problem is admit or deny. If these were in his notebooks would you assume that he did them?"

Howell: "I would assume that he probably did them, but I wouldn't have any evidence that he necessarily did them."

Pincus: "If he testified under oath that he performed these experiments in your lab would that satisfy you?"

Howell: "Yeah. If he testified under oath, yeah. I guess that would be satisfactory."

Pincus: "Okay. And these experiments appear to be associated with the grant we have been speaking of. Isn't that so?"

Howell: "Draw the connection for me in what way that they're related?"

Pincus: "Well, it is involving tritiated thymidine and it is – ."

Howell: "Yes, they involve tritiated thymidine."

Pincus: "Part of your grant involved doing certain – a 100 percent experiment associated with tritiated thymidine. Isn't that correct?"

Howell: "Please provide me with the grant to refresh my memory. This is – which one? This is the – ."

Pincus: "Howell-4 [NIH grant in dispute]. This is your grant application."

Howell: "Well, I submit lots of grants, so this is one of them. So, this is – ."

Pincus: "You don't really seriously mean to tell me that you don't know what grant this suit is all about. Is that what you're telling me?"

Howell: "No. I didn't know which document that you passed to me."

Pincus: "Okay. Well, why would I give you any other grant given the fact – ."

Howell: "You are giving me all kinds of stuff so, God knows. So anyway, this grant, yes, was submitted in 2000. And your question is what?"

Pincus: "That there were a 100 percent experiments that were conducted in regard to tritiated thymidine were there not?"

Howell: "I am going to look because this was dated – that was acquired many moons ago. Let's see here."

Pincus: "That were to be. That were to be performed as part of the grant?"

Howell: "That were to be performed. Well let me look here. Yes. To be performed according to the specific gains would be a 100 percent experiments with tritiated thymidine. That is correct."

Pincus: "I will take that back, you answered my question. Thank you."

All of Lenarczyk's experiments labeled Howell-5 to -9 involved V79 cells. Pincus introduced Howell-10 and Howell-11 and asked Howell to turn his attention to two articles Howell coauthored. He asked him to look at page 91 of the *Rapid Communication* article and Bates stamp 402. "Would I be correct that in this article, Figure 3, is indicative of the survival of V79 cells as a function of cluster activity of tritiated thymidine?" Pincus asked. "Yes," Howell replied. "And that insofar as the 100 percent experiments the results are signified by the use of either of a triangle either shaded or unshaded?" Pincus continued. "Yes," said Howell. "And would I be further correct that insofar as this particular graph the survival results of 100 percent experiments are shown being exponential?" Howell replied yes again. The exchange continued over the graph without much result.

Pincus introduced the *Free Radical Initiated* article – Howell-11 – again drawing attention to data pertaining to survival of V79 cells in the 100 percent experiments. Pincus pressed on the graphics. There was little Howell could offer to explain the exponential decline. "How do you explain the difference in survivals between what is represented in the 100 percent experiments and these two articles and the results of the Lenarczyk experiments that you identified as being biphasic?" Pincus wanted to know. "I don't have an explanation for why they're different. There are many reasons that there could have been, which I went through with the attorneys not with respect – not the attorneys – what are they called? The federal guys," Howell continued, "whatever they're called." Pincus wanted an explanation. Howell didn't have one.

"Not withstanding that you don't have an explanation did you discuss the difference or discrepancy with Dr. Bishayee and/or Dr. Lenarczyk?" Pincus asked. "I don't recall having that discussion with Dr. Lenarczyk," Howell replied. "Or Dr. Bishayee? Pincus rejoined. "Or Bishayee," said Howell. "And then just so we can just come full circle if you don't recall a discussion is it fair for me to conclude you don't recall anything that either one of those individuals may have said to you about that?" Pincus continued his line of questioning. "No, I don't recall," Howell answered.

Further into the deposition Pincus asked Howell to explain the process – the protocol – he used with his postdoctoral fellows, whether it was Bishayee or Lenarczyk as they finished experiments applicable to the NIH grant in dispute, how the completed work came to cross Howell's desk if at all. "Generally, they would perform the experiment, analyze the data. We compare to whatever data that we had on hand and assemble. Once we thought we had sufficient amount of data to obtain a similar result then we would move forward for publication otherwise, we wouldn't," Howell replied. "But generally speaking, can you characterize a time frame in terms of when the experiment the postdocs are doing are concluded when it comes to land on your desk?" Pincus asked again. "That could vary anywhere from days to years," Howell answered. "What about these experiments [Lenarczyk's]. Can you in any manner recall how long it was?" Pincus prompted. Howell didn't remember Lenarczyk's February 19, 2001 experiment ever crossing his desk nor did he recall discussing it. Pincus marked exhibits Howell-15, an experiment dated July 16, 2001, and Howell-16, another experiment dated September 27, 2001.

Pincus: "Dr. Howell, I am going to show you what we marked, as Exhibits Howell-15 and Howell-16. You recognize these documents as experiments that you performed on or about July 16, 2001, and September 27, 2001, respectively?"

Howell: "Yes, these are my writing."

Pincus: "And am I correct that these documents represent experiments that you performed at the request of Dr. Raveché in the aftermath of the complaint that Dr. Hill – ."

Howell: "Not that I recall."

Pincus: "Both of these are 100 percent experiments are they not?"

Howell: "They indicated that they're 100 percent, yes."

Pincus: "And were these – strike that. And insofar as Howell-15, I want to go to page 74555 if you would please?"

Howell: "Okay."

Pincus: "And would I be correct that the minimum survival that you observed insofar as conducting this experiment was – is represented by what appears to be 210 at 0.3623. Is that correct?"

Howell: "That appears to be correct."

Pincus: "And insofar as the chart that appears in the upper right-hand portion – [is it] correct that looking at the points of those charts that the results showed survival to be biphasic?"

Howell: Again, there is one point outside the exponential, perhaps two. So, it is exponential for all but one set of the points. So yeah, I mean maybe you could construe that, but you would need more data points to ensure that it is indeed biphasic.

Pincus: "I am interested how you construe it, however. Can you answer the question?"

Howell: "I can't make a conclusion based on this."

Pincus: "How do the results on this experiment compare with the survival results in the two papers that we reviewed earlier? I am happy to share them with you again if you need be."

Howell: "I'd have to look at that, yes."

Pincus: "I'm going to show you again, Howell-10 and -11 – I'm sorry, -11 and -12, I apologize. So, let's first go to – oh, no, no, no it was Howell-10 and -11."

Howell: "Okay."

Pincus: "So if we go to Howell-10 first and we go to Figure 3 on the document Bates stamped 402, page 91?"

Howell: "Yep."

Pincus: "Would you compare the results of the experiment that you performed on July 16, 2001, with the results of the 100 percent experiment shown in Figure 3 of Howell-10 please?"

Howell: "They're different. They're different axes. The X axes are different on them. One is cellular uptake millibecquerels per cell the other is cluster activity in kilobecquerels.

Pincus: "What about insofar as exponential versus biphasic, the kinetics, so to speak?"

Howell: "The article, which has the survival versus the cluster activity, appears to be a single component exponential. The data in Exhibit 15 has most of the data showing exponential with a single point appearing not to fit in the exponential line. You might construe two points but one for sure."

Pincus: "Well what is the other one that you say one might construe?"

Howell: "I mean if in the absence of the last point, I would fit that with a pure exponential curve. If I look – if I put my thumb over that one point and I were going to pick a function to fit that data, I would pick an exponential function."

Pincus: "So let the record reflect that Dr. Howell has his thumb over the point that appears at the far right, a portion of the chart above the horizontal axis, which is denoted at approximately 3.6. Is that correct?"

Howell: "3.6. Yes."

Pincus: "And that 3.6 by the way corresponds with the minimum survival of that that you identified earlier?"

Howell: "Correct."

Pincus: "But if you include that is it biphasic?"

Howell: "It may or may not be. I mean, you can't say that it definitely is unless you got at least three points on the far side that second phase."

Pincus: "I understand your position. Now let's look if you will please at Howell-16. This is an experiment that you performed on September 27, 2001. Is that correct?"

Howell: "September 27, 2001. Yes."

Pincus: "And this was a 100 percent experiment. Correct?"

Howell: "A 100 percent, yes."

Pincus: "And survivals are shown on 7485. Is that correct?"

Howell: "7485. Yes."

Pincus: "And would you characterize the survival here as exponential or biphasic?"

Howell: "That one, I would characterize as biphasic maybe even triphasic."

Pincus: "So it appears that at least in the majority of the experiments I have shown you and I have asked you to comment on the survivals, the results of survival, were biphasic. Is that fair?"

Howell: "Excuse me?"

Pincus: "Insofar as the majority of the experiments that I – 100 percent experiments that I have reviewed with you here this morning where I have asked you to review and comment upon survivals, you have characterized those survivals as biphasic rather than exponential. Isn't that so?"

Howell: "No, that is not true. I said that – and in fact, I am looking at this one more closely. Again, there is only one data point outside of what appears to be exponential. Indeed, there is a shift of all of the data points relative to unit survival but there is only one data point outside of what would appear to be exponential. So again, even with this one scientifically, one couldn't conclude either way. So, from a purely scientific standpoint, you can't conclude biphasic."

Pincus: "Well isn't the exponential decline at – represented by the far left portion of the graph where it goes from one down to – I am going to point it [out] to you that is the easiest, to that point, which, I guess, is zero-point about eight or so – I am just guessing looking at it upside down."

Howell: "Yes."

Pincus: "And doesn't one then characterize the biphasic nature of the survival as what goes – is represented by the points that occur after that?"

Howell: "Well in this case what is unusual about this curve is there is no points between the first point and the control. So, you don't know what is going on there. So, you have to ask yourself did the whole curve get shifted for some reason by some systematic error or – you know, I can't explain why the first point there starts at .2, .3, .4, .5, .6. That is a little odd."

The bottom line to this exchange came down to whether the September 27, 2001, experiment Howell performed did not replicate the result set forth in the paper marked Howell-11. When he asked Howell if he agreed to this – a simple yes or no question – Howell answered "no." "It did not," Pincus prompted. "No. I would say that it doesn't tell you that. You have two experiments and you're telling me to say that this doesn't replicate that. I can't make that statement." Pincus wanted to know why he couldn't make that determination. "Because if you do any two or three or four experiments and look often, the millibecquerels per cell can change from experiment to experiment. So," Howell continued, "I am not going to conclude, based on the fact that this is two millibecquerels per cell and this D37 is .8 millibecquerels per cell, that I haven't reproduced it. That I am not going to say at all." Pincus pushed back.

"If a principal investigator cannot replicate published results what is your understanding that that principal investigator should do?" Leonard objected to the question. "That is such a general – general question. I can't answer the question. You need to be more specific," Howell replied. "Do you have any understanding insofar as any obligation on the part of a principal investigator to report the inability to replicate results with the entity that afforded them, or granted to them, the grant monies?" Pincus rephrased the question. Leonard objected again. "Do you really think," Howell retorted, "that each and every experiment everyone does, they get exactly the same thing? And each time that they don't, they call and say hey, I didn't get [it]?" The exchange between Pincus and Howell became heated. "I am asking you," Pincus pushed back, "I get to ask the questions. You just get to answer them." Howell didn't back off. "I would say that if somebody doesn't get the exact same results no, they would not pick up the phone every time," he testified. "So, you didn't see any obligation based on the experiments, the 100 percent experiments that have gone through here this morning, to report the inability to replicate the survival results with the NIH?" Leonard objected, asked and answered. "My focus was not as much on the 100 percent experiments with respect to survival. I was looking to see if there was a mutation response so...," said Howell. "So, the answer was that you did not feel an obligation to report these results. Correct?" "No, I did not," Howell replied.

Pincus introduced Exhibits Howell-17, an experiment dated December 26, 2000; Howell-18, an experiment dated January 15, 2001; Howell-19, an experiment dated February 5, 2001; Howell-20, an experiment dated June 14, 2001, and Howell-21, an experiment dated July 5, 2001. "Before we move on to these, when I was discussing the 100 percent experiments to you, I asked whether you felt any obligation to report those experiments to the National Institutes of Health. Did you report these experiments in any way, shape or form to Dr. Baker?" Pincus asked Howell. Leonard objected. "No," Howell replied. "Dr. Putterman?" Pincus prompted. Leonard objected again. "The 100 percent experiments?" Howell asked. "Yeah," said Pincus. "No," Howell answered. "The answer was no?" Pincus wanted to confirm. "Not to my recollection," said Howell. "Let's start looking at these. If you look at Howell-17 this appears to be a 50 percent experiment done with V79 cells by Dr. Lenarczyk on or about December 26, 2000. Is that correct?" Pincus asked. Howell confirmed the experiment as Lenarczyk's work. As with the 100 percent experiments, Pincus questioned Howell about the particulars of the outcomes. He went on to further introduce Howell-22, an experiment dated November 20, 2000; Howell-23, an experiment dated November 28, 2000; Howell-24, and experiment dated February 15, 2001. These were further Lenarczyk 50 percent experiments.

Pincus began to ask about survival as he had previously done with other experiments, asking Howell if what he was looking at in the November 20, 2000, experiment demonstrated the existence of a bystander effect. He directed him to a page Bates stamped 19978. Leonard became concerned. "You know what let me just put on the record, Shelly, that although it is stapled together this looks like the Bates number comes from two different sources," he objected. "So, I am not altogether sure that these pages even go together." Pincus replied that they were the same exhibits that were shown to Lenarczyk during his deposition. "I don't think that is – I think he identified the pages, his name -," Leonard continued. "You can note your objection," Pincus rejoined. Pincus turned back to Howell, pointing out Bates 19978. "One other thing of note it is interesting," observed Howell, "that this date here is crossed out on the first page and 11/20 is penned in and then here it is, 11/20. So that may be the source of your [Leonard's concern]." What began as Leonard's objections and Pincus' pushback during this exchange soon boiled over to what Hill described as plaintiff and defense counsel "nearly coming to blows." What had been tedious question and answer over a lengthy series of experiments to which Howell claimed no firsthand knowledge dissolved into a verbal brawl between legal counsel. Suddenly, it was no longer about the importance of fleshing out the science, establishing the facts and fixing accountability – and what happened next served only to muddy the deposition of a key defendant, the first of the named defendants in Hill's case to be deposed.

> *Leonard*: "Shelly, I just want to make one clarification. We are going through all these documents and we're just flipping through a graph. So long as we are clear that Dr. Howell is not looking at any of the data, how the data – wait a minute. How the data is contrived. You might as well just take what you're asking him and you might as well just take pictures and draw it, whether something that looks like that represents a bystander effect. Is that your question?"
>
> *Pincus*: "That is my question. Whether that chart [from the Lenarczyk experiment] indicates to you that these results indicated the existence of a bystander effect?"
>
> *Leonard*: "Well wouldn't it be quicker to just have him demonstrate what a chart that would show that would look like?"
>
> *Pincus*: "Maybe if you wanted to do it that way when you question him, you are free to. Right now, I am asking him this way."

Leonard: "Well here is the problem. I am not going to let him – I want to be very clear that – and we keep going through this, you keep showing him all these documents and he's not substantiating, nor does he have any firsthand knowledge about the information in them [the Lenarczyk experiments]. If you want him to look at pictures, like is this a picture of the sun? Yes, I think that looks like the sun, that is fine. But he is not rendering any opinion as to what is in these documents. So, if you just want to flip to the last page and stop all the commentary about what Lenarczyk said or didn't say, we could just do that and he can say well no because as I have told you before in order to show bystander [effect] it would have to depict this. But other than that, he is not rendering any opinion or substantiating any of this information."

Pincus: "Well, your objection is noted. I want to proceed the way – ."

Leonard: "Are we in agreement on it?"

Pincus: "I am not agreeing or disagreeing with you. I am questioning him. I want to proceed the way I want to proceed. Are you directing him not to answer?"

Leonard: "Well, I am going to direct him not to guess. If what you're asking for is an opinion on this document, he has no firsthand knowledge. Then, I am going to tell him not to guess as to what any of this means."

Pincus: "We have gone through a whole series of experiments in which you did not, you know, indicate this problem until now. I have this one and a few others to go through and I want to conclude it."

Howell: "That is not true. I will interrupt – ."

Pincus: "No, you will not interrupt. There is no question pending to you, sir. So please with all due respect sit there and do not respond – ."

Leonard: "Shelly, he's allowed – ."

Howell: "May I respond?"

Pincus: "Do not respond until a question is posed to you. I am having a dialogue with your counsel. I am not looking to argue with you."

Leonard: "Shelly, I think this is the problem. You see what the problem is? He thinks – ."

Pincus: "The problem is – ."

Leonard: " – you're looking at pictures."

Pincus: "The problem is, you told him well gee here is the basis for objecting. You're testifying for him."

Howell: "No, that is not true."

Pincus: "I am not speaking to you, sir."

Howell: "Well, I am speaking to you."

Pincus: "Then please – ."

Leonard: "Lower your voice or I will throw you out of here."

Pincus: "Don't throw me out."

Leonard: "Get out."

Pincus: "No, I am not getting out."

Leonard: "Get out. Deps over. Out."

Pincus: "I am not getting out."

Leonard: "This is my building. You will get out or I will throw you out."

Pincus: "You will not – ."

Leonard: "Dep is over. Get out."

Pincus: "If this dep is over then I want to get on the phone with the judge. Let's get on the phone."

Leonard: "I don't care who you call. Do it on your way down the elevator."

Pincus: "Let's take a break and – ."

Leonard: "No."

Pincus: "Why are you concluding this dep?"

Leonard: "Because you're yelling at my witness and it is over. I asked you to lower your voice and you wouldn't."

Pincus: "Let's go off the record and have a discussion outside. Will you please do me the courtesy."

Leonard: "No."

Pincus: "You're not going to do that?"

Leonard: "You're not going to yell at my client."

Pincus: "I did not yell. I respectfully disagree with you –."

Leonard: "You're yelling now."

Pincus: "No, I am not. I am not yelling. My voice tone is much lower than yours."

Leonard: "You're accusing me of telling him – you're letting him – the reality is, he's looking at pictures and what you don't like is, you want him to act as if he is giving you an opinion on these documents and that is not what is happening."

Pincus: "John, number one are you done?"

Leonard: "I am and so are you."

Pincus: "No, no, no. I want this on the record here. Now hear me out and I am surprised that you're taking this position here right now because the fact that we have our first disagreement on the record, you're terminating a deposition? I politely asked your client not to interrupt a dialogue that you and I were having. You, yourself, interjected when he attempted to do that requesting him not to do so. You and I continued to talk about it, he went in, I politely asked him not to do that."

Leonard: "Shelly, you raised your voice. I asked you not to – ."

Pincus: "You raised your voice at me, too. You have, too. So alright if we raised our voices, or whether we didn't raise our voices, there is no basis for you given the timelines here. Are you declaring discovery over?"

Leonard: "I don't even know what you're talking about."

Pincus: "No. If you're terminating this deposition then you expect other depositions to go on before we conclude this one? Because there is not a good reason for you to terminate this deposition based on what occurred. And if you are going to do that before we break, I want to reach out – I want us to reach out to Magistrate [Judge Mark] Falk so that we can have a discussion concerning that. That is what the order requires us to do when we have those discussions.

Leonard: "I don't – ."

Pincus: "Let's go outside please. Can we please go off the record and go outside?"

Leonard: "Off the record. I'll be right back." [A recess was taken.]

When counsel returned, Pincus began to ask Howell about the experiments he'd introduced before the argument with Leonard. One of those was, of course, was Howell-25, a 50 percent experiment that Howell had done on

April 27, 2001, and about which Azzam had observed and about which Azzam had already been deposed. The documents contained Howell's writing on the pages, but he didn't recall the specifics of the experiment. Pincus asked if, looking at what was in front of him, if the experiment showed a bystander effect. Howell noticed, flipping through the pages, that some of the handwriting wasn't his own. "It may be Marek's, but I am not sure." Pincus asked if it might be Bishayee's. "It might be Bishayee's. It could be." Pincus asked if it might be Azzam's. "No, it couldn't be Dr. Azzam's." After some exchanges about the tubes and colony counts, Pincus asked him why Azzam participated in or observed the experiment. "I wanted – I don't remember in this particular one," Howell replied, "But the reason [for] having Dr. Azzam was as a witness that these experiments were being done."

> *Pincus*: "A witness for whom?"
> *Howell*: "For me."
> *Pincus*: "Who told you to do them?"
> *Howell*: "Nobody told me to do them that I recall."
> *Pincus*: "You don't recall being instructed to perform this experiment by Dr. Raveché or any other administrator at the university?"
> *Howell*: "I do not recall being instructed as to do so."
> *Pincus*: "And is it your recollection that – well what is your recollection of what Dr. Azzam did in regards to this experiment?"
> *Howell*: "He – I was at the hood in 451B, which is toward the back of that room. The window is to my right along with a desk. I don't remember if the desk was there at the time and he sat by that, what I think is where the desk was."
> *Pincus*: "That was all he did."
> *Howell*: "Yes."
> *Pincus*: "Did you confer with him in any fashion in regards to this experiment?"
> *Howell*: "In terms of setting it up and so on?"
> *Pincus*: "Setting it up and carrying it out?"
> *Howell*: "Other than to ask him to sit? No."
> *Pincus*: "Is it your recollection that he sat and observed each and every step association with this?"

Howell: "I can't recall whether he saw every step or not."

Pincus: "Do you recall whether he sat and observed any other experiments or just this one?"

Howell: "I believe there was more than one, but I can't recall. It would be indicated by initials."

Pincus: "Is it your recollection that Dr. Azzam initialed the documents associated with any experiment that he observed at your request?"

Howell: "That is what you see right here."

Pincus: "Well, I see it insofar as this particular exhibit – Howell-25."

Howell: "Correct."

Pincus: "But so I am clear. Are you telling me that to the extent he observed other experiments that you performed, he would have engaged in that initialing process?"

Howell: "I believe so, yes."

Pincus: "That is your recollection?"

Howell: "Yeah, that is my recollection."

Pincus: "Fair enough. Now I am going to show you Howell-26, please."

Howell: "Okay."

Pincus: "Do you recall whether you performed this experiment on or about April 19, 2001?"

Howell: "Again, I don't recall which ones were done and weren't done so, I don't recall."

Pincus: "Does this contain your handwriting?"

Howell: "Yes."

Pincus: "Does it assist you at all? If you look at the document Bates stamped B007390, it has your name. See where I referring to, sir, up here?"

Howell: "73 – ."

Pincus: "90 – ."

Howell: "Okay. Yeah."

Pincus: "Does that assist you in terms of identifying whether you performed this experiment or not?"

Howell: "That would indicate – that would indicate that I entered data into the spreadsheet, or that would suggest that I entered data in the spreadsheet to analyze the variety of data that we have here."

Pincus: "The significance of identifying you as the investigator means what?"

Howell: "Means that I would have carried out the experiment. That is normally what I would – normally how I would indicate that."

Pincus: "So if the name Lenarczyk or Bishayee appeared there it would be reasonable to conclude that those individuals carried out the experiment?"

Howell: "Assuming that when they filled out the spreadsheet – because we use the same spreadsheet over and over. Assuming that they had typed in – you know erased what was there and typed in their name that is the case."

The exchange between Pincus and Howell over particulars of the evidentiary experiments continued. "I am going to show you what we marked as Howell-27. Do you recognize this document?" Pincus asked Howell. "Again, I don't recall this particular document, but my writing is on it." The document in question was a 50 percent experiment that was performed on May 3, 2001. "And to your recollection, did Dr. Azzam observe or have anything to do with this experiment?" Recall that Azzam only admitted to one observation of a Howell experiment. "I have to see only – if his initials are on it. I guess the answer would be yes. Otherwise, I don't know," Howell replied. "Well, I don't believe that his initials are anywhere on this document," Pincus added. "So, he may or may not have been part of it. I believe that I had him initialing but I can't be certain," Howell responded. Much of the detail to the experiment in question – what followed – Howell could not recall clearly. Pincus showed him another exhibit, Howell-28, asking again if he recognized the document. "Do you recognize whose handwriting that is?" Howell wasn't sure. "It's probably Marek's. I am not sure."

Pincus: "Did Marek tend to use European sevens in his writing? Do you know what I am referring to?"

Howell: "Yeah. But I would have no idea."

Pincus: "What about Dr. Bishayee? Do you recall whether he did so?"

Howell: "I don't recall."

Pincus: "But in any event let's look at the survivals again here if you would please, as we have been doing. If you got to 73 – I'm sorry – 7430.

Howell: "7430."

Pincus: "What does this tell you in regards to bystander if anything?"

Howell: "Okay. So, if this represents a 50 percent experiment, which it says it is on page 7419, only one of the data points has a survival value below point – no, two. Excuse me, two of the data points have a survival value below .5. And those two data points would indicate perhaps some bystander. The other ones are above .5.

Pincus: "And that – ."

Howell: "Indicating that there doesn't appear to be a bystander – survival bystander or killing bystander.

Pincus: "Do you have any recollection in regards to these last few exhibits that I have just shown you having any discussion about them with Dr. Raveché?"

Howell: "With Dr. Raveché? Which docs now? So, we are looking at what?

Pincus: "These past four experiments that I just showed you, which you identified as having your names on them?"

Howell: "Those docs, I did not discuss with Dr. Raveché.

Pincus: "Did you discuss those documents with Dr. Baker to your recollection?"

Howell: "No."

Pincus: "To Dr. Putterman?"

Howell: "Not the document, no."

Pincus: "To your program director at National Institutes of Health?"

Howell: "No."

Pincus: "Who was that individual?"

Howell: "Program director? Gentleman – what was his name? Paul something. I forget. He is not – oh what was his name? I don't recall. I don't recall. In fact, I am not sure I ever spoke with him during the entire course of the grant. I don't typically give my program directors calls." Struedler (phonetic), Paul Struedler [actually Paul K. Strudler]. It might have been Mahoney. There is a Paul Struedler and Francis Mahoney, but I can't remember what they're two functions are."

Pincus produced Howell-29 and asked Howell if he recognized it. He did. He'd authored it. "Do you recall when you did so?" Pincus asked. "When the attorneys – after the district attorney's office visited my office. I think – was he the district attorney? Who the heck were those guys again?" Leonard stopped him.

Leonard: "The attorney general's office – the United States Attorney's Office?

Howell: "Yeah. Whatever they were. They came to my office. Subsequently, university attorneys got involved and they asked me to provide a summary [Howell-29] of the – of some experiments."

Pincus: "Can you – ."

Leonard: "Shell, can we take a minute? Are you saying you drafted this at the request of the in-house lawyers?"

Howell: "Yes."

Pincus: "It was disclosed to the U. S. attorneys. It's been out here."

Leonard: "I know. But it would be inadvertent production in terms of you guys [the plaintiff]."

Pincus: "Well, unfortunately there has been quite a bit of discussion about this and the other day when you weren't there in another context so…"

Leonard: "I am going to direct him not to answer any questions. I know it was provided to the U. S. attorneys [and] it was done at the behest of in-house counsel so, I'm going to say inadvertent production with respect to the civil litigation."

Pincus: "I have to ask him some other questions surrounding this. Okay? If I may? Okay? Let me ask a question. If you have a problem, you let me know without getting into the substance of the document, just procedure."

Howell: "Now before you do that can I have a word with you outside? I just want to ask you a question. It's not about this document, [it's] about something else. Can I take – I would like to take a break and ask a question or not?"

Pincus: "Let me just ask you a couple of questions and then I will let you do that. Okay?"

Howell: "Sure."

Pincus: "Did you discuss the preparation of this document with Dr. Azzam?"

Howell: "Did I – the preparation of the document? No."

Pincus: "Did you discuss with Dr. Azzam the material that forms – did you discuss with Dr. Azzam the information that is contained within the document?"

Howell: "Some of the information. [Recall that Azzam did not admit to this in his deposition.] I can answer these?"

Leonard: "You can answer the question."

Howell: "So some of the information in the document certainly was discussed with Dr. Azzam. But – well prior to the preparation of the document."

Pincus: "Okay. Am I correct that the information that you discussed with Dr. Azzam involved some of the variables that may have affected the inability to replicate the bystander effect in these experiments?"

Howell: "Yes."

Pincus: "Okay. And the variables that may have changed do you recall whether that information was provided to you by Dr. Azzam?"

Howell: "No that was not. You mean in written form."

Pincus: "No, no, no just in substance again. Did he orally say, you know, and discuss with you – ."

Howell: "We had discussions from which brain it emerged from; I couldn't tell you on a case-by-case basis."

Pincus: "And did you discuss with Dr. Azzam issues regarding the culture medium?"

Howell: "I discussed the culture medium issues perhaps with Dr. Azzam. But obviously there were issues also related to the cells, which involved Lenarczyk. So, who I discussed what with, I don't recall."

Leonard: "Shell let me just say, I have no problem in the world with you asking him these types of questions, but I prefer it not come from a line-by-line reading of the document. So, anything you want to ask him that is fine but, in my view, you shouldn't have that document."

Pincus: "Well, I am not so sure I should, or I shouldn't. And the fact of the matter is, I do. And at this point in the game so to speak it having been out there for as long as it has been, I am not so sure whether in fact there is a waiver to the extent that it is protected as you may suggest. So are you telling me that I can't show – you're not going to permit any questions associated with the document?"

Leonard: "I think that there is a lot of material in the document, and you are hitting upon it that certainly, you know, questions with Dr. Azzam before the preparation. What I don't want is for you to sit with the document, look at it and ask questions based on its contents obviously. But, you know, because that constitutes work product. If he is being told to compile this information [in] this format by the in-house lawyers that is work product."

Pincus: "Are you sure that you did not do this at the request of Dr. Raveché?"

Howell: "No, not at the request of Dr. Raveché.

Pincus: "You did not share this document with Dr. Raveché?"

Howell: "No."

Pincus: "Dr. Putterman?"

Howell: "With Putterman? I don't know if she saw it or not."

Pincus: "Did you share it with Dr. Baker?"

Howell: "No, I did not."

Pincus: "Did you discuss this document with any of those individuals?"

Howell: "This document? No."

Pincus: "Why don't you have your discussion." [A recess was taken.]

Leonard: "I'd like to put this on the record that I have looked at what has been marked as plaintiff's exhibit 29 [Howell-29] and I'm going to allow the witness to answer questions regarding that document."

Pincus: "So, Dr. Howell this document isn't dated but can you be any firmer insofar as a date or time when it was that you caused this document to be prepared?"

Howell: "When I was asked for this, they wanted it within 48 hours or something because I can remember sitting up in bed at midnight typing this up to get it done within the required time frame. So it must have been very soon after the whole thing the other attorneys started, I mean. So, I don't know the date."

Pincus: "Can you identify a year?"

Howell: "Two... When was your first – I think 2004."

Pincus: "She [Hill] can't respond."

Howell: Maybe 2004, 2005, somewhere in there."

Pincus: "Somewhere in there. Okay. Fair enough. And did anybody else have input into the preparation of this document other than yourself."

Howell: "In preparation of the document? As I said, this was done at midnight in my bed and my wife was in the bed and awake so, she would have been there while I was typing. But beyond that, no."

Howell's deposition was broken into two parts on the same day. Certainly, the second half of his deposition turned to the personal, to the people in his lab. Pincus produced exhibits Howell-32 to Howell-39. Howell-34 was the letter Howell wrote to Baker about Hill.

Pincus: "Do you recognize this document?"

Howell: "This document I recognize."

Pincus: "You authored this?"

Howell: "I have initialed it and therefore appear to have authored it."

Pincus: "Is the data accurate insofar as when it was prepared?"

Howell: "I don't know. I assume it is, but I don't know."

Pincus: "Why did you prepare this if you remember?"

Howell: "Why did I prepare this? I don't recall. I was indicating to Dr. Baker some sense of what was happening I presume. Let's see, let me read through."

Pincus: "Take your time. You will let me know when you are done."

Howell: "Yes, this document I prepared after, after, let's see, this was prepared after I learned that Dr. Hill was poking around in the laboratory."

Pincus: "Poking around?"

Howell: "Yes, poking around."

Pincus: "Okay. At that point in time though, she had not filed a formal complaint to your knowledge?"

Howell: "Not at that time."

Pincus: "You indicate approximately a third of the way down of that first paragraph that, 'as I told you this morning, I have requested that my postdoctoral fellow Marek Lenarczyk, Ph.D., repeat some of Dr. Bishayee's experiments as they check on the validity of the experiment.' Do you see where I am referring to?"

Howell: "Yes."

Pincus: "Did you tell Dr. Baker at the time that Dr. Lenarczyk in fact had already done some of the experiments?"

Howell: "I don't recall, and I don't think that that was known to me if he had or hadn't at the time but I don't recall."

Pincus: "So we are clear – ."

Howell: "What experiment are you talking about?"

Pincus: I will go over them with you right now. Did you tell him at the time that you prepared this memo that you were aware Dr. Lenarczyk had conducted an experiment which we have identified as Howell-5?"

Howell: "I believe this document is October 2000. I believe I had told you I wasn't sure that I had even seen it before. So how would I tell him?"

Pincus: "So the answer would be – ."

Howell: "So I don't – not that I recall."

Pincus: "Fine. That's what I need to know."

Howell: "Because I don't recall seeing Howell-5."

Pincus: "Did you share with Dr. Baker the fact that Dr. Lenarczyk may have performed an experiment on or about December 14, 2000, that we have identified as Howell-6?"

Howell: "I shared with Dr. Baker what's in this letter."

Pincus: "Is your answer to my question 'no, I don't recall'? You tell me, I don't want to put words in your mouth."

Howell: "This is December 2000?"

Pincus: "Yes."

Howell: "I don't recall discussing this with Dr. Baker."

Pincus: "May I take that back?"

Howell: "Hold on. No, I don't believe I was discussing that with Dr. Lenarczyk."

The memorandum was important. In it, Howell clearly informed Baker he was having Lenarczyk repeat Bishayee's experiments. "To the extent that Dr. Lenarczyk had, in fact, conducted 50 percent experiments preceding the time you prepared this memo you did not share that fact with Dr. Baker at this time. Correct?" Howell asked. Howell asked the purpose of the 50 percent experiments [prior to Lenarczyk being asked to repeat them]. "I didn't sit and discuss with Dr. Baker what experiments have been done and haven't [been] done," Howell replied. Pincus was skeptical. "You are telling me then that you did not discuss with him what experiments Dr. Lenarczyk had done prior to your giving him his memo. Is that correct?" Howell answered quickly: "That's correct." The memo also stated "'I [Howell] will also personally sign all of Dr. Bishayee's data sheets and Coulter dishes to ensure that they are not tampered with.' Do you see where I am referencing?" Pincus pointed out. Howell didn't recall having actually done this – initialing of Bishayee's evidentiary material.

Pincus showed Howell the minutes of the April 27, 2001, Newark CCRI meeting, marked Howell-36. He went back to questions regarding discussions with Baker. The question from Pincus, prior to showing the minutes was whether Howell had showed Baker any of the documents associated with the experiments. "So, my answer," Howell repeated, "stands. No, I didn't show him those documents." Pincus moved on. "Have you had any occasion to review the minutes that you have before you which was marked as Howell-36. Do you recall?" Howell replied that he believed he was given a copy of the minutes "about a year ago, maybe a little more." Pincus drew Howell's attention to comments Howell made before the committee regarding Lenarczyk. "You indicate, you indicated to this committee that Dr. Lenarczyk had been nonproductive in his nine months of one of your postdocs. Do you recall that statement?" Howell provided this answer: "I don't recall specifically but what, what I can tell specifically is he had been working on the asobel [*sic*][abscopal][145]. The plan was to look at the bystander effect of the abscopal and they had not observed any and they were having various other problems with the Coultering and that was not a portion of the grant. We had embarked on that because of his excellent idea of using the hydrolyson and neolyson resistance cell and we had poured a considerable sum of money to those experiments to no avail."

Pincus wanted to know if there was anything else on which Howell had based his statement to the committee about Lenarczyk. "That was his task, [it] was to do the asobel [*sic*][abscopal] experiments. That's what it is based on." But this subject matter nor the mention of abscopal experiments hadn't come up in Lenarczyk's nor Azzam's depositions. Was it the reason? "Did you base the statement [to the committee] on the fact that his results did not agree with Dr. Bishayee?" asked Pincus. "No, I didn't base it on that. I just said they just weren't getting to the point that they had data which we considered to be publishable," Howell replied. "Are you telling me that you were dissatisfied with his [Lenarczyk] performance in your lab," Pincus pressed. "All I can tell you is that, you know, that we didn't arise, no publications arose from his work and the productivity is publications."

Moving on, Pincus asked if Howell recalled when Bishayee resigned. Howell had difficulty remembering his departure until Pincus showed him an exhibit marked Howell-37, a memorandum dated July 30, 2001, and asked if he recognized it. He did – he'd written it.

Pincus: "Did you prepare this?"
Howell: "Yes, I did."
Pincus: "You did prepare this document?"
Howell: "Yes."
Pincus: "Do you recall giving this to Dr. Bishayee with an instruction for him to sign it?"
Howell: "I gave the document to him after having for months been encouraging him to consider other options with his life."
Pincus: "Explain to me the sequence of these discussions. When did they commence?"
Howell: "As we proceeded, during the summer, I discussed with him the fact that I believe that Dr. Hill would never leave him alone."
Pincus: "The summer of when?"
Howell: "In the months prior to this. And even before that obviously, but – ."

Pincus: "No, tell me right now."

Howell: "In the summer prior to, say, June, July and August – ."

Pincus: "that's what I want to know."

Howell: " – we concluded, I shouldn't say we concluded, I concluded she would never leave him alone no matter what because she had been after him the moment he walked through the door."

Pincus: "What activity led you to conclude that Dr. Hill was looking to get him since the day he walked in the door, which was when, by the way, if you know?"

Howell: "When she walked in the door. I don't know. He walked in well before she did." [Hill had been in the department since 1981. This statement would appear to be related to the specifics of Howell's lab only or Howell misspoke.]

Pincus: "You are saying that from the first day?"

Howell: "I don't know about the first day. Forgive me."

Pincus: "When did you first observe something that led you to believe Dr. Hill was for lack of a better phrase 'out to get' Dr. Bishayee?"

Howell: "She did make statements at some point, I don't remember where they were, in what period, that she had a distrust of Indians and they would do anything to please their boss, that statement she had made. What other documents? She was also angered with him after he used his computer to look at pornographic material. And two weeks after that is when she raised her first claim, if I'm not mistaken, that he was doing things incorrectly. And the timing of that is rather unusual."

Pincus: "Yes."

Howell: "Okay."

Pincus: "Anything else by way of statements?"

Howell: "You know, off the top of my head I would have to sit and think."

Pincus: "Right now you have identified three. Let me ask you some questions about them."

Howell: "Oh, and she requested also that I fire him. She made a specific request that he be fired surrounding that event of the pornography, and I refused. And I fully documented that event with human resources."

Pincus: "Anything else?"

Howell: "What else? You know, offhand you know, I, I would imagine that there's more but right now I can't think of any."

Pincus: "If my questions bring forth statements, if the statements come to mind as a result of my questions, let me know and we'll review them. When did she make the statement she distrusted, people of Indian descent I take it, and that they would do anything to please their boss?"

Howell: "I don't recall the date. I assume that it must have been on or about the time that she first made her allegations, but I don't recall specifically."

Pincus: "What allegations?"

Howell: "That he had done something in the September experiment, but it might have been prior to that."

Pincus: "September of what year?"

Howell: "'98 is it? '99, '99 I think that is. Can we please get the date straight on that."

Pincus: "I'm asking you to the best of your recollection."

Howell: "Okay, I don't remember."

Pincus: "Sitting here it was '98 or '99 to the best of your recollection. Is that fair?"

Howell: "Yes."

Pincus: "Was anyone else present at the time she made this statement to you?"

Howell: "I don't know. But I also know that other people have told me that she has made similar statements or demonstrated similar, similar representations."

Pincus: "Well, you don't recall whether there was anyone present at the time she made the statement to you?"

Howell: "No, I do not recall."

Pincus: "Where were you at the time she made the statement?"

Howell: "I don't recall."

Pincus: "Do you remember – ."

Howell: "I mean we were in the building near in the vicinity of the labs. I don't remember if we were in my office or the hallway or laboratory."

Pincus: "Do you remember what time of the year she made the statement to you?"

Howell: "I just told you I didn't remember the date that she made the statement."

Pincus: "I ask you were there leaves on the trees, on the ground, hot, cold. Does that in any manner refresh your recollection?"

Howell: "Have you been to our laboratory?"

Pincus: "No, I have not."

Howell: "If you are in the area of the offices or in the hallway you may as well be in the basement. If you are in the laboratory, yes, you can see out the windows. Yes, I don't recall."

Pincus: "Did you document as you said you did the issue relating to pornography any of the comments that Dr. Hill made to you about distrust of individuals of Indian descent because they will do anything to please their boss?"

Howell: "I'm not exactly in the habit of memorializing events."

Pincus: "The answer is, no?"

Howell: "The answer is I did not document. Oh, I didn't – ."

Pincus: "You have answered the question."

Howell: "Documented at the time, at the time of that you mean?"

Pincus: "Yes."

Howell: "No, I didn't."

Pincus: "Did you report it?"

Howell: "Did I report it to anyone, no."

Pincus: "You didn't report it to any affirmative action officer in terms of it being an ethnically offensive comment?"

Howell: "No, I didn't."

Pincus: "Did you find it that to be an ethnically offensive comment or comment offensive to one's national origin?"

Howell: "I found it surprising, yes."

Pincus: "But did you report it to anyone?"

Howell: "No, I did not."

Pincus: "You have no documents that would refresh your recollection of the circumstances of this document?"

Howell: "Other than the accusations that he did something wrong. As I said I think it may have been around that time, but I don't recall."

Pincus: "You say others have reported to you that she has made similar comments. Who?"

Howell: "I have heard some statements indicating – ."

Pincus: "I asked you who, not what."

Howell: "Dr. Rameshwar."

Pincus: "Who else?"

Howell: "Dr. Azzam."

Pincus: "Who else?"

Howell: "Those are the two I can, I think of at the present moment."

Pincus: "When was it that Dr. Rameshwar commented to you about such statements?"

Howell: "Sometime well after all of this occurred. Apparently – ."

Pincus: "What's 'this'?"

Howell: "All of this – sometime after 2001."

Pincus: "After 2001?"

Howell: "Yes, sometime after 2001."

Pincus: "You are saying, is it fair to say when you say 2001 was it after the time that Dr. Hill initiated a complaint to the committee?"

Howell: "Yes."

Pincus: "On the scientific integrity?"

Howell: "Yes, after that time."

Pincus: "Where did that conversation take place?"

Howell: "I believe in the vicinity of Dr. Rameshwar's laboratory."

Pincus: "Was there anyone else present?"

Howell: "I don't recall."

Pincus: "Was there anyone else besides the two of you?"

Howell: "No, I wouldn't have such conversations I don't believe in front of other individuals."

Pincus: "Tell me what he said to the best of your recollection."

Howell: "It is a 'she'."

Pincus: "She. Tell me what she said."

Howell: "She had, I don't recall the, I don't recall the words. I just recall the implication."

Pincus: "The implication that you recall was where, was what?"

Howell: "That Dr. Hill has reservations regarding Indian, people of Indian citizenship in research."

Pincus: "Did she share with you any documents or events – ."

Howell: "No."

Pincus: " – that led her to make the statement?"

Howell: "No."

Pincus: "Is she still employed at the university?"

Howell: "Yes."

Pincus: "And Dr. Azzam, when did you have a conversation with him in regards to mistrust of individuals of Indian descent?"

Howell: "No. In his case it was not Indian descent. In his case it was other descent."

Pincus: "Which is?"

Howell: "Which is anything other than, anything other than that, Caucasian, but I don't know in that respect."

Pincus: "When did this conversation take place?"

Howell: "I don't recall."

Pincus: "Was it after the complaint was initiated in 2001 to your recollection?"

Howell: "Yes."

Pincus: "Does that refresh you recollection in so far as 2001, two, three, four?"

Howell: "I don't recall."

Pincus: "Did you make any note of your conversation."

Howell: "No."

Pincus: "What did he say to you as best you recall?"

Howell: "The manner which the, the manner I which what was said to him which I don't recall what [searching for the words] those things were even if he relayed them to me made him feel she felt less of him because of his ethnic background."

Pincus: "You didn't personally observe that, did you?"

Howell: "I did not personally observe that."

Pincus: "Nor did you personally observe the conduct Dr. Rameshwar had reported to you had reported to you. Is that correct?"

Howell: "No, I did not."

Pincus: "Was there anything else that you recall discussing with Dr. Azzam in regard to this subject?"

Howell: "Not offhand, no."

Pincus: "In light of what you said she said to you and in light of what Dr. Rameshwar and Dr. Azzam related to you did you cause it to be reported to the affirmative action officer?"

Howell: "No, it never would have occurred to me to do so."

Pincus: "At that point you were head of the department, were you not?"

Howell: "No, I've never been head of the department."

Pincus: "Were you her supervisor at that point in time?"

Howell: "Whose supervisor?"

Pincus: "Dr. Hill's."

Howell: "No, she did not report to me from a – only in terms of the research data that she was doing on the grant."

Pincus: "Notwithstanding what you said, based on what others told you, you did not report this to the university, did you?"

Howell: "No."

Pincus: "Now, you say that Dr. Hill was angered by the fact that Dr. Bishayee had access to pornography on the university computer. Is that correct?"

Howell: "Repeat that?"

Pincus: "You indicated that Dr. Hill indicated she was angered by Dr. Bishayee having access to pornography?

Howell: "She was extremely angry, yes."

Pincus: "Is there anything wrong with her being angry?"

Howell: "Yes, if one would anticipate that somebody won't be happy that somebody accessed pornography on their computer."

Pincus: "To your knowledge did it violate university policy?"

Howell: "To my knowledge, I haven't read the policy, but I would imagine it does."

Pincus: "You have said you documented all of this?"

Howell: "That's correct."

Pincus: "Do you have any recollection as to whether Dr. Bishayee admitted it?"

Howell: "I believe that he did, and I believe that he apologized for having done so."

Pincus: "Did he suffer any other sanction as a result of that to your knowledge?"

Howell: "Other than that went into his human resources record?"

Pincus: "Yes."

Howell: "Yes, I think he was restricted from the computer in some way for some period of time. I don't remember. I think there was some computer sanction."

Pincus: "In light of his admission in being sanctioned on what basis did you conclude that Dr. Hill being angered by Dr. Bishayee's conduct signified she was out to get him?"

Howell: "If somebody makes a statement or claims against an individual two weeks after this event, it is not too hard to put it together. It seems pretty apparent."

Pincus: "The statement was the distrust statement you and I have identified or some other statement?"

Howell: "You are just talking about the pornography."

Pincus: "The statement about two weeks after the event, what statement are you referring to?"

Howell: "She made claims regarding the veracity of his data two weeks after I believe this event."

Pincus: "Now I understand what you are talking about."

Howell: "The pornography event."

Pincus: "Are there any other acts or statements that you can recall that led you to believe or to conclude that Dr. Hill was out to get Dr. Bishayee?"

Howell: "At what point in time?"

Pincus: "At any time."

Howell: "At any time?"

Pincus: "Yes, at any time. Prior, prior to her filing the complaints with the committee."

Howell: "Yes."

Pincus: "What else?"

Howell: "When I was told that she was in the laboratory poking around. So I had no idea why she was doing that."

Pincus: "When was that?"

Howell: "Right, it is probably around April [20]01."

Pincus: "When, are you referring to what has been documented in these proceedings and she and Dr. Lenarczyk observing what Dr. Bishayee was doing or shadowing?"

Howell: "I didn't know they were shadowing her [him] until after the committee pulled them in."

Pincus: "Is that the event, what you were referring to – ."

Howell: "Yes."

Leonard: "Where did we come up with the phrase shadowing?"

Howell: "That was incorrect. It wasn't after the committee, it was when I was informed that she was doing that or doing something or – I don't know what she was doing."

Pincus: "Let's get back to this, what you have, Howell-37, the letter of resignation. You said you prepared this document?"

Howell: "Yes."

Pincus: "And you were describing for me a discussion or discussions that you said had taken place and commencing in some prior, earlier in the summer that perhaps he should move on. I don't want to put words in your mouth but the substance of what I understood you to say was that. Is that correct? Yes?"

Howell: "Yes, that's correct."

Pincus: "Is that what you said, or did you say something different?"

Howell: "I said we had discussions where I encouraged him to move on because she was not going to leave him alone."

Pincus: "You prepared this document, and did you instruct him to sign it?"

Howell: "I suggested that he move along."

Pincus: "And he signed it? Obviously, you recognize that as his signature."

Howell: "He signed. I did not demand the signature."

Pincus: "Did you tell him that his signature on this document or whether he did not sign it would influence whether you gave him a recommendation or not?"

Howell: "I don't recall."

Pincus: "When you say you don't recall I want to be sure about something. Are you denying saying that to him or is it that you just have no recollection of discussing it?"

Howell: "I have no recollection of the specific discussion."

Pincus: "At the time he signed this document did you instruct him not to come on campus any longer."

Howell: "Did I – excuse me."

Pincus: "Did you instruct him not to come on the campus any longer?"

Howell: "Not that I recall."

Pincus: "Do you recall taking away his keys and ID from him after he signed this letter?"

Howell: "I'm sure that I took the keys, because if you are resigning, although I don't know if I took them then or on August 17, that I don't know. I'm sure I took the keys because a person who is in the lab, or isn't employed or not permitted in the lab shouldn't have a key."

Pincus: "What about [the] ID? Do you have a recollection of taking that?"

Howell: "I don't remember."

Pincus: "Did you tell him not to apply for research positions on the UMDNJ campus?"

Howell: "I suggested him to go elsewhere, again for the purposes [of] getting away from Dr. Hill, as far away from her as possible."

Pincus: "Did you tell him not to apply for positions in radiation biology research anywhere?"

Howell: "Not that I recall."

Pincus: "So I'm clear in terms of 'I don't recall,' you are denying making such a statement?"

Howell: "Yes."

Pincus: "What about Dr. Lenarczyk? Did you evaluate him?"

Howell: "I don't know if I filled out an evaluation. Usually, the evaluations are annual. He was with us, how long I don't recall. I don't recall if I filled one out or not. I mean if I did, I imagine it is on file."

Pincus: "I will take that, please. Thank you. Did Dr. Bishayee indicate to you that he had a discussion with Dr. Hill on or about August 1 regarding his resignation?"

Howell: "I don't recall him mentioning any such thing."

Pincus: "Were you aware that Dr. Bishayee rented apartment space from Dr. Hill?"

Howell: "I believe, yes. I am now and I believe I was at the time."

Pincus: "Were you aware at the time that Dr. Azzam had lived at Dr. Hill's home for a period of time?"

Howell: "Yes."

Pincus: "Were you aware that Dr. Hill had a social relationship with both Dr. Azzam and his spouse, Dr. de Toledo?"

Howell: "What's a social relationship?"

Pincus: "Did she socialize, did they go out and do anything outside of work. Are you aware of that?"

Howell: "I'm not. I have no idea. Dr. de Toledo was in Boston at the time as I recall. Which time are you referring to?"

Pincus: "At any time are you aware that they had a social relationship?"

Howell: "I mean obviously when he was living in their house he must have had one then, yes."

Pincus: "When Dr. de Toledo came to New Jersey were you aware he and his wife purchased a home and resided at this time in Livingston?

Howell: "Yes."

Pincus: "Are you aware that once they began residing in that house that they had a social relationship with Dr. Hill?"

Howell: "Again I don't know what you are defining as a social relationship."

Pincus: "That they were going out for dinner or for a walk, anything outside the workplace?"

Howell: "I don't recall Dr. Azzam telling me they went out for dinner. I don't know."

Pincus: "You didn't observe them outside of the workplace. Is that fair?"

Howell: "Did I observe them outside of the workplace together? No, I can't recall."

Pincus: "You don't recall any interactions outside the workplace between Dr. Hill and Dr. Azzam?"

Howell: "Maybe at department parties and so on I imagine there must have been interaction, but I don't recall there was, whether there was or wasn't."

Pincus: "Sitting here today you can't picture them?"

Howell: "I can't picture an event where I remember them."

Pincus: "Same question with Dr. Hill and Dr. de Toledo."

Howell: "Not that I remember, except there must have been instances on a professional level we were together. I imagine at the Jack Little symposium. I don't remember the dinner, but we must have been together."

Pincus: "Did you observe any hostility between them on those occasions?"

Howell: "Between who?"

Pincus: "Dr. Hill and Dr. Azzam."

Howell: "Not that I recall with the occasion of the Jack Little, but that was prior to this event."

Pincus: "Prior to her complaining, is that what you are saying?"

Howell: "No, I don't remember whether it was or wasn't."

Pincus: "At the symposium was Dr. de Toledo present?"

Howell: "I don't know if she was there or stayed back. I don't remember. Dr. Lenarczyk was there."

Pincus: "What about Dr. Bishayee?"

Howell: "I think he was there."

Pincus: "Did you observe any hostility between Dr. Hill and Dr. Bishayee at the time?"

Howell: "Not that I recall at the time. I mean it is interesting though that I would raise the point here. May I ask a question?"

Pincus: "No. You just respond to my questions, please."

Howell: "Sure. I mean it depends what you consider hostility."

Pincus: "Did you assist, were you aware that Dr. Bishayee received unemployment compensation after his resignation from UMDNJ?

Howell: "I would like to roll back and complete that last comment."

Pincus: "Which comment?"

Howell: "When I said I didn't observe any hostility."

Pincus: "Anything come to mind?"

Howell: "Clearly that's incorrect. I would say putting in a complaint against someone is hostility. No doubt there is hostility."

Pincus: "The fact that Dr. Hill exercised the right afforded to her by university policy to report what she believed to be improper conduct is an act of hostility or maybe shadowing."

Leonard: "Objection to form."

Howell: "All of these events could easily be viewed as hostility. It would be pretty bizarre if they weren't."

Pincus: "I hear what you are saying. My question to you was were you aware subsequent to his resignation from UMDNJ that Dr. Bishayee collected unemployment compensation?"

Howell: "I was not aware of that, no."

Pincus: "Were you asked in any manner to respond to the New Jersey Department of Labor regarding his receiving or not receiving unemployment compensation?"

Howell: "I do not believe I was. No recollection of anything like that."

Pincus: "Did you have any conversation at the time you suggested to Dr. Bishayee that he resign and gave him the letter that we showed a couple minutes ago that if he did, the university wouldn't contest his collecting unemployment?"

Howell [turning to Leonard]: "I just told him I didn't know he collected unemployment. I'm learning this now."

Pincus: "Whether you knew whether or not he collected or not, I'm asking you at the time he discussed the resignation did you make any statements to him about his ability or inability to collect unemployment?"

Howell: "Absolutely not."

Pincus: "Did you make any promises to him in so far as his ability to collect unemployment were he to resign?"

Howell: "I don't remember having any discussion about unemployment whatsoever, unemployment insurance or whatever that is. Can we roll back again?"

Pincus: "If John wishes to question you when I'm done by all means."

Leonard to Howell: "Is this your complete answer?"

Howell: "I just wanted to roll back when he was asking whether I knew he lived in her apartment, whether Bishayee lived in her apartment."

Pincus: "Go ahead, what do you want to say?"

Howell: "I would like to hear her read it back."

Pincus: "I don't want to have it read back."

Howell: "I would like to – ."

Pincus: "If it is something you want to add it is fine."

Leonard: "Don't worry about it."

Pincus: "Do you recall meeting with Dr. Carol [Karen] Putterman on or about April 2, 2002, in your offices?"

Howell: "April 2 – ."

Pincus: "April 16, 2002."

Howell: "In my office, I met one time I believe Dr. Putterman in my office. The date, I don't recall the specific date."

Pincus: "During that meeting do you recall if anyone else was present besides the two of you?"

Howell: "No one else was present."

Pincus: "Do you recall indicating to her that prior to Dr. Bishayee's resignation that you had discussed with him ill will in the lab and uncomfortable working conditions due to the dispute between Dr. Hill and Dr. Bishayee?"

Howell: "I don't remember if I discussed that specifically with her, so I don't recall the specific discussions."

Pincus: "Do you recall telling her that you felt that Dr. Hill would not leave Dr. Bishayee alone, therefore it was best for him to leave the lab?"

Howell: "That's what I just told you that I had discussed with Dr. Bishayee. Did I discuss that specifically with Putterman or not, I don't recall specifically what I discussed with Dr. Putterman."

Pincus: "Let me see if I can refresh your recollection. Do you recall telling her that you had encouraged Dr. Bishayee to find a position outside the university entirely?"

Howell: "I may have said that to her. I don't, I can't tell you specifically if that's actually what I said."

Pincus: "Do you recall telling her because of your encouraging him to find a position outside of the university entirely you did not give him any letter of recommendation for any internal position?"

Howell: "I don't believe that I gave him a letter of recommendation for an internal position."

Pincus: "But do you remember telling Dr. Putterman that was the reason why you didn't give him one?"

Howell: "I don't recall."

Pincus: "Do you recall telling Dr. Putterman that you would recommend him for positions outside the university?"

Howell: "Again, I don't recall the specific conversation with Dr. Putterman."

Pincus: "Were you aware of any e-mail exchanges that went on between Dr. Bishayee and Dr. Hill regarding the payments for the apartment that he was living at?"

Howell: "I have – no."

Pincus: "Dr. Bishayee never discussed that with you at all?"

Howell: "The only thing, no, the only thing that I have seen regarding payments for the apartment was in the docs that were provided to me a year ago indicating she had put them in some trust fund for her granddaughter, that's the only thing I can recall seeing."

Pincus: "You learned about that you are saying to me after these complaints arose?"

Howell: "Way after. 2000 – one year ago approximately."

Pincus: "Do you recall telling Dr. Putterman that no one else in your lab has repeated the bystander experiment?"

Howell: "I don't recall specifically what I said to Dr. Putterman."

Pincus: "Do you recall having any discussion with her regarding the bystander experiment?"

Howell: "Again I know that there was a discussion about bystander. I don't exactly remember the line of discussion."

Pincus: "Is there anything on the basis of the questions that I have asked you that comes to your recollection about what may have been discussed between you and Dr. Putterman between the April 2002 meeting that you have already identified?"

Howell: "Try me again."

Pincus: "As a result of my question is there anything else that you can recall that occurred during the course of the April 2002 meeting with Dr. Putterman that you have not already identified?"

Howell: "I recall there was some discussion regarding the spread and values from one measurement to the next in the Coulter. That I recall. She did mention something that Dr. Hill felt the standard deviation was not anticipated. I do remember that."

Pincus: "What did she say to you as best you can recall?"

Howell: "She said the standard deviations were too small I believe, that I recall."

Pincus: "Did she share any documents with you during the course of this meeting?

Howell: "No."

Pincus: "You have no recollection of her sharing any documents?"

Howell: "No, she did not."

Pincus: "What about when she was discussing this Coulter count issue, did she describe any documents regarding the Coulter count?"

Howell: "No. Only I believe, I believe I pulled out Dr. Azzam's Coulter counts."

Pincus: "Anything else that you can recall that was not identified discussed between the two of you?"

Howell: "That's the main one I can recall."

Pincus: "Nothing else?"

Howell: "Not that I can recall."

Pincus: "Did there ever come a time where you advised Dr. Putterman that you would fax to her any kind of figures or data?"

Howell: "That I faxed to her?"

Leonard: "Objection to form.

Pincus: "Subsequent to your discussion with her, in any discussion with her in April, did you have any further discussions with her whether in person or by telephone or other means?"

Leonard: "Objection to form. About these issues by the way so we are clear."

Howell: "After when?"

Pincus: "After the meeting in April 2002."

Howell: "I don't recall – I know I saw her one or another time after that."

Pincus: "I'm not talking about passing her in the hallway. I'm talking about specifically the issues we are here discussing today. Did she ever request of you, or did you ever indicate to her that you were going to supply her any kind of facts or figures?"

Howell: "I didn't – only at the time when the attorneys requested that thing she may or may not have been in on that, the summary."

Pincus: "You are talking about the summary?"

Howell: "The summary, yes. But other than that, I don't remember her requesting docs – how did you word that?"

Pincus: "Facts or figures."

Howell: "Yes, I don't recall sending her facts or figures."

Pincus: "Did you ever tell Dr. Putterman that you had tried to reproduce Dr. Bishayee's Coulter results?"

Howell: "Not in those particular terms. I recall relating to her that we were having difficulty reproducing the experimental conditions of the experiment."

Pincus: "When did that discussion take place, please, Dr. Howell?"

Howell: "That I couldn't put my finger exactly on when it was."

Pincus: "What about in reference to your meeting with her in April 2002?"

Howell: "2002, I don't recall. I don't recall the timing of when it was."

Pincus: "You are aware of the 1999 article Howell-10 and 2001, article 2000, I'm sorry – Howell-11 have continued to be cited in the literature. Is that correct?"

Howell: "By whom?"

Pincus: "You are familiar with something called the Web of Knowledge [now the Web of Science[146]]?"

Howell: "Never heard of it."

Pincus: "Never heard of that?"

Howell: "No."

Pincus: "Have you yourself cited it in any articles that you have prepared subsequent to that time?"

Howell: "I'm sure that some of the articles that have gone out of the lab have cited it."

Pincus: "And you also recall making reference to the graphs within those two documents that we discussed earlier today in a slide show that you gave at Harvard Medical School in or about 2003?"

Leonard: "Objection to form."

Howell: "2003, Harvard Medical, I don't recall."

Pincus: "You don't recall that?"

Howell: "No. May have, but I don't recall."

Pincus: "Fair enough." [Pincus produced Howell-38, a photocopy of the slides in question.]

Pincus: "Do you recognize this document at all?"

Howell: "These would appear to be slides that were in my collection."

Pincus: "Do you recall presenting these?"

Howell: "I don't recall, I don't specifically recall but I'm sure if it was within my PowerPoint presentation that was delivered to Harvard and it is in there, yes, I presented it."

Pincus: "I'm correct that this graph in the upper right-hand corner comes from the literature that we reviewed earlier today?"

Howell: "From the publication, yes, I imagine so."

Pincus: "I will take that, please." [Pincus introduces Howell-39, a copy of page 35.]

Pincus: "Dr. Howell, I am going to show you a copy of page 35, what I understand to be the grant renewal application that was submitted in or about 2005 concerning this grant."

Howell: "Yes."

Pincus: "I have the whole document here in the event you want to check."

Howell: "Okay. Yes."

Pincus: "I would like to turn your attention to the figure C-1. Am I correct that that comes from the earlier literature?"

Howell: "I don't know just looking at it, but it says, let's see, published experimental data for cesium so I assume, yes."

Pincus: "You caused this grant to be submitted?"

Howell: "I submitted this grant, yes."

Pincus: "I understand that on this grant Dr. Azzam is now – ."

Howell: "Wait a minute, wait a minute – ."

Pincus: "A co-investigator with you?"

Howell: "Roll back again. Go ahead on this that – this is which one of the two grants? This was submitted when."

Pincus: "2005. Okay let's go back to the figure, C-1, and – I was referring you to the top panel.

Howell: "Published experimental data."

Pincus: "You are in agreement with me. Correct?"

Howell: "The, what caught me off guard, theoretical modeling, but it follows with theoretical modeling of the published data."

Pincus: "Do you recall citing those two papers in the course of this grant application?"

Howell: "Yes."

Pincus: "Can I have that back, please."

Howell: "Sure."

Pincus: "Thank you. I had also asked you that insofar as this grant renewal, at that point in time am I correct that Dr. Azzam became a co-investigator on this grant with you?"

Howell: "Dr. Azzam was a co-investigator, was a co-investigator on the renewal, yes."

Pincus: "I understand that you serve as either a principal investigator on another grant besides this with him in which he is the co-investigator or alternatively you are a co-investigator on grants for which he is the primary investigator?"

Howell: "Make that less circuitous, I mean simplify the question."

Pincus: "Are you a co-investigator on any grants on which he is the primary investigator, principal investigator.

Howell: "Before?"

Pincus: "Yes. Identify them, please?"

Howell: "NASA grant."

Pincus: "How much?"

Howell: "I don't know, NASA must be somewhere around [a] million. Department of Energy [DOE] low dose energy grant I think I'm five percent of that. What else does he have? There is a grant from the Cancer Center on bystander studies on animals. I'm a co-investigator. I forget how the layout of that is, but I am participating on that. What else where he is PI [principal investigator], I don't think, he's got a subcontract on a DOE grant held by Iowa. I don't believe I'm part of that from my recollection. I think that's all of them."

Pincus: "Let's flip it."

Howell: "So he is a co-investigator on my current continuation grant that we have been discussing. And he is a co-investigator on a grant that we have from NIAID [National Institute of Allergy and Infectious Diseases]."

Howell's deposition ended shortly after this exchange. The breadth of the deposition revealed remarkable factual and common knowledge inconsistencies between members of the Howell laboratory.

Chapter 7 –
Deposing Bishayee

By the time Anupam Bishayee[147] was deposed on December 22, 2008, he was an assistant professor of pharmaceutical sciences and research assistant professor of internal medicine at Northeastern Ohio Medical University in Rootstown, Ohio. Bishayee's purported scientific misconduct was the catalyst for two university inquiries, a review and report from the National Institutes of Health (NIH) Office of Research Integrity (ORI) and, finally, Hill's *qui tam*. By the time the deposition took place, it had been nine years since he'd performed the experiments at the centerpiece of the federal suit. From the beginning of Bishayee's examination, Hill, who was present, observed that he didn't remember much from that time and made it clear during the proceeding that he had very little understanding of what he had been doing during the experiments. She later wrote that he denied making up media for the experiments but "if not he, then who? The future pharmacy professor and department chair," she continued, "denied knowing what deoxycytidine (one of the building blocks of DNA) or, for that matter, cell synchrony, was." Prior to arriving at Northeastern Ohio University, he was a radiation safety specialist at UMDNJ.

At the beginning of the deposition, it was apparent that Bishayee hadn't carefully reviewed the documents sent to him by defense counsel Scott Flynn, including Marek Lenarczyk's deposition. Sheldon Pincus asked didn't waste much time getting into experiments in question. "The various experiments that were performed, you know, by you in particular, how were the protocols for those experiments constructed? Did you do it together?" Flynn objected to the form of the question. "It really depends on the specificity of the protocol," Bishayee replied. "Some protocol he told me how to do it. Some other things I tried, and I showed it to him and he said go ahead. Like that. It's a constant discussion and, you know, checking and making sure that things are okay." To clarify his next question, Pincus told Bishayee that what he was asking about had to do with Roger Howell's National Institutes of Health grant, not anything else that he may have done. "In these kinds of experiments, you say insofar as the protocols, some he gave to you, some you constructed and shared with him. To use your words, there was 'constant discussion.' Is it fair for me to say that any experiment that you undertook was undertaken with the knowledge of Dr. Howell that you were going to do that?" Bishayee responded yes, that that was correct.

Pincus wanted to know more about the procedures followed by Bishayee, Howell and even Lenarczyk. "It's hard to tell," Bishayee replied, "It's been seven to nine years, depending on start to finish. I cannot remember each and every experiment or every detail, but sometimes he asked me to do certain things and it took me for a week or sometimes, you know, longer than that. Then I went back, and he [Howell] checked. So," he continued, "there is no fixed procedure that each actually happened. It depends on the nature of the experiment or whatever he asked me to do. There is no fixed, you know, format that I took." But was it his practice that at some point during or at the end, Pincus pressed, that Bishayee shared his results and had a discussion with Howell as to what had occurred during the experiment? "Not each and every one. Once again, it's long time so I can't remember the specificity," Bishayee opined. "But sometimes he checked immediately, or I did discuss immediately. Sometimes after one month of experiment we sit together and then we discussed whatever we have found, like that." Were there established times for these meetings, Pincus wanted to know. "No. Sometimes it was structured. We met on a specific day and time. But sometimes like on demand, say, 'Hey, what you got? Let's go and see.' So, it was random." Pincus asked if there was any question in Bishayee's mind that ultimately Howell reviewed each experiment that he performed on the NIH grant. "I would say yes, but once again it also depends on the specific incident or specific nature. I cannot say that it happened all the time," Bishayee explained. If it didn't happen all the time, did it happen the majority of time? "I can't really recall the frequency of that," he continued.

Pincus asked about status reports. Did he prepare any for the experiments performed for the grant. "I documented the experimental finding using a protocol, that kind of format to fill out some of the blanks, and whenever I get some data, I provided him the raw data, and that was the mode of exchange of document," Bishayee told Pincus. "So, the protocols were exchanged, if I understood you correctly," Pincus prompted. "Yes, protocol in protocol book," said Bishayee. "Protocol book is for experiment. I just, you know, print it, a couple of pages, depending on what I was doing that time, and then I recorded whatever I found. And then everything was placed in a book, laboratory notebook, and – sometimes he checked the laboratory notebook but sometimes he didn't but when I got the final result on then I gave it to him, and he analyzed those results."

"You would share that raw data along with the protocol to Dr. Howell, and if I understood you correctly, he would interpret it? Pincus asked. "Right," said Bishayee. Pincus wanted to know if Howell provided guidance in terms of what he wanted Bishayee to do next in an experiment. He wanted the practice and procedure. "Sometimes he say 'Okay, that's fine. Let's move on to a new experiment or think about that,' you know, something different. Sometimes he say just to repeat this. You know, so it depend on the particular incident [he meant experiment]," Bishayee replied. "So obviously different experiments may give rise to different instructions, but you recall that as you went through the experiment and you shared your data with him you looked to him for guidance?" Pincus asked. "Right," Bishayee answered. When Pincus asked if that meant "correct," Bishayee told him "That's correct."

"To your knowledge you didn't do any experiments with regard to the [NIH] grant that Dr. Howell didn't know about, did you?" Pincus rejoined. He had to ask twice. "I don't recall. We did a lot of things in the lab – several work – not only related to this grant but also, I did something which is no way related to this grant, so I'm not clear when I don't know which experiment he put in the grant and which experiment he didn't," Bishayee responded. "As you sit here today, do you have any conscious recollection of secretly doing research that Dr. Howell didn't know about?" Flynn objected. "No, there is no such thing," replied Bishayee. "You said that Dr. Lenarczyk worked in the lab. Do you remember," Pincus asked, "when he came and commenced his work." Bishayee remembered that he came after he did but didn't know exactly when Lenarczyk reported to work in the lab. He recalled only that he thought the two postdoctoral fellows worked together for approximately one year. He also didn't know if Lenarczyk was working on the grant. "Did you have any responsibilities insofar as instructing Dr. Lenarczyk what to do? Pincus asked. "Yes. I showed him the basic lab techniques and a few laboratory experiments that I was asked to." Pincus wanted to know if these experiments were associated with the NIH grant. Flynn objected. "I showed him what I was doing at that time, and that was associated with the NIH grant."

Certainly, it was important to establish the relationship ladder in the lab and most especially who performed the experiments on the NIH grant. "Did you do any experiments together to your recollection?" Pincus asked. "It's hard to tell. Maybe he watched me doing some of the things to better understand what I was demonstrating something," Bishayee speculated, "but principally he had his own project, and I was doing other area, so I don't recall that working together day after day. Maybe occasionally we just learned something together. Basically, he learned," he continued, "from me and I showed him to do something." Bishayee performed "all the different things, like cell culture work," he told Pincus. In large measure Bishayee wasn't certain what was or wasn't in the bounds of the NIH grant. He told Pincus he'd worked with animals but when asked to explain, he didn't know if they were part of the grant or not. "I'm not sure whether it is part of this grant or something different in my various times," he replied. "I did various kinds of experiment, not only cell culture but also other forms of research. So, it would be hard for me to recollect which experiment belonged to which grant."

"Did you work on mutagenesis?" Pincus pushed on. "Briefly, yes," Bishayee said. "Did you work on survival?" asked Pincus. "Yes," he said. What about Lenarczyk, had he worked on mutagenesis? Bishayee said that he had – and survival, too. "When you were working either on mutagenesis or survival work did you know what experiments he was working on and to your knowledge did he know what experiments you were working on?" Pincus wanted to know. "Sometimes yes, but most of the time no," Bishayee replied. "Well, we were given probably the separate part of the study to do by our own, so he joined -," he continued. "By who? You said we were given a separate part of the study. By?" Pincus interjected. "By Dr. Howell, and our background was different so what interest was probably different, too," Bishayee explained. "The only thing that make them working together is that we were working for the same PI [principal investigator] and I don't know what he was doing exactly." Bishayee's ultimate direction, confirmed in this line of questioning, came from Howell. "To your recollection were there joint meetings between the three of you in regards to your experiments and research, the three of you being Dr. Lenarczyk, Dr. Howell and yourself?" Bishayee responded that "we may discuss something together. I don't exclude, but I don't have any recollection of specific event." He didn't know if there were any minutes taken in these meetings. He didn't take notes and he didn't observe Lenarczyk or Howell taking them either.

Pincus began to roll out documents for Bishayee's identification and comment, starting with the controversial September 6, 1999 experiment – Bishayee-1. Shown the pages, Bishayee told Pincus that he saw his handwriting "but I don't remember creating this, but now that you're showing me." Pincus asked him to look harder at the document; there was handwriting on various pages and was all of it his or was some of it not his own. "The handwriting part of this documents seem to be mine. I can recognize my handwriting but there is some other thing which probably I didn't generate, like this chart," he observed, looking more closely at what was in front of him. The chart was on B013921. "I didn't create this chart," he iterated. But he also didn't know who did. The chart pertained, he thought, to the response of the V79 cells against radiation.

Still looking at the September 1999 experiment, Pincus asked if Bishayee recalled a mutagenesis arm of the experiment and whether he performed the same with Hill wherein he did the survival arm and she did the mutagenesis arm. He didn't recall. "Do you have a recollection as to whether you observed Dr. Hill while she was performing the mutagenesis part of the experiment so that you'd be able to do it at a later date?" asked Pincus. "I don't have a recollection of that," he replied. Pincus asked Bishayee about Hill in the lab. "She was doing something that I didn't know [on the grant] and still I don't know which part she was associated with this grant and what was her own experiment," Bishayee explained. "That I have no recollection." He also didn't remember doing any experiments with her. "You have no recollection of doing any experiments with her?"

Pincus pressed. "There may be some things that she showed me, but I don't recall what exactly she showed me at that time." Pincus continued. "So, if she were to testify that in regards to this experiment that you did the survival arm of the experiment and that she did the mutagenesis arm of the experiment you have no knowledge or recollection of that?" "That's correct," he insisted. Bishayee didn't have much to say other than he and Hill did "a lot of things so specifically for this case I don't have any recollection of who did what and what was the purpose." Further, as Pincus asked more specific questions about the result of the experiment and what the chart indicated, he didn't have much to offer. "This document I didn't create so I don't know how it was plotted and how the document was created so it would be hard for me," Bishayee explained, "to interpret that which I didn't create." This included not being able to interpret the chart. "I'm not expert in mutation. I'm not expert in radio biologics so I don't know," he replied when asked to interpret it. "So, whatever was asked me to do, I did, and I provided data to Dr. Howell. I don't know who created this chart and what was the interpretation of the data."

"Based on your knowledge and experience you're telling me that you can't tell me whether this chart appears to show an increase in mutation based upon dose?" asked Pincus. "I see a lot of the things here and I don't know the specificity of this result so I can't tell that, what was happening here." He then showed Bishayee his September 20, 1999, experiment, marked Bishayee-2. He recognized all but the first two pages of the document, though there was some writing that "may not be my handwriting." There wasn't much that Bishayee could offer when asked to interpret the data on the pages in front of him including charted documentation of the number of cells. While he'd performed the experiment, he couldn't tell Pincus what part of the experiment related to the cells that were resuspended just before the radiation. "Say it again?" he asked Pincus.

> *Pincus*: "Can you tell me which page of this data relates to cells that were resuspended just before the radiation by this chart?"
> *Bishayee*: "No, I can't tell."
> *Pincus*: "Can you tell from this chart?"
> *Bishayee*: "I didn't create this chart so it would be hard for me to interpret it."

Pincus: "You can't interpret it?"

Bishayee: "No."

Pincus: "Do you know who created it?"

Bishayee: "No."

Pincus: "Who else could it have been if it wasn't you?"

Bishayee: "I don't know."

Pincus: "Would it have been Dr. Howell?"

Bishayee: "I don't know."

Pincus: "Would it have been Dr. Lenarczyk?"

Bishayee: "Could be anybody, but I don't know who."

Pincus: "You've never seen that chart before?"

Bishayee: "I don't know."

Pincus reminded Bishayee that the documents he was being shown came to plaintiff's counsel as a result of a subpoena issued by the United States Attorney's Office for the District of New Jersey to the UMDNJ for all the laboratory notebooks associated with the experimentation done for Howell's NIH grant. "You have no recollection as to whether these documents comprised your lab notebook associated with this experiment. Is that what you're telling me?" Scott Flynn objected. "No. I'm telling you that I recognize the part that I did, my handwriting is there, but I don't know much about the other documents which are created not by me," Bishayee said. "I'm still trying to figure out given your description earlier insofar as the experiments that you did and what you then did with them, if it wasn't you – and this part of a document associated with this experiment – who else would it have been [who] created this chart on 7891 or 7892?" Pincus pressed Bishayee. Flynn objected again. "I can only tell that I did my part of it and I provided the data to Roger, Dr. Howell, and I don't know what happened after that, who created which chart. That I have no idea." In denying that he created the chart, Bishayee was asking for Pincus to believe that he never charted the results of his experiments. But was that true? Flynn objected. "All I know I did the experiment and I collected the data and I recorded it, and I don't any recollection that I made any chart," Bishayee insisted. "I know that I probably gave the raw data to Roger to analyze and interpret, but I have no recollection in regarding making this particular chart." Pincus wasn't satisfied with the answer. "My question to you is more general. Did you in the course of your experiments chart your results? If not this particular experiment, were there experiments where you undertook to chart your results?" Bishayee conceded that he might have done "some but I don't know which one and I don't know – what I can tell, that this one is not created by me." But how could he tell? Pincus was curious. What was it about that chart that convinced Bishayee that he hadn't prepared it? "I see there are bunch of number of something called at the top of the diagram -," said Bishayee. He was looking at 7891 and there he saw some "specific numbers and I don't know how these numbers were generated and what does it mean."

Pincus tried to walk Bishayee through additional pages of the experiment document. He took him back to page 7900, where he'd identified his own handwriting. "What does this data relate to? Pincus asked. "Counting some colonies," Bishayee replied. "And when you look at that, what does that data signify to you in terms of the results of your experiments when you were counting these colonies? What was the significance of the numbers that are set forth on this chart?" Pincus continued. Flynn objected, telling Bishayee he could answer if he knew the answer. "I see the colonies were all over the place. Some particular tub dilution, the number was high and for some other was low," he told Pincus. "Do these indicate to you there was a decrease in survival or an increase in survival?" Pincus rejoined. Flynn objected again. "That depends," said Bishayee, "which tube you're talking about." But did the data relate to cells that remained as clusters before and during the radiation? Pincus wanted to know. Flynn objected. "I cannot remember," Bishayee replied. At this point it was clear that Bishayee wasn't able, in any way, to interpret the chart in question. Flynn objected again and again. The bottom line was that Bishayee entered data he told Pincus he couldn't interpret. "I recorded whatever I saw but I cannot interpret," Bishayee reiterated. "I'm not a radiobiologist. I'm not specialized in mutation. So, I can't tell for that this is mutant or something different." This confused Pincus. "If you recorded what you saw how did you know what you were looking for if you can't interpret it?" Flynn objected. "Well, probably I was told to just record whatever I see on this plate, but I didn't know, you know, the scientific depth about those issues. So, I just recorded whatever I got and I documented the result and I gave it to Dr. Howell, and it's up to him to interpret and analyze and use the way he want to," Bishayee continued. "Okay. Do you recall any other discussions with [Howell] about this experiment?" Pincus asked. "No," Bishayee said. There was nothing further Bishayee could tell Pincus about the September 29, 1999 experiment.

Bishayee's deposition was comparatively brief compared to those of Lenarczyk, Azzam and Howell. Bishayee couldn't tell Pincus much about any of the protocols, experiment particulars, equipment used and most especially the documents associated with each one, to include the two Newark CCRI inquiries. When asked about an event in March 2001 that involved Hill, Bishayee's responses were brief and largely uninformative – and also illustrative of the tenor of the entire deposition.

Pincus: "Do you recall an event in or about March 2001 when Dr. Hill requested and you indicated that you were going to give her some V79 cells which she can do an experiment with?"

Bishayee: "No, I can't."

Pincus: "Do you ever recall telling her that there were three flasks of V79 cells growing but you were unable to give her the cells?"

Bishayee: "No, I can't."

Pincus: "Do you have any recollection of telling her at or about this time that you had trouble with the Coulter counter and had lost all of the cells?"

Bishayee: "Time to time I had trouble with the Coulter counter, that probably I discussed with many people in the lab, but I don't know the specificity of this incident."

Pincus: "What kind of trouble did you have with the Coulter counter?"

Bishayee: "Clogging."

Pincus: "Do you remember at or about this time having any difficulties with the Coulter counter?"

Bishayee: "Say it again please?"

Pincus: "Do you remember back in March of 2001 having clogging problems with the Coulter counter?"

Bishayee: "No, I can't."

Pincus: "Is that something that you would have memorialized in a document?"

Bishayee: "No."

Pincus: "Does that require, when you have clogging in a Coulter counter, the counter to be serviced in any fashion?"

Bishayee: "No. I asked Roger how to remedy it or what I should do if it happens, and then he showed me."

Pincus: "What did he say or tell you to do?"

Bishayee: "To clean and open it up and just to try to remove the clog and then start again."

Pincus: "How do you do that?"

Bishayee: "I don't know. I don't have recollection about that machine, but it may be spray to remove the nozzle – removal of the obstruction to the nozzle or something that I really forgot."

Pincus: "So do you remember telling Dr. Hill that the trouble with the Coulter counter precludes you from giving her some V79 cells?"

Bishayee: "I cannot connect these two pieces together. I know I had trouble with the Coulter counter."

Pincus: "Is it correct that you maintained or stored your cells in a flask or flasks?"

Bishayee: "Yes."

Pincus: "Your V79 cells in particular?"

Bishayee: "Yes."

Pincus: "Do you remember how many cells you could harvest from one flask?"

Bishayee: "No."

Pincus: "Do you remember what volume you suspended the cells?"

Bishayee: "No."

Pincus: "Do you remember what size aliquot you used to withdraw cells from?"

Bishayee: "No."

Pincus took Bishayee back to when he first joined the lab. "Earlier when I asked you to describe [when he arrived at Howell's UMDNJ's lab] you said 1997 as I recall. That's why I'm – the end of 1997. So, I'm saying you'd have been there for approximately a year at the time [of the November 30, 1998 100 percent tritiated thymidine experiment he performed, labeled Bishayee-6]." Bishayee confirmed that this was correct.

Pincus: "Dr. Howell ever come into the lab and count dishes for you or with you during the course of your experimentation?"

Bishayee: "I don't recall."

Pincus: "Did he ever look at the dishes after you had counted them?"

Bishayee: "Don't recall."

Pincus: "In these two experiments [the other was Bishayee-5 dated November 12, 1998] is there anything set forth in the protocols or the documentation to indicate that deoxycytidine was present in the medium during the overnight incubation with tritiated thymidine?"

Bishayee: "I can't tell."

Pincus: "You can't tell?"

Bishayee: "No."

Pincus: "Do you know what deoxycytidine is?"

Bishayee: "No."

Pincus: "Do you have any recollection of preparing mediums for experimentation associated with this grant, culture mediums?"

Bishayee: "No."

Pincus: "So you have no recollection of every using deoxycytidine?"

Bishayee: "No."

Pincus: "Did you make any attempt to synchronize the cells before you started the experiment?"

Bishayee: "I have no recollection."

Pincus: "Do you know what synchronization is?"

Bishayee: "No."

Pincus: "So is it fair for me to say that if you don't know what synchronization of cells is that it's unlikely that you engaged in any act to synchronize those cells?"

Bishayee: "Can you rephrase the question?"

Pincus: "Sure. You say you don't know what synchronization of cells is. Is that correct?"

Bishayee: "That's correct."

Pincus: "So is it fair for me to say if you don't know what it is you weren't engaged in doing that in regard to any of the experimentation?"

Bishayee: "No. That is not true. I don't know. Maybe I was told to do certain steps and I have no recollection whether I did and why I did."

Pincus: "Okay. So do you have any recollection of Dr. Howell telling you to synchronize cells for any of these two experiments, Bishayee-5 or 6?"

Bishayee: "Don't recall."

Pincus: "Do you have any idea how you go about synchronizing cells?"

Bishayee: "No."

Pincus: "No idea whatsoever? Never read any literature?"

Bishayee: "No."

Pincus: "Do you have any recollection of Dr. Howell instructing you how to synchronize cells?"

Bishayee: "No, no recollection."

Pincus: "Do you have any recollection of Dr. Lenarczyk instructing you how to do that?"

Bishayee: "No."

Pincus: "Dr. Hill?"

Bishayee: "No."

Pincus showed Bishayee the two articles he coauthored with Howell, labeled Bishayee-7 and Bishayee-8. He asked if Bishayee recognized them. "Yes," he replied. Pincus put Bishayee-7 in front of him.

Pincus: "You coauthored this."

Bishayee: "Yes."

Pincus: "Is there any significance that your name is listed first?"

Bishayee: "I don't know."

Pincus: "You don't know?"

Bishayee: "No."

Pincus: "I'm going to show you Bishayee-8. Do you recognize that?"

Bishayee: "Yes."

Pincus: "And you coauthored this also, did you not?"

Bishayee: "That's correct."

Pincus: "Along with Dr. Howell and another individual, a Dr. Rao – ."

Bishayee: "Rao."

Pincus: "R-a-o. Okay."

Flynn: Just to be clear, that had more names on it."

Pincus: "Bishayee-7 shows there are other authors including Dr. Hill, Dana Stein, Dr. Rao, Dr. Howell. Correct?"

Bishayee: "Right."

Flynn: "This is 8. That's what I'm saying. We're getting confused with 7 and 8."

Pincus: "Oh, I'm sorry. Bishayee-8. My apologies. Let me go back, clear up the record. Bishayee-7 has your – ."

Bishayee: "Correct."

Pincus: "Bishayee-8: yourself, Dr. Hill, Dr. Stein, Dr. Rao, and Dr. Howell. Correct?"

Bishayee: "Yes. Right."

Pincus: "You read this and you're familiar with it, this article?"

Bishayee: "Yes."

Pincus: "I want to call your attention to Figure 3 on page 402."

Bishayee: "Okay."

Pincus: "Would I be correct that this figure among other things depicts the 100 percent survivals of V79 cells as a function of cluster activity of tritiated thymidine?"

Bishayee: "We're looking at the legend. It shows that either 50 percent or 100 percent of the cells were radio labeled in multicellular clusters. So it's 100 percent and 50 percent, not only 100 percent."

Pincus: "And this figure shows that with regard to the 100 percent survivals, the survivals were exponential, isn't that so?"

Flynn: Objection to form.

Bishayee: "I can't interpret the data here and I don't see the top part very clear so I can't tell."

Pincus: "You wrote this, did you not?"

Bishayee: "No. I just did the experiment, and I provided data to Dr. Howell, and he probably wrote this thing."

Pincus: "To your knowledge were all of the 100 percent experiments that are reported in this paper done by you?"

Bishayee: "I don't know which data he used. I did the experiment, and I gave data to Dr. Howell and that's all I know, but I don't know which data he used to put in this paper."

Pincus: "Are you aware of any other 100 percent experiments that were used to prepare this paper other than Bishayee-5 and 6 that I showed you a few moments ago?"

Bishayee: "Once again, I did the experiment, and I provided data to him. I don't know what he did next."

Pincus: "So are you aware of performing any other 100 percent experiments prior to the time this paper was performed?"

Bishayee: "No, I have no recollection."

Pincus: "And to the extent there were such 100 percent experiments done, if I understood your testimony regarding your practice and procedure those experiments would have been placed in your lab notebook. Correct?"

Bishayee: "Say it again?"

[Pincus had it read back.]

Bishayee: "Well, whatever experiment I did I have recollection that I recorded the data, and that's all I can say. If I did an experiment, I would record that. But I don't know what was in the paper, which data, and those data came from which experiment. I have no – ."

Pincus: "My question to you was, if you performed 100 percent experiments as we've been describing them here, those experiments would have been placed by you in your lab notebooks?"

Bishayee: "I think so."

Pincus: "Okay. And other than the two experiments that we have identified here are you aware of any other 100 percent experiments that you performed for Dr. Howell before such time as this article was published?"

Bishayee: "I have no recollection."

Pincus went on to question Bishayee about more of his lab experiments though many of Bishayee's responses were the ubiquitous "no," "I can't tell" or "I don't recall." Moving on, Pincus introduced exhibits 11, 12 and 13 (also Bishayee-11, -12, and -13). Bishayee-11 was the 50 percent experiment dated April 29, 2001 [in the deposition excerpts that follow the reference is to the April 19, 2001 experiment]. Before he questioned Bishayee about this test, Pincus went back to exhibit 10. "Do you have any recollection of Dr. Howell sharing with you the results of this experiment?" Bishayee replied "yes." "Really?" Pincus responded. "When did that occur?" Bishayee replied "I don't know."

Pincus: "Okay. This experiment, were you employed at the time that he shared this with you [the results]?"

Bishayee: "I don't know whether this experiment was done while I was there or I left."

Pincus: "Going by the date of the experiment – September 27, 2001 – can you see where I'm looking in the upper right-hand corner?"

Bishayee: "Yes."

Pincus: "Based on the time frame that you outlined to me you were no longer employed in the lab by this point in time, September of 2001. Is that correct?"

Bishayee: "September – I don't know exactly when I resigned. If you refresh my memory – ."

Pincus: "I'm going to represent to you it was about July 31 or August 1 of 2001."

Bishayee: "Okay. If my resignation date is prior to this date then I would not know that Dr. Howell discussed with me or not. When I was there, I know there was some experiment was done that didn't match what I go so I cannot tell about specific – this experiment or some prior experiment."

Pincus: "But you recall that an experiment was done that didn't match with the results of your experiment."

Bishayee: "Correct."

Pincus: "Correct? Now let's first talk about your experiment. Do you recall whether it was a 100 percent experiment or 50 percent experiment that we're talking about?"

Bishayee: "I can't tell."

Pincus: "Do you recall whether your experiment showed an exponential decline in survival as opposed to a biphasic decline in survival?"

Bishayee: "All I can recall that some of my experiment couldn't be reproduced the way I did but I don't recall the specificity of the particular experiment."

Pincus: "And who was doing the other experiments? That was Dr. Howell?"

Bishayee: "Which experiment?"

Pincus: "You say you learned that the results of an experiment or experiments that you had performed could not be replicated. Is that fair?"

Bishayee: "Right, right."

Pincus: "Somebody else had tried to do that. Who was that person? Was it Dr. Howell?"

Bishayee: "I don't know who did the experiment. Could have been Dr. Howell, could be Lenarczyk. I don't know which two we are talking about."

Pincus: "But what you do recall, if I understood you correctly, was that their results were different from your results?"

Bishayee: "That's right."

Pincus: "And you don't remember whether their results are biphasic or exponential?"

Bishayee: "No."

Pincus: "How about a time frame? Do you remember when it was that you learned that somebody was trying to replicate your experiments? When was that?"

Bishayee: "I don't know."

Pincus: "Do you remember whether if we go back to Bishayee-11 – I don't think that's right in fairness to you, sir. If we go back to Bishayee-9, do you recall whether this was the exhibit that was discussed with you, or the experiment that was discussed with you, or again are you not sure?"

Bishayee: "I can't tell which particular one was discussed with me."

Pincus: "Alright. I'm going to show you Bishayee-11. This is a 50 percent experiment as opposed to 100 percent experiment?"

Bishayee: "Correct."

Pincus: "Right?"

Bishayee: "Right."

Pincus: "That appears to have been performed on or about April 19, 2001. This doesn't, on the face of the document 7383, indicate an investigator, but if you go back to 7390 you see Dr. Howell's name listed as an investigator."

Bishayee: "7390, okay."

Pincus: "Sir?"

Bishayee: "Yes."

Pincus: "See where I'm referring to?"

Bishayee: "Yes, I can."

Pincus: "What is this page? Can you identify what this is as part of this experiment data?"

Bishayee: "You're talking about this page, B007390?"

Pincus: "Yeah. Do you recognize this?"

Bishayee: "No."

Pincus: "Have you ever seen any documentation associated with these experiments?"

Bishayee: "No."

Pincus: "Did you ever use an Excel program in terms of setting forth or plotting or describing the data in any of your experiments?"

Bishayee: "No, I can't."

Pincus: "You can't, or you didn't? Did you use an Excel program?"

Bishayee: I don't recall whether I did or not."

Pincus: "Do you recall having any discussions about a 50 percent experiment that Dr. Howell performed on April 19, 2001?"

Bishayee: "No, I don't."

Pincus: "Does this refresh your recollection at all insofar as whether Dr. Howell advised that he was trying to replicate your 50 percent experiment as opposed to 100 percent experiments?

Bishayee: "No."

Pincus: "Okay. If you look at page 7394, the upper – can you tell me whether the data there is, by way of survival, shows an exponential versus a biphasic survival?"

Flynn: Objection to form.

Bishayee: "Which figure are you talking about?"

Pincus: "The upper one, this one here. Can you tell now?"

Bishayee: "I can't tell."

Pincus: "You can tell?"

Bishayee: "I can't tell."

Pincus: "You can't?"

Bishayee: "Yes."

Pincus: "There's a line here. I thought that might help you."

Bishayee: "It depend on how you draw a line. I can make it biphasic; I can make it exponential. It depend on how you draw the line."

Pincus: "What you mean by 'I can make it biphasic' or 'exponential'?"

Bishayee: "Can I show you on a piece of paper or directly on this document?"

Pincus: "No. I don't want to write on that."

Bishayee: "Okay."

Pincus: "You can use my pen."

Bishayee: "It might not be accurate representation. I can – it could be both curve. If I draw a line, then the line would be like this. If I draw the line the way I did, so you can get two curves."

Pincus: "Okay. [Pincus had Bishayee's handwritten diagram marked Bishayee-12.] Go back on Bishayee-11 if you would please. Do you see where it has Coulter counts here?"

Bishayee: "Yes."

Pincus: "Is that your handwriting?"

Bishayee: "No."

Pincus: "Do you recognize whose that is?"

Bishayee: "Possibly Dr. Howell's."

Pincus: "Fair enough. [Pincus then introduces Bishayee-13, a 50 percent experiment dated May 3, 2001.] Dr. Bishayee, I realize some of this is tedium and I apologize, but I got to go through it. Alright?"

Bishayee: "Do it. We're all here to do our job."

Pincus: "I'm going to show you Exhibit Bishayee-13. It appears to be a 50 percent experiment performed by Dr. Howell on May 3, 2001. Do you recognize this document?"

Bishayee: "No, I don't."

Pincus: "I'd like you to look through the various pages and tell me if you can identify any of the handwritten comments that are yours, if any?"

Bishayee: "I haven't seen any."

Pincus: "Okay. Do you have any recollection of Dr. Howell sharing this experiment with you?"

Bishayee: "No."

Pincus: "You were working in the lab at this point in time still. Correct?"

Bishayee: "The date shows, yes."

Pincus: "You were still employed in the lab. Correct?"

Bishayee: "That's correct."

Pincus: "Do you have any recollection of Dr. Howell discussing that the results of this experiment differed from the results that were reported in the publications?"

Bishayee: "Not specifically that, but there was some discussion about some of the experiment by either him or somebody else that he didn't match what I got."

Pincus: "What did he tell you in that respect? Tell me the discussion as best you can recall."

Bishayee: "Okay. Some of the experiment he did, or somebody else, didn't show the same result that I observed, so."

Pincus: "Did he tell you that he had been ordered to perform those experiments?"

Bishayee: "I'm sorry?"

Pincus: "Did he indicate to you why he was performing these experiments?"

Bishayee: "No."

Pincus: "Did he tell you that he had been ordered by anyone in the university to do that?"

Bishayee: "No."

Pincus: "Did he ever tell you that Dr. Raveché – do you know who Dr. Raveché is?"

Bishayee: "Yes."

Pincus: "Who is she?"

Bishayee: "She's one of the faculty at UMDNJ."

Pincus: "Did Dr. Howell ever indicate to you that he had been instructed by Dr. Raveché to perform any of these experiments?"

Bishayee: "No."

Pincus: "So you don't recall discussing this experiment or the results of the experiment?"

Bishayee: "That's right.

Pincus: [He then introduced Bishayee-14, a 50 percent experiment dated June 28, 2001.] "Dr. Bishayee, I'm showing you what I've marked as Bishayee Exhibit 14 it appears to be a 50 percent tritiated thymidine experiment done by Dr. Howell on or about June 28, 2001. Do you recognize this document?"

Bishayee: "I see it now, but I don't recall."

Pincus continued to question Bishayee about this experiment but got nowhere. Bishayee continued to reply "no" and "I don't recall." Then Pincus took him back to the January 11, 1999 50 percent tritiated thymidine experiment marked Bishayee-15. Bishayee had performed that experiment and recognized his own handwriting. He asked Bishayee if he could tell him what the results of the experiment were by way of survival. "To your knowledge was it exponential or biphasic?" Bishayee replied that he couldn't tell. "Where were you looking to determine that? Pincus asked. "Survival fraction." Bishayee indicated B012289. "For purposes of seeking to determine whether survival was exponential versus biphasic the data you relied upon was the column S. F. or survival fraction?" Pincus rejoined. "That's correct," said Bishayee. "And you can't tell?" Pincus asked again. "No, I can't," Bishayee retorted. "Do you recall learning of the fact that there came a point in time where Dr. Hill and Dr. Lenarczyk had been observing an experiment you were performing in about March of 2001?" Pincus redirected his questioning. "No," said Bishayee. "You never learned of that?" Pincus appeared surprised. "I learned later but I didn't know that at the time," Bishayee replied. "Not at the time. When did you learn of it?" Pincus wanted to know. "I don't know exactly when," Bishayee told him. "Did you learn of it after such time as a complaint was lodged with the Committee on Scientific Integrity [first Newark CCRI], or was it before that?" Pincus asked. "I don't know the time sequence," Bishayee insisted. And so do you recall the experiment that you were advised that they had observed you performing, what it was about?" Pincus pressed. "No," Bishayee replied. "You don't have any recollection of it?" Pincus continued to push. "Not for any particular experiment. Somebody's – I didn't have any recollection about others observing me doing any particular experiment," Bishayee replied.

> *Pincus*: "Do you recall there being a question about certain [Helena] tubes being left in an incubator?"
> *Bishayee*: "Yes."

Pincus: "What do you recall about that?"

Bishayee: "As part of the university investigation and the committee I appeared before they asked me about some question about the tubes in the incubator."

Pincus: "And they questioned you about the fact that when – apparently the tubes that you were using during your experiment, you were going to take them to a machine known as a FACS, F-A-C-S, for separation. Correct?"

Bishayee: "Right."

Pincus: "And that apparently when you went to get those tubes you related there was still a rack of your tubes in the cold incubator. Correct?"

Bishayee: "Right."

Pincus: "Am I right that's your recollection?"

Bishayee: "I didn't – actually I don't really recall saying me – that was part of the discussion. I don't recall what are the specific questions were asked."

Pincus: "But you do recall the subject coming up at the committee?"

Bishayee: "All I recall, that there is some issues with the tubes and the cold temperature."

Pincus: "You told them that you believed those tubes were for another experiment. Correct?"

Bishayee: "I don't recall that."

Pincus: "And do you recall there being a question as to where the protocol for that experiment was and that you were unable to produce that? Do you recall anything of that nature."

Bishayee: "No, I don't."

Pincus: "You have no recollection of that?"

Bishayee: "No recollection it was asked."

Pincus: "If there is a reflection of this dialogue between you and the committee in the minutes of the investigation sitting here today you have no other recollection of that discussion other than what might be set forth in those minutes. Is that fair to say?" [Flynn objected to the form of the question.]

Bishayee: "I don't recall anything. Once again, if I am given more information maybe I can tell."

After a brief recess Pincus introduced Bishayee-16, a 50 percent experiment he performed on March 19, 2001, and asked if Bishayee recognized it. "It seems my handwriting so it appear it is documented. I don't recall about doing it," Bishayee told Pincus. "But the handwriting that's on this document is yours. Is that correct?" Pincus asked. "That's correct," Bishayee replied. Pincus put aside the document. "Now we were talking about your leaving Dr. Howell's lab earlier today. I indicated to you that that took place on or about July 31 or August 1 [2001]. Do you recall at or about the time that you determined – it was determined that you were leaving Dr. Howell's lab having lunch with Dr. Hill in a church?" Pincus continued. "Yes," Bishayee confirmed. "What's the name of the church?" asked Pincus. "Priory," said Bishayee. "And was anyone else present other than you and she?" asked Pincus. "No," he replied. "Did she ask you to go to lunch with her or did you ask her to go to lunch?" Pincus pressed. "She asked me to go to lunch," Bishayee answered. "You didn't feel uncomfortable going to lunch with her?" Pincus pushed back. "Initially I did and then I thought it would be nice to go and see what's happening," Bishayee explained.

> *Pincus*: "Okay. So, what do you recall discussing with her during that lunch?"
>
> *Bishayee*: "Well, the discussion was what is going on, what I was going to do, resign from the lab, and discussion about my future, what I should do. [Bishayee speaking in third person] Dr. Hill indicated Roger fired you so we should – you should take action against him, so Roger is a bad guy, you should bring something against him."
>
> *Pincus*: "So let's go over this again. Tell me again please as best you can recall?"
>
> *Bishayee*: "She asked me about what happened in the lab. I said that I resigned, and she indicated that Roger didn't like me to be in the lab so he used me and then he's kicking me out from the lab, and I should bring something against him."
>
> *Pincus*: "Okay. So, Dr. Hill asked you what had happened?"
>
> *Bishayee*: "Yes."
>
> *Pincus*: "And you said you had resigned?"
>
> *Bishayee*: "Yes."

Pincus: "And then you said that Dr. Hill said that Roger didn't like you?"

Bishayee: "Yes. Roger didn't want me to be in the lab so he's kicking me out from the lab, and I should be doing something."

Pincus: "You should be doing something in terms of what?"

Bishayee: "In terms of – that the way he behaved was not right so I should do something."

Pincus: "Do you recall whether the decision to resign was yours or Dr. Howell's?

Bishayee: "It was mine."

Pincus: "Why did you resign?"

Bishayee: "Well, the environment was not very good for me. I realized that tenured faculty was fighting left and right, and as a young investigator who just started the career it was not very pleasant environment to do the research and to do the job. So, the atmosphere was very bad and I decided to leave to get rid of this kind of hostile environment."

Pincus: "And who was the atmosphere bad between?"

Bishayee: "Between Dr. Hill and Dr. Howell."

Pincus: "Did you know the reason or reasons why – ."

Bishayee: "No."

Pincus: " – they had bad feelings?"

Bishayee: "No."

Pincus: "Did Dr. Howell ever tell you that the atmosphere was bad because of you?"

Bishayee: "No."

Pincus: "Did he tell you that the atmosphere was bad because of anything that Dr. Hill was doing to you?"

Bishayee: "I recall that he said that Dr. Hill doesn't like me to be in the lab."

Pincus: "That she doesn't like you being in the lab?"

Bishayee: "Yes."

Pincus: "Did he go into any details or explain the basis on which – the factual basis on which he was making that statement?"

Bishayee: "No."

Pincus: "Okay. So, he didn't indicate any event or events that led him to believe that [Dr. Hill] didn't want you in the lab?"

Bishayee: "Yeah. He didn't give me any specificity or what – the overall situation was not pleasant, so it was – it was very clear to me. I didn't need his explanation to understand."

Pincus: "You say it was very clear to you. What was very clear to you?"

Bishayee: "The environment was not good for me."

Pincus: "Was it very clear to you that Dr. Hill did not want you in the lab?"

Bishayee: "I was told about that. I didn't hear directly from her."

Pincus: "Who told you that?"

Bishayee: "Dr. Howell."

Pincus: "But you never heard Dr. Hill state to you directly that she didn't want you in the lab. Correct?"

Bishayee: "That's correct."

Pincus: "She never said any words to that effect to you. Correct?"

Bishayee: "No, not directly."

Pincus: "Did she always treat you politely?"

Bishayee: "Yes, she did."

Pincus: "Did she manifest any behavior toward you that you found to be hostile?"

Bishayee: "I'm sorry? Say it again."

Pincus: "Did she manifest any behavior, physical, verbal or otherwise, to you that you found to be hostile?"

Bishayee: "That I found she was hostile to me?"

Pincus: "Yeah."

Bishayee: "Okay, no."

Pincus: "She never behaved that way to you?"

Bishayee: "No."

Pincus: "Did she ever make any comments about your national origin or ethnicity to you?"

Bishayee: "No."

Pincus: "Is it fair for me to say if she had made such a comment or comments about your national origin or ethnicity to you, you would remember that?"

Bishayee: "Yes, I would."

Pincus: "Did anybody relate to you that Dr. Hill had made such comments about you?"

Bishayee: "I'm sorry? Say it again?"

Pincus: "Did anybody ever say to you that Dr. Hill made comments about your national origin or ethnicity?"

Bishayee: "No."

Pincus: "Dr. Howell never related that to you?"

Bishayee: "No."

Pincus: "Did Dr. Azzam ever say anything to you about that?"

Bishayee: "No."

Pincus: "Did doctor – are you familiar with an individual by the name of Rameshwar?"

Bishayee: "Yes."

Pincus asked Bishayee what position Dr. Rameshwar held at the university. "At the time *he* [emphasis added] was a faculty with UMDNJ," Bishayee responded. "Did Dr. Rameshwar ever indicate to you that Dr. Hill had made statements about your national origin or ethnicity?" Pincus asked. "I can't recall," said Bishayee. [Dr. Pranela Rameshwar is female, not male, as indicated in Bishayee's comment; he either misspoke or had never met Rameshwar to know that he should have been referring to "her," not "he."] Pincus turned back to the luncheon Bishayee attended with Hill at the church. "Do you recall during the course of the luncheon saying to Dr. Hill that you were sorry that you had hurt her?" "Yes, I said that," Bishayee replied. "Why did you say that to her? Pincus wanted to know. "What were you talking about?" Bishayee replied: "Yes. I was a little surprised when she called me for the luncheon and she said that we need to talk. At that time, I was not sure about her motive, what she was trying to do to me. I heard that she doesn't want me in the lab," he continued, "and now she's calling me for luncheon, so I was very worried. So, I was trying to get some information to know what's coming for me next." Pincus encouraged him to continue. "So, I didn't want to argue anything; rather, I wanted to become very polite and to understand what she's trying to do and what she wants me to do," Bishayee explained. "So, I understand that," said Pincus, "when you say that you recall saying to Dr. Hill that you were sorry that you hurt her. In what way did you think you had hurt her?" Bishayee replied: "I didn't hurt her in any way but the whole atmosphere was very unfriendly, so I said if I did anything wrong you, you know, just as a nice gesture, to be friendly with her and know what's in her mind." Pincus wasn't satisfied with the answer. "I'm not there, you are. So, I hear a statement that you admit making and I'm trying to understand what was your thinking behind making that statement," said Pincus. "Thinking was just to behave as friendly and to get as much information as I can of because I didn't know what she was thinking," Bishayee elaborated further, still not making it clear why he was apologizing for "hurting Dr. Hill."

Pincus produced a copy of Bishayee's resignation letter dated July 30, 2001, and asked him if recognized it. He did. "Now, do you recall in the course of this luncheon at the Priory with Dr. Hill telling her that Dr. Howell handed you this letter of resignation and told you to sign it?" Pincus asked. "I don't recall exactly saying that, but it was part of a discussion that I resigned from the lab," Bishayee replied.

Pincus: "Did you prepare this letter of resignation?"

Bishayee: "No."

Pincus: "Who did, to your knowledge?"

Bishayee: "Dr. Howell did."

Pincus: "So tell me as best you can recall the discussion you had with Dr. Hill during this lunch about your letter of resignation. What did you tell her? What did she have to say, if anything?"

Bishayee: "I don't recall exactly the conversation in detail. What I was trying to accomplish there, to know what she was up to, and I told that I resigned from the lab and that's it. It was part of the discussion."

Pincus: "Did you tell her that you had been afraid not to sign the letter lest Dr. Howell would not give you any recommendations?"

Bishayee: "I don't recall that."

Pincus: "Do you have any recollection at all of discussing recommendations with Dr. Hill during the course of this luncheon?"

Bishayee: "I don't recall."

Pincus: "By the way, had Dr. Howell told you that he wouldn't give you a recommendation unless you signed this?"

Bishayee: "I don't recall."

Pincus: "No recollection whatsoever of any discussions with Dr. Howell about letters of recommendation. Is that your testimony?"

Bishayee: "Yes. I don't have any recollection about the specific."

Pincus: "At the time you signed this letter of recommendation did Dr. Howell – ."

Bishayee: "I'm sorry. It's letter of resignation."

Pincus: "Resignation. Did I say letter of recommendation? I'm sorry. I apologize. Did you tell Dr. Hill that you'd been instructed by Dr. Howell not to come on campus after resigning."

Bishayee: "I don't recall."

Pincus: "Do you remember telling Dr. Hill that Dr. Howell had taken your keys away?"

Bishayee: "I don't recall."

Pincus: "Do you remember telling Dr. Hill that Dr. Howell had taken your ID away?"

Bishayee: "I don't recall."

[Flynn objected to the form of the questions.]

Pincus: "So do you have a recollection of saying any of those statements?"

Bishayee: "No, I don't have a recollection."

Pincus: "Do you have a recollection of telling Dr. Hill that Dr. Howell told you not to apply for research positions on campus, campus being UMDNJ?"

Bishayee: "No, I don't have."

Pincus: "Had Dr. Howell told you not to apply for research positions on campus?"

Bishayee: "I don't recall."

Pincus: "Do you have a recollection of telling Dr. Hill that Dr. Howell told you not to apply for research positions in radiation biology anywhere, that would be UMDNJ or any other campus or facility?"

Bishayee: "I don't recall whether that was part of discussion?"

Pincus: "Did Dr. Howell ever have discussions with you about not applying for research positions on UMDNJ?"

Bishayee: "I don't have a recollection of that, he said that."

Pincus: "Do you have any recollection of Dr. Howell telling you not to apply for research positions in radiation biology anywhere?"

Bishayee: "No, I don't."

Pincus: "Do you recall discussing your visa status with Dr. Hill during the course of the luncheon?"

Bishayee: "I don't recall."

Pincus: "Do you remember what visa you were on at that point in time?"

Bishayee: "Probably H1-B."

Pincus: "Do you recall having any discussions about changing that visa from H1-B to visitor's or tourist visa?"

Bishayee: "No."

Pincus: "Did you in fact ever undertake to do that?"

Bishayee: "No."

Pincus: "Do you recall at or about this time that you were living in her son's apartment in Newark?"

Bishayee: "Right."

Pincus: "What was the address? Do you remember?"

Bishayee: "It was 25 Manor Drive."

Pincus: "Do you recall having discussions with Dr. Hill during this lunch as to whether you could continue to live in that apartment?"

Bishayee: "I don't recall."

Pincus: "Did you continue to live in that apartment for a period of time?"

Bishayee: "Yes."

Pincus: "For what period of time?"

Bishayee: "I would say – maybe beginning of 2002 until I got married and brought my wife, and we stayed there shortly and then we found an apartment and we moved out."

Pincus: "When did you move out to the best of your recollection in 2002?"

Bishayee: "2002, let's see, March – maybe April, maybe May."

Pincus: "May? So for the better part of almost another year you continued to reside in that apartment, from the end of July of 2001 through the following May of 2002. Is that correct?"

Bishayee: "Yeah. Whatever the period was, yes, I was there."

Pincus: "Fair enough. She never told you you needed to get out, did she, Dr. Hill?"

Bishayee: "No pressure to move out."

Pincus: "You were welcome to stay as long as you wanted, isn't that so?"

Bishayee: "Right."

Pincus: "Do you recall telling Dr. Hill during this luncheon that up until that time all of the evaluations that Dr. Hill had prepared regarding your performance had been good?"

Bishayee: "I don't recall."

Pincus: "Do you recall having any discussions with Dr. Hill during this luncheon regarding any of the bystander effect experiments?"

Bishayee: "I don't recall."

Pincus: "Do you recall telling Dr. Hill during the course of the luncheon that some of the experiments that you performed showed it and some of it did not?"

Bishayee: "I don't recall."

Pincus: "That being the bystander effect so we're clear."

Bishayee: "Okay."

Pincus: "You don't recall?"

Bishayee: "No."

Pincus: "Do you recall telling Dr. Hill that in regards to the bystander, you know, experiments that Dr. Howell had picked and chosen from among the experiments for purposes of reporting?"

Bishayee: "I don't recall, but I know that I did the experiment, and I gave it to him and it was up to him to use whatever he thought reasonable."

Pincus: "Did you ever give him any experiments that you recall that reflected results different from that which was reported in the literature that we reviewed here earlier, the two articles?"

Bishayee: "Say it again?"

Pincus: "Do you recall giving Dr. Howell any experiments' results which results differed from the results reported in the two articles?"

Bishayee: "It's the other way. I did the experiment and gave it to him, and he used those data to produce this publication, so I was not sure which one he used and which one he didn't."

Pincus: "Okay. So, you're telling me that the data that was chosen to be put in those two articles, Bishayee-7 and -8, to your knowledge was chosen by Dr. Howell and not by yourself?"

Bishayee: "That's correct."

Pincus: "Do you recall telling Dr. Hill that you had been pressured to produce data by Dr. Howell?"

Bishayee: "No."

Pincus: "Were you pressed to produce data by Dr. Howell?"

Bishayee: "No."

Pincus: "Do you recall telling Dr. Hill that Dr. Howell was always pleased when your data fit his mathematical models?"

Bishayee: "I don't recall exactly."

Pincus: "I don't understand your response."

Bishayee: "What I said whether she asked me the specific question and I answered, I don't recall that."

Pincus: "You recall the question being asked of you?"

Bishayee: "No, I don't recall this question."

Pincus: "That's what I wasn't sure of. So, if you would again tell me why it was exactly that you left the lab, Dr. Howell's lab?"

Bishayee: "The environment was not pleasing for me as I realized that a lot of things is happening, not very healthy environment for me to be there as it's kind of animosity between two senior faculties – tenured faculty and being a very junior postdoctoral fellow, it was not pleasant for me to stay and survive."

Pincus: "Were there any other reasons?"

Bishayee: "No."

Pincus: "That was it?"

Bishayee: "That was it."

Pincus: "So it was the relationship that was in the laboratory?"

Bishayee: "Yeah."

Pincus: "Do you recall being questioned by Dr. Putterman regarding the allegations, Dr. Karen Putterman?"

Bishayee: "There was a part of the procedure."

Pincus: "And do you remember when it was that you were questioned by her?"

Bishayee: "When?"

Pincus: "Yeah. Do you recall when?"

Bishayee: "During this time. I don't recall the exact timing."

Pincus: "And do you recall telling her that a reason that you left the lab or decided to leave the lab was the relationship with Dr. Hill and the allegations she brought against you which made you feel uncomfortable working in her proximity?" [But he had stated these things to Putterman.]

Bishayee: "I don't recall that."

Pincus: "You don't recall making that statement?"

Bishayee: "No."

Pincus: "Do you recall telling Dr. Putterman that there were any professional reasons or career reasons why you chose to leave the lab?"

[Again, he had made a statement to this effect to Putterman.]

Bishayee: "No, I don't recall."

Pincus: Do you recall telling Dr. Putterman that you were desirous of upgrading your technical skills in basic molecular biology?" [This is on record in the transcripts of the university inquiry.]

Bishayee: "Yeah. Probably I said that, that I would like to learn new things and try something different."

Pincus: "Did you tell her that you couldn't accomplish that by working for Dr. Howell?" [He did make such a statement.]

Bishayee: "I don't recall that."

Pincus: "Do you remember telling Dr. Putterman that you wanted to further your development and career with a goal of becoming an independent researcher?" [He made this statement also.]

Bishayee: "I don't recall that."

Pincus: "Do you recall telling Dr. Putterman about working – the desire to work in Dr. Pain's lab?"

Bishayee: "That I was talking to Dr. Pain for a possible position, too."

Pincus: I want to just be clear. I've heard everything you said and the fact that Dr. Howell prepared the letter of resignation. Do you feel by Dr. Howell preparing the letter of resignation that he was asking you to leave the lab?"

Flynn: "Objection. Go ahead."

Bishayee: "It was my decision more or less to leave the lab because that's the only thing I could do as a favor to myself, to be out of this kind of environment."

Pincus: "Why couldn't you write your own letter? I'm confused. Why did Dr. Howell write the letter as opposed to you writing the letter?"

Bishayee: "I don't know."

Pincus: "Alright. Was it your understanding that Dr. Howell wrote any letters of recommendation for you?"

Bishayee: "Say it again?"

Pincus: "At or about the time you resigned was it your understanding that Dr. Howell wrote any letters of recommendation for you?"

Bishayee: "I don't know."

Pincus: "Did he ever provide you copies of letters of recommendation?

Bishayee: "No."

Pincus: "Did you ever ask for a recommendation for Dr. Pain from Dr. Howell?"

Bishayee: "No."

Pincus: "How would you describe your relationship with Dr. Howell at the time you left the lab?"

Bishayee: "It was okay."

Pincus: "Okay? Any bad – any positive – what were the positives and what were the negatives, if you could tell me, if any?"

Bishayee: "Nothing. You know, I just took it as friendly way that you work hard for him, and we have good times and now the situation is not that good so I must leave."

Pincus: "Now, what efforts did you then undertake to learn molecular biology?"

Bishayee: "I understand that I need to upgrade my basic technique about scientific research, and of staying in the lab for that long you only learn a few things but a lot of things that I should learn to expand my horizon."

Pincus: "You needed that job for purposes of maintaining your H1-B visa, isn't that so?"

Flynn: "Objection."

Pincus: "If you know?"

Bishayee: "No, I don't know that."

Pincus: "You don't know that?"

Bishayee: "No."

Pincus: "You had recently been promoted to research associate?"

Bishayee: "Recently? What time?"

Pincus: "Prior to your resigning you had recently been promoted by Dr. Howell to research associate. Isn't that so?"

Bishayee: "Yes."

Pincus: "And along with that promotion had come an increase in salary, isn't that so, to your knowledge?"

Bishayee: "I can't recall, but there may be some increase that goes by year."

Pincus: "Were you aware whether that promotion came with a salary increase?"

Bishayee: "Possibly, yes, but I can't tell you exactly when that was effective."

Pincus: "But notwithstanding that you had been promoted and that you were in line for a salary increase you were willing to resign and get along without any income for a period of time. Isn't that so?"

Bishayee: "Say that again."

Pincus: "Notwithstanding you had been promoted and were due to receive an increase in salary you determined to resign and get along without an income for a period of time. Isn't that so?"

Flynn: "Objection to form. You can answer."

Bishayee: "Again, if you could please rephrase the question again?"

Pincus: "Notwithstanding that you had been promoted and were due an increase in salary, you determined to resign. Correct?"

Bishayee: "Yes."

Pincus: "And you resigned knowing that by doing so you would do without income for a period of time until you obtained other employment. Isn't that so?"

Bishayee: "That's right, that's right."

Pincus: "By the way, did you sign up for any courses in molecular biology?"

Bishayee: "Not full-time courses for molecular biology.

Pincus: "Part-time?"

Bishayee: "No."

Pincus: "None?"

Bishayee: "No."

Pincus: "Did you purchase any molecular biology textbooks?"

Bishayee: "No."

Pincus: "So if I understood you correctly, you left the lab because there was this environment, but you could not identify anything specifically that Dr. Hill had said or done to you. Correct?"

Flynn: "Objection. Go ahead."

Bishayee: "Not directly to me, but once again, the environment was not suitable for me."

Pincus: "Did you ever tell Dr. Putterman that you were uncomfortable working in close proximity with Dr. Hill?" [He had done so.]

Bishayee: "I don't recall that."

Pincus: Do you recall whether you and she carried on certain e-mail correspondence in terms of your payment of rent?"

Bishayee: "Correspondence between whom?"

Pincus: "Dr. Hill and yourself."

Bishayee: "I don't recall."

Pincus: "Do you recall ever using the terms 'how's the weather' in any e-mail for purposes of ascertaining whether Dr. Howell had left the lab so that you could pay the rent [to Hill]?"

Bishayee: "I don't recall the specific word, but I recall that I needed to come to give her a check as the rent."

[Pincus produced Bishayee-18, copies of e-mail correspondence between Bishayee and Hill.]

Pincus: "Dr. Bishayee, if you would please take a moment to look at these pages comprising various e-mails. My question to you is, do you recall being one of the parties who authored or received these e-mails?"

Bishayee: "Yes."

Pincus: "Do you recognize these?"

Bishayee: "Yes."

Pincus: "Any references to Anupam or e-mails – ?"

Bishayee: "I see my name."

Pincus: "Do you recall writing to her and she writing to you?"

Bishayee: "I don't recall exactly but I recall the document which is part of it."

Pincus: "Do you know what the basis was for your communicating with her by e-mail?"

Bishayee: "You see this is to giving her the money for the rent."

Pincus: "Do you remember what period of time you communicated with her this way?"

Bishayee: "No, I don't know how long I did."

Pincus: "So do you recall in the fall of 2001, you know, subsequent to your resignation, receiving unemployment compensation?"

Bishayee: "Yes."

Pincus: "Can you tell me how you were able to do that when you resigned from your employment with Dr. Howell?"

Flynn: "Objection. You can answer if you know."

Bishayee: "I don't know. I resign and then I realized that somebody mentioned, I don't recall who, and I learned that there is some mechanism that I can be qualified, be eligible, for unemployment benefit because I pay taxes while I was employed, unemployment insurance. So, I checked with the human resources whether I am eligible or not, and I also checked with the Department of Labor. The Department of Labor told me that they would check – they need to check with human resources to make the decision about my case, and so that I get the name – the information."

Pincus: "You provided the name of who?"

Bishayee: "Provide the name of the department, human resources, and let them figure it out, and subsequently it was approved and I got a check."

Pincus: "Is it your recollection that you told [the] unemployment [office] you had resigned?"

Bishayee: "I didn't say what I did. I told them to verify from human resources whether I am eligible for any compensation or not. My understanding is that they talked to each other. Department of Labor verified the fact with the human resources, and then they decided to grant me the unemployment benefit."

Pincus: "Do you recall telling unemployment that you had been terminated by the university or Dr. Howell?"

Flynn: "Objection. Go ahead."

Bishayee: "I don't recall what I said but I recall that I told them to verify whether I am eligible or not, and that's what they did."

Pincus: "Did Dr. Howell assist you in regards to your unemployment application?"

Bishayee: "No."

Pincus: "To your knowledge did the university ever contest your application to receive unemployment benefits?"

Flynn: "Objection."

Bishayee: "I don't know."

The claim for unemployment is unfounded if Bishayee resigned from the lab. The *New Jersey Unemployment Handbook*[148] is clear on this point. According to the handbook "you cannot receive New Jersey unemployment benefits if you quit your job. In all-capitalized letters, this advisory publication published by a private service to educate New Jersey residents to the law, the handbook cautions the following:

IF YOU HAVE A JOB, DO NOT QUIT IT

IF YOU QUIT YOUR JOB, DON'T EXPECT TO COLLECT UNEMPLOYMENT

IF YOU HAVE A JOB, AND ARE LOOKING TO CHANGE JOBS, DO NOT QUIT YOUR JOB UNTIL YOU HAVE A NEW ONE

"In order to be eligible for unemployment benefits in New Jersey, you have to have lost your job through no fault of your own," per the manual. "This means that if you quit your job, you can't get unemployment." Further, it goes on to note: "Be sure that you [did not] quit your job because you didn't like it, or it was too hard, or because it didn't pay enough, or because the hours didn't work for your schedule – and still expect to collect unemployment." If a state does allow you to collect unemployment when you've quit your job – the requirement to do so will be *very* difficult to meet.

The State of New Jersey Department of Labor and Workforce Development[149], on its web site, is also clear on this point: "If you voluntarily quit your job without 'good cause connected with the work,' or if you voluntarily retire, you may be disqualified for benefits. 'Good cause connected with the work,' means that your reason for leaving must be directly related to your job and be so compelling that you had no choice but to leave the job. For example, a person quits work to move out of the area. While this is a good personal reason to quit, the reason for quitting is not connected with the work and the person would be disqualified." The site notes exceptions but none of these qualified in Bishayee's case: "There are exceptions to this disqualification. One exception may apply when the separation was related to or due to domestic violence. Another exception may apply when a spouse or civil union partner of an active military member leaves work to move with the military member who is transferred outside of the state." Certainly, it would be memorable, even to Bishayee, had he been scheduled for a claims examiner interview. "If you quit your job, or if you voluntarily retire," according to the state's labor workforce development site, "you will be scheduled for a claims examiner interview. The examiner may request certain documentation as supporting evidence of your separation. The examiner will determine if you are entitled to benefits based on unemployment insurance laws and regulations."

There is another hurdle that would have been equally memorable to Bishayee had he merely resigned and was seeking unemployment benefits: "To remove a disqualification for voluntary leaving, you must return to work for at least eight (8) weeks, earn at least 10 times your weekly benefit rate, and then become unemployed through no fault of your own. The new work must be in employment covered under the unemployment compensation law."

Moreover, according to LSNJLAW's New Jersey Find Law[150] site, "if you quit for a good, work-related reason, it is important to show that you tried to *work out the problem before you quit* [emphasis added]. If you quit because of harassment, did you first complain to a supervisor to try and fix the problem? If your working conditions were unsafe, did you let management know so they could try to correct the situation?" Ultimately, it goes on, "if you didn't try to work out the problem, can you show that it would have been impossible to work it out?" Thus, a lot of documentary proof would have had to transpire for Bishayee to receive unemployment payments. Bishayee did not document a hostile work environment in the resignation letter. If he was claiming discrimination, New Jersey law is clear: "Various laws make it unlawful to discriminate on the basis of race, sex, pregnancy, religion, national origin, age (over 40), disability, marital status, and sexual orientation." But, as state law applies: "If someone at work was discriminating against you, that can be a good, work-related reason for quitting." The individual is thus required "to write down any discriminatory actions or comments as well as who witnessed them. In the cases of disability and religion, the law may require the employer to take extra steps to accommodate you or to help you to do your job. You may also file a claim with the New Jersey Division on Civil Rights or take other legal steps to challenge the discriminatory treatment."[151] What was clear in the Bishayee case is that there was absent a trail of documentation of any such discrimination having occurred to include the witnesses being on the record as supporting such a claim. Pincus' line of questioning is telling because this kind of engagement to obtain unemployment benefits for a foreign national who voluntarily quit his job would have left a document trail given the difficulty of meeting eligibility gateways.

New Jersey law is also quite clear that personality conflicts that don't involve unlawful discrimination are not good cause to quit a job. Workers are expected to do their best to work together, but when an employee is subjected to intentional harassment, vulgar, abusive language, or threatened violence, working conditions can be so intolerable so as to be good cause to quit. It takes a lot, though, to prove a hostile working environment. Many situations that workers feel are hostile will not amount to hostile environments for the purposes of obtaining unemployment benefits.[152] From Pincus' and the Newark CCRI's questioning of Bishayee – by his own admission for the record – he was not in what is clearly defined as a hostile work environment and not subject to any of the aforementioned intolerable working conditions.

Pincus switched his line of questioning to a different subject.

"You were kind enough to share the fact that you had an impending wedding. Correct? You had a wedding coming up at this point in time [after he had quit]? You went to India to get married, right?" Pincus asked. "Right," Bishayee replied. "When was that exactly?" Pincus wanted to know. "Oh boy. I should remember my marriage date," said Bishayee, searching for the date. "I think it was February 28 is my marriage."

Pincus: "Of 2002?"
Bishayee: "2002."
Pincus: "And you had a reception to celebrate your wedding?"

Bishayee: "Right."

Pincus: "Yes?"

Bishayee: "Yes."

Pincus: "Where was that?"

Bishayee: "That was one of the restaurants in Montclair."

Pincus: "Did you invite Dr. Hill?"

Bishayee: "Yes, I did."

Pincus: "Why did you invite her?"

Bishayee: "I invited everybody whom I worked, so Dr. Azzam, Dr. Howell, all my associations, social gathering, and I was still staying with her apartment. It was a nice gesture to introduce her to my newly wife, bride."

Pincus: "Is it fair to say that had she done anything that caused you to pause and say gee, I don't want to invite her, or she shouldn't be invited to this day of happiness – ."

Bishayee: "I was upset about the atmosphere."

Pincus: "You might not have invited Dr. Howell, too, right?"

Bishayee: "In the end I invited everybody regardless how I feel I guess about these people."

Pincus: "So regardless of how you felt about Dr. Howell or Dr. Hill or the atmosphere you undertook to invite them both. Correct?"

Bishayee: "Yeah. I think it was maybe it's last time we see each other and I was prepared to leave the department and go somewhere else, maybe new job, so."

Pincus: "She gave you a gift to your knowledge?"

Bishayee: "I'm sorry?"

Pincus: "Did she bring you a gift to your knowledge?"

Bishayee: "Yes, she did."

Pincus: "Was she accompanied by her husband to your recollection?"

Bishayee: "Probably she came alone."

Pincus: "Have you ever met her husband?"

Bishayee: "Yes, I did."

Pincus: "What is his name?"

Bishayee: "George."

Pincus: "Do you have any recollection of George attending your wedding reception at that restaurant in Montclair?"

Bishayee: "No, I don't."

Pincus: "You do recall sending her an invitation?"

Bishayee: "Right."

Pincus: "And do you recall sending her a thank you for attending?"

Bishayee: "That I did for everybody."

The deposition ended shortly thereafter.

Chapter 8 –
Taking Responsibility

Department of Radiology chairman[153] Dr. Stephen R. Baker's December 29, 2008 deposition was by the shortest of all those recorded in the discovery phase. In sharp contrast, Baker, who had previous experience having his deposition taken, provided direct and concise yet complete responses to the questions he was asked. Baker established the line of accountability to his office when he informed plaintiff's counsel that he signed off on the grants, including Howell's to NIH, that he was responsible for the ethical behavior of faculty, fellows and researchers in his department, and when he acknowledged that Roger Howell hadn't informed him that Anupam Bishayee's experiment results couldn't be replicated. Further, it was Baker who went on record that Howell had not followed up on his promise to report back to him after postdoctoral fellow Marek Lenarczyk tried to repeat Bishayee's experiments. But what was inordinately clear after his deposition was taken, said Hill, "is that he had not taken a proactive role in the [entire] affair."

In his capacity as department chairman, Baker had supervisory responsibility over Howell. Pincus asked him to describe what that supervision entailed. "Well, like any other physician in my department, I'm responsible for them to maintain their position in an ethical manner. I'm responsible," he continued, "for evaluating them at least annually. And I'm responsible for making sure that they obey the rules and regulations of the institution. I'm also in charge of deciding whether they would be promoted or punished." Pincus wanted to know if he was correct that Howell's title had been recorded correctly as the chief of the division of radiological research. Further, he asked Baker to briefly describe the hierarchy in terms of how Howell fit in the organizational structure of the department. "Yes, he's responsible for all matters relating to radiation research, which is one of the components of the department. And, therefore, reports to me both in his capacity as a researcher and his capacity as chief of that particular division of our department," Baker explained. Pincus asked how many other chiefs were in the department. "There's vice chairs," said Baker, "I have three vice chairs." What was the difference between someone who is designated a vice chair versus a chief, Pincus asked Baker. "There's no strict relationship there," he replied. "It's, the three vice chairs all have responsibilities separate and distinct in terms of the disciplines they supervise than does Dr. Howell." The terms vice chair and chief are not synonymous. "They're not synonymous," Baker confirmed, "but they really don't differ substantively regards to what his responsibility is versus someone who is responsible for the clinical work."

The radiological research division was not divided into sections. "Well, yes, we do have people," Baker explained, "- let me try to answer that question the way I perceive it. They have different research interests so that there would be an individual who would be involved in radiation research in terms of physics and others in terms of biology and others in terms of functional imaging, and another one in terms of issues related to the juncture of psychiatry and radiology. So, they don't have specific titles, administrative titles, they have specific interests to, and they are now in that section." Pincus asked if Baker was aware of the fact that at one point in time Helene Hill served as head of the section of cancer biology. "Yes," he replied, without elaboration. "And was that prior to the time that Dr. Howell became chief of the division of radiation research to your knowledge? Pincus asked. "To my knowledge, that is probably the case. I don't recall." Curious still about the departmental hierarchy, Pincus asked, when Hill was head of the cancer biology section, were the sections synonymous with what Baker had previously described in terms of those "interests." In other words, did Baker know how the hierarchy worked at that time. "It would be synonymous in terms of interest not in terms of strict stratified administrative responsibilities," Baker replied. Pincus asked if Baker could tell him the responsibilities of a principal investigator at UMDNJ in the context of a grant. "What is your understanding in terms of the responsibilities of that individual?" he asked. "The responsibilities functionally in terms of managing the grant – in terms of the reward system," said Baker, "which is an important point of that, they would be the grantee of that award, and that would devolve upon their responsibilities, of course, but on the recognition of those responsibilities by everyone else who evaluates the performance of that individual as a researcher."

"Insofar as grants that are provided from the National Institutes of Health [NIH], do you have responsibility insofar as the application process for that grant?" asked Pincus. "I sign off on those grants," Baker replied. "Yes, I have to sign off because a chair must sign off on each grant." Pincus followed up, asking if Baker knew the purpose of that requirement. "Well, I think for accounting purposes and oversight in case there are issues related to those grants, then there has been, it's been looked at by a, someone who is not the author of the grant before it's submitted. And it makes good sense administratively." Baker admitted that he wasn't intimately familiar with the rules and regulations that the National Institutes of Health promulgates per the grant process. "I'm not intimately familiar with it. All my grant proposals are looked at by the administrator in my department who has a long experience, and then they're looked at by a grant department in the medical school," Baker explained. "I have an administrator of my department who has a long history of becoming familiar with some of the mechanical issues related to grants and their appropriate filling out of the forms, that's Ms. Karel Campbell, who is my administrator." The UMDNJ, at that time, also had an associate dean for research who looked over each grant as well. But ultimately it was Baker who signed off on the grant.

Pincus asked Baker insofar as the submission of the grant, if the individual who's looking to serve as the principal investigator has to certify that the application is true and complete and accurate to the best of his or her knowledge. "I believe that's correct," Baker said. "Was it your understanding that they [the principal investigators] are made aware of the fact that any false, fictitious or fraudulent statements or claims may subject the principal investigator to criminal, civil or administrative penalties?" Pincus pressed. Defendant attorney John Leonard objected to the wording of the question. Despite that objection, Baker answered "I'd agree with that [Pincus' statement about the penalties]." Following up, Pincus asked if it was Baker's understanding that the principal investigator in submitting the grant agrees to accept responsibility for the scientific conduct of the project. Leonard objected again. But Baker replied yes. Pincus asked if the principal investigator, in submitting the grant, agreed to provide progress reports if the grant is awarded as a result of the application. "Yes," Baker answered.

> *Pincus*: "During the time that the subject matter of this suit was transpiring, do you know how many postdocs Dr. Howell had working with him on this grant?"
>
> *Baker*: "No."
>
> *Pincus*: "Have you ever met Dr. Marek Lenarczyk?"
>
> *Baker*: "I'm sure I have. I met all of his postdocs."
>
> *Pincus*: "So you met Dr. Lenarczyk?"
>
> *Baker*: "If that was one of them, I don't recall all the names. Unless I have something else to jog my memory, but if it was a postdoc, I would have met him."
>
> *Pincus*: "Same question. Do you recall meeting Dr. Anupam Bishayee?"
>
> *Baker*: "Yes."
>
> *Pincus*: "And I believe you knew Dr. Hill working in his lab?"
>
> *Baker*: "Yes, I knew Dr. Hill."
>
> *Pincus*: "What was your understanding insofar as what Dr. Hill was doing insofar as the grant that is the subject matter of this suit?"
>
> *Baker*: "I am not sure."

Moving on, Pincus showed Baker a copy of the grant application at the center of the lawsuit involving Drs. Howell and Bishayee and the university, marked Exhibit Baker-5. Baker told him that he couldn't say that he'd seen it, "it's a long time ago." Pincus drew his attention to the bottom right-hand corner of the first page of the document [Bates stamped 000094]. "Below what appears to be Dr. Howell's signature is someone else's signature. Is that your signature," Pincus asked. "Doesn't look like my signature," said Baker. "You identified earlier that you have as part of your duties and responsibilities, you sign your approval on an application for a grant before such time in this case as this would have been submitted to the National Institutes of Health, do you recall testifying to that effect?" Pincus followed up. "Yes," Baker replied. "When you sign off your approval, is it on the grant itself or is that a separate document?" Pincus wanted to know. "Separate document that doesn't look like this," Baker explained. Was there a name or title of such a document? "I recognize it when I see it. I don't remember the title of it," Baker told Pincus. "I look to where I'm supposed to sign. I don't look at the top." Pincus asked if it wasn't procedure that a copy of the document must be kept in either Baker's office or the university as a paper trail pertaining to grant. "Absolutely," Baker responded. "So that approval that you signed off on this grant would still exist?" Pincus asked him. "Yes," Baker responded. "And Ms. Campbell would have that in her possession?" Again, he answered yes.

> *Pincus*: "Now, while we're here, I'd like you to turn to page 29 of the grant, and in particular what is denoted as figure seven, and that would be the document Bates stamped 0001227?" [Baker searched for the page.]
>
> *Pincus*: "To your recollection, did Dr. Howell ever inform you that the results depicted in this figure, and in particular in the B-graph of this figure, conflicted with results from a similar experiment that had been performed earlier by Dr. Hill?"
>
> *Baker*: "No."
>
> *Leonard*: "Objection to form."
>
> *Baker*: "No, I don't recall he ever informed me about that."
>
> *Pincus*: "And would your response be the same whether I asked the question whether he informed you before the submission of this grant or after, after the submission?"
>
> *Baker*: "It was a very specific issue related to the science of this case. I don't think I was informed about it one way or the other that I can recall."
>
> *Pincus*: "Fair enough."
>
> *Baker*: "I meant to say grant not case."
>
> *Pincus*: "Grant, okay, by all means. Did he ever tell you at or about the time this grant was being submitted that Dr. Hill had observed certain dishes and that she had suspected, dishes used during experimentation, and that she had concerns that the postdoc, Dr. Bishayee, had fabricated the results that were ultimately presented in figure seven?"

Leonard: "Objection to form."

Baker: "No, I don't recall that."

Pincus: "Do you recall having any conversations with Dr. Howell in which he conveyed to you suspicions about fabricated data that had been related to him by Dr. Hill?"

Baker: "No."

Pincus introduced exhibits he marked Baker-6 and -7. Baker-7 was Hill's memorandum. "Dr. Baker, I'm going to show you what I've just marked for identification as Baker-6. Have you ever seen this document before? Did you have an opportunity to look at it?" Pincus asked. Leonard tried to correct the phrasing of the question. "Do you remember seeing this document?" Pincus rephrased. "I don't remember seeing this," Baker replied. "I just want to orient you by way of time," Pincus explained. "This is an experiment, so you know, what has been described to us as one that was performed by the postdoc, Dr. Bishayee, commencing on or about September 20, 1999. It's event from looking at the protocol," he continued, "for the experiment that is on the document that is Bates stamped 13911 in the upper right-hand corner. Do you see where I'm referring to?" Baker found it. "Now, if you then jump ahead two pages to 913 – B013913 – all the way down at the bottom. There's a reference to colony counting on October 11, 1999, do you see where I'm referring?" Baker saw it. "I'm going to show you what's been marked as Baker-7 [this is Hill's memorandum]. Have you seen this document?" Baker told Pincus he didn't recall seeing it. "I'll represent to you that this sets forth her observation with regard to the experiment that I placed before you involving Dr. Bishayee and which has been identified as Exhibit Baker-6. Does this refresh your recollection," he asked Baker, "in any fashion insofar as any discussions that you may have had with Dr. Howell regarding suspicions that were related to him by Dr. Hill?" Baker, without elaborating, answered: "This does not refresh my memory."

The document in front of Baker – the Hill memorandum – set forth Hill's belief and position that her having worked a similar experiment in which she obtained conflicting data or data that conflicted that which Bishayee got from performing the same experiment. "What's your understanding insofar as the process that Dr. Howell should have followed or employed where there are conflicts involving the same experiment conducted by two individuals within the same laboratory?" asked Pincus. Leonard objected again, telling Pincus he didn't understand the question. "Do you understand the question?" Pincus repeated to Baker. "Not really," he said. "I want you to assume that Dr. Hill and Dr. Bishayee performed two experiments, each of them using the same protocols, you with me so far?" Pincus asked, rephrasing the question. Baker was with him. "And having performed those two separate experiments they came up with data that conflicted with one another. What was your understanding of Dr. Howell's responsibilities in terms of resolving the conflict between the data of those two experiments?" Pincus continued. "You're asking two different questions in the question," said Baker. "The question you asked, what we do in terms of the conflict. The conflict to me indicates that there is going to be some contest. What you have is two different results. So, I really don't know how to perceive from your question when you get two different results, because the possibilities, each of which may initiate a different action, depends upon a lot of other factors besides what's included in the generality of your question." Undeterred by the respondent's circular answer, Pincus pressed Baker. "What if the data in one of the experiments was used to support that figure seven in the grant application that I place before you, which was Baker-5, and the other data which was at variance with that as opposed to conflicting with it? Under those circumstances what was your understanding of Dr. Howell's responsibilities in resolving this?"

Leonard: "I'm going to object to this line of questioning. You're asking Dr. Baker to assume two identical experiments and all variables were exactly the same."

Pincus: "Yes, I am."

Leonard: "I think that's a scientific impossibility."

Pincus: "Your objection is noted. Can you answer?"

Baker: "My answer would be as specified. I don't know if these are identical experiments, under identical conditions with identical precursors. There's not enough information to assume that two experiments done at different times realizing different results were in effect in conflict, which was in your question. So, I don't know how to answer that question."

Pincus: "I understand your response. But do you not recall Dr. Howell relating such circumstances that existed regarding two experiments, one of which had been performed by Dr. Hill in or about October 1999, the other which had been performed by Dr. Bishayee in or about October 1999?"

Baker: "No, I don't recall that."

Pincus then introduced Howell's April 6, 2001, memorandum, marked as Exhibit Baker-8. He showed it to Baker and asked him to take a moment to review it.

Pincus: "Do you recognize this document?"

Baker: "I'm not sure I understand the question as do I recognize the document."

Pincus: "Do you recall receiving this document from Dr. Howell on or about April 6, 2001?"

Baker: "I don't recall receiving it. It doesn't mean that I didn't receive it. I just don't recall that."

Pincus: "In the second sentence, I believe, of Dr. Howell's memo to you, there's a statement that, 'as I mentioned to you, I first became aware of Dr. Hill's concerns about six to nine months ago when she brought this to my attention.' Do you see where I'm referring to?"

Baker: "Yes, I do."

Pincus: "Do you recall him discussing with you the concerns that Dr. Hill had raised with him?"

Baker: "All I recall was that there were concerns that were raised by Dr. Hill. I didn't know the specifics."

Pincus: "At the time that Dr. Howell spoke with you, did he provide to you any documentation relating to the concerns of Dr. Hill that are referred to?"

Baker: "Not that I recall."

Pincus: "Did you require or request of him any additional follow-up insofar as setting forth or articulating of the concerns of Dr. Hill of which he notes?"

Baker: "No, I didn't, because my knowledge, if I can recall, was that there was a dispute. I didn't know the specifics of that dispute."

Pincus: "And your knowledge was that there was a dispute, what was the dispute that you knew of at that time?"

Baker: "As I said that I was aware there was a dispute between the two, that's all that I knew."

Pincus: "So, you didn't know what the substantive aspects of the dispute was?"

Baker: "I didn't know the substantive aspects of the dispute."

Pincus: "When had you first learned of the so-called dispute?"

Baker: "My recollection it's about 2000, the year 2000."

Pincus: "And how did you learn about it?"

Baker: "I think Dr. Howell told me that there was a dispute."

Pincus: "Do you recall anything else that he told you back in 2000?"

Baker: "That's all that I recall."

Pincus: "Do you recall whether in 2000 he shared with you any documentation relating to the dispute?"

Baker: "I don't recall that."

Pincus: "When he told you there was a dispute in the year 2000, what, if anything, did you do?"

Baker: "Well, Dr. Howell was the director of the section. And if I recall, the dispute was to me, as I remember it, more over lab space and two people working together. I didn't recall that there was a specific issue about science. I told him he should resolve the differences not knowing that it was one specifically related to an issue of integrity in science."

Pincus: "Okay, so your recollection is it related to lab space, number one, correct?"

Baker: "Vaguely remember that."

Pincus: "And that the time frame that you seem to recall was preceding the date on which you received this memorandum in or about the year 2000?"

Baker: "Yeah, but my memory isn't sharp on this because that's a long time ago."

Pincus: "So, you're not certain of the date, is that what you're telling me?"

Baker: "I'm not certain of the date."

Pincus: "Are you telling me that it could have occurred after this discussion or is it your recollection that you're certain that the – ."

Baker: "I'm not certain of anything. The specifics are that it seemed to be about that time."

Pincus: "About that time?"

Baker: "About 2000, somewhere in that regards, a long time ago now."

Pincus: "I understand."

Baker: "And I have a lot of people dispute with a lot of people in my department, that's the nature of things."

Pincus: "If I were to represent to you that Dr. Hill did not make a formal complaint about Dr. Bishayee and Dr. Howell until April 20, 2001, does that in any manner refresh your recollection insofar as when it was that Dr. Howell first related to you that there was a dispute or describe the substance of that dispute?"

Baker: "No, sir."

Pincus: "If you go down approximately two-thirds of the way in the first paragraph, there is a statement, 'as I told you this morning, I have requested that my postdoctoral fellow, Marek Lenarczyk, Ph.D., repeat some of Dr. Bishayee's experiments as a check of the validity of the data.' Do you see where I'm referring to?"

Baker: "I do."

Pincus: "Did he inform you in this meeting or any prior meeting that Dr. Lenarczyk had attempted to replicate Dr. Bishayee's results in eleven experiments without success?"

Baker: "I don't recall that."

Pincus: "There's also a statement about two sentences following that, 'I will also personally sign all of Dr. Bishayee's data sheets and culture dishes to insure they are not tampered with.' Do you see where I'm referring to?"

Baker: "I do."

Pincus: "Do you recall [how] the discussion regarding his signing the data sheets and culture dishes came about?"

Baker: "No, I just remember that there was some issue with Dr. Bishayee, but I don't remember the specifics."

Pincus: "I'm going to show you, Dr. Baker, a series of documents from a prior deposition that was taken of Dr. Marek Lenarczyk on or about, I believe it was October 28, 2008, and we've marked those separately, so rather than take the time to remark them as separate exhibits for your review. These exhibits span a period of the year 2000 into 2001, all of which precede the date of the memo you have before you dated April 6, 2001, which was marked as Baker Exhibit 8, I believe."

Baker: "Yes."

Pincus: "And I had asked you a moment ago, and you said you don't recall Dr. Howell informing you in your meeting, this meeting or any prior meeting, that Dr. Lenarczyk had been unable to replicate Dr. Bishayee's results in eleven experiments without success. Did he ever show you, I'm going to identify the exhibits for the record, Lenarczyk-14, -15, -19, -20, -21, -25, -26, -27, -30, -31 or -32."

Baker: "What are those numbers again, please?"

Pincus: "Lenarczyk-14, let's take them one at a time. Did he ever show you this or share with you this exhibit?"

Baker: "No, I wouldn't have seen this?"

Pincus: "Okay, you can just flip them over and I'll put them back. What about Lenarczyk-15?"

Baker: "Similarly, no."

Pincus: "Never saw that, never provided to you by Dr. Howell?"

Baker: "No."

Pincus: "Lenarczyk-19, did you ever see that or was it ever provided to you by Dr. Howell?"

Baker: "No, I would have never seen this." [Baker denied having seen any of the Lenarczyk documents.]

Pincus: "Similarly, I'm going to show you two papers that are marked for identification during the course of Dr. Lenarczyk's deposition, that were authored in part by Dr. Bishayee and Dr. Howell. I show you Lenarczyk Exhibit 11 and Lenarczyk Exhibit 12. Have you ever seen these documents?"

Baker: "I'm aware of the bystander effect; I'm aware that people in my department publish articles. I'm not, I do not look at these articles. So, I'm not aware of them specifically."

Pincus: "I understand that. Did Dr. Howell ever relate to you or discuss in any way that the results set forth in either of the two papers, Lenarczyk-11 or -12, are at variance with any of the eleven experiments that I just shared with you?"

Baker: "I don't recall that at all."

Pincus: "So we're clear, when you say you don't recall, I just want to be certain that beyond any oral discussion, you were provided no document that related that the results were at variance?"

Baker: "I don't recall a discussion or that presentation of documents related to that. I don't recall any of things."

Pincus: "Okay, I'll take those back. Again, I want to first orient you to what we marked as Exhibit Baker-8 and the date of it being April 6, 2001. I'm going to show you an additional five experiments, and I'm going to represent to you that the dates of these experiments took place subsequent to the date of the memo, Baker-8, okay? And I want to go through each one and just, the question to you is did Dr. Howell at any time share with you a copy or the results of the experiment – I'm sorry, there's got to be a fourth. I missed one, let me go back. This is on top. I'm going to share with you six experiments, not five, that were performed subsequent to the date of the memo, Baker-8. Did Dr. Howell at any time share with you a copy of what has been marked as Lenarczyk-33?"

Baker: "This is Lenarczyk-18."

Pincus: "So you need 33. Thank you for correcting it. I'll take that one back."

Baker: "Maybe it's in here, I don't know."

Pincus: "No, no, thank you for correcting me. There's 33, so I'm showing you Lenarczyk-33."

Baker: "I never saw this that I can recall."

Pincus: "Were you aware of any actions that he took insofar as a Dr. Azzam? Do you know Dr. Azzam?"

Baker: "Yes."

Pincus: "Were you aware of any actions of Dr. Azzam undertook insofar as observing the conduct of this experiment?"

Leonard: "Objection to form."

Baker: "The question you asked me, was I aware of he, and then you said Dr. Azzam, does the 'he' refer to Dr. Howell or Dr. Azzam?"

Pincus: "He refers to Dr. Howell."

Baker: "Can you repeat the question?"

Pincus: "Sure, first, you said are you familiar, you've not seen this document?"

Baker: "I have not seen this document that I can recall."

Pincus: "Were you aware of any requests that Dr. Howell made of Dr. Azzam to observe the conduct of this experiment?"

Baker: "I'm not aware of that."

Pincus: "Have you had any conversation or discussion with Dr. Azzam regarding this experiment?"

Baker: "No."

Pincus: "Or any experiments that were undertaken in Dr. Howell's lab regarding the bystander effect?"

Baker: "Dr. Azzam's main area of interest is the bystander effect. So, I'm sure that there had been conversations between Dr. Howell and Dr. Azzam about the bystander effect."

Pincus: "My question was did you have any conversation with Dr. Azzam, not Dr. Howell?"

Baker: "About the bystander effect?"

Pincus: "The experimentation."

Baker: "These experiments?"

Pincus: "Yes."

Baker: "No."

Pincus: "Now, let's go to Lenarczyk-34 next. Is that the right document, just want to be sure?"

Baker: "Yes."

Pincus: "Have you ever seen that one?"

Baker: "No."

Pincus: "Do you recall any discussions with Dr. Howell regarding the results of this experiment that was undertaken by Dr. Lenarczyk or himself?"

Baker: "I don't have any recollection of any specific experiments relating to the bystander effect or this component of it."

Pincus: "I want to go back. This was an experiment as well as the preceding one, which was conducted by Dr. Howell, not Dr. Lenarczyk. Did he discuss with you the fact that he undertook these series of six experiments subsequent to his providing you the memo of April 6?"

Baker: "No, I'm not aware of that."

Pincus: "Okay, so let's just be sure, you don't recall seeing Lenarczyk-34, which was the second document that you just flipped over. What about -35, will you answer, please?"

Baker: "I'm not aware of that."

Pincus: "Lenarczyk-36?"

Baker: "I'm not aware of that one."

Pincus: "Lenarczyk-22, I believe?"

Baker: "I'm not aware of that one."

Pincus: "And last, Lenarczyk-23?"

Baker: "I'm not aware of this one."

Pincus: "So, if I understand you correctly, at no time did Dr. Howell relate to you that in any of the, that any of the experiments that Dr. Lenarczyk had performed were unsuccessful in replicating the results that Dr. Bishayee and Dr. Howell had reported in the two papers that I showed you, you have no recollection of such a discussion?"

Baker: "No recollection of such a discussion."

Pincus: "Do you have any recollection of him discussing with you the series of six experiments that I just shared with you and the fact that in understanding them they were unable to replicate the results of Dr. Bishayee as reported in those two papers?"

Leonard: "Objection to form."

Baker: "I'm not aware of that."

Pincus: "Did they indicate to you that they were unable to replicate the results in the experiment as reported in the grant application?"

Leonard: "Objection to form."

Baker: "I'm not aware of that."

Pincus: "You were aware that there came, were you aware that there came a point in time where Dr. Lenarczyk left Dr. Howell's lab and returned to Poland?"

Baker: "Now that you refresh my memory, he no longer works there. I didn't know where he would return to. I know that he was there and then I knew he wasn't there, that's all I know."

Pincus: "So, you knew that he left?"

Baker: "Yes."

Pincus: "Fair enough."

Baker: "I don't keep tabs on where they go unless I happen to know them personally."

Pincus: "I understand. Did Dr. Howell indicate to you that before he left the lab, that being Dr. Lenarczyk, he had performed five yet additional experiments, none of which were able to replicate the results that had been reported in the two papers or the grant application?"

Leonard: "Objection to form."

Baker: "I'm not aware of that."

Pincus: "I'm going to show you, let me just state the numbers, Lenarczyk-16, -17, -18, -28 and -29. And just as we have, let's take them one at a time, please. Lenarczyk-16, did Dr. Howell at any time provide you a copy of this exhibit or discuss it."

Baker: "No, sir, not that I can recall."

Baker's responses remained the same for each of the Lenarczyk documents. He answered "not that I can recall" to all," recorded as "same response" in the deposition transcript as Baker tried to prompt his recollection. Pincus took Baker back to Howell's April 6, 2001 memo and asked what kind of follow-up had occurred in the wake of the issues it raised. "No, I do not," Baker responded, indicating that he recalled no follow-up from Howell. Pincus then showed Baker his own memo, marked Exhibit Baker-9. "Do you recall authoring this?" Pincus asked. "Yes," Baker replied. "Do you recall whether this was authored shortly after the decision had come down from the committee on scientific integrity that there was insufficient evidence of scientific misconduct as alleged by Dr. Hill?" Pincus pressed. "I don't recall the chronology," Baker responded. "Okay, in this letter you appoint Dr. Howell the chief of the division of radiation research, am I correct? Pincus asked. "Right," Baker rejoined. "Am I further correct that in doing so you abolished the section of cancer biology?" Pincus pushed back. "Yes," he answered. "And this effectively made Dr. Howell Dr. Hill's supervisor, is that correct?" asked Pincus. Leonard objected again. "Yes, that's correct," said Baker.

Pincus: "Did Dr. Hill ever relate to you the fact that Dr. Howell called her into his office in or about July, early July of 2001, and told her he wanted nothing to do with her?"

Baker: "Again, I know that there was a dispute between the two of them. That dispute had been simmering, and I don't know what the specifics of that dispute were except that my perception of it was there [were] disagreements. So, I don't recall that specifically, and I would not react to a statement when someone doesn't want something to do with someone. That's not a very productive statement to be made to me."

Pincus: "Is it a statement that you would expect from, of an individual who you had appointed in a supervisory capacity within your department?"

Leonard: "Objection to form."

Baker: "It's a statement I would not expect from anybody who works for me or who is a colleague of mine. So, it's not just a statement made by someone who's taking on a supervisory capacity, it's a hostile, nonproductive statement."

Pincus: "Did you come to have knowledge of the fact that Dr. Hill was then locked out of access to shared laboratory space within the division sometime later that month?"

Leonard: "Objection to form."

Pincus: "July 2001?"

Leonard: "Objection to form." [He directs Baker to answer if he can.]

Baker: "I don't recall, I just recall that there was bad blood."

Pincus: "And by bad blood what are you referring to?"

Baker: "There was a lot of disputes, and I'm not referring to anything specifically."

Pincus: "And when you learned that there was this bad blood and/or these disputes, what, if any, actions did you undertake to seek to resolve them?"

Baker: "I told Dr. Howell that there needs to be a modus Vivendi among the two of you."

Pincus: "And how did you – ."

Baker: "Between the two of them."

Pincus: "And how did you monitor whether, in fact, that occurred?"

Baker: "I didn't monitor that, I didn't hear either of the individuals speak to me further about it."

Pincus: "So, on the basis of silence did you presume that there was no longer a problem?"

Baker: "Well, I presumed that there was no longer an acute issue, I'm sure that there was a problem."

Pincus: "So, if you were sure that there was still a problem, did that cause any concern to you?"

Baker: "Of course it causes concern, but I have a lot of people who may not get along with a lot of other people."

Pincus: "What was your concern that you had at that point in time as regards Dr. Hill and Dr. Howell?"

Baker: "That question's kind of vague, what is my concern?"

Pincus: "Well, you said it caused you a concern. I'm asking you what was the nature of your concern?"

Baker: "I regretted that two people couldn't work together. I know they had different research focus, and I would hope they would pursue them."

Pincus: "So, did you take any other steps insofar as seeking a resolution of those concerns?"

Baker: "No, I took no other steps."

Pincus showed Baker a document he marked Baker-10 and asked him if he recognized it. Baker said no. Pincus wanted to know if Baker had had any discussions with either Howell or anyone else listed in the document regarding assignment of laboratory space. Baker again said no. "If you look at the reassignments that are set forth in regards to offices [or] laboratory, would you agree that this announced that Dr. Hill was going to be locked out of the division shared spaces in MSBF451A and –B?" Leonard objected to the question. "Nothing here that says she'd be locked out of anyplace. It says there's a room assigned to her, both in the office and in the laboratory," Baker replied. "Did you ever come to learn that, in fact, she had been locked out?" Pincus wanted to know. Leonard objected again. "No, I was not informed by either Howell or by Hill about this matter," said Baker. "Were you ever made aware of any grievances that Dr. Hill filed in regards to this assignment of laboratory space?" Pincus asked Baker. "No, nor did I ever hear from Dr. Hill that she was aggrieved."

> *Pincus*: "So, you've not seen any documents associated with any grievances associated with this?"
> *Baker*: "She never came to me and talked to me that there was [an] issue, which I would have expected what she would do if there was an issue to which she was aggrieved."

Pincus pulled out Exhibit Baker-11, the Howell letter. He asked Baker to review it. Then he asked if he recalled seeing it.

> *Baker*: "I don't recall, but I was [copied], so I probably received it."
> *Pincus*: "Am I correct that at the time Dr. Howell authored this, h had been appointed chief of radiation research by you, correct?"
> *Baker*: "Correct."
> *Pincus*: "So, does this refresh your recollection at all insofar as Dr. Howell denying or restricting access to certain laboratories to Dr. Hill?"
> *Baker*: "*It states he restricted her access* [emphasis added]."
> *Pincus*: "I see you reading that, but does that refresh your recollection insofar as any further involvement or discussions that you had regarding this subject matter?"
> *Baker*: "No."
> *Pincus*: "Can you think of any reason or reasons why Dr. Howell at the time would have restricted access to the laboratory spaces that are denoted to Dr. Hill?"
> *Baker*: "You're asking me to speculate?"
> *Pincus*: "I'm asking do you know of any reasons why he did that?"
> *Baker*: "Could you rephrase your question then?"
> *Pincus*: "Sure, do you know of any reasons why Dr. Howell denied access to Dr. Hill as set forth in this memorandum?"
> *Baker*: "I don't know any reasons."

Pincus then produced Exhibit Baker-12, the Dr. Robert A. Saporito[154] correspondence. He again asked Baker to take a look at the information in front of him and whether he had seen it prior to the deposition. Baker replied that he hadn't seen it and he wasn't copied on it. "I've never seen this document."

Pincus: "Do you recall having any discussions with Dr. Howell regarding the correspondence that he had with Dr. Saporito?"

Baker: "No, I don't recall that."

Pincus: "You recall having any conversations with Dr. Saporito regarding this correspondence?"

Baker: "I did not have any conversations with Dr. Saporito, involvement in this issue."

Pincus: "Do you have any knowledge why this issue regarding lab space would have gone to Dr. Saporito as opposed to you for resolution?"

Baker: "I don't know why that happened."

Pincus: "Do you have any other knowledge regarding the issue of lab space that you haven't already identified?"

Baker: "Yes, one clarification, Dr. Howell was seriously ill about this time."

Pincus: "Okay."

Baker: "I'm not sure what year it was, but it seems about this time."

Pincus: "So, if he was ill – ."

Baker: "There was several weeks when he wasn't physically present."

Pincus: "And how does that relate to the circumstances?"

Baker: "I don't know where these, whether these letters were penned by him as he was recuperating or he was physically present. I don't know. But it seems to me that in my mind there was at least two events were happening close to each other, but I'm not sure."

Pincus: "What would his illness have to do with his authorship of these documents?"

Baker: "Good question, they probably have little to do with it, but he may not have been physically present at that time. I don't know, you can find out from him exactly. Just to tell you I don't know whether he was there, or he was not there, seems to be about the same time."

There were at least two problems with Baker's response, not the least of which was his deflection of the question as to why the issue was sent to

Saporito rather than himself for resolution. Howell's presence – or not – has nothing to do with the matter about which Baker was asked. Further, whether Howell was ill is a matter of record within the department and the university and is remarkable enough in the seriousness of his illness that providing a record of his absence, bookending the event with dates, would not have been difficult to do. Suggesting Howell's absence in this context was a negative response to the question and, in fact, irrelevant. As it was, Pincus changed the line of questioning, going back to Bishayee.

> *Pincus*: "How, if at all, were you informed of Dr. Bishayee leaving the laboratory in or about the end of July or early August 2001?"
> *Leonard*: "Objection to form."
> *Baker*: "How was I informed?"
> *Pincus*: "How did you learn that Dr. Bishayee left Dr. Howell's laboratory at the end of July, beginning of August 2001?"
> *Baker*: "I believe he told me."
> *Pincus*: "He told you?"
> *Baker*: "Yes."
> *Pincus*: "Where did that take place, if you recall?"
> *Baker*: "Someplace in the confines of the Newark campus. It could have been in my office, it could have been in the hall, it could have been in the cafeteria. I don't know."
> *Pincus*: "Was it just the two of you or do you recall anyone else being present?"
> *Baker*: "It was a conversation only the two of us would have heard, wherever it took place."
> *Pincus*: "What did he tell you?"
> *Baker*: "I don't recall that he was leaving."
> *Pincus*: "Did he tell you that Dr. Howell had prepared the letter of resignation for him to sign?"
> *Baker*: "Did who tell me?"
> *Pincus*: "Dr. Bishayee."
> *Baker*: "You're asking did I have a conversation with Dr. Bishayee?"
> *Pincus*: "Yes."
> *Baker*: "Oh, I thought you asked me if I had a conversation with Dr. Howell about Dr. Bishayee."

Pincus: "Alright, then, so your answers to the preceding few questions about a conversation about Dr. Bishayee leaving the laboratory, you were referring to a conversation that you had with Dr. Howell, not Dr. Bishayee is that correct?"

Baker: "Yes, in all respects I thought that's the way you asked me the question."

Pincus: "I'm glad you asked for clarification, so we have an accurate record then. The conversations to which you were referring related to a conversation that you had with Dr. Howell and not with Dr. Bishayee, is that correct?"

Baker: "That's correct."

Pincus: "Did he tell you that, Dr. Howell that is, that he had prepared the letter of resignation for Dr. Bishayee to sign?"

Baker: "I don't recall that."

Pincus: "Did he describe anything else by way of the basis or circumstances surrounding Dr. Bishayee leaving the laboratory?"

Baker: "I know there was another issue with Dr. Bishayee."

Pincus: "That issue was what?"

Baker: "A question in whether it was appropriate looking at images in his computer."

Pincus: "And do you know when in reference to his resignation the issue relating to, looking at things on his computer had occurred?"

Baker: "I don't know, it's probably approximate, but I don't know."

Pincus: "So, was it your understanding that that served as a basis for his leaving the lab?"

Baker: "I don't know the answer to that question."

Pincus showed Baker a staff information adjustment form, which he marked as Exhibit Baker-13. As before, he asked Baker to take a look and tell him about the form and for what purpose it was used. "That's [used] anytime someone comes on or leaves or goes on leave, there needs to be a form commemorating that change in status. That's available as a full time record, as a record, for resolution of that issue in terms of salary, benefits and just a historical record of employment or any change thereof," Baker explained.

Pincus: "Is that your signature or authorized signature at the bottom?"

Baker: "It's [an] authorized signature."

Pincus: "That's Ms. Campbell?"

Baker: "That's the 'KC,' yes."

Pincus: "So, other than the conversation that you identified with Dr. Howell a few moments ago, do you recall any other discussions that you may have had with him regarding Dr. Bishayee's termination?"

Baker: "I don't recall any other."

Leonard: "Objection to form, Shelly. Just, you're saying termination, this just designates termination as leaving, but it says resignation on the document."

Pincus: "Termination and/or resignation, okay?"

Leonard: "Well, it's not and/or, it says type."

Baker: "No, the type of termination of his employment is resignation, not a termination."

Pincus: "You understanding was he resigned?"

Baker: "Yes."

Pincus: "Were you aware that after Dr. Bishayee resigned, he collected unemployment compensation from the New Jersey Department of Labor?"

Baker: "I would not be aware of that at all."

Pincus: "Is it you understanding that an individual employed within your department who voluntarily resigns from their position of employment is entitled to receive unemployment compensation?"

Leonard: "Objection to form, you can answer it."

Baker: "That's an HR issue. I'm not aware of the specifics of human resources policy."

Pincus: "So, that being said, do you have any knowledge of the fact that Dr. Bishayee on resigning from UMDNJ, and specifically Dr. Howell's laboratory, collected unemployment compensation for a period of time thereafter?"

Leonard: "Objection, asked and answered."

Baker: "I would have no way of knowing."

Pincus: "To your knowledge, would you have had to sign any documentation relating to Dr. Bishayee's ability or inability to collect unemployment compensation?"

Baker: "I would never have had that opportunity as far as I can recollect for people who have resigned voluntarily or been asked to resign. So, I have no knowledge of that."

Pincus: "What about any of your employees, and Ms. Campbell for one, would she to your knowledge have any responsibilities in regards to any documentation?"

Baker: "That's an HR responsibility. That's a university HR department would have that responsibility, not the individual department."

Pincus directed Baker's attention Putterman's correspondence, identified as Exhibit Baker-14. He again asked Baker to review it and, also as before, had he seen it prior to being deposed. Baker told Pincus he'd never seen it.

Pincus: "Do you know who Dr. Putterman is?"

Baker: "Yes, I certainly do."

Pincus: "What position did she hold at the university in or about April 2002, if you know?"

Baker: "She was, I'm reading the wrong title, but I'll give you the right effect. She was the deputy to the vice president for academic affairs. So, she was the second [in] command in that university function."

Pincus: "So, did she have any discussions with you regarding any of the items that are identified within this document?"

Baker: "She had one discussion with me about the issue that this would be an investigation into Dr. Bishayee sometime in 2001 in the first half of the year."

Pincus: "What do you mean this would be an investigation?

Baker: "There would be an investigation because Dr. Hill had raised issues."

Pincus: "And beyond that?"

Baker: "Beyond that there were no discussions."

Pincus: "Was there an exchange of any documents between you?"

Baker: "There was no exchange of any documents."

Pincus: "That would include e-mails, things of that nature."

Baker: "There were no e-mails. She called me, I called her. I think I got a cell phone call. I was at Union Station, and I had a discussion with her on the phone. That was the only discussion or only exchange we had."

Pincus: "Did you have any involvement insofar as the investigation?"

Baker: "I had none."

Lastly, Pincus showed Baker a memorandum marked Exhibit Baker-15. He asked if he'd review it and whether or not it looked familiar to him. "It's not authored. Can you tell me who wrote this?" Baker opined. "I believe this was authored by Dr. Howell," Pincus replied. Baker hadn't seen the document prior to his deposition. "I didn't mean to interrupt your review of it, but other the other question is if you have not seen the document, did you have any discussions with Dr. Howell regarding the substantive aspects of this memo?" Pincus continued. "I did not," said Baker. The deposition ended shortly thereafter.

Chapter 9 –
Box 6

Defense counsel John Leonard deposed whistleblower Dr. Helene Hill on January 23, 2009. Hill's deposition, like Marek Lenarczyk's, was lengthy. She would later opine that the defendants' attorney "did a masterful job of twisting my words and forcing me to say things that I really did not want to say." Like Stephen Baker, Hill had prior experience with giving a deposition and like all of those previously deposed, she was under oath. Hill's prior deposition had been taken four to five months prior and involved the female faculty at the UMDNJ suing for equal pay; it was a class action case in the New Jersey civil courts, though she wasn't positive of that fact when Leonard asked her about it. Hill was a named member of the class. Other than that deposition, she'd never been deposed.

Hill's deposition in the *qui tam* case was in many respects an exercise in futility for her and for the defendants' counsel. She was well versed in her own complaint and the discovery up to the point of her own deposition. Just as her attorney, Sheldon Pincus, had done in prior depositions, Leonard showed Hill exhibits but unlike Pincus, he drew them from prior depositions to avoid having yet another copy of the same documents used from one session to the next. This paper-saving measure would also have been welcomed by the court, cutting down the redundancy of exhibits, had the case gone that far.

While Leonard went back to the beginning of Hill's career, relevant information did not apply until she got to the UMDNJ on September 1, 1981, and even then, pertinent didn't come into play until Leonard began asking questions that applied to the *qui tam* case. Leonard asked Hill if she had obtained any grants of her own since 1999. She told him no. "In 1999 I was 70," she explained. "I decided that I didn't want to apply for grants anymore. I had had it. You write about ten grants for every one that you get."

Leonard asked Hill if she was tenured at UMDNJ. "Yes, I am," she replied. "When did you become tenured?" he asked. "The way the rules went when my husband [George, a medical doctor] and I came here, you had two and a half years to get tenure. And so we came in 1981, and I got tenure I think in 1983." She told Leonard to refer to her curriculum vitae, a document that he had, and letters of appointment, all of which spelled out the information he had asked. When he asked if she was granted tenure the first time she was up for the appointment, Hill told him yes. "You have the letters, so you know," she told Leonard, "That there was a little bit of an argument there, but things went through the first time." He wanted to know what she meant about the argument. "The members of the department were supposed to vote – the tenured members – and there were only three of them. And one of them abstained and two voted against me. So," she continued, "the current chairman was not going to support me, but then he decided that he would. He supported me and I went to the committee known as FCAP." She informed Leonard that she was still a member of FACP (Faculty Committee on Appointments and Promotions) at the time of the deposition. "I went through, actually I'm told, with flying colors. I have very good credentials."

Leonard was curious why two colleagues were unwilling to support her tenure. "[There] was a little bit of infighting," Hill told him. "My husband was the chief of surgical oncology and one of those members was the former chairman of the department of radiology [who] believed that my husband had been responsible for him no longer being the chair." She identified that individual as Dr. John T. Mallams, M. D., a radiation oncologist. The other was Dandamudi V. Rao, Ph.D., who published a substantive number of papers with Roger W. Howell, Ph.D., and was notably a coauthor of the 2000 Bishayee et al. (to include Howell) paper at the heart of Hill's *qui tam*. "When I came to the New Jersey Medical School, Dr. Mallams told me that I was to be in charge of the research,"

Hill told Leonard. "And Dr. Rao had, unbeknownst to me, just gotten an NIH grant and RO1 grant. So Mallams put me in charge of Rao and Rao didn't like that. So I think that he would have preferred that I not get tenure." Hill's chairman just then was convinced to support her tenure after a dean – Lawrence A. Feldman, Ph.D. – spoke to him.

> *Leonard*: "What did you do when you first found out that two of the three members had voted against you and the third abstained?"
> *Hill*: "I was pretty annoyed."
> *Leonard*: "What did you do about it?"
> *Hill*: "Nothing."
> *Leonard*: "Did you talk to anybody?"
> *Hill*: "Oh, I'm sure I did."
> *Leonard*: "Anybody in administration?"
> *Hill*: "I don't remember. I think I probably went and talked to the dean. Yeah, I think I did."
> *Leonard*: "Do you remember who the dean was at that time?"
> *Hill*: "He's dead now – Vincent Lanzoni."[155] [Vincent Lanzoni, M.D., was New Jersey Medical School dean from 1975 to 1987. He died on June 28, 2007.]

The time between the time Hill found out that her tenure was in jeopardy to the department chairman changing his mind was two days. Leonard wanted to know what Hill's working relationship was like with Mallams and Rao after she got tenure. "Well, they had their project. Mallams was a clinician, and he was mad at everybody and left very shortly after that," Hill explained. "And Rao had his own grant, and he was working on that. His interest was in nuclear medicine and my interest was in cancer biology. So, our interests didn't intersect." Up to that time, Hill told Leonard, she never felt the university had discriminated against her in any manner. But in hindsight, Leonard wanted to know if her opinion of that had changed. "No, other than, you know, the salary business. I recognize looking back at the records and so forth that I probably was not being offered a competing salary. The theory is, and I'm sure you know it, is that women don't argue, they don't bargain."

Leonard: "You never felt that you were being paid less than your male counterparts [at UMDNJ]?

Hill: "I didn't know."

Leonard: "Even given your experience [with the same unequal pay] at Colorado [University of Colorado] and Marshall [University] and the work you did there to do comparative studies [of pay inequality between male and female faculty], did you not have inquiries as to UMDNJ?"

Hill: "A the time that I came to UMDNJ, I just wanted to do good work and I had enough of this political stuff. Put my nose to the grindstone and do some good science."

Leonard: "So it wasn't that you weren't aware of it, you just – ."

Hill: "I wasn't aware of it but I didn't want to get into it either."

Leonard: "But certainly you came to UMDNJ with an understanding that from your experience, women in your field were being discriminated against in terms of salary with respect to their male counterparts?"

Hill: "That's correct."

Pincus: "Objection, asked and answered."

Hill: "That's correct. In the global sense we're talking about."

Leonard: "In your personal experience."

Hill: "Oh, okay."

Leonard: "At two universities prior to UMDNJ you not only were aware of it but helped investigate it."

Hill: "Yeah, right."

Leonard: "So in 2000 is the last time you applied for a grant?"

Hill: "Yes. Maybe not even that. I'm not sure when I hung up my grant writing pen."

Leonard showed Hill the complaint – her complaint. The complaint spoke for itself, but Leonard proceeded to ask Hill questions about her claim that Bishayee and Howell had, in simplest terms, defrauded the United States government in conduct tantamount to scientific misconduct. The complaint was filed under seal, meaning just then that she wouldn't be allowed to talk about it and the named defendants wouldn't know about it. But that's not what happened. "I believe that they were told something about it but not the details," Hill recollected. "Do you believe they were told you filed a complaint?" Leonard asked. "I think so, yes." Leonard then wanted to know if she understood that information contained in the complaint would not be made available to the public. "I don't understand your question," she replied. "When you filed this under seal, you had the understanding that nobody was going to be able to read the contents of the complaint?" Leonard explained. "Right," she answered. "Until when? What were you waiting for, do you know?" Leonard asked her. "The attorney general had to release the seal." But in fact, the United States attorney general's office had to conduct an investigation and make a determination whether it wanted to purpose the case. Hill understood this and more. The government investigated the case for years. FBI agents interviewed all persons involved and reviewed voluminous amounts of information. At the end of the review, the government elected to not participate in the case, a move that is not at all unusual in *qui tam* suits. In most of these cases, the government largely elects the relator to take the case. If Hill's case had no merit, the government would have informed all parties that the suit had no merit – it didn't elect to do that and informed Hill she could proceed as relator. Leonard tried to press her on this point, but she replied that the letter received from the United States attorney further stated that the federal government could come back into the case at any time it chose to do so. Another data point that could and should have been raised was the fact that in the year 2009, the same year that discovery heated up in Hill's case, the United States Department of Justice recovered some $2.4 billion in settlements and judgments from False Claims Act cases, about $2 billion of which was the result of *qui tam* actions.[156]

Leonard wanted to know Hill's thoughts as to why the government didn't jump on her case. "At just about the time that this case [her case] was getting their interest and they finally got the documents that had been subpoenaed and so forth, the $80 million cardiology case came in from UMDNJ and I believe that they felt that that was more bucks and that they needed to devote their energies and their personnel to that case [harkening back to the aforementioned note that the UMDNJ was a troubled institution before being absorbed by Rutgers University] rather than to my case. I think my case was triaged." In other words, the United

States Attorney's Office decided that the cardiology scandal was more important. When Leonard asked her why she thought the other case was more important than her own, she replied "because $80 million is a whole lot more than $1.4 million, and many people [were] involved." When Leonard tried to suggest that she couldn't know if that case wasn't concluded years before, Hill told him that she didn't believe that was true. "The case is still pending." She was correct.

In September 2009 the UMDNJ agreed to pay $8.3 million to settle charges that it paid illegal kickbacks to cardiologists. Justice Department documentation indicates, in brief, that in 1995 the UMDNJ was having trouble finding enough patients to perform the minimum number of cardiac procedures required to maintain its government funding and accreditation as a level-one trauma center. Everything turned bad when the UMDNJ created part-time employment contracts with community cardiologists who, in turn, referred patients to the facility. The Department of Justice thus alleged that the part-time contracts were, in truth, masked kickbacks to cardiologists for patient referrals. The six cardiologists involved in the scheme settled with the government as part of the deal struck with prosecutors. This was not, however, the first time the Justice Department had settled with the UMDNJ. In the years just prior, a federal monitor found that the school had double-billed Medicaid for $5 million in procedures and was forced to settle that claim with a $2 million payout. In 2005 UMDNJ paid $4.9 million to the federal government and the state of New Jersey as part of a deferred prosecution agreement.[157]

Leonard wanted to know what supplemental – or new – information Hill sent to the offices of United States attorney Susan Steele and FBI agent Mary Beth Gardocki, assigned to the case, and that Hill subsequently presented as a PowerPoint in Steele's office. He wanted to know if the information contained in the PowerPoint was something that the United States attorney and the FBI didn't have previously. She told him it was new data. "That summer I spent a great deal of time in the FBI offices and the office of the United States attorney going through the documents that had been subpoenaed and I discovered," said Hill, "at that time that the results of experiments that had been done, the repeat experiments that had been done by Dr. Howell, which were entirely at odds with reports done by Dr. Bishayee; that Bishayee's results were scientifically impossible, and that there was a very good scientific explanation for the results that Dr. Howell and Dr. Lenarczyk had gotten."

Leonard asked Hill how much time she spent reviewing the documents in Steele's and Gardocki's offices. "A lot of time. I went down probably – I could keep a record. I don't know whether I kept a written record of the number of times I went to the FBI office. Probably five sessions in the FBI office. But then more like 10 or 12 sessions in the U.S. Attorney's Office, and I would spend maybe two or three hours each time." She further explained that her time in the federal offices was spent in closely monitored space. "Mary Beth Gardocki had to watch me very carefully. So, she was always with me." She was always provided an escort in the FBI spaces. "When I was in the U.S. Attorney's office they gave me a conference room." She was permitted to move unescorted in Steele's office spaces. "What exactly were you looking for? Did you know what you were looking for when you went there?" asked Leonard. "I knew that Howell had not been able to repeat the experiments," Hill explained. "I also had a Zip drive from Lenarczyk, and I knew I had seen his experiments. There are two types of an experiment – there's the 50 percent experiment and there's the 100 percent experiment. I had only focused on the 50 percent experiment because that involved this so-called 'bystander effect.' Well," she continued, "why was I doing it? They asked me to do it, that's why I was doing it." The "who" was the United States Attorney's Office and the FBI, she told Leonard. "There were 11 books that they had subpoenaed. The first time they called me in they asked me just to prioritize the documents. They hardly told me," she iterated, "what was in any of them, but I prioritized them as best I could. And then there was one box, Box 6, which really seemed to contain the copies of most of the notebooks that would have been important, and they asked me to go through them." She did. Leonard asked again what she was looking for when she was there. "That's what I'm saying," said Hill, "is that I didn't really know what I was looking for. They asked me to go through the documents, and at their request, I was going through the documents. Basically, what I thought that I was looking for was the actual data that had been generated by Bishayee."

Hill got more specific. "I already knew from the ORI that the Coulter counter, this particle counter that was used to count the cells, were – how shall I say – out of whack. Bishayee Coulter counts were not consistent with random distribution, which they should have been. So," Hill continued, "I thought I was looking for more evidence of Coulter counts being not consistent with random. So, I was focusing on the Coulters, but I also looked at the data as well." She was coming to the end of looking at Box 6 when she came upon repeat experiments performed by Howell. "And I had not looked at the 100 percent experiments that Lenarczyk had sent me because I had not suspected that there was anything wrong with Bishayee's 100 percent experiments. But what I saw," said Hill, "I came upon the stuff from Howell's notebook, and there were two experiments there that were 100 percent experiments. [Then] I went through the first one and I saw," she reiterated, "that the data went down like that and then plateaued." She drew an illustration to illustrate her point for Leonard.

What Hill had seen in this box – Howell's 100 percent experiments – stunned her. There was a many-fold difference in the survival between Bishayee's results and Howell's results in the 100 percent experiment. "I lay awake at night because I thought this [what she'd seen of his experiments in the box and documented in Howell's notebook] was a 50 percent experiment and I had to go back through and look at the beginning; it wasn't a 50 percent experiment; it was a 100 percent experiment. And I had believed Bishayee's 100 percent survival, the exponential decline," she explained, "and now I'm looking at Howell's 100 percent results and it's plateauing at 50 percent, and I couldn't believe it. I lay awake at night [again] thinking 'what's going on,' what's going on.' And I'm a biochemist, I'm a radiation biologist, and I know that thymidine blocks the cell cycle. The explanation for Howell's results is that the tritiated thymidine was blocking the cell cycle. Cells are only going to be killed if they go into the phase of DNA synthesis." Then Hill realized that the explanation for Howell's results was that half the cells in the population weren't going into DNA synthesis; they were being blocked, and they had to have been blocked by tritiated thymidine. "So, I knew then," she continued, "that that was what the explanation was, and that Bishayee's results were impossible. And that's what I told the U.S. Attorney."

Leonard challenged Hill's conclusion about Howell's 100 percent experiment. He suggested that she was just speculating since she'd not performed the experiments she critiqued, nor had she spoken to those who'd done them. "You know, I'm a scientist. And one of the things that we have to do is interpret data. And nothing in biochemistry or radiation biology is 100 percent, but it can be very close to 100 percent," Hill told Leonard. "And I can tell you as we're sitting here that this is very, very close to 100 percent that that is what is going on; that the thymidine, the tritiated thymidine was blocking the cell cycle, cells are not entering the "S" phase, they're not being killed, and that explains Howell's results." Further, she continued, "If Bishayee is doing his experiments in exactly the same way that Howell is, which he is, because the protocols are exactly the same, that Bishayee could not be getting the results that he was getting."

Leonard wanted to know what, before looking at the documents in the possession of the federal authorities, prompted her concerns. Much of this information was covered in the other depositions but he was prompted to ask after Hill's discussion of the findings, most especially those in Box 6. "Everything began in 2000 – well, it really began in 1999 – but it kind of really began in 2001 when Dr. Lenarczyk came to my office and tried to get me to go into the lab to see what Dr. Bishayee had been doing in the course of the experiment he was doing. And we [she and Lenarczyk] shadowed him," Hill replied. "We believed that he was fabricating the results of his experiments." In truth, she told him, it went back to 1999, "which was two years before this, looking back – well, the 2001 experiment brought me to tell Dr. Howell about that, and you have the documents that were involved in that, and to go to the [Newark CCRI] to report those results. The 1999 experiment was another experiment in which I had observed that Bishayee was making up results in another experiment. So," she continued, "two observations in which Bishayee was coming up with results that were incompatible with my observations."

Hill reiterated that the Newark CCRI had largely looked into Bishayee's second experiment. "But then I told them about the first experiment, and they looked into that somewhat," Hill told Leonard. "They came up with the conclusion that there was not enough data there to rule on research misconduct." Importantly, she would testify that at the time in the spring of 2001 that that first Newark CCRI reviewed the case, the university policy stated that the CCRI was supposed to determine whether there was some indication of research misconduct and then a new committee should have been formed, had the CCRI determined misconduct, that would have included outside members and experts in the applicable field spelled out in the complaint. The Newark CCRI had, just then, no experts in the field, consisting instead of deans and administrators whose job it was to protect the university, a point Hill stressed in her deposition, and which was certainly true. "So, in my mind," she explained, "the committee was not properly constituted and then, rather than referring the case to a second committee, which would have had some experts, they decided that there wasn't enough evidence for research misconduct. So, they kicked it out, basically." The wagons had been circled within the UMDNJ, but Hill was just beginning to form that picture as she fought back to get a proper vetting of the case in a court of law.

After being rebuffed the first time by the Newark CCRI, she went to the United States Public Health Service (PHS) Office of Research Integrity (ORI). "I sent everything that I had given the research integrity committee at the university – I sent them everything – and then they took another year to go through all of that," Hill told Leonard. "But in the meanwhile, I talked with Dr. Fields, who was charge of the case and she told me about the Coulter counts. Actually, she kind of told me at the very last minute before the committee was going to meet and then I went and got some more Coulter counts and sent them to her. These are all," she continued, "in the disclosure or in the documents that accompanied the disclosure." What she discovered was that there was even more evidence of nonrandomness in Bishayee's numbers.

Going back to what Leonard asked about additional information Hill pointed out to federal investigators that were not part of the original complaint, she offered this: "So I thought when I was in the United States Attorney's Office that what I was looking for was more evidence of Coulter counts that were amiss and I found this other thing. But I also found, because what I was looking for, and already had the results of Lenarczyk's 50 percent experiments, that the 50 percent experiments look almost like that (she indicated this on paper)," she said. "Again, Bishayee has got this exponential decline in survival and Howell and Lenarczyk – you could put Lenarczyk here as well – Howell and Lenarczyk, their results plateau. Actually, in the 50 percent experiments, they plateau at about .7 or 70 percent." To confirm her testimony, Leonard asked again if nothing regarding what she'd found in Box 6 formed the basis of anything that was in the original complaint. "That's correct. I didn't see Box 6 until after the complaint."

When Leonard asked if Hill could sit in the deposition, under oath, and state to a certainty that Bishayee fabricated data, Hill's response was spot on: "No, I can't. We're in the 99 percent level here." Leonard didn't understand the reference. "We're at the 99 percent level here." He still didn't get it. "That I can't say for sure, but I am very convinced [99 percent certainty] that he did." When Leonard asked if she had firsthand knowledge that Bishayee had fabricated data, Hill replied: "I have the firsthand knowledge of that first experiment in 1999 that I observed with a set of dishes that had no colonies on them; I have that in my head. I know that those dishes were empty, and I have the observations that Lenarczyk and I made of the experiment in 2001 where tubes were left in the cold incubator that should have been processed and were not."

Leonard drew Hill's attention to page six of the original complaint, filed in 2003. The passage read: "It is, therefore, critical that patients not be misled about the results of the research." Leonard wanted to know, now nearly six years beyond the date the complaint was filed, what patients had been misled. Leonard's question was designed to make the ubiquitous "patients" vague and unaffected recipients of the bad science that Hill was trying to expose. "Dr. Howell's experiments and research involves setting standards for nuclear medicine," Hill explained. Nuclear medicine uses radioactive tracers to both treat cancer and to detect cancer. Dr. Howell's premise is that there's a nonrandom distribution of radiation so that the isotopes go to different parts of the body and some parts are irradiated heavy and some parts of the body are not irradiated heavily. This is very important research and it's very important to understand the kinds of doses patients are receiving."

Hill continued, explaining to Leonard that in radiology she and her colleagues deal with very large numbers. "Like, for example, a chest x-ray. Does a chest x-ray cause cancer? Yes. But if you went for a chest x-ray today, would that cause cancer? I couldn't say that it caused cancer. Even if you got lung cancer, I couldn't say that it caused your lung cancer," but, she explained, "I can say that I can make a calculation over the population, say 200 million people in the United States, and maybe how many of them are getting chest x-rays every year, maybe a million. I can tell you – and I've done this calculation – I'm telling this off the top of my head, so I don't know whether this is actually true or not, but I could tell you that of all those chest x-rays, those million chest x-rays, let's say, that are being given in the United States in 2008, that perhaps one or two people will get lung cancer resulting from the radiation exposure that they have gotten from those chest x-rays." Further, she explained, "I can't tell you who, but I can tell you that statistically a chest x-ray, if you integrate over a large population of people, will cause cancer. Leonard dismissed this answer, going back to the ubiquitous "patients" to ask if she knew any particular patients affected by Howell's and Bishayee's miscalculations.

"No," said Hill. "But it's very important for people who are working in nuclear medicine to understand the mechanism of the isotopes that they're using and also to understand what safe doses are and Howell is also involved in determining what the safe doses are."

The wrinkle in Hill's deposition came first when Leonard interrupted her response regarding Howell and dosing. Sheldon Pincus, who'd been quiet, asked that Leonard let her finish. "Okay. I will say this, though," Leonard retorted. "Dr. Hill, you seem like a very nice person and I'm indulging your responses, sometimes for five or ten minutes. Typically, I would ask a question and you would answer it and not go on a diatribe. So, I'm trying to sort of reign that in a little bit. Had you finished your answer?" She said no, she hadn't. "What else would you like to say?" Leonard asked. Certainly, what she had done up to that point did not meet the definition – anyone's definition – of a diatribe. Hill was actually providing informative, intelligent responses to the questions defendants' counsel had asked. The adage that most applied in this deposition, evident from its beginning: "If you don't want to hear the truth, don't ask the question."

Hill finished the answer she'd begun before Leonard interrupted her. "That the NIH gave Dr. Howell this grant, they renewed his grant because they considered that the research that he was doing was critically important. The NIH," she continued, "doesn't give grants to just anybody. The competition is very stiff and there has to be a good reason. And this grant is coming from the National Cancer Institute, so the National Cancer Institute feels that this research is important enough to support it." Leonard's response to her, in part: "Frankly, what you just said isn't responsive or doesn't relate to that [original question about patients being affected] at all." He moves on, clearly agitated by her answer. He draws her attention to the complaint, to page 7, where Hill documented a conversation with Howell. Leonard wanted to know if she remembered what she told him about Bishayee.

"Well, the first experiment, in the 1999 experiment, I informed Howell about it, and he brushed it off. He used the results of the Bishayee experiment that I believed had been fabricated in his grant application." Leonard asked about the passage that followed: "As a result of their actions, Hill and Lenarczyk concluded that Bishayee had, in fact, fabricated the experiment's data and engaged in scientific fraud." He noted that she had attended Lenarczyk's deposition and heard him deny all of this under oath. She couldn't answer the question. She'd also never shown Lenarczyk the complaint language that applied to him prior to filing it in federal court. Lenarczyk's deposition, however, was rife with contradictory statements that were troubling without Leonard's suggestion that she should have made a full disclosure to her former colleague prior to using such strong language to suggest he supported her findings.

Just below the statement enjoining Lenarczyk to Hill's finding of fabrication, was another that Leonard called into question, also involving Lenarczyk, another postdoctoral fellow with much to lose not only during his employment with the UMDNJ but any future fellowships and research associate jobs if he said the wrong thing in his deposition, which would eventually become public record: "After the Committee's report, Howell proceeded to terminate the employment of both Bishayee and Lenarczyk." Leonard wanted to know the basis of the statement. "I was told by Bishayee that Howell had written a memo [resignation letter], which he told Bishayee to sign, in which was written 'I resign my position as,' whatever his position was, and he signed it." She'd also sat in on Bishayee's deposition. Leonard noted that she'd heard Bishayee testify that Howell hadn't terminated him. "He testified that Dr. Howell wrote that memo." He pressed on the word "terminate."

Bishayee didn't admit to being fired but his collection of unemployment – if true – would have required it. If he and Howell persisted in claiming Bishayee "resigned" it is thus an open admission that Bishayee presumably and illegally collected unemployment compensation from the state of New Jersey, which is clear on the point that a resignation is not grounds to qualify for unemployment. Bishayee's deposition makes it clear he "resigned" and was not "fired" but the contradiction is just as clear in several responses in which he insists he qualified for unemployment compensation suggesting that it would be improbable for both circumstances to be true, most especially for a foreign national with an H1-B visa working under finite grant funding, otherwise known as soft money.

As a point of fact, the Bishayee and Lenarczyk depositions point the finger at a truth the defendants in Hill's case would have most assuredly drawn unwelcome discussion not in their favor: the reality that postdoctoral fellowships in the United State are flooded by foreign Ph.D.s and data about them is sorely lacking. More often than not, their stay is predicated on short-term performance evaluations and the availability of grant funding to keep them here. These often newly minted Ph.D.s are paid low salaries funded via soft money and know that the funding line determines whether they stay or go on an H1-B visa. The foreign postdocs coming to the United States to polish their skills are vulnerable to being fired for any number of infractions and poor performance and, importantly, scientific misconduct where they are either guilty as charged or the easiest, most vulnerable candidate in the lab to be blamed and sent home or passed along to the next research institution. They remain on a constant hunt for the next open fellowship or research associate position, a point made abundantly clear in Lenarczyk's deposition and the number of postdocs and research associate positions he has held since first coming to Howell's lab in 2000 with returns to his native Poland sprinkled in-between.

As [the NIH] started delving more deeply into the data on foreign postdoctoral fellows, it became clear that that lack of reliable information about the postdoctoral fellow population in the United States could be attributed to several factors. "First and foremost, we do not collect much information about foreign-trained Ph.D.s who come to the United States to do a postdoc, and we have no idea how long they stay or how many leave after their training," wrote Sally J. Rockey, Ph.D., of the NIH in June of 2012. "These foreign-trained postdocs comprise about two-thirds of the total postdoc population. In addition, postdocs have many titles, and some institutions require they change their titles after a certain number of years."[158]

Leonard's attempt to disable the messenger, in this case Hill, stood to backfire in the long game. These foreign postdoctoral fellows, Bishayee and Lenarczyk included, come to the United States to work for several reasons, among them the fact that there is an oversupply of Ph.D. scientists in their home country to include reduced opportunities for them to attain permanent positions in university teaching and research.[159] Whether in their own country or the United States, foreign postdoctoral fellows often spend a substantial period of time in low-paid posts prior to entry into their field, where they are *treated as temporary workers* with *reduced and nonexistent benefits.*[160] The leverage held by American colleges and universities, their various departments, divisions and faculty who oversee their work, over foreign postdoctoral fellows creates an undeniable conflict of interest when they are called to testify, under oath, about controversial matters whether the circumstance is a deposition or the matter lands in a court of law. Even if the postdoctoral fellow testifies to the truth – full disclosure of all circumstances he either participated in or witnessed – at one university, the next university to which he applies might just pass on employing that individual concerned that should a similar matter arise on their own campus, the postdoctoral fellow who blew the whistle the first time might just do it again. Being a whistleblower has a downside, and for a postdoctoral fellow implicated by one, the result can lead to any number of outcomes from being fired and sent home, poor employment opportunities going forward to being ostracized. With little to no benefit to them, foreign postdoctoral fellows are particularly sensitive – and wary – to being party to controversies in the workplace. Hill, whether she had shown the complaint to Lenarczyk or not, would have almost assuredly gotten the same result during his deposition. Bishayee was simply unwilling at every level of testimony, to provide full disclosure, most especially after it went to federal authorities for investigation.

Hill never stated that Bishayee was fired. She said he was terminated. Leonard questioned the semantics. "He was terminated. You can look at the personnel action form," Hill replied. "He was terminated because according to the personnel action form he resigned. That's a termination. The form is called a termination." Leonard wanted to know what impression Hill was trying to convey by stating, in the complaint, that after the Newark CCRI issued its report, Howell had moved to terminate the employment of both Bishayee and Lenarczyk. "That Howell encouraged the termination of both of these people," Hill answered. Leonard tried another approach, suggesting to Hill that Howell couldn't resign for Bishayee. "Dr. Howell wrote the memo." The bottom line is this: if you write the resignation letter for your employee – which Bishayee stated Howell did – and you hand it to the individual, you have not only crafted the narrative but are, in a passive-aggressive manner, compelling that individual to submit to the terms written into the document along with the date of termination. What muddied the story for all concerned was that Howell and Bishayee continued to call Bishayee's termination a "resignation" and that was a obfuscation of what took place.

Accordingly, it is well documented in human resources that if your employer doesn't want you around any longer, that employer can force you to resign or terminate you if you don't agree to resign. In most cases, that employer probably wouldn't be violating any labor or employment laws[161] but as the employee presented with the letter of resignation, a forced resignation deprives the recipient of unemployment benefits because the termination is recorded as voluntary with state employment offices. What disadvantaged Bishayee and Lenarczyk in this situation and most especially Bishayee as the unfortunate receiver of the resignation letter Howell crafted, was his foreign status and being unfamiliar with and nor entitled to, the protections, benefits and services of state employment law. Bishayee didn't know that he couldn't be forced to sign a letter of resignation if he absolutely refused to put his signature on a letter he didn't write nor wanted to write. Recall that he had just been promoted to research associate and hadn't yet seen the pay raise windfall when Howell handed him the letter to sign.

Bishayee, however, was further disadvantaged as the point of blame for his supervisor's inability to replicate data in a critically important NIH grant and arguably as the object of fault, he'd put Howell and Howell's reputation at jeopardy if Hill proved Howell had gone forward with data Bishayee fabricated and hadn't moved to retract two published papers and set the record right with the National Cancer Institute, which was support-ing Howell's grant. Thus, what Hill suggested – that Howell terminated Bishayee and Lenarczyk – wasn't a false statement in her complaint as Leonard tried to paint it. Hill was up against defendants who weren't well acquainted with the terms of full disclosure, including a medical school that could ill afford another costly court battle and settlement, especially since it was already engaged in several other whistleblower cases at the time Hill's came up to the plate.

Leonard wanted to know what kind of retaliatory action Howell had committed against Hill. "We had a laboratory we all shared. After the [Newark CCRI] had finished its work, my chairman [Baker] chose to make Howell the chief of the division and Howell, in his wisdom, decided to change the locks so that I was no longer able to get into the shared laboratory," Hill explained. "Howell told me to my face even though he had just been appointed the division chief, that he wanted nothing more to do with me. Howell has not spoken to me, except in public, since that time. He's my division chief. He has not spoken to me [again], except in public, since that time. He has shunned me for how many years? 2001 to 2009 – going on eight years." Leonard wanted to know what she meant by "shunned." "Divisions are supposed to have meetings," Hill replied. "I have never been invited to a division meeting in the eight years since this happened. This is supposed to be a collegial organization. There is absolutely nothing collegial about the way I have been treated since then." Leonard wanted to know more about the "when." "Since 2001, when he [Howell] called me into his office and told me that he wanted to have nothing further to do with me." Further, Howell reassigned the laboratory space so that the lab that had been Hill's was shared with Edouard Azzam. "And then, Dr. Azzam, in his wisdom, started moving my stuff out and putting it elsewhere and not telling me where it was." Leonard asked if there was anything else. "Dr. Azzam has not over the years treated me with – after Howell said he wanted nothing more to do with me, I told Azzam that if he wanted me to, I would do experiments for him and I did. Then in April of 2002 he came into my office and he insulted me and he told me that I was a bad person, because if Dr. Howell's postdocs found out that Dr. Howell had done a bad thing, or if Dr. Howell were to lose his grant – they were afraid that Dr. Howell would lose his grant because of the misconduct – that if Dr. Howell lost his grant, that the postdocs would lose their jobs, and that would be my responsibility, because I would have forced the postdocs to lose their jobs. If Dr. Howell's children found that Dr. Howell had done a bad thing," Hill continued to relate

Azzam's confrontation, "that would be a terrible thing for his children. And so I was a bad person. And after that, then I couldn't work with Dr. Azzam anymore."

Leonard tried to parse the difference between two people having a conflict and someone being retaliated against. "Retaliation is something that makes a person feel very uncomfortable and I feel very uncomfortable with these guys," Hill told him. "Well, it's more than that. Retaliation is action taken in response to something," Leonard said and whether he knew just then or not, he hadn't covered all the bases by opening the door with "in response to something," which in Hill's case it was retaliation for her pursuit of inquiries with the Newark CCRI that questioned Howell's competency. Legally, retaliation includes any adverse action taken against an employee for filing a complaint or supporting another employee's complaint under a variety of laws. The most common type of retaliation claim involves an employee who alleges that she was first harassed or discriminated against and later punished for making a complaint to her employer or a relevant federal agency. Employers also are prohibited from punishing employees for exercising other rights assured to them. For example, employers aren't supposed to be able to retaliate against an employee who exercises those rights. Recently, the United States Supreme Court has also ruled that the protections against discrimination and harassment guaranteed in Section 1981 of the Civil Rights Act of 1866 and the Age Discrimination in Employment Act (ADEA) extend to retaliation. It's also illegal to retaliate against an employee for opposing an illegal employment practice, or for testifying, assisting, or participating in any manner in an investigation, proceeding, or hearing under some statutes.[162] Hill's circumstance most certainly qualified as retaliation.

The United States Supreme Court's objective test for retaliation complaints was set forth in *Burlington Northern and Santa Fe (BNSF) Railway Co. v. White,* 548 U.S. 53 (2006), which stated in the simplest terms: If an employee engages in some protected activity – for example like blowing a whistle or filing a complaint – and then that employee's job is changed in some way that is not good, to include a *materially adverse* change, and if the employee and other reasonable people take that as a warning or even a disincentive to complain, it's retaliation. Paul G. Mattiuzzi, Ph.D., wrote that in a retaliation complaint, "the law is not concerned about whether or not it was meant to be a threat or meant to chill free speech. The *objective test* has to do with how people experience or respond to the events they observe." He further remarked that how people would respond is knowable, measurable and a matter of common sense and common experience. "That is what makes it *objective*. Intuitively, people know what retaliation looks like. Everyone knows that if you complain, there is a good chance you are going to be punished, and everyone knows about pretend excuses from management."[163] Lawyer and employment law expert Lisa Guerin echoes Mattiuzzi's analysis. "Retaliation occurs when an employer punishes an employee for engaging in legally protected activity. Retaliation can include any negative job action, such as demotion, discipline, firing, salary reduction, or job or shift reassignment. But retaliation can also be more subtle."[164]

In Hill's case there was outright retaliation to include being locked out of the lab for up to two weeks, Howell's and Azzam's demeanor changes, and her exclusion from division meetings. But there were also subtleties. "Sometimes, it's hard to tell whether your employer is retaliating against you," wrote Guerin. "For example, if you complain about your supervisor's harassing conduct, his attitude and demeanor may change. But if the change means he acts more professionally towards you, that isn't retaliation even if he isn't as friendly as he once was. Only changes that have an adverse effect on your employment are retaliatory." Hill had many of the aforementioned happen to her. "On the other hand," continued Guerin, "if something clearly negative happens shortly after you make a complaint – like firing or demotion – you'll have good reason to be suspicious. And remember, not every retaliatory act is obvious or necessarily means your job is threatened. It may come in the form of an unexpected and unfair poor performance review, the boss micromanaging everything you do, or sudden exclusion from staff meetings on a project you've been working on."[165] Hill's case fit every definition by law of retaliation.

Leonard wanted to know if department chairman Baker had retaliated against her. "As soon as the committee had finished their deliberations and sent out the memo that there wasn't enough evidence of scientific misconduct, I was chief of the section of cancer biology and Baker abolished the section and made Howell my division chief and sent a memo to the whole department inviting Howell to be the division chief. I believe that that was a slap in the face for me because he had never sent a memo about appointing any other division chief and division chiefs come and go and never had there been a blanket memo like that about appointing someone as a division chief. I think he did that," she Hill explained, "to humiliate me." Leonard wanted to know why she thought he'd want to humiliate her. "Why would he want to humiliate me? Because I made the department look bad. He even said that," Hill continued. "When I went down to report to him about the results of our observations in March of 2001, I took the documents that I had assembled to Dr. Baker and, and Dr. Baker scolded me for making the department look bad." What he said, Hill told Leonard, was something she'd not soon forget. "'You made the department look bad. It's all over the medical school,' he said." By the negative job action – demoting Hill from section chief and abolishing the cancer biology section immediately following the Newark CCRI report – Baker's action and his admonition of Hill could readily be construed as retaliation and certainly fell into the descriptives used by Mattiuzzi and Guerin.

Asked if Howell humiliated her, at that point it could be considered obvious. Not only had Howell told her he didn't want anything more to do with her but her office was moved about two years after that [by then she was in the middle of a full-blown complaint in the federal court]. I'm in a different hallway – in the F level. But I have to get my mail from the laboratory that Dr. Azzam and I are supposedly sharing. I find it's very humiliating to have to go over there and get my mail," she explained. "I find I feel very humiliated when I am confronted with those people, that they have treated me the way they have treated me, that they have shunned me for all these years. I feel very uncomfortable in that whole environment." Leonard wanted to know if she bore any responsibility for that environment. "When I blew the whistle, I was following the instructions of the university. We have received memos telling us if we observe – ." "That's not my question, Dr. Hill," Leonard interrupted. "She's answering your question. Let her respond," said Pincus. "I was following the guidelines of the university. We were told if we observed misconduct we were supposed to report it. I did report it," she continued. "And we actually had a card that we were supposed to wear along with our ID that basically told us that we had to report these things to our supervisor. If we observed misconduct, we were supposed to report it to our supervisor. I was following the rules of the university when I did what I did."

So, what about the "shunning" Leonard asked. Did Hill feel any responsibility for being spurned by her own division chief and colleagues? "Do I feel any responsibility for my being shunned? Hill responded, incredulous. "I don't think that a person should be shunned because they were doing what the university instructed them to do because they were doing what was right, and that was to report that misconduct had occurred." Back to the conversation with Howell, Leonard asked her when Howell said he wanted nothing more to do with her, was it an accurate characterization of what transpired. "That's what he said," Hill replied. "Are those his exact words?" Leonard rejoined. "Those are his exact words." No one else was present when he said them to her either. Leonard asked why she was in his office at the time he said what he did to her. "Because he called me in." Had he said anything else? "That was about it," she replied. "Oh, he told me that Bishayee was leaving at the end of the month and that Lenarczyk was leaving [also] at the end of the month." Presumably neither was scheduled to leave at the end of the summer. Bishayee had just been recently promoted to research associate and Lenarczyk's soft money funding hadn't run out just yet.

Leonard questioned Hill's October 24, 2002 document prepared for Dr. Karen Putterman, pointing to the passage in which she wrote that Howell claimed that the cells had changed over time since the original experiments were done or that the bystander effect he purported to have observed was dependent on serum that is a component of growth medium. She explained to Putterman that these were possible but unlikely explanations for the non-reproducibility of the data. He asked if she remembered the gist of that passage written to Putterman. She did. So, he wanted to know, under oath, if she stuck by the analysis. "I have further information as a result of documents that we have received," she told Leonard. "And Dr. Howell states that he got the serum that Bishayee had used in his experiments, and that he used that same serum and he was still unable to replicate Bishayee's results. So, the serum is not an answer to his non-reproducibility." When Leonard pressed, asking again if under oath that she could rule out the serum, she replied: "You're asking me a question that as a scientist, I have to answer that I cannot absolutely rule out serum. But as a scientist I can tell you that it is highly, highly, highly unlikely that serum would have any effect on these results." He pressed again regarding Howell's claim that the cells changed over time. "I can come very close to ruling that out," Hill explained, "because Dr. Howell said that they lost their liquid nitrogen freezer, so they had to go back to freeze-downs from 2,000 and that means that they were actually going back to using cells that were closely related to the cells that Bishayee had used." Leonard repeated his question, asking if under oath she could definitely say that Dr. Howell was wrong that the cells changed over time. "As a scientist, I have to say that I can't say definitely. But I can say to you as we sit here today that it is [again] highly, highly, highly unlikely that the inability of Howell to replicate these results is due to the cells having changed," Hill replied. "And furthermore, I can tell you that the biochemistry, the biomedical explanation for the shape of Howell's curve is a universal effect; that the thymidine is blocking the cell cycle because it's blocking certain enzymes, and those enzymes are present in organisms from bacteria to elephants. Biochemistry has a unity," she continued. "When you're breathing, your oxygen is going through something called the Krebs cycle[166]. The Krebs cycle is present in organisms from the very earliest organisms to elephants." Leonard continued to try to compel Hill to admit under oath that she couldn't know if what Howell stated about cell changes was true or not. "It's the same sort of thing," she replied. "I can tell you that it's highly, highly, highly unlikely that the cells have changed." When Leonard continued to push for the answer he wanted and not what she had already explained, Pincus objected as asked and answered. Leonard insisted he hadn't gotten his yes or no answer and as Hill began to respond again, he interrupted her, and it took Pincus telling her to finish her response. "Your legal way of saying things is so unscientific," said Hill. She was right.

Further into Hill's deposition, Leonard asked Hill, after several attempts to get a yes or no answer to the same question, if Azzam had participated in the cover up or not. Each time Hill provided a lengthy response. "You paint things in black and white," she told Leonard. "I don't paint things in black and white. I think that Azzam knew that something was going on and he didn't say anything about it. He told me he knew something was going on because he said to me if Dr. Howell's children find out that Dr. Howell has done something bad, that will be a terrible thing for Dr. Howell. If Dr. Howell's postdocs lose their jobs because Dr. Howell has done something bad, that will be terrible. So, he knew that something was going on. He may not have known exactly what was going on, but he knew that something was going on." Leonard wouldn't let that answer stand. He pressed again for the yes or no. Pincus objected, asked and answered. "I want a yes or no," Leonard retorted. "I think she's saying she can't answer it yes or no. She explained," said Pincus, "the facts on which she relied. You're asking the same question over and over and over again. She's answered it three times the same way. I don't think it's going to change." Tension building Leonard asked sharply, "Yes or no."

Hill: "Do I go to jail if you don't [get an] answer to your question.

Leonard: "You don't go to jail, but you lose credibility insofar as you're writing documents to people making that accusation [that they are involved in a cover up], but now you won't say it under oath."

Pincus: "Objection to the form."

Leonard: "That's the problem."

Pincus: "My objection is noted."

Leonard: "Your objection stands. I get it. With respect to the document you wrote to Dr. Putterman, isn't it true that you accused Dr. Azzam of doing that [being part of the cover up]?"

Hill: "I did accuse Dr. Azzam of doing that."

Leonard: "But as you sit here today, you can't say yes or no whether you believe that to be true?"

Hill: "I will say I believe it to be true."

Leonard: "See how easy that can be?"

Hill: "No, it wasn't easy."

Leonard moved on. He asked Hill if she had problems with Bishayee prior to the experiments that had gone awry. "Like what?" she asked. "Like did you have a conflict with him because you thought he was looking at pornographic material on your computer?" Leonard asked. "I gave him permission to use my computer," Hill explained. "And I came in the next morning, after he had used my computer and I found this e-mail. I didn't know this at the time but when e-mail went out from my computer, it went out under my name. He didn't know it either. So, I don't know where he got these names but anyway, he was sending e-mail out to Cutie, Baby and Honey Doll and goodness knows who all, and I was pretty shocked," she continued. "And then Dr. Howell came in and I didn't even know that you can see what people have been doing on the Internet, and Dr. Howell found that he had gone to some pretty salacious Internet sites, and he was totally outraged. And I was outraged for maybe a day or so, and then I thought about it. I thought well, here's this guy, he's far from home and he's a man and he's looking for some fun or girls or something like that." She told Leonard that Howell wrote up the incident as a memorandum that went into Bishayee's personnel file. "And I wasn't, you know, I wasn't upset about it for very long. I thought he's a lonely guy and I would hope my son wouldn't behave like that, but that's the way it was." Leonard suggested that the first complaint Hill filed against Bishayee was just a couple of weeks after the pornography incident. "It was a lot more than a couple of weeks. It was about two and a half months," she told him.

Leonard: "Two and a half months?"

Hill: "Right. Dr. Howell said it was a couple of weeks, but it was a lot longer than that. And I had totally forgotten about it by then."

Leonard: "So at that time you still were not upset with him for having used your computer to visit those sites?"

Hill: "No. And he didn't get to use my computer anymore. He did it in the lab."

Leonard: "When you first started going to Dr. Howell with what you believed to be problems with Dr. Bishayee, was it your intent to get Dr. Bishayee terminated or fired?"

Hill: "No. You don't fire a postdoc. I'm sure I never used the word 'fire.' What you do is you help them find the next job. You show them the door." [That's exactly what Howell did.]

Leonard showed Hill a document she wrote before going to the Newark CCRI. He drew her attention to the second to the last paragraph that started, "On Monday, March 26."

Leonard: "About four sentences down it says 'it was my hope that by careful documentation I would be able to present a convincing case to Dr. Howell regarding Dr. Bishayee's incompetence and that I could then persuade Dr. Howell to terminate his appointment.' Do you see that?"

Hill: "Yes."

Leonard: "Does that refresh your recollection as to whether you wanted Dr. Bishayee fired?"

Hill: "I told you I [didn't] use the word 'fire.' Terminate his appointment would mean that, you know, it was time for Anupam to move on. He had been there for three years or something like that. And if I could make Howell realize that Anupam was doing things that were not scientifically correct, that Howell would say, 'Well, Anupam, you've been here for three years. Now it's time for you to get another job.' I don't use the word 'fire' for a postdoc. What you do is you say, you know, it's time to find another job. And what I was hoping was that Howell would do that."

Further into the deposition, Leonard asked about Lenarczyk's employment. He asked about Hill's statement that Howell "...made it impossible for Lenarczyk to extend his visa." To which Hill replied, "Well, Lenarczyk told me, he didn't say it in his deposition, but he told me that he couldn't stay because he wasn't getting enough money, and he wanted to bring his wife over. So," she continued, "that's basically the background for that statement."

Leonard subsequently returned to the issue of individuals who leave the university and the fact that there were, among them, those to whom she would not provide a recommendation. "Not say anything bad about is kind of what I said," Hill replied. "I will tell the truth. You know, [what's] very important to me is the truth. So, I'll say what's true. You know," she continued, "Bishayee never worked for me. I never would have to give a recommendation for him. I think I wouldn't if I had to. But other people that I've had that weren't very good, I said what they did; I said that they were nice, I said things like that. But I wouldn't say that I think this was a terrible worker because I really don't like to destroy people. I hope that they will be able to get along and get a boss who is as easygoing as I am, who will really put the screws on them and make them work right." Leonard asked if anyone had ever worked for Hill and engaged in suspicious activity who left the university had asked for her to provide a recommendation.

Yes. I had, I guess she was a postdoc, and she was doing a procedure. We were measuring DNA strand breaks in gels. We scientists do an awful lot of stuff in electrophoresis and gels and so forth. Her results were not coming out right and I knew something was wrong. I said, 'Let me run the experiment and I'll see if I can figure out where it is that something isn't working.' It was a pretty complicated experiment, and you had to cut out the gels into little pieces and then you had to digest them with acid. So, I went through the experiment and when I was finished, then you have to count the radioactivity that's coming from these gels. And I counted my radioactivity, and it was gone, and I had put in lots. You can see from Howell's experiments, there's lots of counts. And this was astounding. And I did it again and the same thing happened. And I thought this is really weird.

Then I realized when you digest the gel with the acid, you keep a bottle of concentrated hydrochloric acid on the bench, and you just take a pipette, and you take that, and you put it into each one of the vials to digest the gels. I think actually what she was doing was not getting them thoroughly digested. But I think what she did then to sabotage my experiment was that she took the hydrochloric acid and she put it into every single one of the tubes and it just destroyed everything. I have no proof. But anyway, that's what I figured, that she was sabotaging my experiment. Because she didn't want it to work for me if it couldn't work for her.

So, I went to the university lawyer, and I said, 'Look, I think this woman has been sabotaging my experiment, or sabotaged my experiment when I tried to [do] it, and what do I do?' Things were a lot looser in those days, and the lawyer said, 'Go back to the lab and tell her to pack up her things and to leave and give her 15 minutes.' And I did that. And she was supposed to be taking her off the payroll. They didn't actually take her off the payroll for another two or three months or something like that. And then they called me, and they said she's getting another job in Connecticut, and they needed[ed] a letter of recommendation. I said I really don't want to write a letter of recommendation. They said, 'you have to.' So, I wrote her a letter of recommendation and I did just what I said I would do, I said that she's an attractive person and she worked. I tried to say what was true and nothing more. And so that's the way that was handled.

Toward the end of her deposition, Leonard asked Hill, after covering so much ground, if there was anything else she wished to say. "I suppose what I want to say is that I don't take this lightly. Obviously," she observed, "I have expended a great deal of my time, a great deal of my family's finances, that I'm sad and disappointed in the university and Dr. Howell. Bishayee, I don't really blame him because I think he grew up in a place where in order to get ahead you had to make up things. So, he's kind of – he's really the cause of it all. But I think if he had been properly educated when he came to the United States and it was explained to him that we don't do things this way, maybe he wouldn't have done what he did." She went on to note, too, that she knew people were being hurt and that she was very sad about that as well. "So why do I do it? I think it is an obligation and a duty. I think that science has to be as true as we can possibly make it and when we find out we've made mistakes, I think we have to be willing to stand up and to say 'I made a mistake, and I have to retract.' I would have given anything to have not had this happen," she continued. "When Marek called me in to see what Anupam had done that fateful night, March 26, I believe it was, of 2001, I kept saying, 'No, go away, I don't want to hear you. I don't want to see it.' And finally he persuaded me and that was a terrible moment. So, I guess that's really all that I can say. I'm sorry. But once I started down this slippery slope," she opined, "I felt that I had an obligation to continue. So, I guess that's it."

Just before the deposition came to an end, Leonard asked if Hill had ever tried to confront Bishayee directly. "You know, I never did. I thought that was Howell's job," she replied. "And when you and Dr. Lenarczyk observed or shadowed Dr. Bishayee, you didn't go back to Dr. Howell, though, did you?" Leonard asked. "Marek and I shadowed and when the experiment was completed, then I put everything together in three notebooks. And I didn't think that Marek should have any further role because I'm the senior person and I had to take responsibility for it. So, I took the three notebooks," she explained, "and my obligation really was to report to Baker, but I thought out of courtesy I had to report to Howell, and so I did. And then he and I went together to Baker." Leonard wanted to clarify that Hill had informed Howell prior to going to Baker. "I gave him [Howell] the notebook. I don't know whether he looked at it or not. But what happened was, I was worried because I knew that Anupam was going to take contaminated cultures into the FACS lab, the fluorescent activated cell sorter. That's a delicate instrument," Hill replied, "and I didn't know whether putting bacterial contamination or yeast or whatever it would be into that, whether that would mess it up, make a real mess of that instrument or not. I was worried about that." Thus, she determined that it was better to approach the head of the lab quickly. "I told the boss of the lab that I was worried that these contaminated cultures might be coming in on Friday. And I asked him – I told him I was telling him in confidence – but then he broke my confidence, and he told Howell. And then actually I think Howell told Bishayee to 'clean up your act.' Then Bishayee got the little tubes out of the freezer and threw them away. Where he threw them, we don't know because remember we couldn't find them," she explained. "But in any case, because [Thomas N.] Denny, [M.Sc., M.Phil.,] the head of the lab, spread the word, I guess then Howell went and told Baker what had been going on, that I had been making these observations," she continued. "And then, when we went to Baker, Baker said, 'It's all over the medical school. The dean knows about it. And why are you making the department look bad,' and so on and so forth. So that's kind of the way it all began." Leonard wanted to know why Hill didn't go to Howell the day that Lenarczyk first approached her. "Because we both knew," Hill told him, "that Howell thought Bishayee was wonderful and really believed in him. We thought if we told Howell then that Howell would tell Bishayee to stop the experiment, and we would never have the proof that we would need to show that Bishayee had been carrying out an experiment with contaminated cultures. So, we decided to carry the experiment on through the very end." Leonard asked if it wouldn't have been better to have gone to Howell and let him stop the experiment. "We wanted to convince him that Bishayee was carrying out an experiment through that he knew was contaminated," she responded. "You know, it didn't occur to me at that time that he would actually make up the numbers for the results. But that was our reasoning, that we should carry it through to the end, and then we'd show how the flask that contained, he called it the garden, because it had so much contamination in it. We thought we'd have more of an impact if we did it that way." But had this experiment gone awry wasted resources associated with the grant? "It was just wasting incubation time, not a big deal," said Hill. "It wasn't big money at that point. The experiment, by the time I started in on it, by the time Lenarczyk called me, which was like Wednesday, the experiment had started on Monday, and everything was set up. It was not a big deal from that point on."

Leonard had a difficult time understanding why, when Lenarczyk came to Hill, if it was Howell's NIH grant, he wasn't brought in on what was happening immediately so that he might be afforded the opportunity to participate in what he dubbed "righting the wrong." Hill replied that it was because she and Lenarczyk didn't think he would set it right. "We thought we really had to knock it home for him. We had to go through to the very end and then show him at the end of the experiment. Then what absolutely shocked us was the realization that Anupam had taken the clean cells that he had gotten from Marek and substituted them."

Chapter 10 –
The Experts Weigh In

ill eventually enlisted the expertise of two witnesses, Joel H. Pitt, Ph.D., a statistician and chairman of the Department of Computer Science and Mathematics at Georgian Court University in Lakewood, New Jersey, and Michael E. Robbins, Ph.D., a radiation biologist and associate director, Brain Tumor Center of Excellence, section head, radiation oncology-radiation biology and professor of radiation oncology-radiation biology at Wake Forest School of Medicine in Winston Salem, North Carolina. The Round Table Group had recommended Robbins to her and while she'd initially put aside his credentials, Hill realized that Robbins was exactly who she needed. Both of them filed reports with the court and were deposed. The defense elected not to deal with the statistics and only employed one expert who was provided just the Robbins report to examine and was never privy to Pitt's statistician's report. "Unlike science, where one hopes to have all of the facts available," said Hill later, "the law can be selective and need not present all of the information. This is one very important reason why the courts are no place for science to be judged." The expert reports were due after all of the principals had been deposed.

In reviewing Dr. Anupam Bishayee's data for his report Pitt used three different techniques. First, he determined the relative frequency with which each of the digits 0-9 appeared as least significant digit in Bishayee's data (a standard technique used by the Public Health Service's Office of Research Integrity). Using appropriate control data to confirm assumptions about the expected relative frequencies it was determined the probability that non-fabricated data could result in such frequencies is considerably less than one in one hundred billion. Second, data from the experiments critical in supporting non-replicated results was organized in groups of three measurements. Pitt examined the frequency with which one of the three measurements were close to the average of the three measurements and found that the frequency and pattern of closeness in Bishayee's data was completely at variance with the pattern in control data from various sources and computer simulation data. Although he could not assign a specific probability to the results, the distinctive pattern evident in Bishayee's data would lead any reasonable observer to conclude that Bishayee repeatedly and deliberately invented one value in each triad to force his data to conform to the experimental results he wished to report. Third, Pitt determined the relative frequency with which the two least significant digits in Bishayee's measurements were equal. Based on reasonable assumptions about the likelihood that the terminal digits of a non-fabricated measurement would be equal, assumptions that were borne out in Pitt's control data, he found the probability that the relative frequency of such incidents diverged from the expected frequency as much as they did in Bishayee's case is one in ten million.

In considering any claim that the probability of an outcome with certain anomalous or distinctive characteristics is miniscule it is absolutely essential, Pitt reported, to understand the assumptions on which calculation of the given probability value is based. The mere unlikelihood of an event certainly does not imply that it cannot have honestly occurred by chance. After all, he pointed out, lotteries regularly return winners despite the significantly low probability of winning. Nonetheless, the staggering improbability that the pattern evident in Bishayee's data would occur in the ordinary course of research left Pitt quite certain that much of it had, indeed, been fabricated. When their statistical results were considered in combination with direct observation of scientific misconduct by Bishayee and the irreproducibility (and apparent impossibility of reproducing) his results the conclusion that he had committed fraud was inescapable.

Pitt had quantified cell counts from the laboratory using the Coulter counter and found that Bishayee's counts were far from random while those of everyone else in the laboratory were consistently random. He made the same determination for colony counts and in like manner showed that Bishayee's counts were far from expected uniform or random while everyone else's were, as expected, consistent with uniform or random. Pitt further determined that the frequency with which the rightmost digit in Bishayee's Coulter counts was equal to the next to rightmost was also significantly different from the expected 10 percent or random as compared to the terminal pairs of others that occurred at a frequency of 10.1 percent. Pitt also found that the average of the three triple colony counts recorded for each of Bishayee's data points appeared as one of the individual counts at extremely high frequency, in stark contrast to the triple counts of others that did not contain the average at an unusually high frequency. He and his team had done their homework.

In his conclusion, Pitt wrote that "although one hope that the incidence of the use of fabricated data in research is low, it is clear that however low it is, it does exist. Researchers may choose to fabricate data as an expedient to justify scientific results that they believe (or would like to claim) to be true, or they may simply do so to relieve themselves of the burden of honest toil. In either case," he continued, "the consequences are seriously damaging to science."

Pitt invoked the work of James Mosimann et al.[167] who pointed out in 2002 that "a useful way to assess questioned data is to examine inconsequential components of data sets that are not directly related to the scientific conclusions of the purported experiment [...] [if] the allegation is true and the data are falsified, the falsifier typically devote[s] attention to numbers that establish the desired scientific outcome. Properties of the numbers that are not directly related to the desired outcome are less likely to receive consideration by the falsifier." In Pitt's study of Bishayee's experimental data he and his team found ample indications of such a failure to pay attention to the "inconsequential components" of his data sets. Having done so certain patterns of regularity that were observed to be present in the comparable control data sets that Pitt studied were absent in the data that Bishayee presented. These regularities included:

1. the non-significant low order digits of data were ordinarily uniformly distributed: control from multiple sources and settings displayed this regularity; it was easily seen from graphs of Bishayee's that it did not display this regularity. Applying the standard

statistical Chi-square goodness-of-fit test to Bishayee's data showed that the probability of obtaining the actual distribution of low order digits that occurred in his data assuming they were truly uniform was less than one in one hundred billion.

2. when data is collected in triads with elements having essentially the same distribution, the median value of the triad is equally likely to be as close to the smallest or the largest as it is to the mean of the two: thus data from Pitt's controls and a computer simulation clearly showed this regularity while in Bishayee's data the median was extremely close to the mean in more than sixty percent of his trials – a pattern that seems clearly to indicate that Bishayee was deliberately inventing results to justify an assumed result.

3. when the two lower orders of digits are non-significant, they will be equal in about ten percent of data values: Pitt's control data exhibited this regularity while in a test of the null hypothesis that Bishayee's data did he and his team rejected the null hypothesis with a p-value that was less than one in ten million.

Ultimately, Pitt considered, to reiterate, the staggering improbability that the patterns evident in Bishayee's data would occur in the ordinary course of research in combination with the known direct observations of scientific misconduct by Bishayee and the irreproducibility (and apparent impossibility of reproducing) his results the conclusion that he fabricated his results and has committed fraudulent research "is inescapable." He couldn't emphasis his findings enough but the defense, perhaps already realizing how much Pitt's report could hurt their case, elected not to present it to an equivalent expert for rebuttal.

Michael Robbins' task was to analyze the radiobiological data produced by Bishayee, Marek Lenarczyk, Roger Howell and Hill. In Robbins' report, he concluded that Bishayee's exponential killing of cells by tritiated thymidine (3H-TdR) was impossible because tritiated thymidine only kills cells when they are in the DNA-synthesis phase of the cell cycle (S) and no attempt was made to synchronize the cells indicating that all of them would have been S during their exposure to tritiated thymidine. Furthermore, tritiated thymidine blocks cells from entering S unless deoxycytidine is present and there is no evidence that it was present in any of the questioned experiments. Robbins pointed out that, in contrast to Bishayee's experiments, Howell's and Lenarczyk's biphasic results were entirely consistent to expectation and with other reports in the literature when no deoxycytidine was present. He further concluded that the Helena tubes used by the workers in the lab were hypoxic, in contrast to Howell's conclusion in his grant application that hypoxia was minimal. Robbins went point by point through Howell's list of possible reasons why he and Lenarczyk could not reproduce Bishayee's results and argued that there was no evidence for any one of them.

Robbins outlined the experiments in question, collectively called either 100 percent or 50 percent experiments, both of which followed two similar protocols. In the 100 percent experiments, V79 hamster lung fibroblasts, a cell line that has been used extensively in radiation biology studies over the past forty years, were placed in tubes and incubated overnight with varying amounts of tritiated thymidine. During this time the cells that are synthesizing DNA in a portion of the cell cycle called the S-phase (synthesis phase), take up the 3H-TdR and incorporate this into their DNA. Cells that are in other phases of the cell cycle (G1, G2 and M) will not take up the 3H-TdR. The following morning (in the protocol), the cells are washed with medium to remove any of the 3H-TdR that has not been taken up by cells and then placed in narrow 400 µL (the aforementioned Helena tubes) and then gently centrifuged (spun down) to form "clusters" of cells. The Helena tubes are then to be incubated for three days at ~ ten degrees Celsius to allow the 3H to decay without any cell replication taking place. Those cells that have incorporated 3H-TdR into their DNA are irradiated. The type of radiation produced by the 3H-TdR, called beta (β) radiation, will only travel a short distance, and therefore only kill cells that contain the 3H-TdR. Thereafter, the cells are placed in new larger tubes, washed and counted for both cell number (performed by the Coulter counter) and radioactivity, diluted, and then placed in tissue culture dishes containing growth medium. The cells are diluted such that each plate contains between 50 and 250 colonies. Once plated, each cell divides several times forming a colony of cells that can be counted about one week later. The number of colonies present is thus used to determine the surviving fraction of cells following treatment with 3H-TdR. This is determined by dividing the average number of colonies in the dishes containing cells exposed to tritiated thymidine by the average number of colonies in control dishes that were not exposed to the same.

In the 50 percent experiments, half the tubes are incubated overnight, Robbins explained, without the radioactive 3H-TdR; these will be the bystander cells. These cells are subsequently mixed with an equal number of cells treated with tritiated thymidine overnight prior to the three-day incubation. The rest of the experiment is performed as described in the 100 percent experiment. No mystery there.

The results of the experiments in question were reported in a successful grant application submitted by Howell to the NIH as well as in two published papers written by Bishayee, et al., though Bishayee disavowed having written either, putting authorship on Howell, during his deposition. These 100 percent experiments showed that the cells were extremely sensitive to irradiation. There was an exponential decline in cell survival (seen as a decrease in surviving fraction on a log [surviving fraction]-linear [dose of radiation] plot) down to around three logs or, in other words, a surviving fraction of around .001 or .1 percent (99.9 percent of the cells had been killed). These results are impossible to generate due to the following three reasons that Robbins documented in his report:

1. Tritiated thymidine blocks the movement of cells through the various phases of the cell cycle. Thus, cells that are not in the S-phase of the cell cycle during the overnight incubation with 3H-TdR cannot enter the S-phase, will not incorporate 3H-TdR into their DNA, and will not be killed by the subsequent radioactive decay of the 3H.

2. No deoxycytidine (dCyd) was present in the medium at the time the cells were exposed to 3H-TdR. Previous studies have shown that the inclusion of dCyd in the medium prevents the 3H-TdR from blocking cell movement through the cell cycle leading to an exponential decrease in cell survival.

3. No attempt was made to synchronize the cells into the same phase of the cell cycle prior to their treatment with 3H-TdR. If all the cells were in the same phase of the cell cycle then there is a possibility that they would all have been in the S-phase of the cell cycle at the time the tritiated thymidine was added;

however, as he exposed in his final report, Robbins wrote that special experimental procedures are required to ensure that the cells are synchronized, and these were not used in the experiments performed by Bishayee.

Robbins' work provided critical perspective. He reiterated that additional 100 percent experiments performed by Marek Lenarczyk and Roger Howell failed to confirm data generated by Bishayee. In these studies no dCyd was added to the medium, nor were the cells synchronized prior to addition of the tritiated thymidine. In marked contrast to the surviving fraction of .001 reported by Bishayee et al., these latter studies noted a surviving fraction of .3 (30 percent or, in other words, 70 percent of the cells were killed) or less, a value 300 times greater than reported in the experiments performed by Bishayee, and a value entirely consistent with the experimental conditions used and published in the literature.

For the 50 percent experiments, Bishayee et al. reported that the surviving fraction of the cells declined exponentially down to a value of .01. Subsequent attempts by Lenarczyk and Howell to reproduce Bishayee's findings were singularly unsuccessful. These investigators were unable to demonstrate any bystander effect, although one would have expected to see such a response. The likely explanation for the absence of any bystander effect in these cells is the presence of hypoxia in the Helena tubes. The radiological basis for these arguments in support of the allegations of fraud were further spelled out in Robbins' report. But one of the critical points that cannot be overemphasized is that there was no evidence of deoxycytidine added to the medium in any of the protocols used in the Howell lab. The only way to prevent the effects of tritiated thymidine on blocking cell cycle progression is to add dCyd to the medium. Since this was not added, all the deleterious effects of 3H-TdR on the cell cycle did occur in the experiments carried out by Bishayee, Lenarczyk and Howell.

The failure of Howell to duplicate the results generated and published by Bishayee could reflect differences in the experimental protocols used. This is clearly not the case, Robbins reported. Howell experiment number five dated July 16, 2001, had the same protocol with minor variations as that used by Bishayee in his experiment number two dated November 30, 1998, in 100 percent cluster experiments except for the range of the

3H-TdR concentrations used. The uptakes were comparable; however, the experiment performed by Bishayee claimed an exponential decrease in surviving fraction, to a value less than .5 percent. In contrast, the data generated by Howell showed, as predicted, a biphasic decrease in surviving fraction, with a minimal value of some 50 percent, a level of survival 100 times greater than Bishayee's. This marked discrepancy between the data generated by Bishayee and Howell was evident for the 50 percent experiments as well. Howell experiment number three dated May 30, 2001, had the same protocol as Bishayee experiment number one dated December 29, 1998, except for the range of tritiated thymidine concentrations used. The uptakes were different but at 10 mBq per cell, Bishayee's survival had fallen to .009 (.9), while Howell's had fallen to .6 (60 percent), a level of survival 67 times greater than Bishayee's. These findings indicate that the protocols used by Bishayee and Howell were the same, and thus the pronounced differences in the level of cell kill observed between the two investigators could not reflect differences in experimental procedures.

So, what about the comparison of experiments performed by Lenarczyk and Howell, many of which became the centerpiece of the Lenarczyk deposition? The 100 percent experiments performed by Lenarczyk and Howell completely failed to duplicate the results published by Bishayee et al. in 1999 and 2001. To illustrate this, Robbins looked at four experiments using the V79 cells performed by Lenarczyk, those dated October 2, 2000; May 3, 2001; May 21, 2001, and June 21, 2001. In each of these, the surviving fraction showed a biphasic decline with increasing dose that plateaued at a value of about .6 (60 percent). Two experiments performed by Howell, namely experiment number five dated July 16, 2001, and experiment number six, September 27, 2001, also showed a biphasic decline in surviving fraction with a plateau value of approximately .3 (30 percent). Finally, an experiment performed by both Lenarczyk and Howell dated December 14, 2000, also demonstrated a biphasic response, with a minimal value of about .2 (20 percent). Averaging the data from these seven experiments indicated the following: survival decreased in a biphasic manner with increasing dose, and the average break point in these survival curves was .5 (50 percent) at greater than .3 mBq per cell.

Robbins found that in three additional experiments dated November 10, 2000; November 28, 2000, and Lenarczyk exhibit 21 dated February

19, 2001, which Lenarczyk performed using a different Chinese hamster cell line (CHOK1-A1), a biphasic decreased in survival was again clearly evident. The average break point with this cell line is .29 (29 percent). Thus, in ten 100 percent experiments performed by Lenarczyk and Howell, the survival curves seen when cells were treated with tritiated thymidine (again, 3H-TdR) gave the expected biphasic response, reflecting the lack of synchrony and the lack of dCyd present in the medium to prevent the tritiated thymidine mediated cell cycle block. These results were completely at odds with those generated by Bishayee et al., who reported an exponential decline in cell survival down to levels less than .1 percent. Based on the radiological principles outlined by Robbins, it seemed clear that the data generated by Bishayee from his 100 percent experiments could not have been generated without falsification of the data.

The 50 percent experiments performed by Lenarczyk and Howell similarly completely failed to duplicate the results published by Bishayee et al. To illustrate this point Robbins looked at experiments using V79 cells performed by Lenarczyk, labeled in deposition as exhibits 25, dated December 26, 2001; exhibit 26, January 15, 2001; exhibit 27, February 5, 2001, and exhibit 28, June 14, 2001. These data showed a biphasic response with a break at a surviving fraction of approximately .7 (70 percent). An additional experiment, performed by both Lenarczyk and Bishayee, tagged as exhibit 29, dated July 5, 2001, provided essentially identical results. Three experiments performed by Howell, namely experiment two, dated April 19, 2001; experiment three, dated May 3, 2001, and experiment four, dated June 28, 2001, also demonstrated a biphasic decline in surviving fraction, with a plateau value of approximately .7 (70 percent). Combining the data from these eight experiments indicated the following: survival decreases in a biphasic manner with increasing dose, and the average break point in these survival curves was .7 (70 percent). Thus, in the ten 50 percent experiments, half of the tubes were incubated overnight without the radioactive 3H-TdR; these were the bystander cells. These cells were subsequently mixed with an equal number of cells treated with 3H-TdR overnight prior to the three-day incubation. The rest of the experiment was performed as aforementioned for the 100 percent experiment. The expected survival of the 100 percent labeled cells was approximately fifty percent. These made up 25 percent of the survivors in the 50:50 clusters present in the 50 percent experiments. Considering the bystander cells, if all survived then 75 percent of the cells in the clusters would be survivors. In contrast, if none survived, then 25 percent of the cells in the clusters would be survivors. Thus, the predicted bystander effect would be evident between 25 and 75 percent survival.

In the 50 percent experiments carried out by Lenarczyk and Howell, survival appeared to be in the order of 70 percent, indicating little or no bystander effect. These results were completely at odds with those generated by Bishayee et al., who reported an exponential decline in cell survival down to a survival of .01 (1 percent). Based on the radiobiological principles outlined in his report, Robbins noted that it seemed clear that the data generated by Bishayee from his 50 percent experiments could not have been generated without falsification of the data.

But why did Lenarczyk and Howell fail to observe the bystander effect? Robbins noted that experiments that had been published by Rudranath Persaud et al.[168] in 2005 did indicate a bystander effect, with survival of the bystander cells appearing to be around 40 percent. What is the difference between these studies? There is a major difference, Robbins iterated, in the protocols used by the two groups. Persaud et al. used microfuges tubes with 100 μL of air present above the cells. In contrast, Lenarczyk and Howell used Helena tubes in which no air was present above the cells. In this situation the cells would have been hypoxic, a condition in which they were much more resistant to the killing effects of radiation. In Bishayee experiment one, performed on September 7, 1998, a colony forming assay was obtained using V79 cells cultured in Falcon tubes, in which the cells were aerobic and therefore radiosensitive. The cell survival curve indicated a marked decline in surviving fraction, as expected. These data were in agreement with well documented studies of the survival data for V79 cells. These data further showed the average dose at which a surviving fraction of .1 (10 percent) was observed. The remaining plots were from cell survival assays performed by Bishayee, Lenarczyk and Bogdan I. Gerashchenko, M.D., Ph.D., using Helena tubes and irradiating the cells as clusters or in suspension in all of these experiments. In all of these experiments, the survival curves were much shallower, reflecting radioresistant cells due to their being hypoxic. Thus, given the experimental protocol used in the Howell lab, the cells were hypoxic when cultured in the Helena tubes, and would be radioresistant. Further, the lack of a bystander effect observed by both Lenarczyk and Howell was the result of the cells being hypoxic. Bishayee's conditions and protocols were the same as those used by Lenarczyk and Howell and thus the cells would have been hypoxic. The marked radiosensitivity of the cells on the 50 percent experiments indicated that his results could not have been generated without falsification of the data.

Attempts to reproduce the data generated by Bishayee and published by Bishayee et al. in 1999 and 2001, as well as in the NIH applications, failed completely. These failures reflected the experimental conditions present: tritiated thymidine-mediated cell cycle block; lack of dCyd in the medium at the time the cells were exposed to 3H-TdR[169]; lack of any attempt to synchronize the cell into the same phase of the cell cycle prior to their treatment with 3H-TdR, and the presence of hypoxia in the Helena tubes used for the 50 percent labeling studies prevented the anticipated bystander effect.

In response to the failure to show bystander effect, Howell (who is not a radiology biologist) presented a document that summarized the experiments, in which he proposed a number of possible factors that might explain the differences in the data generated; however, as Robbins documented, these fail to provide any evidence that might explain the marked differences in the experimental data generated by Bishayee compared with that of Lenarczyk and Howell.

1. Variable source of microfuges tubes: It seems highly unlikely that "contamination" of the microtubes used with trace elements would explain the failure to replicate Bishayee's findings. No experiments were performed by Howell to substantiate this "claim."
2. pH of media: This is not a valid concern. If the pH changed during the course of the experiment, then it would be noted by the color of the phenol red in the medium; any acidification due to contamination would lead to stopping the experiment.
3. Level of trace elements in the water: Without any evidence to indicate that this wither occurred or would have any significant impact on the data generated, this response provided no explanation.
4. Wetting agents on filter apparatus: Without any evidence to indicate that this either occurred or would have any significant

impact on the data generated, this response, again, provided no explanation.

5. Methods used to clean bottles: It is not clear how this might have impacted the studies or served as a potential explanation for the inability to duplicate Bishayee's data.

6. Sodium bicarbonate product changed: As long as the final concentration of chemicals used was the same, there would be absolutely no difference in the experimental conditions.

7. Different incubators: Laboratories change incubators on a regular basis as the equipment ages or breaks down. As long as the gas concentrations and temperature are maintained, the incubator is kept clean, then no difference would be seen in the cells being cultured.

8. Fetal calf serum: Without any evidence to indicate that different sources of media used in the Howell lab led to differences in cell survival ranging from .1 to 70 percent, this point is invalid.

9. Different V79 cells used: A review of the literature reporting cell survival for V79 cells after irradiation indicates very little difference in the radiation response of these cells over several decades and being cultured in numerous laboratories in Europe and the United States. There was no data showing the 100-fold difference in survival of V79 cells following irradiation noted between Bishayee, Lenarczyk and Howell.

None of these variables, Robbins wrote, explained the inability of Lenarczyk and Howell to replicate Bishayee's findings. The data generated by Lenarczyk and Howell were exactly what was expected based on the experimental protocols used and the radiobiological principles he laid out for the court. The inability to reproduce Bishayee's findings reflected the fact that these data were falsified, he concluded.

The third expert, for the defense, stunned Hill. "I was dumbfounded when I learned who the expert witness for the defense was to be Ludwig Feinendegen, M.D., a well-known physician specializing in nuclear medicine, the field in radiation biology that all of this was about," Hill observed. "I knew – or thought I knew – Ludwig well. We had worked together on a grant application to the Department of Energy, and he had been a guest in our house for the better part of a week in 2001 or 2002. More than that," she continued, "he had been our guest at the Explorers Club annual dinner, one of the most prestigious (and expensive) social affairs in New York City each year. I had considered asking him to serve as an expert for me, but rejected the idea because I knew that he was also friendly with Howell and I thought he would be too much of a gentleman to accept. How wrong I was."

Feinendegen took nearly five months to rebut the Robbins report; he was originally given a month to do so. Feinendegen never saw the Pitt analysis. He took issue with Robbins conjecture that tritiated thymidine would block the cell cycle because, he argued, its thymidine concentration was too low due to its high specific activity (amount of radioactivity relative to the amount of chemical thymidine) and in such case, deoxycytidine would not have been necessary to counteract the cell cycle blocking effect. He calculated the slope of Bishayee's survival curves and said that it was as predicted based on radiobiological principles. This could certainly have been true except that when those experiments were analyzed statistically, it was evident that the results were extremely at variance with any expectation of randomness. In fact, there were in all 2,048 tritiated thymidine experimental Coulter counts by Bishayee with a p-value for random or uniform terminal digits of 4.7 times 10 to the minus fifty-eighth power (57 zeros after the decimal point). In all, there were 1,380 tritiated thymidine experimental colony counts with a p-value for random or uniform terminal digits of 3.6 times 10 to the minus eleventh power (10 zeros after the decimal point). In all, there were 465 tritiated thymidine experimental triples of which 389 or 84 percent contained the mean as one of the triples. But Feinendegen was totally unaware of any of these determinations. In his report to the court, Feinendegen went through Howell's list of possible explanations for the inability to repeat Bishayee's experiment, justifying each one without finding any evidence for roles in any of Howell's experiments.

The last phase of discovery deposed Pitt, Robbins and Feinendegen. Hill's experts were deposed in Sheldon Pincus' law office, and Feinendegen was deposed on September 17, 2009, at a motel on Long Island, located near the Brookhaven National Laboratory, where he spent a great deal of his time, even in retirement. Pitt was deposed on September 2, 2009. Hill recalled later that he was "calm, cool, collected and spoke convincingly and sincerely. He was unflappable. The defense attorney was completely unable to find any flaws in his testimony. Too bad it was totally lost on the court." Robbins wasn't deposed until January 7, 2010.

During his deposition, Pitt told defense counsel Scott Flynn that he'd looked at all the data he got from Howell's laboratory. "I looked at Dr. Hill's data, I looked at Dr. Howell's data, I looked at Dr. Lenarczyk's data. And in looking at that data I said, "Well, did their last digits look random? They are all doing roughly the same kind of experiment, they are all pulling out roughly the same kind of thing," he continued, "and I don't see that kind of nonuniformity. Then you can argue that Dr. Hill and Dr. Lenarczyk to some degree are the people of interest in this group. I actually looked at everybody else's separately. Still uniform. But another question," he told Flynn, is what happens at other labs? That is why I actually asked Dr. Hill to contact other people to see whether I could get data from other Coulter counters from roughly similar experiments." Pitt received data from Case Western University and University of Texas Southwestern Medical Center, Dallas-Fort Worth to get that extra data from similar experimentation.

Pitt explained to Flynn the point of simulation, which is to obtain some idea of what random mechanisms produce, an important aspect of what he considered in analyzing Bishayee's data. "One of the interesting features of the way in which randomness works is that even though things are random when you take them one at a time, but when you do them over and over again there are patterns," Pitt testified. "This is what is called the law of large numbers. So, if I perform a simulation five thousand times of something, in general if I do – if I sort of pick new random numbers and do the same thing again, I am going to see something which is pretty close."

Flynn challenged Pitt's statement "when our statistical results are considered in combination with," and Pitt lists the "direct observation of scientific misconduct," Flynn asked what he meant by "direct observation. Was that Dr. Hill's observations? Flynn wanted to know. Pitt said yes. "And the irreproducibility and apparent impossibility of reproducing Dr. Bishayee's results –," Flynn pressed. Pitt was on point. "By the way, I think that is actually just a mild statement of my position. I frankly – looking at just the numbers [from the various Bishayee experiments], I believe they are fabricated. I believe it is inescapable, just as I believe that if the molecules – if you told me the molecules accumulated in that corner, I wouldn't believe you," he told Flynn. "But I was trying to actually state this in, you know, what I felt was a reasonable fashion. In other words, you look at the whole picture. To me," he continued, "the whole picture spells it out. But I suspect, I don't know [...] whether I should have written it that way." Flynn asked him to clarify if that meant he meant the need to keep "in combination with" because Pitt felt that his conclusions stood alone. "I do feel it stands alone [his conclusion of fabrication]." He didn't need Hill to help him come to it.

"There was a great deal of back and forth regarding Robbins' and Feinendegen's depositions," Hill observed. "Robbins was supposed to be deposed first but Howell wanted to attend, and Howell's brother was being operated on that week [of the planned deposition] for an ocular melanoma. In this situation, proton radiation is used and since this was close to Howell's expertise, he wanted to be there. So Feinendegen was scheduled to go first." Hill and Pincus arrived at the deposition to find Feinendegen waiting in the meeting room. "When I walked in, he came over, threw his arms around me and gave me a big bear hug," she wrote of that day. "I was so shocked that I didn't push him away. If our paths should, heaven forbid, ever cross in the future, I will be sure he never touches me again. Both university lawyers and Howell were also in attendance. Howell made a big show of taking notes on his computer." Notably, Feinendegen admitted to having retired sixteen years before and that he had spent thirty-six hours preparing his report. He had not seen Robbins' but not Pitt's findings. Pincus explained to Hill that Feinendegen was a rebuttal witness only for Robbins' report.

Feinendegen kept wanting to make statements but was not permitted to do so, Hill subsequently recalled. "He brushed off three papers in the literature as irrelevant although high specific activity tritiated thymidine, no deoxycytidine and biphasic killing had been observed. He even claimed that one of those papers was irrelevant while admitting he had not read it. He [also] admitted he did not look at the lab protocols. 'I relied on the accuracy of the published papers,' [Bishayee et al. 1999 and 2001] he said. He claimed to have found many papers in which there was exponential survival without deoxycytidine, but he had not cited any of them [in his final report]." Further, Feinendegen had shared an advance draft of his report with Howell, who recommended nips and tucks to the final product. Bishayee was not consulted by Feinendegen or, for that matter, Pitt or Robbins.

Pincus asked Feinendegen if he acknowledged in his review of the Robbins report that Robbins compared the protocols and results of Bishayee versus those of Lenarczyk and concluded that the protocols were the same? "That I cannot answer the question," Feinendegen began. "I only know that this is a scientific question which is of great importance, and it has nothing to do with the accusation of falsification. It is of enormous importance scientifically," he continued, "it's fascinating to see this differences coming up and I have come – let me just add that, in my own career, similar situations where I just failed to reproduce data and I have paper here that I even cited that in my report, a slight change of conditioning now, it may be the page [probably pH], it may be a similarity, it may be the concentration of buffer, it may be the type of buffer, all this influences an experimental outcome, and the lack of reproduction of the second group is a scientific question of great interest, but it has nothing to do with the accusation of falsification." Pincus let him ramble. "It's of great scientific interest but it does not refer to whether the first data were – and that was the starting point of it all, was falsified or not. That is a very fascinating topic you are opening up here and it is a fascinating scientific issue and I spent, to finish my answer, to make you aware of what that means, I spent more than the whole year once in order to find out why I could in my own laboratory not reproduce data, and then after a whole year of very precise efforts I found the conditions, what I called conditions, as you referred to that, and then we could reproduce and then we were off." Feinendegen, at one point in his deposition, referred to and defended his use of a dead html link to support one of his conjectures, and then he defended Howell for not reporting his inability to replicate Bishayee's results in twenty-two attempts, according to his deposition.

> *Pincus*: "You cite on page 17 [of his report] a link, the hyclone.com link, you see where I'm at?"
> *Feinendegen*: "Where is it?"
> *Pincus*: "On page 17 of your report."
> *Feinendegen*: "Where is it?"

Pincus: "Below those four items."

Feinendegen: "Yeah."

Pincus: "Are you aware that that's a dead website?"

Feinendegen: "No."

Pincus: "Do you have a copy of the document?"

Feinendegen: "No, I did not look at that website. I cited it but I did not look at that website; I tried to but I didn't look at it. I thought it was not available anymore, so I took it as it is and the document asks, 'actually supported by recommendations regarding the testing of disposable laboratory supplies.' So this – I took that to be expressed in this figure here."

Pincus: "You went to look at the link, when you looked at the link it was not available?"

Feinendegen: "Yeah. I should have crossed it out perhaps."

Pincus: "You don't have a copy of what those recommendations were that you allude to in your report, correct?"

Feinendegen: "What?"

Pincus: "You don't have a copy of what those recommendations were that you cite in your report?"

Feinendegen: "No. Here we are, there is a general recommendation by every manufacturer, any manufacturer who supplies tissue culture tools, they all more or less say the same thing, it is just like a – let's say a recommendation how you cross street, you should first look, then look left, then right, then see if cars are coming or not, these are generally valid. [He continues to compare this to the handling of tissue cultures and nothing to do with falsification of data. He is generally unresponsive to the question Pincus asked.]

Pincus: "I just want to be clear; you never actually had the opportunity to review – ."

Feinendegen: "That website."

Pincus: " – that website?"

Feinendegen: "Correct."

Pincus asked Feinendegen, given his experiences, having worked in controlled laboratories for years, did Feinendegen think that in the event there was an issue or a problem regarding reproducibility, did Howell have an obligation to report it to his chairman [Baker]? "No," said Feinendegen. "Here we come to the question of science. I put myself now in the position of Dr. Howell, who I didn't know at that time. If that would happen to me, I get an experiment and then I try to repeat the experiment and don't succeed, an alarm bell goes off. Why? There must be a reason, and this reason is a scientific question. What is the reason," he continued. "It could be – I could name off not only just that we discussed the vials, cleanliness, it could be the composition of the culture medium unknown, it happened to me from the same supplier, suddenly our cells didn't respond anymore as they used to do and we found out eventually the hard way it was a change in the media supplied by the same company, my lab, and so this – and the search, as I understand, is still going on why is there a discrepancy but that is a scientific question." Pincus pressed, asking as a group leader, would Feinendegen expect his subordinates to report such problems to him? "No," he said again, repeating "it's a scientific question. I mean, depending on to whom I report. If I am an independently working scientist, I'm responsible to the scientific community, that's not a question of reporting to the superior, it's a scientific question, what did we do, what is wrong." Pincus didn't let it go. "If you had associates working on a grant with you in your lab and they had difficulties reproducing results, you would have no expectation that your subordinate would report to you?" he asked Feinendegen. At no point in this exchange did it appear Feinendegen answered the question Pincus posed, which, in truth, went to the ethics of Feinendegen's laboratory practices. "That question is not quite complete," Feinendegen responded, "in that it was in certain way reproducible because many other people saw the thing, same thing, and it was reproducible in a way but suddenly it changed. It is not so as you now implied that there was one experiment and then the others didn't follow, there was a number, a large – I gave you the data before, I don't know how many, they all showed this exponential survival curve, so it was something strange and suddenly something that was reproducible becomes not reproducible anymore. It was not one experiment," he continued, "that could not be reproduced, it was a whole series of experiments [Feinendegen's work] which were in line actually with observations other people have made, myself included."

Feinendegen had still not answered the ethical question Pincus posed. Pushing ahead, Pincus asked if Feinendegen was aware that by the time the first Newark CCRI convened that Howell and/or Lenarczyk had performed eighteen experiments that did not reproduce Bishayee's results. "No," Feinendegen replied. "I don't know anything." "Do you think he [Howell] had an obligation to report that to the committee?" Pincus asked Feinendegen. "What?" he replied. "That they were unable to reproduce the results of Dr. Bishayee in those eighteen experiments?" Feinendegen again replied no, that it was "a scientific question." Pincus didn't let up. "I'm asking you to answer it." Leonard objected, claiming the question had been asked and answered. But it had not been answered.

Pincus: "Your answer is they had no obligation?"

Feinendegen: "To whom?"

Pincus: "To the committee?"

Feinendegen: "Which committee?"

Pincus: "The university committee on scientific misconduct."

Feinendegen: "I don't know what the issue was."

Pincus: "You don't know?"

Feinendegen: "I don't know. I don't know the circumstances. I cannot answer this."

Pincus: "Have you ever submitted reports to the National Institutes of Health regarding grants?"

Feinendegen: "No."

Pincus: "You never have?"

Feinendegen: "No, I work there." [Feinendegen had been retired from academia for sixteen years and was an assignee and program manager for the DOE from 1994-1998, a Fogarty Scholar, NIH from 1998-1999, and served as a consultant only to NIH from 2000-2008].

Pincus: "Were you aware that he [Howell] submitted annual reports to the NIH without ever mentioning the inability of members of his lab to reproduce the Bishayee results?"

Feinendegen: "I can't answer that question, it goes far beyond – in order to answer that question honestly, I must know the details."

Pincus: "You don't know any details [of the crux of the case to which he was asked to provide the report], is what you're telling me?"

Feinendegen: "No. I have the report by Dr. Robbins, okay. I examined the report by Dr. Robbins, and I accept the probabilities."

Pincus: "You limited yourself to the Robbins report. I understand what you're saying."

Feinendegen: "Yes."

Pincus: "Let me ask you this question: Are you aware that the two papers by Bishayee that are in *Radiation Research* contain results that have been cited numerous times in the literature since then?"

Feinendegen: "Yes, because they weren't confirmed with other people's observations, it was nothing strange."

Pincus: "Do you believe that a scientist has an obligation to the scientific community to inform a journal that results in the papers could not be reproduced?"

Feinendegen: "No. I'm sorry. If you do an experiment, you do that in your own laboratory many times and they did that and they could reproduce it; it's not just one experiment that could not be reproduced, there's a whole series of experiments that you have to – then they are ready. Now the new experiment was done and different results come up and then you ask why, and now it's a question in subsequent publications to – that is being done, why is that so."

Pincus: "Are you telling me that it's your understanding that Dr. Howell was able to reproduce the Bishayee results?" [Howell was never able to reproduce Bishayee's results.]

Feinendegen: "Initially, yes. There was a whole series of experiments; they all showed the same data, and then suddenly it didn't work anymore, so that is a scientific – ."

Pincus: "In other words, the experiments that you're referring to that all showed the same thing were the ones that Bishayee performed?"

Feinendegen: No, Harapanhalli [Ravi S., Ph.D.] for example."

Pincus: "I'm confining my question right now to the experiments that were done within the Howell lab itself, nothing external, you with me?"

Feinendegen: "Yea. Harapanhalli was at this place [he had a postdoctoral fellowship at UMDNJ from 1989-1994 and had nothing to do with Howell's laboratory or the experiments in question.]. You see, it's not so – I mean, as I understand that what was in front of me, right, I have no other further detail. I didn't visit the lab. I didn't ask people. I don't know. I didn't even ask you what really happened [in Howell's lab]. I just got that what I have, and my question was, is this accusation – the suspicion of fraud justified… [he repeats himself at that point, telling Pincus there was no misconduct and that everything is a 'scientific problem.']"

Pincus: "Are you aware that Dr. Howell is a member of the Medical Internal Radiation Dose Committee?"

Feinendegen: "No. I guess he should be."

Pincus: "In the spring – we're almost done here. In the spring of 2001, am I correct that yourself, Dr. Hill, Dr. Azzam, Dr. Howell, Dr. Hei of Columbia and a Dr. [F. Avraham] Dilmanian of Brookhaven were engaged in writing a grant application together?"

Feinendegen: "Yes, very much so." [The grant was submitted to the United States Department of Energy.]

Pincus: "You developed a friendly relationship with members of this group. Is that fair to say?"

Feinendegen: "We became friends. I stayed at Lanie Hill's [Dr. Hill's] house and enjoyed very much taking walks with the dogs in the morning."

Pincus: "In March of 2002 did Dr. Hill arrange for you to give a seminar at the New Jersey Medical – ."

Feinendegen: "Yes, I enjoyed this very much."

Pincus: "Did you receive an honorarium for that seminar?"

Feinendegen: "I don't even know."

Pincus: "You've been a guest in the Hill house?"

Feinendegen: "Yes."

Pincus: "Did they take you to dinner?"

Feinendegen: "Yes."

Pincus: "Did they take you as a guest to the Explorers Club banquet?"

Feinendegen: "Yes, I enjoyed that."

Pincus: "Did they take you to the reception the next day?"

Feinendegen: "Did we? Lanie, what was it, did we go to the reception the next day? I think we went home that night."

Pincus: "Did you see any basis in those experiences to recuse yourself in this proceeding?"

Leonard: "We're done, right?"

Pincus: "Not quite."

Feinendegen: "What?"

Pincus: "Did you see any basis in which to recuse yourself from this proceeding based on your experiences?"

Feinendegen: "No, this is strict science, that has nothing to do with personal relationships or friendships. We are arguing here science, and I got a report here [just the Robbins report] that was a very interesting question and also a scientific question and I tried to do the best and I didn't see any – I felt sorry that – ."

Pincus: "On page four – ."

Leonard: "Let him finish his answer. He felt sorry what?"

Pincus: "He's giving the same speech over and over again."

Leonard: "You felt sorry. You felt sorry about what?"

Feinendegen: "That our personal relationships was there but I thought I could clarify and help by going through the Robbins report very carefully and did the best I could to come to some kind of an understanding for this happening."

The importance of Pincus' line of questioning and the point of asking why he hadn't recused himself harkened to the relationship Feinendegen had with all the players. Hill had tried to determine, early on in her lab investigation of Bishayee, how he had purportedly "discovered" the bystander effect. "Roger [Howell] had said many times that it was a case of serendipity: Anupam had done the experiment that showed it but Roger, not being up on the radiation biology literature, was at a loss to explain it until he told Ludwig Feinendegen about it when he saw Ludwig at a meeting and Ludwig said that it must be the bystander effect," Hill recalled later. "This is all very mysterious because Roger's statements imply that there would have been sometime between the first 'bystander' experiment(s) and the subsequent ones. As it happens, the bystander experiments that were presented in the first paper [1999] on the subject followed one on top of another without any gaps."

Pincus ended Feinendegen's deposition with a query regarding the thymidine acronym:

> *Pincus*: "Are you aware of the acronym IUPAC-IUB [International Commission on Biochemical Nomenclature] as it being an authority on biochemical acronyms?"
>
> *Feinendegen*: "Would you spell that?"
>
> *Pincus*: "Sure. IUPAC-IUB."
>
> *Feinendegen*: "Yeah."
>
> *Pincus*: "Do you recognize that as an authority on biochemical acronyms?"
>
> *Feinendegen*: "Yeah."
>
> *Pincus*: "Are you aware that that body abbreviates thymidine as T small d capital R?"
>
> *Feinendegen*: "Listen – ."
>
> *Pincus*: "Just answer my question."
>
> *Feinendegen*: "My answer is the following: There are many ways of acronyming thymidine. My point is and I think you are referring to that, I criticize as kind of a negligence to stick to one acronym. I don't care which one, in the report but don't use different acronyms in one report in order for those who are not familiar with it, I am familiar with it, I don't care, but it is just a slight oversight which I pointed out." [He was incorrect in his criticism of the Robbins report on this point.]
>
> *Pincus*: "I wanted to understand the basis for your comment."
>
> *Feinendegen*: "Yes, I don't care whether you call it TdR or ThR [this is wrong and not the proscribed abbreviation for thymidine. The proposed alternative, not the proscribed, is dThd and is listed by the IUPAC-IUB.][170]

With that last exchange, Feinendegen's deposition was over.

Howell did not attend Robbins' January 2010 deposition. Defense counsel John Leonard did his best to discredit Robbins by emphasizing that he has never worked with tritiated thymidine or even with thymidine; that he had never worked with the bystander effect, and that he never spoken to Lenarczyk about the case or visited the laboratory to try to replicate the experiments in question. Leonard couldn't play the "have you ever spoken to" card when it came to Bishayee because his rebuttal expert, Feinendegen hadn't spoken to him either. But Robbins would not have been much of a radiation research section head, noted Hill later, had he not been familiar with these topics in radiation biology, his specialty at Wake Forest University School of Medicine. At the time of the deposition, Robbins had taught residents for seventeen years. He was also a member of the Radiological Society of America and had been involved in the board's educational subcommittees for several years. Robbins also had a grant from the Nuclear Regulatory Commission to develop teaching modules in radiation biology just then. "Plus," he told defense counsel John Leonard, "I've been doing this for thirty years." He informed Leonard, too, that he did not study bystander effect directly because he had his own research focus that involved giving doses of radiation that kill cells, which is not based on the bystander effect, which up until recently [in terms of the deposition] was only looking at low dose effect. But since he taught radiation biology of bystander effect in his course work, he was aware of the topic and had kept up to date with the subject matter. "He twisted Robbins words as he had done with mine. Robbins kept his cool," Hill observed of his deposition. Once Robbins' deposition was complete, discovery was over.

The crux of Robbins' testimony came when Leonard questioned Robbins's responsibility when looking at all the factors of the case to determine whether fraud had, in fact, occurred. "The question – the reproducibility is the fundamental consequence or fundamental concept of what we do [as scientists]," said Robbins. "Biology, I say, by its nature is variable. Therefore, I do not expect to see exactly the same thing exactly the same every single time. That's not the issue," he continued. "But the issue is a I need – when I generate information, if I'm going to publish that or present it in a – in some proposal and I'm going to build up a case for saying I need more money to pursue this [as in Howell's grant], I have to believe that that really is correct, that it's true." He went on to explain: "If I'm presented with a whole lot of evidence that those data are very questionable, it's my responsibility to respond to that – not to apparently ignore it." Leonard asked how Robbins would respond to it. "I would try – I would try and replicate that. If I couldn't replicate it, I would then say, okay, we have to do something about this," he replied. "In my own circumstance I have had a situation where I got an NIH grant, one of the aims was to do one thing; it didn't work, but I reported that to the NIH. Every year," he explained, "you write a noncompetitive renewal in which you state progress and future direction. If things aren't working out correctly, I notify people. I say, you know, we can't do this, we're going to do something different. I'm not misleading people. If I publish something and it's in the literature, it's for real." Leonard wanted to know what made Robbins think Howell didn't believe "this is for real." "The data that is presented in my report," Robbins replied, "clearly shows that those data are impossible to generate unless they're fabricated."

Asked if he had anything further to offer regarding his report for the court, Robbins had much to say. "I think I'd just like to add [...], the reason for my opinion and my support [to Hill's case] is essentially for concern over, to my mind, being faced with a situation in which, if I have something that is purported to be reality, has been published, and I'm presented with over twenty experiments that failed to reproduce that, that even I am doing and I can't reproduce that, that to me is a big red flag. I have to do something about that," he reiterated, "I have to report that, report that to the scientific community, report that to the funding body. And it is something that, you know, to report to the funding body, to my mind, is no big deal. If you tell NIH up front there's no problem, you know, I think they will help you deal with it. Part of the process," he explained, "for them and for us is really to help the principal investigators achieve success. Nobody's out to nail anybody. But the failure to do that, the failure to recognize what strongly I believe is fraud, really concerns me. And that's why I'm here."

Chapter 11 –
Court v. Science

O nce discovery was over, counsel for Helene Hill and the defendants had the option to file a motion to dismiss, also known as a dispositive motion. Both sides chose to do this before the deadline of May 25, 2010. On Hill's behalf, lawyer Sheldon Pincus filed seven documents: the notice for summary judgment, dated April 15, 2010; Hill's statement of undisputed material facts of the case, certifications provided to opposing counsel and the court, Hill's brief (basically the evidence and other cases cited), the order to grant the summary judgment, Pincus' proof of service and cover letter, all filed on May 25, 2010, to federal district court judge Dennis M. Cavanaugh. The defense's undisputed facts brief was also filed with the court on May 25. Each side then had the opportunity to respond once again. Pincus filed a reply brief before the June 21, 2010 deadline set by the court and the defense filed their reply brief in support of the defendants' motion for summary judgment after the deadline, which should have given Hill and Pincus a win but that's not what happened. In his reply brief, Pincus focused on the strength of the science and the experts' opinions and, as Hill opined, "he would have done well in a court of science."

The defendants' opposition to the plaintiff's motion for summary judgment failed to comply with Local Rule 56.1(a). The defense, as Hill noted later, and her counsel wrote in his reply brief, failed to file with the clerk a responsive statement to the plaintiff's statement of undisputed material facts in a timely manner. As a consequence, the plaintiff's statement of undisputed material facts was thus deemed to be undisputed for the purposes of Pincus' summary judgment motion on Hill's behalf. The filing of the statement, as a matter of law, is required at the time opposition papers are due to be filed with the clerk per Local Rule 56.1(a) and Federal Rules of Civil Procedure, cited as Fed R. Civ. P. 5(d)(2) and (3). The deadline date for the parties' opposition papers in Hill's case was midnight, June 21, 2010. The defense's 5:33 in the afternoon filing on that date (ECF Document Number 50) did not include a responsive statement. The defense's responsive statement wasn't filed until 9:53 the following morning, June 22, as ECF Document Number 51. The statement was untimely filed and its acceptance worked in prejudice to the plaintiff because the defense sought to dispute facts that the rule deemed to be undisputed (since the defense didn't file a dispute on time with the court).

The defense took a position in their reply brief that Hill described as sarcastic and dismissive. "They claimed I had a personal vendetta against the defendants," she recalled later. "This was a scientific disagreement that only dealt with one small part of the NIH grant, a dispute regarding methodology and interpretation of data. They emphasized that Bishayee had denied falsifying the data (what did they expect?), in contrast to [Eric T.] Poehlman, [Ph.D.], who falsified data on several [NIH] grant applications and ultimately admitted it. As for Dr. Joel H. Pitt's analyses, they brush these off as 'certain statistical anomalies.' They claimed," she continued, "that I failed to prove that the defendants acted 'knowingly, recklessly or with deliberate ignorance.' I failed to demonstrate that the data was false, or that, if it were false, it would have made a difference in the funding decisions. They emphasized that the CCRI on two occasions unanimously concluded there was insufficient credible evidence, that the ORI supported this decision and that the United States Attorney declined to pursue the case, especially after I made my PowerPoint presentation purporting to present new evidence. They referred to the 'questionable and self-serving reports of [my] experts, that Howell's only knowing was based on my self-serving suspicions. Furthermore," she concluded, "[they wrote that] I had failed to file a formal complaint in 1999 after my first observations. Lastly, I had failed to demonstrate that the data I claimed was false was material to the funding decision."

The defense was playing a calculated game of Russian roulette raising the Poehlman case. Like too many of a lengthy and growing roster of scientific misconduct cases, the years that elapsed between Poehlman's first transgressions and comprehensive criminal, civil and administrative settlement for falsifying and fabricating research data in numerous federal grant applications and in academic articles from 1992 to 2002, took many years of fits and starts to ultimately force an admission to his misconduct and crimes in a federal court. Poehlman, a former tenured research professor at the University of Vermont College of Medicine in Burlington, Vermont, would make numerous denials before an admission of guilt. The National Institutes of Health funds meritorious biomedical research with, according to its spotlight piece on Poehlman's misconduct, "the full expectation that its funded researchers will meet the highest standards of research integrity. NIH considers the scientific misconduct admitted by Dr. Poehlman deplorable and an affront to the American people and the vast community of scientists who conduct their research in an honorable and ethical manner."[171] Poehlman's work was supported by NIH grants from the National Institute of Aging (NIA), the National Institute of Diabetes and Digestive and Kidney Diseases (NIDDK), and the National Center for Research Resources (NCRR). What Poehlman had done proved one of those cases that shared a number of similarities to Hill's at UMDNJ.

A March 17, 2005 press release[172] promulgated by the Department of Health and Human Services' (HHS) Office of Research Integrity (ORI) laid out all that Poehlman had done but it took the concerted effort of the United States Attorney's Office for the District of Vermont, the HHS Office of Inspector General (OIG) and ORI to end Poehlman's research misconduct and, in truth, his career. According to court documents filed on the date of the press release, Poehlman agreed to plead guilty to making material false statements in a research grant application in April 1999, upon which the National Institutes of Health (NIH) paid $542,000 for Poehlman's research activities. In addition, Poehlman agreed to pay $180,000 to settle a civil complaint related to numerous false grant applications he filed while at the University of Vermont College of Medicine. Further, he was required to pay $16,000 in attorney's fees to counsel for Walter F. DeNino, a research assistant whose complaint of scientific misconduct spurred an investigation by the university. Also, Poehlman agreed to be barred for life from seeking or receiving funding from any federal agency in the future, including all components of the Public Health Service, and to submit numerous letters of retraction and correction to scientific journals related to his scientific misconduct. Poehlman also agreed to be permanently excluded from participation in all federal health care programs. In these agreements, Poehlman admitted that he acted alone in falsifying and fabricating research data and filing false grant applications.

"Preserving the integrity of the grant process administered by the Public Health Service is a priority for the Department of Justice," said United States Attorney David V. Kirby. "This prosecution demonstrates that academic researchers will be held fully accountable for fraud and scientific misconduct. Dr. Poehlman fraudulently diverted millions of dollars from the Public Health Service to support his research projects. This in turn siphoned millions of dollars from the pool of resources available for valid scientific research proposals. As this prosecution proves, such conduct will not be tolerated."[173]

Acting assistant secretary for Health, Cristina V. Beato, M.D., acknowledged just then the invaluable assistance of the Department of Justice in bringing the Poehlman case to a conclusion and upholding the high standards for research integrity in research supported by the Public Health Service. HHS actions against Poehlman included a lifetime debarment from receiving Public Health Service research funds and an agreement to retract or correct ten scientific articles due to research misconduct. Dr. Beato added that "while criminal charges against research scientists are rare, the egregiousness of Dr. Poehlman's conduct in this case fully supports the actions of the U.S. Attorney's Office and the administrative actions taken by HHS."[174] Through ORI, HHS is authorized to investigate and oversee institutional investigations of allegations of research misconduct in order to protect the integrity of Public Health Service funded research.

Poehlman was subsequently arraigned and pleaded guilty to the criminal charge filed also on March 17, 2005, in federal district court for the District of Vermont. Poehlman was sentenced to a year and a day in federal prison on June 28, 2006, for making a false statement on a federal grant application in 1999. The action by U.S. District Judge William Sessions in Burlington, Vermont, ended the most extensive case of scientific misconduct in the history of the National Institutes of Health (NIH). Experts say the case marked the first time an American scientist served jail time for research misconduct not linked to fatalities.[175] The

Poehlman case was just the beginning of what would be several like it to follow. So what had he done?

From 1987 to 2001, Poehlman held various research positions as an assistant, associate, and full professor of medicine at the University of Vermont College of Medicine in Burlington, Vermont (1987-1993 and again 1996-2001), and the University of Maryland in Baltimore, Maryland (1993-1996). In these academic positions, Poehlman conducted research on human subjects related to exercise physiology and other topics that was funded primarily by grants from federal public health agencies and departments, including the NIH, the United States Department of Agriculture (USDA), and the Department of Defense.

From in or about 1992 to 2000, Poehlman submitted seventeen research grant applications to federal agencies or departments that included false and fabricated research data. In these grant applications, Poehlman requested approximately $11.6 million in federal research funding. In most cases, he falsified and fabricated research data in the "preliminary studies" sections of grant applications in order to support the scientific basis for and his expertise in conducting the proposed research and in this respect, he was on par with what Hill had accused the defendants in her case of doing at UMDNJ. Reviewers of Poehlman's grant applications relied on the accuracy of the "preliminary studies" to determine if a grant should be recommended for award. While many of the grant applications were not awarded, NIH and USDA expended approximately $2.9 million in research funding based on grant applications with false and fabricated research data.

Poehlman falsified and fabricated research data in grant applications and research papers related to several topics including his study of the impact of the menopause transition on women's metabolism (referenced as "the Longitudinal Menopause Study"), his study of the impact of aging in older men and women on a wide range of physical and metabolic measures ("the Longitudinal Study of Aging"), and his proposal to study the impact of hormone replacement therapy (HRT) on obesity in post-menopausal women ("the Prospective HRT Study"). Poehlman also presented falsified and fabricated data in grant applications and academic papers related to his study of metabolism in Alzheimer's patients and the effect of endurance training on metabolism.

In the Longitudinal Menopause Study, beginning in 1994, Poehlman began presenting research data to academic colleagues that was purportedly from a study of women who were tested for basic metabolic characteristics before and after their transition through menopause. Poehlman presented the findings of the purported Longitudinal Menopause Study in an academic paper entitled "Changes in Energy Balance and Body Composition at Menopause: A Controlled Longitudinal Study," which was published by the *Annals of Internal Medicine* in 1995 ("the 1995 *Annals* article"). The purported Longitudinal Menopause Study and the 1995 *Annals* article were based almost entirely on falsified and fabricated research data. Poehlman represented in the 1995 *Annals* article that he had tested thirty-five healthy women for basic metabolic characteristics and retested the same women six years later for the same characteristics; in fact, Poehlman falsified and fabricated test results for all but three of the thirty-five women in the study. From 1995 to 2000, Poehlman used the false and fabricated research data from the Longitudinal Menopause Study and the *Annals* article in nine NIH grant applications, two USDA grant applications, and an additional six published academic papers, which were often cited in the grant applications.

In 1996, Poehlman initiated a new project where he planned to recruit subjects who had previously been tested at his University of Vermont medical school from 1987 to 1993 and retest them for the same and additional physical and metabolic characteristics over time. Beginning in 1999, Poehlman began presenting falsified and fabricated data from this Longitudinal Study on Aging to seek additional federally funded grants for his research projects. In these grant applications, Poehlman exaggerated the number of subjects tested and changed the values for the physical characteristics of the subjects and the test results for these subjects (often reversing the values from the initial test and the retest) in order to create trends during the aging process that were not reflected in the actual research data. In 1999 and 2000, Poehlman presented the same or similar false research data in three grant applications to the NIH and to the USDA.

Beginning in 1996, Poehlman was a co-investigator on a study at University of Vermont College of Medicine designed to evaluate the effect of hormone replacement therapy (HRT) on twenty post-menopausal women over several years' time. Among other things, the study had double-blind controls so that the researchers and the patients were unaware of any test results (or even who received HRT or the placebo) until the study was completed in 2002. In 1999 and 2000, Poehlman used data purportedly from the Prospective HRT Study in two grant applications to NIH seeking federal funding for additional HRT studies. But, in fact, Poehlman did not have access to the data and just fabricated the preliminary test results in the grant applications.

Beginning in December 2000 and continuing until April 2002, the University of Vermont College of Medicine conducted an investigation of scientific misconduct by Poehlman while at their medical school. The investigation began in response to allegations of scientific misconduct by Walter F. DeNino, one of Poehlman's research assistants. During the course of this investigation, Poehlman destroyed electronic evidence of his falsifications and fabrications, presented false testimony, presented falsified documents, and influenced other witnesses to provide false documents to the investigating authorities, and herein lies defense counsel in Hill's case misstep in introducing the Poehlman case. In Hill's case, Howell held on to the lab notebooks and other materials in the case, per discovery in her *qui tam*, in his university office. The Coulter counter Bishayee purportedly used was no longer in the lab by the time Hill could bring her case to court, and thus it was no longer available for inspection.

Back to Poehlman. In September 2001, he resigned from University of Vermont College of Medicine and moved to Montreal, Canada, to work as an academic researcher. Following the university's investigation, the matter was referred to the United States Attorney's Office and the ORI within the Public Health Service. Scientists at ORI working with investigators at OIG-HHS and the United States Attorney's Office then conducted a far-reaching investigation into all of Poehlman's grant applications and scientific publications. As a result of this investigation, Poehlman agreed to enter into the comprehensive criminal, civil, and administrative settlement filed in federal district court on March 17, 2005. In concluding its case against Poehlman, the United States Attorney's Office also acknowledged the important role that individual scientists have in identifying and responding to research misconduct. "ORI depends largely on the assistance of honest research scientists in the lab in discovering and reporting suspected misconduct, as occurred in this case," read the federal prosecutor's press release. "Without their assistance, ORI and HHS would have great difficulty in taking appropriate actions to protect the public health."[176]

Cavanaugh filed his decision on October 18, 2010, and in doing so wrote in his opinion that there had been no violation of the False Claims Act. The Act has three parts, the first of which states that the government must have been defrauded (money had to be involved and in Howell's case, there was). The fraud clause was satisfied in that a federal grant was issued based on the grant application filed for the November 1, 1999 deadline. The Act's second requirement is that the defrauder had to have known that he or she was defrauding the government. The law uses quaint language that the defrauder had to be scienter, a legal term that refers to intent or knowledge of wrongdoing. This means that an offending party has knowledge of the "wrongness" of an act or event prior to committing it. Lastly, the False Claims Act requires that the fraudulent action had to be material.

The judge ruled that Howell was not scienter because he did not know on the date that he filed the grant that the government was being defrauded. Since the expert reports came along after the date of filing, they were immaterial because Howell did not know about them at the time he submitted his materials to the NIH. The judge did not consider that the renewal of Howell's grant had any role as a new application, even though Howell might have been knowingly at the time of the renewal, that did not matter. By legal definition and according to provisions in the Model Penal Code, an individual is deemed to have acted knowingly in regard to a material element of an offense when: in the event that such element involves the nature of his or her conduct or the circumstances attendant thereto, he or she is aware that the conduct is of such nature or that those circumstances exist; if the element relates to a result of the person's conduct, he or she is conscious of the fact that it is substantially certain that the conduct will precipitate such a result. When the term knowingly is used in an indictment, it signifies that the defendant knew what he or she was going to do and, subject to such knowledge, engaged in the act for which he or she was charged. But Cavanaugh brushed aside what Howell knew at the time that he filed the grant: that Hill had reported to Howell her suspicion that the dishes material to the data in figure 7 of the grant application were devoid of colonies and that Hill's results performing the same protocol did not agree with those of Bishayee.

Cavanaugh concluded, in basic terms, that failure to replicate didn't imply any wrongdoing had taken place. He hit on Hill's lack of absolute certainty about what she'd seen regarding the lack of colonies on the dishes she examined. He called Hill's battle "a quest of Quixotic proportions that ultimately must be put to rest," but that she had not acted out of malicious intent in filing the *qui tam* in federal court. This removed the possibility of the defendants coming back at Hill and suing her for frivolous action. When it came to Bishayee, the judge ruled that he was neither scienter nor material, the latter in spite of the fact that Bishayee had produced the majority of all the preliminary data that went into the grant application. The judge exonerated Howell for not informing the NIH of his inability to replicate Bishayee's results, reasoning that even if the NIH had known that the reported results had been falsified, it would have made no difference regarding their granting decision.

After Cavanaugh rendered his opinion that there had been no violation of the False Claims Act, Hill decided to go back to Newark CCRI and the ORI with the new information she had gleaned from discovery. Perhaps there was no violation of the False Claims Act, but she believed just then that with all the new information in her possession, she might be able to get a scientific misconduct ruling from the university and ORI, and better still, get the eight papers coauthored by Bishayee retracted.

Hill sent identical cover letters and information to the Newark CCRI and the ORI. In the letters, dated October 29, 2010, she emphasized that the enclosed material was new, distinct from documents either organization had previously examined. To make it easy, she included a flash drive that contained the .pdf files she'd accessed via discovery; an Excel spreadsheet documenting the Coulter counter terminal digit data; copies of the eight papers Bishayee coauthored with Howell (and others); expert reports promulgated by Joel Pitt, Michael Robbins, including Robbins' figures; the memorandum from Howell to Stephen Baker; notes that Karen Putterman kept on the case that demonstrated that Howell had concealed from her the fact that Lenarczyk had been unable to replicate Bishayee's experiments; the summary of experiments document written by Howell in 2004 acknowledging ten unsuccessful attempts to repeat Bishayee's results; a preprint from Hill summarizing the statistical and radiobiological findings that are covered in the two expert reports, and finally copies of pages from Howell's grant renewal showing data that could not be replicated and noting that he cited at least twenty times in two *Radiation Research* papers that contained the unsubstantiated data.

Hill had no success with the ORI. Donald Wright, M.D., M.P.H., the acting director of the ORI, wrote Hill on December 2, 2010, in simplest terms telling her she'd brought them nothing new:

> The Office of Research Integrity (ORI) has received your letter of October 29, 2010, along with additional documentation of your concerns regarding possible scientific and research misconduct on the part of Drs. Anupam Bishayee and Roger W. Howell. Some of these concerns first came to ORI's attention in 2000-2001, when it was determined that there was insufficient evidence for ORI to pursue findings of misconduct. You subsequently filed a *qui tam* suit which, on October 18, 2010, was dismissed by Judge Cavanaugh.

> ORI has reviewed Judge Cavanaugh's Opinion, and notes that he has ruled on all of the issues you are asking this office to reconsider. Your request to ORI does not present substantive new information concerning Dr. Bishayee's research, or Dr. Howell's reporting of it, and no party has ever concurred with you that the evidence of misconduct is sufficient to make findings. For all these reasons, DIO declines to pursue this matter further.

The ORI's turn down hadn't considered the fact that Cavanaugh had limited information considered by the court to November 1, 1999, the submission date of Howell's grant application, a maneuver that thus allowed the judge (and ORI) to disregard information obtained during discovery. Cavanaugh, in his opinion, barely acknowledged expert reports. From the opinion, Cavanaugh wrote that "the fact that the results could not be replicated is proof only that the results could not be replicated" and the statistical analysis done Joel Pitt had been done after the ORI investigation (though it was performed legitimately as part of discovery for Hill's case in Cavanaugh's court), and therefore could not have contributed to defendant Howell's knowledge as of October 1999. Further, although "it was bio-chemically and radiobiologically impossible for the outcome of Bishayee's 100 percent experiments to occur" these "were likewise not known to Dr. Howell when he applied for the NIH grant," Cavanaugh determined.

Cavanaugh's opinion stood to have enormous implications pertaining to interpretation of the False Claims Act. For example, suppose a contractor builds a bridge for the federal government that subsequently collapses, kill-ing several people and it becomes known that the contractor had at some point in time become aware that a subcontractor had excessively watered down the concrete, perhaps the number one reason that greatly reduces the strength of concrete on a job site. Cavanaugh's decision would absolve both the contractor and the subcontractor from responsibility because the contractor was not scienter (did not know about the excessively watered down concrete when the bridge was constructed) and the subcontractor was neither scienter (did not know he was cheating the government), nor material (did not sign the original contract). Neither one would be held responsible for violating the False Claims Act.

Rebuffed again by the ORI, Hill met on January 7, 2011, for several hours with Drs. Pranela Rameshwar and Vivien Bellofatto, then serving as members of the Newark CCRI. Both were full professors, Rameshwar in the department of Medicine and Bellofatto, in the department of micro-biology, biochemistry and molecular genetics. "I thought at last I would be getting a fair evaluation of my findings," Hill opined. "How wrong I was." They questioned Hill at length about her charges and the findings, and while it seemed to her that they were coming around to supporting the findings she'd put in front of them, something wasn't quite right.

The session with Rameshwar and Bellofatto was recorded and, on the tape, Rameshwar closed the meeting noting that there was a significant discrepancy and in summary that they believed what Hill presented warranted further investigation, "a full investigation." Nearly two months later, on March 1, Rameshwar called to tell Hill that the CCRI had, again, decided not to go forward with the investigation. "Rameshwar had turned 180 degrees," Hill recalled later. "Even though I had new information, how could it be used to show fraud? They were concerned that there were no printouts of the Coulter raw data. [Rameshwar] said that the CCRI had met with the university lawyer after they had interviewed me. I expect that that was what turned the tide." The third Newark CCRI hadn't interviewed Howell or Hill this time, nor had the committee called in a statistician to provide an outside opinion of the data Hill provided. "I asked how she had voted, and she said they did not take a vote." The same day, Hill passed Leroy R. Sharer, M.D., a pathologist and member of the third Newark CCRI to rule on her case, in the hallway. Hill observed that his head was down, and he avoided eye contact. About a month later, Hill received an informal letter from Rameshwar, that read, in part, that on February 7, 2011, the committee met pursuant to the university policy on research misconduct number 001-01-05-10:15 section V.G.I. ("Policy"):

> The Committee reviewed the statements and supporting data you presented in support of your request. After the Committee's thorough review, it determined that, pursuant to the Policy, the allegation was not sufficiently credible and specific so that potential evidence of research misconduct may be identified. As a result, the Committee members unanimously decided that an Inquiry is not warranted.

Hill also had another option after Cavanaugh issued his opinion: she could appeal. Pincus filed prepared and filed the appeal in the United States Court of Appeals for the Third Circuit in Philadelphia on November 8, 2010. Even though she knew it was going to be an exercise in futility, Hill had to give it one last try. Documents submitted were repeated from what had been provided to the federal district court. The defendants relied, again, on the findings of the two Newark CCRIs and the ORI rulings. Defense counsel was afforded the opportunity to reply to Pincus's filing on Hill's behalf, and then Pincus would have to reply again to defense's filing. Everything boiled down to more sit-and-wait, and at this point it had been a decade's long back and forth for Hill. Oral arguments took place on September 13, 2011. Each side was given fifteen minutes, some of which could be reserved for rebuttal. The court asked the questions, and the lawyers responded. Presiding were appellate judges Dolores K. Sloviter, Anthony J. Scirica, and David Brooks Smith. Pincus went first and his fifteen minutes started on the wrong footing when Sloviter pronounced that she didn't see that any law had been broken. "We're just judges," Sloviter said, in open court, "I never had a science course in my life."

Much of the hearing discussion focused on whether Howell was scienter. Pincus tried to point out that Howell had concealed from the first CCRI his and Lenarczyk's inability to reproduce Bishayee's results. Scirica wrote the opinion brief for the appellate court affirming Cavanaugh's original ruling and filed it on October 20, 2011. The result was not unanticipated. "I was informed that at this point the only course of action would be to appeal to the United States Supreme Court, which would be a futile exercise. As I saw it," Hill wrote later, "my only course of action would be to publish the statistical analysis of the raw data that we had amassed, including the arguments made in Robbins' report that Bishayee's tritium experimental results were impossible because tritiated thymidine blocks the cell cycle and no deoxycytidine was present during the experiments to abrogate this effect," she continued. "I thought this next phase was going to be easy."

Hill and Joel Pitt, who continued to collaborate after her case was over, summarized their experience from Hill's *qui tam* case in a poster presentation at the biennial meeting of the American Society for Photobiology held in Montreal, Canada, from June 23-27, 2012. In this presentation, they'd use the method published by the ORI's statistical forensics advisory to check rightmost digits for uniform distribution.[177] This is advisory is posted to the ORI's website, last updated on April 20, 2011:

> Numbers are often recorded beyond the repeatability of the experimental procedure. When counts or measurements are recorded to higher precision than can be repeated in replications of an experiment, the rightmost digits of the recorded numbers have little biological meaning. Consider a count of radioactivity for biological preparation, for example, 5179. In a recount of the sample, or in a replication

of the assay, it is unlikely that the rightmost digits will be the same. Thus, with three repetitions, 5179, 5118 and 5134 could be expected.

The rightmost digits of these three numbers differ. Thus xx 79 differs from xx 18, and, in turn, both differ from xx 34.

In large samples of numbers, such rightmost digits often occur with the same frequency, like lottery digits where each of the digits 0, 1, 2, . . ., 9 has the same expectation. Statistically speaking, rightmost digits are approximately uniformly distributed in many circumstances.

In one ORI case, the respondent's notebook contained fabricated counts as well as un-fabricated counts. For the fabricated counts the radioactive spots on the experimental sheets had not been excised and hence could not have been counted in the scintillation counter. The unfabricated counts were supported by counter tapes.

Investigators from ORI's Division of Investigative Oversight (DIO) compared rightmost digits of fabricated and un-fabricated counts. The fabricated digits differed significantly from uniform. The unfabricated digits did not so differ. (The respondent accepted voluntary exclusion from receiving federal funds for three years.)

In another case, one column of a published table of numbers was not supported by notebook data. DIO investigators found that the rightmost digits of the un-supported column differed significantly from uniform. The rightmost digits of the supported columns did not so differ. (The paper was retracted, and in a related Department of Justice settlement, the government recovered over $1 million from two universities.)

To succeed in fabricating data, the fabricator must make the leftmost digits exhibit the desired biological magnitudes. Rightmost digits, given little thought, may be subject to personal preferences of the moment, and hence not uniform. Even when instructed to "make up" numbers with uniform digits, many subjects appear unable to do so. (See Mosimann, James E., Ph.D., Claire V. Wiseman, Ph.D., M.A., and R.E. Edelman, Ph.D., "Data Fabrication: Can people generate Random Digits?" *Accountability in Research*, Volume 4, 1995, pages 31-55).

In cases of scientific misconduct, unscientific details, like rightmost digits, are worthy of attention.

While members at the photobiology conference expressed interest in Hill and Pitt's poster, the reaction wasn't universal. Hill submitted an abstract for a similar poster to be presented at the annual meeting of the Radiation Research Society, to be held in San Juan, Puerto Rico, from September 30 to October 3, 2012, it was rejected. "I believe it is very rare for a presentation by a member in good standing to be turned down," Hill lamented. "Curious."

The scientific atmosphere over the past quarter century has undergone a paradigm shift. Over the centuries there have been a number of cases of fraud but until recently, these have been considered to be rare occurrences. Since the early 1980s, the scientific community has become more aware of instances of scientific misconduct, and articles and advisories have appeared in popular newsstand magazines and newspapers, scientific journals, online advisories and retraction sites like *Retraction Watch*, and NIH and ORI government websites. While retractions once seemed a rare occurrence, by the time Hill concluded her *qui tam* case, they had become almost commonplace, expectant and unwelcome reminders that the line between fabrication, fraud and misconduct and scientific breakthroughs is fragile. The increase in retractions doesn't indicate that all are the result of fraud or gross misconduct, but the truth is, many of them are. What divides the scientific community is what to do about it when it happens.

With Pitt, Hill wrote a paper they titled, "Statistical detection of potentially fabricated numerical data: a case study," which both believed would be welcomed as an addition to growing list of misconduct revelations that had now enveloped the scientific community. She was wrong. The paper was rejected by nine journals, only three of which made it beyond the editorial office. In her first submission to *Science* on June 15, 2011, Hill made all parties in her case anonymous; it was rejected on July 6, 2011. In the second submission to *Nature* on July 13, Hill received a quick rejection from the editor the following day. Setting her sights on *PLoS ONE*, Next, I set my sights on *PLoS ONE*, Hill decided to get the advice of an attorney that specialized in defamation law. The attorneys pointed out that the case was obviously transparent and that there had been no point to making all the parties anonymous. They advised presenting the facts and letting the readers draw their own conclusions. She rewrote the paper on their advice, sought editorial and statistical critiques and submitted the paper to *PLoS ONE*, which received it on January 22, 2012. The editorial office rejected it on February 10. Though she appealed the decision, she was rejected again.

Hill kept trying to publish the article, this time in the journal *Accountability in Research*. By its title, Hill's article – and her case study – would have seemed a perfect marriage. She sent it directly to one of the editors with whom she'd corresponded during her search for an expert in ethics. Again, the article was rejected.

Since four of the eight papers coauthored by Bishayee and Howell had been published in *Radiation Research*, Hill decided to apply to the editor, also mentioning that she wanted her name removed from the paper she'd coauthored with Bishayee et al. In his rejection letter, the editor stated that Hill would be allowed to withdraw her name as an author on one of the papers, but that Howell would have to be afforded the opportunity to respond to the request. Hill had no objection.

Hill's next stop was the *Annals of Applied Statistics*. She submitted the paper on March 13, 2012; it was rejected three days later. This was the journal that published the statistical analysis of chemotherapy studies at Duke University that led to multiple retractions, several resignations and a *60 Minutes* segment titled "Deception at Duke" that aired on March 5, 2012, right in the middle of Hill's many attempts to place her paper.

The *Annals'* editor provided Hill with constructive commentary, enough to encourage Hill and Pitt to revamp the statistics and develop a better method to analyze Bishayee's findings. Hill resubmitted the revised paper to *Annals* and was rejected again, this time with permission to resubmit (a second time). "We conscientiously rewrote, resubmitted and were re-rejected," Hill opined. "I appealed to the editor-in-chief who politely declined to reconsider."

Bruce Ames, famous for the Ames test for mutagenesis/carcinogenesis, coauthored[178] a series of exchanges in the Life Sciences Forum of the *FASEB Journal*, the official publication of the Federation of American Societies of Experimental Biology, in 2009 in which they questioned the unusual clustering of coefficients of variation in published articles from a medical biochemistry department in India. The authors of the questionable data were given the opportunity to respond in the forum. This, to Hill, seemed fair. "We would certainly be happy to have Bishayee and Howell respond to anything that we might publish," she observed. Hill submitted the required initial inquiry documents – an abstract and cover letter – on December 17, 2012. Within hours, Hill and Pitt's paper was rejected. They moved on, submitting to *Statistics in Medicine* on December 29, 2012; it was rejected on January 7, 2013. The rejection read, in part, "Although the problem you are working on is interesting, the degree of methodological innovation is not at the level of development that would receive a high enough priority for publication in *Statistics in Medicine.*" Since one of the papers Bishayee and Howell coauthored had been published in *Acta Oncologica* they decided to give that one a try. The paper, by now titled "Forensic statistics used to analyze preclinical survival data in nuclear medicine," was first rejected on March 6, 2013. Hill convinced the editor to reconsider. He did. But on further review and with extensive comment from reviewers, *Acta Oncologica* rejected the paper again on May 2, 2013.

While discouraged, Hill and Pitt didn't give up. They decided to contact the journal editors directly, starting with *Micron*. But the July 8, 2013, letter to *Micron* had some rather unexpected repercussions. Shortly after it was sent, Hill received a letter dated July 13, 2013, a mere five days after sending her query to *Micron*, from Rutgers New Jersey Medical School (formerly UMDNJ)[179] dean Robert L. Johnson, M.D., F.A.A.P., threatening her with dismissal if she did not "cease and desist." The letter read:

> The Dean's Office has received a complaint regarding your most recent actions directed at Dr. Roger Howell and Dr. Anupam Bishayee concerning their findings and publication of research data relative to bystander effects caused by non-uniform distributions of DNA-incorporated [note] I. Having reviewed the long history of your dispute with Drs. Howell and Bishayee regarding the efficacy of this data, our Office has found that your allegations involving research integrity against Drs. Howell and Bishayee were reviewed repeatedly, and dismissed uniformly as unfounded, by both scientific and non-scientific neutral bodies, including the Federal District Court of New Jersey and the Third Circuit Court of Appeals. A copy of the District Court's decision and the Third Circuit's affirmation are attached. [note]

A footnote on page one of Johnson's letter stated that his office's "review has revealed that you have brought this issue before the following bodies, and your allegations were unfounded by all of them." He listed the following: UMDNJ Campus Committee on Research Integrity [the Newark CCRI], NIH Office of Research Integrity, UMDNJ Campus Committee on Research Integrity (second time), U.S. Federal courts all the way to the Third Circuit of Appeals, NIH Office of Research Integrity (second time), UMDNJ Campus Committee on Research Integrity (third time), *Nature* (Journal), *PLoS ONE* (Journal), *Radiation Research* (Journal), *Annals of Applied Statistics* (Journal), *Accountability in Research* (Journal). The letter continued:

> As you are well aware, your participation and membership in an academic profession carries with it special responsibilities. To that end, your bargaining unit, the American Association of University Professors [AAUP], has issued many policy statements over the years to guide

its members. Chief among these is the AAUP's *Statement on Professional Ethics*. I quote from the *Statement* at Paragraph 3 [italics added by Johnson]:

As colleagues, professors have obligations that derive from common

> *membership in the community of scholars. Professors do not discriminate against or harass colleagues. They respect and defend free inquiry of associates. In the exchange of criticism and ideas, professors show due respect for the opinions of others. Professors acknowledge academic debt and strive to be objective in their professional judgment of colleagues.*

In the same vein, the AAUP's *Statement of the Association's Counsel: Freedom and Responsibility* contains the following language:

> *Membership in the academic community imposes on students, faculty members, administrators, and trustees an obligation to respect the dignity of others, to acknowledge their right to express differing opinions, and to foster and defend intellectual honesty, freedom of inquiry and instruction, and free expression on and off campus. The expression of dissent and the attempt to produce change, therefore, may not be carried out in ways that injure individuals or damage institutional facilities or disrupt the classes of one's teachers or colleagues.*

Your continued pursuit of this disputed issue for so many years and to so many different tribunals, every one of which has rejected your claims, compromises the precepts cited above, and is, moreover, inconsistent with the conduct expected of NJMS faculty and an AAUP member. Therefore, I am issuing to you this formal Letter of Reprimand. Moreover, because your actions towards your colleagues also represent conduct unbecoming a member of the faculty of the University you are directed that this harassing and disruptive behavior must now cease; and, you are advised that further incidents of this unprofessional and non-collegial conduct will subject you to additional disciplinary action, up to and including termination of your employment at Rutgers New Jersey Medical School.

Hill did not cease and desist. She responded to Johnson on August 14:

This is in response to your letter dated July 19, received by me on July 25. Pursuant to Article IV of the AAUP Agreement, I ask that this response be attached to and retained with your letter in my Permanent Personnel File. I respectfully disagree that my conduct has been unbecoming a member of the faculty of this university. My intent is not to be disruptive and harassing but to fulfill my obligation as a faculty member with a right to voice my academic views on the theories of other scientists. My goal is to protect the medical school and university.

You appear to threaten that my continuing to raise my academic views of research in my field will subject me to additional disciplinary action, up to and including termination. Please let me point out that Regulation 5(a) of the AAUP's *Recommended Institutional Regulations on Academic Freedom and Tenure* [Hill emphasis added] provides that "adequate cause for a dismissal will be related, directly and substantially, to the fitness of faculty members in their professional capacities as teachers and researchers. Dismissal will not be used to restrain faculty members in their exercise of academic freedom or other rights of American citizens." "**The burden of proof that adequate cause exists rests with the institution and** will be satisfied only by clear and convincing evidence in the record considered as a whole." [Hill emphasis added] Id. at 5(c)(8).

The AAUP's 1994 *Statement on the Relationship of Faculty Governance to Academic Freedom* directly addresses the issue of academic freedom in research and criticism: "The academic freedom of faculty members includes... the freedom to express their views on academic matters... in the conduct of research...on issues of public interest generally, and to do so even if their views are in conflict with one or another received wisdom...**good research requires permitting the expression of contrary views in order that the evidence for and against a hypothesis can be weighed responsibly.** [Hill emphasis added] In the case...of issues of public interest generally, the faculty member must be free to exercise the rights accorded to all citizens.

"Protecting academic freedom on campus requires ensuring that a particular instance of faculty speech will be subject to discipline only where that speech violates some central principle of academic morality, as, for example, where it is found to be fraudulent [Hill emphasis added] (academic freedom does not protect plagiarism and deceit)." At pages 141-143.

The AAUP's *Statement of Professional Ethics* (revised in 2009), states that "Professors...respect and defend the free inquiry or associates, **even when it leads to findings and conclusions that differ from their own.** [Hill emphasis added] Professors acknowledge academic debt and strive to be objective in their professional judgment of colleagues."

Regarding the bodies you mention in your footnote: the first and second Campus Committees on Research Integrity did not have access to nor review the data we analyze in the Micron letter, data which fails multiple statistical tests of legitimacy. With respect to the third committee, in the recording of my meeting on January 7, 2011, with Drs. Pranela Rameshwar and Vivian Bellofatto, Dr. Rameshwar states "…there is a big discrepancy. In summary, we believe this warrants further investigation." This did not, however, occur and, for that matter, I never received an official final report from that committee so I must assume that an investigation according to the guidelines was never undertaken. The ORI refused to consider the new information that I sent them, saying incorrectly, that they had already dealt with the material.

I point out that the False Claims Act requires that the respondent knowingly defrauded the government and that Judge Cavanaugh drew the line at October 1999 [Bishayee's first experiment, included in Howell's NIH grant application submitted that November], when the grant was submitted, thereby bypassing any possibility of "knowing" after that date. He does not understand that for scientists "Replication – the confirmation of results and conclusions from one study obtained independently in another – is considered the scientific gold standard" (Jasny, B. R., et al. *Science* 334:1225, 2001). Judge Cavanaugh is clearly not a scientist: "the fact that results [of tritiated thymidine experiments] could not be replicated is proof only that the results could not be replicated." He goes on to say "the statistical analysis done by Plaintiff's expert, Dr. Joel Pitt, was done subsequent to the ORI investigation, and therefore could not have contributed to Defendant Howell's knowledge as of October, 1999." True enough, but all the more important that it should be considered now. The appeals court affirmed Cavanaugh's decision. However, in initiating the oral arguments on September 13, 2011, Judge Dolores Sloviter said, "We are just judges…I never had a science course in my life." Unfortunately, such ignorance does not preclude judiciary from venturing into the scientific arena.

As regards the journals that you mention in your letter, our submissions never made it beyond the editorial office (no scientific review) in all but the Annals of Applied Statistics, which decided that the paper would not be of interest to its readers. There are many other statistical journals out there and it is an appropriate academic activity for me to continue to seek publication of my work with Dr. Pitt.

I seek to make the system fair for all concerned and I seek to participate in academic discussion of research in my field without threat to my position. I hope that the information above will allow you to reconsider and review the views stated in your letter.

Hill had no intention of backing off.

On October 10, 2013, she and Pitt submitted a reedited, revamped version of their paper to *The American Statistician*. After four months of waiting, Hill finally corresponded with the secretary who informed her that the paper was back from the referees and the outcome looked favorable. Then more time passed, and Hill queried the editor-in-chief, Ronald Christensen. Two referees and one associate editor, from their comments, wanted to publish the paper. Reviewer 1 wrote: "This is an interesting paper and overall, it shows an interesting approach, particularly to analysis of triplicate counts." Reviewer 2 offered this:

This is a well-written paper that addresses an important problem: the apparently increasing prevalence of falsified or fabricated data in scientific publications. The authors present a convincing case study that the data from one researcher in a laboratory appear to have been fabricated. Their novel methods of testing (1) whether the rounded mean is included among triplicate values or (2) whether the mid-ratio of triplicates lies in a small interval around 0.5, have wide potential applicability even beyond the colony assays and Coulter counts that they present, since many biological experiments use triplicates for their basic measurements.

The journal's associate editor commented: "This is a well-written paper on an interesting and particularly relevant topic. There have been news stories recently on the public's lack of trust in scientists, and I can't imagine that some of the highly visible reports on data fraud have helped this perception."

Enter Charles W. McCutchen, Ph.D., who wrote to Hill on December 6, 2013, having seen an article about her in *Nature*. "When the UMDNJ decided it was less embarrassing to deny that Bishayee had done fraud than to punish him for it, you were supposed to shut up. People who challenge those in power must be punished. Otherwise," he wrote, "who knows what might happen?" He also told her that "the punishment [of Hill] must continue as long as you live because cessation might suggest that you should not have been punished in the first place." McCutchen doubted she'd get many opportunities to rehash her case. "The honor, ho-ho-ho, of the establishment is at stake, but you will keep fighting because you can." She was. "Sympathy from working scientists will be limited because many (most) have had to pretend that black is white, so they resent your refusal to do the same," he opined.

"Numbers that detect and prove fraud are hard to derail with farfetched explanations, so HHS' departmental appeals board in the Imanishi-Kari case, having heard a dishonest defense by Terence [P.] Speed, [Ph.D.], Imanishi-Kari's expert, 'toughed it out,' concluding that 'ORI's statistical analyses of the June subcloning data and other data are not evidence from which it is reasonable to infer fabrication here. In general, ORI's statistical analyses were flawed.' The science press," McCutchen wrote, "which has fully as much freedom as *Pravda* in the Stalin era, pretended to take that statement seriously. To whistleblowers it meant, 'That is how we treat evidence, so think what we could do to you." *Statistician* editor Christensen, Hill noted, didn't want to publish the paper so he sent it over to a reviewer, Terence Speed, who tried to destroy the work of NIH statistician James Mosimann, Ph.D., on the Imanishi-Kari case and whose work Hill and Pitt cited in their paper; Christensen turned down the paper on March 18, 2014. Less than a month later[180], McCutchen wrote Hill, "Why does the [bioscience] establishment want to render statistical evidence unacceptable? Because it can be incontrovertible and would limit the establishment's ability [to] decide who is guilty and who [is] innocent." With that said, he told Hill, "Dr. Christensen knows what is expected of his magazine" in turning down the article.

While waiting to hear from *The American Statistician*, Hill and Pitt decided to try another tack. Reading the instructions to authors of the BioMed Central Journals *BMC Medical Research Methodology*, it actually looked as though it might be possible to submit to them the same paper that had been submitted to *The American Statistician*, so they gave it a try, being absolutely open with them about what we were doing. This, too, was unsuccessful. But the editors of the publication did inform them in their rejection notice that if it did not work out with the other journal, if *The American Statistician* rejected it, their editorial office would like another opportunity to review it. Turned down again by the *Statistician*, Hill and Pitt submitted their paper to BioMed Central Journals, where it wasn't published either.

In January 2014, Hill learned that R. Grant Steen, Ph.D., well known for his interest in scientific retraction; research misconduct; medical misinformation; retraction as a proxy for misconduct; neuroepistemology, and cognitive biases associated with misinformation, was organizing an online special issue of the MDPI AG *Publications* entitled "Misconduct in Scientific Publishing." Addressing the special issue, Steen wrote that "scientists believe – or at least profess to believe – that science is a process of iterative approach to objective truth. Failed experiments," he continued are supposed to serve as fodder for successful experiments, so that clouded thinking can be clarified. Observations fundamentally true should find support, while observations flawed in some way are supplanted by better observations. Why then would anyone think that scientific fraud can succeed?"[181] Steen further observed of the recent sea change in the scientific community that "there has been an alarming increase in the number of papers in the refereed scientific literature that have been retracted for fraud." For example, Steen offered, data fabrication and data falsification. "Do fraudulent authors imagine that their fraud will not be exposed?" he wrote. "Do they see the benefits of fraud as so attractive that they are willing to risk exposure? Or do some scientists doubt the process itself, believing themselves to be immune to the failure to replicate?"

As an ethicist, Steen declared that while "it may be true that most scientists who fabricate or falsify data believe that they know the 'right' answer in advance of the data and that they will soon have the data necessary to support their favored answer. It may therefore seem legitimate to fabricate," he continued, and "such scientists may believe that they are simply saving time by cutting corners. They may even believe that they are serving science and the greater good by pushing a bold 'truth' into print. But humans," he observed, "are so prone to bias that the process of scientific discovery has been developed specifically to insulate scientists from the malign effects of wishful thinking. Measurement validation, hypothesis testing, random allocation, blinding of outcome assessment, replication of results, referee and peer review, and open sharing of trade secrets are keys to establishing the truth of a scientific idea. When those processes are subverted, scientific results become prone to retraction."

After reading Steen's introduction, Hill and Pitt composed a new article entitled "Failure to replicate: a sign of scientific misconduct," and submitted it to MDPI AG by the February 28 deadline. In the abstract to their submission, they wrote that repeated failures to replicate reported experimental results could indicate scientific misconduct or simply result from unintended error. Experiments performed by one individual involving tritiated thymidine, published in two papers in *Radiation Research*, showed exponential killing of V79 Chinese hamster cells. Two other members of the same laboratory were unable to replicate the published results in fifteen subsequent attempts to do so, finding, instead, at least 100-fold less killing and biphasic survival curves. These replication failures (which could have been anticipated based on earlier radiobiological literature) raise questions regarding the reliability of the two reports. Two unusual numerical patterns appear in the questioned individual's data, but do not appear in control data sets from the two other laboratory members, even though the two key protocols followed by all three were identical or nearly so. This report emphasizes the importance of 1) access to raw data that form the background of reports and grant applications; 2) knowledge of the literature in the field, and 3) the application of statistical methods to detect anomalous numerical behaviors in raw data. Furthermore, journals and granting agencies should require that authors report failures to reproduce their published results. The paper was subjected to six reviews before finally being accepted and published online on September 1, 2014. "So, at long last, we have published some of our analysis." After fourteen years of trying to publicly expose her case, Hill was making progress.

Chapter 12 –
The Dark Side of the Moon

"The troubling truth is that scientists who lie to their peers are rarely exposed," wrote Daniel Engber in a recent *Popular Science* article. A February 2015 investigation revealed that most scientific fraud uncovered by inspectors for the Food and Drug Administration (FDA) never gets identified as just that in the academic literature. Further, in the *PubMed* database of biomedical research, only one in 10,000 recent papers – or 0.01 percent – has been retracted – most often for fraud.[182] Engber found that other measures of misconduct hinted at much higher rates. Daniele Fanelli, Ph.D., a senior researcher at Stanford University who now studies the nature of science itself and the misconduct of scientists, collected data from eighteen surveys, only to find that nearly two percent of scientists admitted to fabricating or falsifying their work or manipulating data. When asked whether they'd ever seen misconduct among colleagues, fourteen percent admitted that they had. Comparing survey responses between decades to mark changes in how misconduct is reported (or not), he discovered that scientists have become less likely to admit misconduct but also no less likely to report the misbehavior of their colleagues.[183]

Engber offered that perhaps fraudsters today may be cagier to avoid getting caught, or misconduct has become subtler and harder to assess. To find out more about this "softer form" of fraud, Fanelli examined bias in the scientific literature. Published data is now more likely than ever, he concluded, to show positive results that support a scientist's hypothesis. So, what does that mean? Researchers may toss out data points that don't seem right or run experiments many times and publish only the best results. By now this observation should ring familiar. What's driving the trend? Patterns tell a story. Fanelli documented bias that varied between fields. For example, psychology and psychiatry demonstrated the most significant bias but astrophysics showed very little. Papers out of American laboratories tended to give stronger or more positive results than the same research queries from Europe or Canada, perhaps due to greater competition or pressure to publish, he found. Growing awareness of misconduct might also have made the rogue scientists savvier, wrote Engber of Fanelli's findings. "They now know where to stop before what they do is considered intentional fraud," Fanelli observed.[184]

Fanelli's research has been groundbreaking on the subject matter of scientific misconduct, calling into question the overworked image of scientists as objective seekers of truth that is periodically rocked by the discovery of major scientific fraud. In writing about his work in the May 29, 2009 *PLOS One*, he cited significant cases of misconduct, to include Hwang Woo-Suk[185]'s faked stem cell lines and Jan Hendrik Schön[186]'s fabricated and falsified data in seventeen published papers, both cases spotlighting just how easy it can be for a scientist to publish fabricated data in the most prestigious journals in the world, and how this can lead to a significant waste of money and human resources in addition to the risk to human health. "A popular view propagated by the media," he wrote, "and by many scientists sees fraudsters as just a 'few bad apples.' This pristine image of science is based on the theory that the scientific community is guided by norms including disinterestedness and organized skepticism, which are incompatible with misconduct." But increasing evidence suggests, he observed, that known frauds are just the "tip of the iceberg," and that many cases are never discovered. The debate within the scientific community has shifted to defining the forms, causes and frequency of scientific misconduct.[187]

So why did Helene Hill refuse to give up on the scientific misconduct case at UMDNJ? She would argue, as have many others, that the way to protect the integrity of science is to look beyond falsification, fabrication and plagiarism to a far wider range of questionable research practices, ones that continue to jeopardize, at an alarming rate, public support for and trust of what goes on in the lab once scientists have been painted with the broad brush of misconduct. "Serious misbehavior in research is important for many reasons, not least because it damages the reputation of, and undermines public support for, science. Historically," argued Brian C. Martinson, Ph.D., et al., professionals and the public have focused on headline-grabbing cases of scientific misconduct, but we believe that researchers can no longer afford to ignore a wider range of questionable behavior that threatens the integrity of science."[188]

Martinson, senior research investigator at HealthPartners Research Foundation, and colleagues Melissa S. Anderson, Ph.D., professor of higher education, with affiliate membership in bioethics, and associate dean of graduate education at the University of Minnesota, where her work over the past thirty years has been in the areas of scientific integrity, research collaboration, and academy-industry relations, with particular attention to the research environment, and Raymond de Vries, Ph.D., codirector of the Center for Bioethics and Social Sciences in Medicine at the University of Michigan and a professor in the department of learning health sciences and the department of obstetrics and gynecology there, surveyed several thousand early- and mid-career scientists in the United States and funded by the NIH, and asked them to report their own behaviors. Their findings revealed a broad range of questionable practices that were what they described as "striking in their breadth and prevalence." The 2002 survey that served as the basis for their study and cited here from commentary offered by the principles published in the June 9, 2005 *Nature*, was the first time such behaviors had been analyzed quantitatively, so it became the benchmark for all studies to follow.

As Fanelli discovered later, evidence in Martinson et al.'s research suggested that mundane "regular" misbehaviors present greater threats to the scientific enterprise than those caused by high-profile misconduct cases such as fraud.[189] They pointed to the negative aspects of the research environment that feed the misconduct, from intense competition, to difficult, often untenable regulatory, social and managerial demands imposed by federal, state and institutional mandates. This, they observed, creates abundant possibilities for the compromise of scientific integrity that goes well beyond fraud, fabrication and plagiarism.[190]

The Martinson et al. study was based on large, random samples of scientists drawn from two databases maintained by the NIH Office of

Extramural Research. The mid-career sample of 3,600 scientists received their first research project (R01) grant between 1999 and 2001. The early-career sample of 4,160 NIH-supported postdoctoral trainees received either individual (F32) or institutional (T32) postdoctoral training during 2000 or 2001. Of the 3,600 mid-career scientists, 3,409 were deliverable and 1,768 yielded usable data – a fifty-two percent response rate. Of the 4,160 surveys sent to early-career scientists, 3,475 were deliverable, yielding 1,479 usable responses, a response rate of forty-three percent. Study investigators acknowledged that misbehaving scientists were perhaps less likely than others to respond to the survey, largely due, quite possibly, to the belief they might be discovered and sanctioned. "This, combined with the fact that there is probably some underreporting of misbehaviors among respondents, would suggest that our estimates of misbehavior are conservative."[191]

Interestingly, the construct of Martinson et al. survey questions came from six focus-group discussions held with fifty-one scientists from several top-tier research universities who informed them which misbehaviors were of greatest concern on their campuses, all sanctionable conducts that stood – and still stand – to compromise the integrity of research. To further narrow the topics of their research, they took the conversation further, consulting six compliance officers at five major research universities and one independent research organization located in the United States. They asked the compliance officers to assess the likelihood that each behavior, if discovered, would get a scientist into trouble at the institutional or federal level. The top ten behaviors were the most serious, to include: falsifying or "cooking" research data; ignoring major aspects of human-subject requirements; not properly disclosing involvement in firms whose products are based on one's own research; relationships with students, research subjects or clients that may be interpreted as questionable; using another's ideas without obtaining permission or giving due credit; unauthorized use of confidential information in connection with one's own research; failing to present data that contradict one's own previous research; circumventing certain minor aspects of human-subject requirements; overlooking others' use of flawed data or questionable interpretation of data, and changing the design, methodology or results of a study in response to pressure from a funding source. All the officers, it was noted, judged the top ten behaviors as likely sanctionable, and at least four of the six officers determined them to be very likely sanctionable.

Alternative bad behaviors, documented in the Martinson et al. study include publishing the same data or results in two or more publications; inappropriately assigning authorship credit; withholding details of methodology or results in papers or proposals; using inadequate or inappropriate research designs; dropping observations or data points from analyses based on a gut feeling that they were inaccurate, and inadequate recordkeeping related to research projects. This was, by no means, an exhaustive list of those alternative bad behaviors. Fanelli, too, would pick up on these behaviors in his studies of misconduct.

The results of the Martinson et al. study were telling. Scientists were asked if they themselves had engaged in a specific misconduct. For six of the aforementioned top ten behaviors, under two percent reported that they'd participated in wrongdoing to include falsification and plagiarism; however, the frequencies for the remaining behaviors were five percent or above, most exceeding ten percent. Overall, they reported, thirty-three percent of the respondents stated that they had engaged in at least one of the top ten behaviors during the previous three years – 1999-2001 – the time frame that Hill observed and reported misconduct in Roger Howell's radiation research laboratory. Among the mid-career respondents, the data showed that thirty-eight percent had engaged in questionable behavior, and in the early-career group, it was twenty-eight percent. This, it was documented, proved a significant difference. For each behavior where mid- and early-career scientists' percentages differed significantly, the mid-career respondents' numbers were higher than their postdoctoral trainees. The reasons for this difference, they believed just then, was – and is – the underreporting of misbehaviors by those in relatively tenuous, early-career positions. Over half, fifty-one percent, of the mid-career respondents had positions at the associate professor level or above, whereas fifty-eight of the early-career participants were postdoctoral fellows. [192] Martinson et al.'s survey hinted, too, that much scientific misconduct is the result of frustrations and injustices built into the modern system of scientific rewards. The findings of this study would have profound implications for efforts to mitigate misconduct, demanding more focus on fixing systemic problems and less on identifying and weeding out individual scientists gone bad. "Science has changed a lot in terms of its competitiveness, the level of funding and the commercial pressures on scientists," Martinson observed. "We've turned science into a big business but failed to note that some of the rules of science don't fit well with that model."[193]

Since a single, universally accepted definition of research misconduct doesn't exist among various professional societies, scientific journals, government agencies and regulatory bodies concerned with the issue, it should be noted that fabrication, falsification and plagiarism are so egregious that all definitions implicitly and explicitly include these practices,[194] according to Stephen L. George, Ph.D., of the Duke University Department of Biostatistics and Bioinformatics. George cited the United States Public Health Service's definition of research misconduct specifically limited to these practices:

> Research misconduct means fabrication, falsification, or plagiarism in proposing, performing, or reviewing research, or in reporting research results.
>
> (a) Fabrication is making up data or results and recording or reporting them.
> (b) Falsification is manipulating research materials, equipment, or processes, or changing or omitting data or results such that the research is not accurately represented in the research record.
> (c) Plagiarism is the appropriation of another person's ideas, processes, results, or words without giving appropriate credit.
> (d) Research misconduct does not include honest error or differences of opinion.

George further noted that National Institutes of Health (NIH), National Science Foundation (NSF), American Psychological Association (APA), and other such science-based organizations use identical or nearly identical definitions of misconduct. "This is a very narrow definition, covering only the most serious unethical behaviors," he wrote. "For clinical investigators, the United States Food and Drug Administration uses a much broader definition of investigator misconduct, targeting practices that might create a safety risk for patients." The FDA definition reads:

> Failure to report serious or life-threatening adverse events; serious protocol violations, such as enrolling subjects who do not meet the entrance criteria because they have conditions that put them at increased risk from the investigational drug, or failing to carry out critical safety evaluations; repeated or deliberate failure to obtain adequate informed consent, including falsification of consent forms or repeated or deliberate failure to disclose serious risks of the investigational drug in the informed consent process; falsification of study safety data; failure to obtain IRB review and approval for significant protocol changes; failure to adequately supervise the clinical trial such that human subjects are or would be exposed to an unreasonable and significant risk of illness or injury.

Taking a broader perspective, the Council on Scientific Editors defined research misconduct as "behavior by a researcher, intentional or not, that falls short of good ethical and scientific standard."[195] The broader the definition, the greater the opportunity for researchers to come up with myriad reasons they've failed to uphold scientific and ethical standards in the conduct of their work and commit acts that fall between the cracks (other serious deviation), beyond the terms used throughout this narrative, to include fabrication, falsification and plagiarism – the big three. The findings of the Martinson et al. study had already laid bare the fact that American scientists engaged in a range of bad behaviors that extended far beyond fabrication, falsification and plagiarism, all of which stood to irreparably damage the integrity of science. "Attempts to foster integrity that focus only on fabrication, falsification and plagiarism therefore miss a great deal," Martinson et al. concluded. "We assume that our reliance on self-reports of behavior is likely to lead to underreporting and therefore to conservative estimates, despite assurances of anonymity [of the study]. With as many as thirty-three percent of our survey respondents admitting to one or more of the top ten behaviors, the scientific community can no longer remain complacent about such misbehavior."[196]

The PHS definition of scientific misconduct is tighter but if not enforced it means nothing. Even when the NSF and PHS reached a definition of misconduct, a debate ensued over the inclusion of the words "other serious deviation," opponents arguing it was far too broad a term to pin down those doing wrong and might well catch up those who were merely undertaking innovative, groundbreaking science.[197] Certainly, in Hill's case, the report published by federally appointed Commission on Research Integrity in 1995 would have been well known. In it, the commission concluded that "research misconduct is [a] serious violation of the fundamental principle that scientists be truthful and fair in the conduct of research and the dissemination of research results."[198] But further, the HHS-sponsored commission also identified other forms of misconduct, some of which clearly went to the crux of what Hill exposed at UMDNJ:

(a) Obstruction of Investigations of Research Misconduct: The Federal Government has an important interest in protecting the integrity of investigations into reported incidents of research misconduct. Accordingly, obstruction of investigations of research misconduct related to federal funding constitutes a form of professional misconduct in that it undermines the interests of the public, the scientific community, and the Federal Government. Obstruction of investigations of research misconduct consists of intentionally withholding or destroying evidence in violation of a duty to disclose or preserve; falsifying evidence; encouraging, soliciting or giving false testimony; and attempting to intimidate or retaliate against witnesses, potential witnesses, or potential leads to witnesses or evidence before, during, or after the commencement of any formal or informal proceeding.

(b) Noncompliance with Research Regulations: Responsible conduct in research includes compliance with applicable federal research regulations. Such regulations include (but are not limited to) those governing the use of biohazardous materials and human and animal subjects in research. Serious noncompliance with such regulations after notice of their existence undermines the interests of the public, the scientific community, and the Federal Government and constitutes another form of professional misconduct.

The misrepresentation of research results remains the most common form of scientific wrongdoing.

Broader definitions of misconduct, according to George's findings, are vaguer than the PHS definition, leaving open the question of the definition of what makes "good ethical and scientific standard" or "research and scholarship standards" and, in his observation, "what exactly constitutes falling short of those standards." Questionable research practices other than the aforementioned big three, George wrote, "nevertheless result in unreliable results and other serious problems." He listed them, to clarify, grouping the possible list of transgressions into several categories: design, conduct and analysis, to include the use of improper design or analysis techniques, selective reporting, data-dredging, study weakness (not described), misrepresentation of the methodology used; publication and authorship, for example, failure to publish, agreement not to publish, gift authorship and redundant publications; patient safety, to include failure to follow proper protocol safety requirements, failure to obtain proper informed consent, and failure to report adverse events; and other practices, such as misuse of funds, conflicts of interest or refusal to share data.

George characterized data fraud in biomedical experimentation as two ends of the spectrum, the extreme ends of it ranging from the inevitable honest errors to data fraud at the other with misunderstandings, incompetence and sloppiness somewhere in the middle. Should misconduct, by definition, include more of the spectrum, most especially when there is clear intent? "Other sources of data errors are regrettable," George noted, "but data fraud involves a deliberate intent to deceive or 'intent to cheat,' a qualitatively different source of data errors." He has argued that in aggregate more damage is caused by the less serious forms of questionable research practices and from sloppiness or incompetence than from data fraud – largely because these other sources of data errors are more common.

Janis Costello Ingham, then a professor of speech and hearing sciences at the University of California at Santa Barbara, and Jennifer Horner, an associate professor just then in the College of Health Professions, Medical University of South Carolina, in their ethics and research article in the March 2004 *ASHA Leader* describe various circumstances of misconduct, those that run the gamut of possible egregious acts. They describe a well-known researcher, highly successful in obtaining extramural research funding and having a long list of publications, who insisted that her name be included as an author on all publications emanating from her laboratory; a clinical scientist, testing the effects of an experimental treatment, who compared scores from a treated group of participants to scores from a group of individuals from whom treatment was withheld; a university professor serving as a reviewer of a manuscript submitted for publication in a research journal, who provided copies of the manuscript to graduate students so that they could prepare their own critiques, as an educational exercise in peer review; the ambitious graduate student, when analyzing data collected for his master's thesis, changed the numbers in two experimental conditions to make the results fit his hypothesis; a young assistant professor, eager to bolster her publication record in time for her tenure review, cut corners and included published findings from another scientist in a manuscript submitted for publication and reported them as her own, and a busy senior scientist, mentoring several doctoral students and postdoctoral fellows, who neglected to monitor their methods of data collection.[199] What do these cases have in common, they asked. "All of them raise questions about breaches of research ethics or of research misconduct," they wrote. Why should we be concerned about research misconduct, they postulated. "The consequences of violations of research ethics can be far-reaching," Ingham and Horner continued. "On moral grounds, society expects individuals to lead their lives with honesty and the utmost respect for the well-being of others. These expectations are highest for those in positions of trust, who are responsible for the health, safety and welfare of others – such as religious leaders, educators, and government officials. Ethical transgressions within these groups are considered especially abhorrent – and so it is with scientists, whose life's work is fundamentally about seeking truth (fact) and developing an understanding of natural and human phenomena." Certainly, disregard for intellectual honesty, they would agree with Fanelli, Martinson et al. and many others, jeopardizes the credibility of science.

Richard Smith, former editor of the *BMJ* (formerly the British Medical Journal) for thirteen years, who has written extensively about research misconduct, questioned:

> Why does research misconduct happen? The answer that researchers love is 'pressure to publish,' but my preferred answer is 'Why wouldn't it happen?' All human activity is associated with misconduct. Indeed, misconduct may be easier for scientists because the system operates on trust. Plus scientists may have been victims of their own rhetoric: they have fooled themselves that science is a wholly objective enterprise unsullied by the usual human subjectivity and imperfections. It is not. It is a human activity.[200]

Smith, among many others who have spent years studying scientific misconduct, considers it inevitable but not to read "the cost of doing business" in scientific research. Ingham and Horner observed that confirmed cases of scientific misconduct illustrate the practical consequences to the public. In these cases, they wrote, fraudulent research activities have derailed development of a vaccine against hepatitis C, interfered with progress in understanding metastasis in prostate cancer, produced misleading data regarding auditory processing in Broca's aphasia, and misrepresented findings aimed at understanding chemical phenomena active in the edema associated with traumatic brain injury. The consequences of misconduct, they argue, in one particular case, in which a graduate student falsified and fabricated data, were described as "adversely and materially affecting the laboratory's ongoing research [...] by creating uncertainty about all his experimental results, necessitating verification and repetition of experiments, preventing the reporting of results for publication, and preventing the principal investigator from submitting a competitive renewal application for an NIH grant."[201] The fallout of research misconduct is impactful and far-reaching, going well beyond the walls of the laboratory, where the dark side of the moon, as Paul Brookes, Ph.D., has called it, robs all concerned of breakthroughs and hope.

As Congressman John D. Dingell's subcommittee investigators uncovered in the 1980s and 1990s, evidence of misconduct was being covered up or whitewashed by institutions apparently more interested in the appearance of integrity than in the reality of it. By 1988, Dingell expressed shock when he discovered that the National Institutes of Health relied completely on the universities to investigate themselves. The conclusion he reached just then was the proverbial fox guarding the chicken coop. Truer still, if the institutions were policing misconduct, it put campus whistleblowers in the crosshairs for retaliation from their own colleges and universities. After all, the only one who is going to blow the whistle has to be close enough and with knowledge of the research to know if wrongdoing has been committed. By now widely acknowledged is the small percentage of whistleblowers continues to hobble an accurate count of the true number of scientists committing scientific misconduct despite provisions for their protection laid out by the Commission on Research Integrity in 1993 was the inclusion of a whistleblower's bill of rights. The greatest pushback against the commission's inclusion of this provision came from scientists, who, though already well protected by law, objected vigorously to potential accusers' having any rights at all. "Perhaps the most persistent, extraordinary and revealing criticism of the report, and, indeed, of any proposed regulation, was the continuing allegation [from these scientists] that regulation would impede scientific progress, because truly original science might easily be labeled misconduct. In hundreds, even thousands of cases, this has never happened,"[202] wrote Drummond Rennie[203], M.D., M.A.C.P., F.R.C.P., who described his first brush with misconduct coming just four months after he became editor-in-chief of the *New England Journal of Medicine* in September 1977, and C. Kristina Gunsalus, J.D., a nationally recognized expert, speaker, author and workshop presenter on matters of research integrity, ethics and professionalism in academia.

In the whistleblower's bill of rights, whistleblowers, going forward, were to be free to disclose lawfully whatever information supports a reasonable belief of research misconduct as it is defined by PHS policy. An individual or institution that retaliates against any person making protected disclosures engages in prohibited obstruction of investigations of research misconduct as defined by the commission. There were, of course, requisite caveats for the whistleblowers, including that they must respect the confidentiality of sensitive information and give legitimate institutional structures an opportunity to function. Should the whistleblower elect to make a lawful disclosure that violates institutional rules of confidentiality, the institution may thereafter legitimately limit the whistleblower's access to further information about the case. But conversely, the institutions were thereafter duty-bound to protect whistleblowers from retaliation. Per the report, "institutions have a duty not to tolerate or engage in retaliation against good-faith whistleblowers. This duty," written into the rules, "includes providing appropriate and timely relief to ameliorate the consequences of actual or threatened reprisals and holding accountable those who retaliate. Whistleblowers and other witnesses to possible research misconduct have a responsibility to raise their concerns honorably and with foundation." The commission also supported the 1989 recommendation by the Institute of Medicine of the National Academy of Sciences stating that universities should provide mediation and counseling to faculty, staff and students who wish to express concerns about professionally questionable training practices.

Accordingly, under fair procedures, the commission concluded that institutions have a duty to provide fair and objective procedures for examining and resolving complaints, disputes, and allegations of research misconduct. In cases of alleged retaliation that are not resolved through institutional intervention, whistleblowers should have an opportunity to defend themselves in a proceeding where they can present witnesses and confront those they charge with retaliation against them, except when they violate rules of confidentiality. Whistleblowers have a responsibility, the commission observed, to participate honorably in such procedures by respecting the serious consequences for those they accuse of misconduct, and by using the same standards to correct their own errors that they apply to others.

Per procedures free from partiality, the commission determined that institutions have a duty to follow procedures that are not tainted by partiality arising from personal or institutional conflict of interest or other sources of bias. Whistleblowers have a responsibility to act within legitimate institutional channels when raising concerns about the integrity of research. They further have the right to raise objections concerning the possible partiality of those selected to review their concerns without incurring retaliation.

In terms of information, the commission found that institutions have a duty to elicit and evaluate fully and objectively information about concerns raised by whistleblowers. Whistleblowers may have unique knowledge needed to evaluate thoroughly responses from those whose actions are questioned. Consequently, a competent investigation may involve giving whistleblowers one or more opportunities to comment on the accuracy and completeness of information relevant to their concerns, except when they violate rules of confidentiality.

Addressing timely processes, the commission dictated that institutions have a duty to handle cases involving alleged research misconduct as expeditiously as is possible without compromising responsible resolutions. When cases *drag on for years*, the issue becomes the dispute rather than its resolution. Whistleblowers have a responsibility to facilitate expeditious resolution of cases by good faith participation in misconduct procedures.

But what about vindication? Is there any such thing for the whistleblower? The commission was clear that at the conclusion of proceedings, institutions have a responsibility to credit promptly – in public and/or in private as appropriate – those whose allegations are substantiated. Every right carries with it, the commission reported, a corresponding responsibility. In this context, the whistleblower bill of rights carries the obligation to avoid false statements and unlawful behavior.

To a greater degree than she anticipated, Hill ran into the war of legal semantics and proof of intent as it pertains to fabrication and falsification. "There can be little doubt," wrote Fanelli, "about the fraudulent nature of fabrication, but falsification is a more problematic category. Scientific results can be distorted in several ways, which can often be very subtle and/or elude researchers' conscious control."[204] Like Martinson et al., Fanelli addressed cooked data, a process which he noted mathematician Charles Babbage in 1830 defined as "an art of various forms, the object of which is to give to ordinary observations the appearance and character of those of the highest degree of accuracy." Such data can be mined, Fanelli opined, just to come up with a statistically significant relationship that is then presented as the original target of the study. Further, it can be selectively published only when it supports one's expectations. Taking it one step further, it can also conceal conflicts of interest. "Depending on factors specific to each case," he continued, "these misbehaviors lie somewhere in the continuum between scientific fraud, bias, and simple carelessness, so their direct inclusion in the 'falsification' category is debatable, although their negative impact on research can be dramatic."[205]

The Martinson et al. study provided just one estimate of the prevalence of scientific misconduct. Based on the number of government confirmed cases in the United States, fraud is documented in about one in 100,000 scientists, or one in every 10,000 per a different counting.[206] Moreover, Fanelli documented that paper retractions from the *PubMed* library due to misconduct had, more recently, demonstrated a frequency of 0.02 percent, which led just then to speculation that between 0.02 and 0.2 percent of papers in the literature were fraudulent. Eight out of 800 papers submitted to the *Journal of Cell Biology*, he noted, had digital images that had been improperly manipulated, suggesting a one percent frequency. But ultimately, the routine data audits conducted by the FDA between 1977 and 1990 revealed deficiencies and flaws in ten to twenty percent of studies, and led to two percent of clinical investigators being judged guilty of serious scientific misconduct.[207] Yet, Fanelli, concluded, all of the aforementioned estimates were calculated on the number of frauds that had been discovered and reached the public domain. "This significantly underestimates the real frequency of misconduct, because data fabrication and falsification are rarely reported by whistleblowers, and are very hard to detect in the data," Fanelli wrote. "Even when detected, misconduct is hard to prove, because the accused scientists could claim to have committed an innocent mistake. Distinguishing intentional bias from error is obviously difficult, particularly when the falsification has been subtle, or the original data destroyed. In many cases," he continued, "therefore, only researchers know if they or their colleagues have willfully distorted their data."

To get at the actual percentage of scientists gone bad, Fanelli conducted the first systematic review and meta-analysis of survey data on scientific misconduct. He employed direct comparison between studies made possible by calculating, for each survey question, the percentage of respondents who admitted or observed misconduct at least once, and by limiting the analysis to qualitatively similar forms of misconduct – specifically on fabrication, falsification and any behavior that could distort scientific data. Fanelli's meta-analysis yielded mean pooled estimates that were higher than most previous estimates of scientific misconduct. Meta-regression analysis identified key methodological variables that affected the accuracy of the results and suggested that misconduct is reported more frequently in medical research.[208]

What Fanelli found across the continuum of studies was that about two percent of scientists admitted to having fabricated, falsified or modified data or results at least once – a serious form of misconduct – and up to one third admitted a variety of other questionable research practices including what he described as "dropping data points based on a gut feeling," and "changing the design, methodology or results of a study in response to pressures from a funding source." In surveys asking about the behavior of colleagues, fabrication, falsification and modification had been observed, on average, by over 14 percent of respondents, and other questionable practices by up to 72 percent – a big number.

Over the years, Fanelli observed that the rate of admissions declined significantly in self-reports but not in non-self-reports. Citing Martinson et al., which was one of the largest and most oft cited surveys on scientific misconduct published up to the time of Fanelli's study, he opined that the earlier work was conservative because without it the pooled average frequency with which scientists admitted that they had committed misconduct would jump to nearly three percent.[209] In fairness, Martinson et al. also acknowledged the conservative results of their own study. What stands out about both studies is that self-reporting systematically underestimates the real frequency of scientific misconduct, leading to Fanelli's observation that it can be safely concluded that data fabrication and falsification – let alone other questionable research practices – are more prevalent than most previous estimates have suggested.

"Anyone who has ever falsified research is probably unwilling to reveal it and/or to respond to the survey despite all guarantees of anonymity," Fanelli reported in his findings. "The opposite – scientists admitting misconduct they didn't do – appears very unlikely [also]. Indeed, there seems to be a large discrepancy between what researchers are willing to do and what they admit in a survey. In a sample of postdoctoral fellows at the University of California at San Francisco," he continued, "only 3.4 percent said they had modified data in the past, but 17 percent said they were 'willing to select or omit data to improve their results.' Among research trainees in biomedical sciences at the University of California at San Diego, 4.9 percent said they had modified research results in the past but 81 percent were 'willing to select, omit or fabricate data to win a grant or publish a paper."

Hill v. UMDNJ et al. served to further test the limits of how scientific misconduct is defined but it also pushed back at the way in which courts struggle through the adjudication process when it comes to holding researchers and institutions liable for fraud, where fraud is understood in the legal sense of the word. "If a researcher makes a false claim on a grant application or files a false report to the government, this could be construed as fraud under the common law or under the False Claims Act (FCA)," wrote David B. Resnik, J.D., Ph.D., a bioethicist at the National Institute of Environmental Health Sciences (NIEHS) and a former full professor of medical humanities at the Brody School of Medicine at East Carolina University (ECU) and an associate director of the Bioethics Center at ECU and University Health Systems from 1998 to 2004.[210] "Many scholars and scientists who discuss dishonesty in research use the term fraud to refer to many of the misdeeds captured by the phrase 'scientific misconduct.' However," Resnik observed, "fraud has a specific meaning in the law that makes the phrase scientific fraud a poor substitute for scientific misconduct." Resnik has proposed a different definition of misconduct:

> Misconduct is a serious and intentional violation of accepted scientific practices, commonsense ethical norms, or research regulations in proposing, designing, conducting, reviewing, or reporting research. Punishable misconduct includes fabrication of data or experiments, falsification of data, plagiarism, or interference with a misconduct investigation or inquiry. A person who commits to a form of punishable misconduct may receive a sanction proportional to the seriousness of the misconduct. Misconduct

does not include honest errors, differences of opinion, or ethically questionable research practices.[211]

What many of the studies of misconduct reveal is the tenfold increase in the number of retracted papers since the 1970s, and a number of these are attributable to fraud or suspected fraud. An investigation of retractions from the aforementioned biomedical scientific literature database *PubMed*, published by the Proceedings of the National Academy of Science (PNAS), revealed more recently that 63.2 percent of health- and life-science related retractions were due to fraud, suspected fraud or plagiarism, with honest error retractions in a slim minority.[212] Then there are the stories, wrote Sylvia McClain, Ph.D., a biophysicist at the University of Oxford who also writes a blog about science, science policy and philosophy of science. In her October 12, 2012 *Guardian* article, she introduced the story of Marc Hauser, Ph.D., a psychology professor dismissed by Harvard University after being found solely responsible for a series of six misconduct cases in 2012. Hauser invented results, she reported, to support his idea of a biological foundation for cognition in monkeys, specifically if they could recognize changes in sound patterns like human babies can do. She pointed further to Hauser's bestselling book – *Moral Minds: How Nature Designed Our Universal Sense of Right and Wrong* – in which he ironically argued that "policy wonks and politicians should listen more closely to our intuitions and write policy that effectively takes into account the moral voice of our species." This seemed to work out in his case, she observed, since he was busted for scientific misconduct. Hauser's book also pitched the notion that "our ability to detect cheaters who violate social norms is one of nature's gifts." But, as she concluded, "nature's gifts or not, his students and research assistants blew the whistle" on his wrongdoing.[213]

Harvard's eighty-five page report details instances in which Hauser changed data so that it would show a desired effect.

Further, it revealed that he more than once rebuffed or downplayed questions and concerns from people in his laboratory about how a result was obtained. The report also describes "a disturbing pattern of misrepresentation of results and shading of truth" and a "reckless disregard for basic scientific standards."[214] According to an article in the May 30, 2014 *Boston Globe*, a three-member Harvard committee reviewed forty internal and external hard drives, interviewed ten people, and examined original video and paper files that led them to conclude that Hauser had manipulated and falsified data. Their report was sent to the federal Office of Research Integrity in 2010, but it was not released to the *Globe* by the agency until the week prior to the article's appearance in the Boston newspaper nearly a full four years later. The documents offered the public a rare window into one of the highest-profile cases of scientific research fraud in recent years. After the *Globe* first reported that Harvard was investigating his underlying research for potential misconduct, according to its reporting, "Hausergate" became a portal into the mechanics of research fraud and the tense internal politics of a university that was investigating one of its best-connected public figures.

In 2005, according to the *Globe* and touched upon by McClain, Hauser and colleagues did a statistical analysis of an experiment in which monkeys responded to two artificial languages. In a later statistical analysis, an unnamed individual using the raw data got very different results. Unlike Hill's experience with three Newark CCRIs looking into her allegations, the Harvard committee in the Hauser case painstakingly reconstructed the process of data analysis and determined that Hauser had changed values, causing the result to be statistically significant, an important criterion showing that findings are probably not due to chance. For example, after the data from one experiment were analyzed in 2005, the results initially were not statistically significant. After Hauser informed a member of his lab of this by e-mail, he wrote a second e-mail: "Hold the horses. I think I [expletive] something up on the coding. Let me get back to you."

But after correcting for that problem, Hauser concluded that the result was statistically significant. According to the Harvard report, five data points had changed from the original file, and four of the five changes were in the direction of making the result statistically significant. In a second, related experiment, a collaborator asked to be walked through the analysis because he or she had obtained very different results when analyzing the raw data. Hauser sent back a spreadsheet that he said was simply a reformatted version, but then his collaborator produced a spreadsheet highlighting which values had apparently been altered. Hauser then wrote an e-mail suggesting the entire experiment needed to be recoded from scratch. "Well, at this point I give up. There have been so many errors, I don't know what to say. . . I have never seen so many errors, and this is really disappointing," he wrote.

In defending himself during the investigation, Hauser quoted from that e-mail, suggesting it was evidence that he was not trying to alter data. The committee strongly disagreed. "These may not be the words of someone trying to alter data, but they could certainly be the words of someone who had previously altered data: having been confronted with a red highlighted spreadsheet showing previous alterations, it made more sense to proclaim disappointment about 'errors' and suggest recoding everything than, for example, sitting down to compare data sets to see how the 'errors' occurred," the report states.[215]

McClain observed that in Hauser case, among others, when their work is published there is an expectation of some peer review prior to publication, a mechanism that should build trust that what we are about to read has some truth to it. "When a paper is reviewed [for publication], no one goes for a visit to the laboratory where the work was done – there has to be some degree of trust," McClain wrote. "If someone is falsifying data they usually get found out. No one else can repeat the experiments, no one else can find any evidence for the authors' claims. This is actually a standard part of science. Scientific experiments must be repeatable," she argues. "Theories or hypotheses must [be] verifiable in some manner to stand the test of time. If a scientific work cannot be verified or repeated then it tumbles off into oblivion, like most scientific theories." McClain cautions that despite steering for the iceberg that threatens "to sink modern science in a wave of fraud and deceit, science is the same as it ever was. Errors in the science literature, whether a result of fraud or honest errors or faulty equipment, ultimately will fall off into obscurity or get found out,"[216] just like Schön, Hauser and many others.

But how does the public find out what's going on behind lab doors when the institutions tasked to protect the public health from misconduct purposefully withhold the information? A recent *Journal of the American Medical Association* (*JAMA*) study revealed that the FDA remains silent on matters of scientific misconduct and fraud. In at least fifty-seven trials, the FDA found evidence of one or more of the following problems: falsification or submission of false information, problems with adverse events reporting, protocol violations, inadequate or inaccurate recordkeeping, failure to protect the safety of patients or issues with informed consent. Only three of the seventy-eight publications that resulted from these trials made any mention of these problems, and there were largely no corrections made, retractions issued, or concerns listed.[217]

In an article published for the February 9, 2015, *Slate*, study author Charles Seife, M.S., wrote that the FDA routinely buries evidence of fraud in medical trials. He and his students uncovered it.[218] "Agents of the Food and Drug Administration know better than anyone else just how bad scientific misbehavior can get," he wrote. "Reading the FDA's inspection files feels almost like watching a highlights reel from a *Scientists Gone Wild* video. It's a seemingly endless stream of lurid vignettes – each of which catches a medical researcher in an unguarded moment, succumbing to the temptation to do things he knows he really shouldn't be doing." He rattled off the faked x-ray reports, forged retinal scans, phony lab tests, secretly amputated limbs. "All done," he observed, "in the name of science when researchers thought that nobody was watching. That misconduct happens isn't shocking. What is: When the FDA finds scientific fraud or misconduct, the agency doesn't notify the public, the medical establishment, or even the scientific community that the results of a medical experiment are not to be trusted." Seife discovered that, to the contrary, and for more than a decade, the FDA has shown a pattern of burying the details of misconduct, going back to what Dingell observed, the FDA was more interested in the appearance of integrity than in the reality of it. As a consequence, Seife noted, nobody ever finds out which data is bogus, which experiments are tainted, and which drugs might be on the market under false pretenses.

"The FDA has repeatedly hidden evidence of scientific fraud not just from the public, but also from its most trusted scientific advisers, even as they were deciding whether or not a new drug should be allowed on the market. Even a congressional panel investigating a case of fraud regarding a dangerous drug couldn't get forthright answers," Seife explained. "For an agency devoted to protecting the public from bogus medical science, the FDA seems to be spending an awful lot of effort protecting the perpetrators of bogus science from the public."[219] In at least one case, falsified data in a trial comparing chemotherapies led to a patient's death. In another, a stem cell treatment trial, the FDA called it successful because it worked for twenty-six patients but put aside the fact that one patient later had to have a foot amputated.[220]

To illustrate his point, Seife detailed how he and his students were able to uncover evidence of FDA hiding research misconduct, largely accomplished through following the extensive paper trail the FDA creates when tipped off to misconduct. At that point, picture the FDA dispatching several investigators to put a lid on the problem, still generating a lot of paperwork in the process – starting with the agency's form 483 and, if the transgression is particularly egregious, a warning letter. Seife and his students pored over nearly 600 clinical trials, and while much of each one was redacted, to include information describing what drug the researcher was studying, the name of the study, and exactly how the misconduct affected the quality of the data, they were able, by cross-referencing documents, to piece together about 100 cases, identifying the study, drug and which pharmaceutical company was involved. "For the other 500, the FDA was successfully able to shield the drug maker (and the study sponsor) from public exposure."[221]

But as Seife was just as quick to note, it is not just the public that's in the dark – it's researchers and your doctor. As Seife described in the April 2015 *JAMA*, he and his students were able to track down those seventy-eight scientific publications resulting from a tainted study – a clinical trial in which FDA inspectors found significant problems with the conduct of the trial, up to and including fraud. To reiterate, in only three cases did they find any hint in the peer-reviewed literature of problems found by the FDA inspection; the remaining other publications weren't retracted, corrected, or highlighted in any way. "In other words, the FDA knows about dozens of scientific papers floating about whose data are questionable – and has said nothing, leaving physicians and medical researchers completely unaware. The silence is unbroken even when the FDA itself seems shocked at the degree of fraud and misconduct in a clinical trial."[222] Unfortunately, the FDA isn't the only dysfunctional federal agency on the block.

When David Wright[223], Ph.D., director of the Office of Research Integrity (ORI), resigned by letter to Howard K. Koh, M.D., then assistant secretary for health (ASH) for the United States Department of Health and

Human Services (HHS), he cited the dysfunctional atmosphere not only of the ORI but science in general. The letter, dated February 25, 2014, was provided by an anonymous source to the media[224], which released it in its entirety. In the letter, Wright informs Koh that his job had been at once the best and the worst job he'd ever had. "The best part of it has been the opportunity to lead ORI intellectually and professionally in helping research institutions better handle allegations of research misconduct, provide in-service training for institutional research integrity officers (RIOs), and develop programming to promote the responsible conduct of research (RCR). Working with members of the research community, particularly RIOs, and the brilliant scientist-investigators in ORI has been one of the great pleasures of my long career," he continued. "Unfortunately, and to my great surprise, it turned out to be only about thirty-five percent of the job."

Wright explained that the rest of his role as the ORI director had been the very worst job he'd ever had, and it took up sixty-five percent of his time. "That part of the job is spent navigating the remarkably dysfunctional HHS bureaucracy to secure resources and, yes, get permission for ORI to serve the research community. I knew coming into this job about the bureaucratic limitations of the federal government," he opined to Koh, "but I had no idea how stifling it would be. What I was able to do in a day or two as an academic administrator takes weeks or months in the federal government, our precinct of which is OASH [office of the assistant secretary of health]." He explained that there were – and are – a number of reasons for this:

> First, whereas in most organizations the front-line agencies that do the actual work, in our case protecting the integrity of millions of dollars of PHS-funded research, command the administrative support services to get the job done. In OASH it's the exact opposite. The op-divs [operational divisions], as the front-line offices are called, get our budgets and then have to go hat-in-hand to the administrative support people in the "immediate office" of OASH to spend it, almost item by item. These people who are generally poorly informed about what ORI is and does decide whether our requests are 'mission critical.'

He described one occasion in which he'd been invited to give a talk on research integrity and misconduct to a large group of American Association for the Advancement in Science (AAAS) fellows. "I needed to spend $35 to convert some old cassette tapes to CDs for use in the presentation. The immediate office denied my request after a couple of days of noodling," he informed Koh. "A university did the conversion for me in twenty minutes, and refused payment when I told them it was for an educational purpose.

> Second, the organizational culture of OASH's immediate office is seriously flawed, in my opinion. The academic literature over the last twenty-five years on successful organizations highlights several characteristics: transparency, power-sharing or shared decision-making and accountability. If you invert these principles, you have an organization (OASH in this instance), which is secretive, autocratic and unaccountable.

In one instance, by way of illustration, Wright urgently needed to fill a vacancy for an ORI division director. "I asked the principal deputy assistant secretary for health (your deputy) when I could proceed. She said there was a priority list. I asked where ORI's request was on that list. She said the list was secret and that we weren't on the top, but we weren't on the bottom either," he told Koh. "Sixteen months later we still don't have a division director on board."

Clearly frustrated, Wright recounted another occasion in which he asked Koh's deputy why there wasn't an evaluation by the operational divisions of the immediate office administrative services to try to improve their performance. "She responded that that had been tried a few years ago and the results were so negative that no further evaluations have been conducted.

> Third, there is the nature of the federal bureaucracy itself. The sociologist Max Weber observed in the early twentieth century that while bureaucracy is in some instances an optimal organizational mode for a rationalized, industrial society, it has drawbacks. One is that public

bureaucracies quit being about serving the public and focus instead on perpetuating themselves. This is exactly my experience with OASH. We spend exorbitant amounts of time in meetings and in generating repetitive and often meaningless data and reports to make our precinct of the bureaucracy look productive. None of this renders the slightest bit of assistance to ORI in handling allegations of misconduct or in promoting the responsible conduct of research. Instead, it sucks away time and resources that we might better use to meet our mission. Since I've been here, I've been advised by my superiors that I had 'to make my bosses look good.' I've been admonished: 'Dave, you are a visionary leader but what we need here are team players.' Recently, I was advised that if I wanted to be happy in government service, I had to 'lower my expectations.' The one thing no one in OASH leadership has said to me in two years is 'how can we help ORI better serve the research community?' Not once.

Finally, Wright believed there was another important organizational question that deserved mention: Is OASH the proper home for a regulatory agency such as ORI? "OASH is a collection of important public health offices that have agendas significantly different from the regulatory roles of ORI and OHRP," he wrote. "You've observed that OASH operates in an 'intensely political environment.' I agree and have observed that in this environment decisions are often made on the basis of political expediency and to obtain favorable 'optics.' There is often a lack of procedural rigor in this environment," he continued. "I discovered recently, for example, that OASH operates a grievance procedure for employees that has no due process protections of any kind for respondents to those grievances. Indeed, there are no written rules or procedures for the OASH grievance process regarding the rights and responsibilities of respondents. By contrast, agencies such as ORI are bound by regulation to make principled decisions on the basis of clearly articulated procedures that protect the rights of all involved. Our decisions," he iterated, "must be supported by the weight of factual evidence. ORI's decisions may be and frequently are tested in court. There are members of the press and the research community who don't believe ORI belongs in an agency such as OASH and I, reluctantly, have come to agree."

In closing, Wright concluded that his twenty-six months of service as ORI director had still, despite the high negatives, been a remarkable experience. "As I wrote earlier in this letter, working with the research community and the remarkable scientist-investigators at ORI has been the best job I've ever had. As for the rest," he observed, "I'm offended as an American taxpayer that the federal bureaucracy – at least the part I've labored in – is so profoundly dysfunctional. I'm hardly the first person to have made that discovery, but I'm saddened by the fact that there is so little discussion, much less outrage, regarding the problem. To promote healthy and productive discussion, I intend to publish a version of the daily log I've kept as ORI director in order to share my experience and observations with my colleagues in government and with members of the regulated research community."

Wright made no mention of a recent letter to ORI from Iowa Republican senator Charles "Chuck" Grassley, who had complained vigorously that ORI was not tough enough on an AIDS researcher at Iowa State University who faked data to obtain nearly $19 million in NIH funding. Though ORI barred the researcher, Dong-Pyou Han, Ph.D., then a professor of biomedical science, from participating in PHS-funded research for three years, Grassley wanted to know why ORI did not make him return federal grant money or impose harsher sanctions.[225] Grassley's letter, dated February 10, elicited a canned six-page response from acting director Donald Wright, M.D., M.P.H., nearly one month later, on March 5, that included long citations of federal regulations that didn't directly answer many of the senator's questions. Grassley spokeswoman Jill Gerber cited the fact that the ORI response didn't say whether federal officials ever try to recoup money in such cases, or what the agency intends to do with unspent grant money that was awarded, for example, to the Dong-Pyou Han research team at Iowa State. "They just parroted the regulation at us," she opined, "and then they referred it to NIH."[226]

In short, Han admitted in 2013 that he faked the results of his AIDS vaccine study. According to media reports, among other misconduct, Han spiked rabbit blood with human antibodies to make it appear that the vaccine was sparking a response from the rabbits' immune systems. Han resigned after his fraud was exposed. Despite the particularly egregious nature of his misconduct, Iowa State administrators publicly admitted in a March 11, 2014 *Des Moines Register* article, they'd not seen the need to inform law enforcement about the incident. The ORI had no comment, but it was clear to Grassley and his staff from the ORI's March 5 letter that they'd not notified, in this case, federal law enforcement (due to the appropriation of NIH federal funding). Grassley had also asked ORI if the Han case had been referred to any other government agencies. The only agency mentioned by ORI was the NIH, the funding agency, by letter on December 13, 2013, but there weren't any others who told of Han's misconduct.[227] The Iowa senator wasn't the only one incensed by Han's initial slap-on-the-wrist punishment. A number of research ethicists suggested that criminal fraud charges should be considered in Han's case. Shortly after Han's fraud came to light, in fact, James A. Bradac, Ph.D., chief of the NIH's Preclinical Research and Development Branch Vaccine Research Program, who oversees the organization's AIDS grants, called it "the worst case of research fraud he'd seen in his more than two decades at the agency," according to an article published in the December 27, 2013 *New York Daily News*.

Grassley's attention to the Han case put ORI in his bull's-eye. He wanted to know how ORI holds accountable senior faculty or laboratory supervisors for researchers who engage in research misconduct while under their supervision, a question certainly applicable to *Hill v. UMDNJ et al.* In response, the ORI cited 42 C.F.R. Part 93 that requires that ORI prove by a preponderance of the evidence that the accused respondent knowingly, intentionally, or recklessly falsified, fabricated, or plagiarized data in the research record associated with a PHS-funded project. Therefore, the evidence must show that the senior faculty or laboratory supervisors were directly culpable for the creation or publication of the false data. Because of the *mens rea* requirement in the regulation, ORI does not automatically, "impute responsibility for misconduct by subordinates to their supervisors." Nonetheless, ORI has made, Donald Wright wrote, "research misconduct findings against senior faculty of a laboratory if the senior individual knowingly, intentionally, or recklessly reports and/or submits data that are known to be false. This usually occurs," he explained to Grassley, "when a junior person is responsible for creating the false, fabricated, or plagiarized data, but the senior person instructed, coerced, or knew that the data were false when including the data in a grant application or research paper or as part of the research record." He added that ORI had observed that in most cases in which senior faculty members oversee scientists engaged in research misconduct, it is usually the result of poor mentorship or supervision and not intentional research misconduct on their part."

Grassley also asked how many cases had ORI encountered where a researcher intentionally falsified data or research results? He asked for elaboration of any such cases. In response, acting director Wright wrote that of ORI's 276 cases involving research misconduct findings from 1992 through 2014, 258 cases included intentional falsification or fabrication of data or results; 18 cases involved only plagiarism. In a handful of cases, the findings involved knowingly reporting false data. Asked if any of these were repeat offenders, he informed Grassley that there had only been one – a mentally unstable laboratory technician – and findings of research misconduct were made in 1989 and 2002.

The wheels of justice often move at very slow speed. Though unhappy as Grassley and prominent ethicists were that Han hadn't faced criminal charges soon after his fraud was discovered, it didn't take long for federal prosecutors to build their case against the former Iowa State researcher. In June 2014, as a result of his receiving grant money due to falsified results, Han was ultimately indicted on four federal felony counts of making false statements. The case was played out in federal court, where Han pleaded guilty in late February 2015. As part of his plea agreement, Han admitted that his subterfuge cost the federal government (taxpayers) $7 million to $20 million. In return for his guilty pleas to two felony charges of making false statements, prosecutors dropped two other charges. At the time of his allocation, he faced up to ten years in prison, which would be a rare punishment for academic fraud, those monitoring the case observed. But Han's case wasn't normal. He was sentenced to fifty-seven months imprisonment for fabricating and falsifying data in HIV vaccine trials on July 1, 2015. He was also fined $7.2 million as well as being subject to three years of supervised release after his federal prison term ended.

Writing of the Han debacle prior to his sentencing, *Retraction Watch*'s Adam Marcus and Ivan Oransky, M.D., suggested that Han was driven to wrongdoing by the need for impressive lab results to help his team at Iowa State move forward with its work on an AIDS vaccine – "and to continue receiving millions of dollars in federal grants."[228] Criminal charges such as those leveled at Han are uncommon and that he was convicted and put in jail for it fell into yet another inordinately rare category. While research misconduct is clearly far from rare, according to a study from 2013 cited by Marcus and Oransky, "most investigators who engage in wrongdoing, even serious wrongdoing, continue to conduct research at their institutions."

As part of their reporting, Marcus and Oransky have written about multiple academic researchers who have been found guilty of misconduct and have gone on to work at pharmaceutical giants. Iowa State, they opined, had agreed to reimburse the federal government roughly $500,000 to cover several years of Han's salary and the NIH opted to withhold another $1.4 million that it promised the university as part of the grant. Han was subsequently fined $7.2 million. Examples of other recent scientists who committed misconduct and who were fined and sentenced to prison terms include Poehlman, previously discussed, and also Scott S. Reuben, an anesthesiologist, who spent six months in federal prison starting in 2010 for faking data in many of his studies. Reuben, they noted, was further ordered by the court to repay more than $360,000 to Pfizer as restitution for misusing the drug maker's grant money. "Dr. Han may have remained one of the hundreds of fraudster scientists who faced little punishment if it weren't for the attention of [Senator Grassley]."[229] Grassley has seized upon the simple fact that the ORI lacked the teeth to get the job done. Just two of the eleven misconduct cases brought before the agency in 2013 resulted in outright bans, according to Marcus and Oransky. "The [ORI] never recovered from its case against Thereza Imanishi-Kari, [the] Tufts University researcher [also previously discussed] accused of fraud in her work with a Nobel laureate, the biologist David Baltimore. In 1991," they explained, "investigators at the ORI – then called the Office of Scientific Integrity – found Dr. Imanishi-Kari guilty of misconduct and lying to cover up her actions, but in 1996 they were overruled by panelists for its parent agency, the Department of Health and Human Services, who concluded that the [ORI] had failed to prove its case." From that time forward, scientists could cite the Imanishi-Kari case as the poster child for government oversight "run amok" and more than that, the wiggle room needed to get a slap on the wrist versus a prison term.

In his pursuit of ORI and the misconduct that had come to cloak the biomedical research community, Grassley had the help of a reporter-turned-congressional investigator named Paul D. Thacker, whom *Nature*'s Meredith Wadman[230] declared had disrupted the careers of several top researchers with lucrative industry ties. Thacker, who specialized in science, medicine, and environmental reporting, in April 2007 went to work on the Senate Finance Committee for Republican senator Chuck Grassley investigating medical research conflicts of interest, a position he held until 2010. In that time, he became Grassley's point person behind a far-reaching probe into the financial reports required of biomedical researchers who receive federal funding. According to Wadman, Thacker quickly crossed paths with "a rising academic star" named Charles B. Nemeroff[231], M.D., Ph.D., the chair of psychiatry at Emory University, until he was forced out of his position in December 2008 as a result of Grassley's investigation. As it happened, Thacker acquired documents from the pharmaceutical industry that showed Nemeroff had violated NIH rules by failing to declare to the university at least $1.2 million in income from drug makers.[232] But Nemeroff wasn't the only one in Thacker's sights. He'd identified seven other physician-researchers, Wadman reported, those who violated NIH rules by failing to disclose large payments from drug and device companies, and in several cases, the researchers were engaged in human trials of the products made by some of the very companies with which they had financial relationships.

The Grassley investigation has had a profound impact on the biomedical research community. Steven E. Nissen, M.D., a cardiologist and chairman of the Robert and Suzanne Tomsich Department of Cardiovascular Medicine at Cleveland Clinic's Sydell and Arnold Miller Family Heart and Vascular Institute in Ohio, which in December 2008 began requiring its physicians to publicly disclose industry payments of more than $5,000 per year, declared that Grassley's intense scrutiny "has changed the practice of medicine."[233] He credited Thacker as the "bulldog leading the charge." Grassley's – and Thacker's – work to expose misconduct in biomedical research wasn't without its critics. Some worried that they were wrongly putting all researchers into the "bad apple" basket. Institutions and the federal government, in the interest of fueling drug discovery, encourage researchers to cultivate industry collaborations, an art form, Wadman observed, Nemeroff had perfected. But Nemeroff's case, which has been painted as "this black and white thing and it's not," opined Joseph Cubells, M.D., Ph.D., a neurogeneticist and associate professor in the Emory University School of Medicine Department of Genetics and a former colleague of Nemeroff. "It's nuanced and it's subtle and it's difficult. And there are a lot of aspects to this whole issue where there are no right answers."[234]

When Thacker first joined Grassley's staff, the senator had already looked into how drug-industry payments might be influencing advisers on government drug approvals, Wadman wrote. Thacker took up the baton, moved forward by an article he'd seen in the *New York Times* that described payments from drug companies to psychiatrists in Minnesota, the only state that at that time required companies to make public their payments to physicians, according to Wadman's account of Thacker's activity. The *New York Times* had, in fact, discovered that psychiatrists taking the most money and perks from pharmaceutical companies were likelier to prescribe a controversial class of anti-psychotic drug to children.

What got Thacker's attention in the *Times* article was a comment by Melissa DelBello[235], M.D., subsequently named the Dr. Stanley and Mickey Kaplan professor and chair of the University of Cincinnati Medical Center's Neuroscience Institute/Department of Psychiatry and Behavioral Neuroscience on July 1, 2014. DelBello, a nationally recognized expert on child and adolescent mood disorders, and who has lectured and published extensively on bipolar disorder and served as principal or coinvestigator of several NIH grants, was quoted in the story as refusing to disclose how much money she'd made from the drug maker AstraZeneca. She'd published, Wadman noted, a company-funded study in 2002 that reported positive results for AstraZeneca's anti-psychotic drug Seroquel – quetiapine – in adolescents with bipolar disorder. There was a direct correlation between her published study and the twenty-seven percent uptick in the drug's sales in the year after publication. "Trust me, I don't make much," she told the *Times* just then. In her own defense, DelBello later stated she'd been misquoted, that she only meant how much she made per lecture.[236] But by then, Thacker had already locked and loaded on two words – "trust me." By Wadman's account, he walked down to Grassley's chief of investigation's office. There, he told Emilia DeSanto[237] that the article, including DelBello's comments, read "like nonsense." Since he knew researchers had to file a financial-interest disclosure with their respective universities, he sent a letter to the University of Cincinnati requesting copies of DelBello's disclosure forms under Grassley's signature. What Thacker soon learned was that AstraZeneca had declared payments in excess of $180,000 in 2003 and 2004, the first two years after the Seroquel study was published. The money they'd paid DelBello was, according to the filing, for lectures, consulting fees, travel expenses and other services (unspecified).[238]

According to Wadman, it took only a few days for DelBello and the university to call Grassley to the carpet for his remarks on the senate floor regarding DelBello's AstraZeneca income. The university stated that Del-Bello had been completely forthcoming about her income and other perks, filing timely disclosures. Thacker, again, detected something was "off" with DelBello's overall claim. He attached a question, taking a chance, in a letter to AstraZeneca from another Grassley staffer on a separate issue, Wadman reported, asking the company for records of how much it had paid DelBello in recent years. While he waited for the reply, Thacker pondered why Grassley was so interested in a matter that seemed to boil down to a private arrangement between DelBello and AstraZeneca. The answer wasn't hard to find. Recipients of NIH grants – DelBello had and has been a principal or coinvestigator on several of them – are compelled under federal rules in place since 1995 to disclose to their institutions any financial interests of more than $10,000 in cash or five percent (or more) equity in a company. She was taking taxpayer dollars – and considerable money from AstraZeneca. There, Thacker found the conflict of interest and the misconduct. With that understanding, Thacker informed Wadman, he started identifying NIH-funded researchers like DelBello who were likely earning top dollar from drug or device companies, and he did it many times following media reporting and information gleaned from whistleblowers.

Back to the DelBello query Thacker had earlier slipped into correspondence to AstraZeneca. In April 2008, armed with information that came back from the company, Grassley once again hit the senate floor with another revelation. AstraZeneca had paid DelBello $238,000 between 2005 and 2007, during which she had only reported receipt of $100,000 from all sources outside the University of Cincinnati. This led Grassley to opine that the university had engaged in the practice of "trust but did not verify."[239] DelBello had underreported the income to the university but as Thacker soon found out, she wasn't alone. Wadman reported that in June 2008 Thacker discovered that three psychiatrists at Harvard University had each done the same, underreporting hundreds of thousands of dollars in payments. That same month, she continued, Thacker also reported that Alan F. Schatzberg, M.D., a professor of psychiatry and behavioral sciences at Stanford University School of Medicine in Palo Alto, California, and chair of the department from 1991 to 2010, who received multiple national and international awards for his work as an investigator in the biology and treatment of depression, owned equity then worth $6 million in the drug company Corcept Therapeutics, in nearby Menlo Park. Schatzberg had cofounded the company to develop the drug mifepristone for treatment of psychotic depression. Thacker discovered, and passed to Grassley, that Schatzberg had only declared $100,000 "plus investments" in Corcept Therapeutics. At the same time, he was the beneficiary of the equity from his company, Schatzberg was the principal investigator on an NIH-funded study, which included a component that tested the drug Schatzberg's company had developed. "Grassley's report triggered a media storm," wrote Wadman, "and Stanford removed Schatzberg as principal investigator, pending an investigation. The university said that although it believed that Schatzberg and Stanford had not broken any rules, 'we can see how having Dr. Schatzberg as the principal investigator on this grant can create the appearance of conflict of interest.'" In other reporting, interestingly, it was stated that Schatzberg "voluntarily and temporarily stepped down"[240] as principal investigator from a related grant, implying that the university had not "removed him."

Schatzberg, removed from the study of his own drug, declared that Stanford and the NIH had preapproved his service as the study's principal investigator and that he hadn't run the mifepristone trial. Stanford subsequently reinstated him as principal investigator in late July 2009 after the NIH wrote to the university informing them that no rules had been broken – but by that time, Wadman noted, the mifepristone part of the study had ended. Of note, despite the controversy that enveloped Schatzberg just then, he was elected the one hundred thirty-sixth president of the American Psychiatric Association (APA) from 2009 to 2010.

Armed with information that was being provided by Thacker, and with considerable media attention, Grassley went after then-NIH director Elias A. Zerhouni[241], M.D., to better police grant participants, Wadman reported. "NIH oversight of the extramural program is lax and leaves people with nothing more than questions," Grassley wrote in a letter to Zerhouni that he inserted into the June 4, 2008 *Congressional Record*.

Back to Nemeroff. Grassley's letter to Emory University about Nemeroff's financial conflict of interest hadn't come as a surprise to the university's administrators. Over a period of sixteen years, he'd built the psychiatry department into one of the best in the country. But at the same time, Wadman discovered, he'd developed substantial financial ties to many companies and a history of failing to disclose them to the university.[242] "Seen from one perspective, the connections to industry," she wrote, "were just what universities and the federal government wanted; in fact, Nemeroff had won a $3.95-million grant from the National Institute of Mental Health (NIMH) in 2003 that would test at least five new antidepressants developed by the British drug giant GlaxoSmithKline, whose advisory board Nemeroff chaired." Further, in soliciting grant proposals, the NIMH had asked researchers to involve industry as a means to accelerate the development of new drugs – a well-known fact at that time. Nemeroff, however, had made a number of mistakes in his juggling of these industry relationships, according to Emory administrators. Wadman documented that in late 2003, officials there had grown so concerned about his network of industry connections that they convened a conflict-of-interest committee to investigate. The committee issued a fourteen-page report in May 2004, concluding that there were serious lapses in Nemeroff's reporting of his relationships in nineteen companies. The university still didn't curtail Nemeroff's extramural activities. Within two years of the university's internal investigation, his name was mentioned prominently in a *Wall Street Journal* article in which he'd favorably reviewed an implantable device as a treatment for depression without, of course, disclosing that he had financial ties to the company that made the device; he was also the editor of the journal *Neuropsychopharmacology*, where the review[243] he'd written was published. Though the journal informed the *Wall Street Journal* that Nemeroff had filled out author disclosure forms, it did not report his financial ties in the manuscript, as the journal would require of anyone else. Thacker was paying attention to these kinds of discrepancies, Wadman noted.

As for Emory University, it had tried unsuccessfully to control Nemeroff, to rein him in. "I can't remember when I have gotten so many complaints about the action of one or more of our faculty from inside and outside the institution," one Emory administrator wrote in an e-mail to Nemeroff in the wake of the controversy that followed the *Neuropsychopharmacology* article.[244] That was an internal, private communication. Publicly, the university wasn't doing anything to preclude Nemeroff's activities and, in fact, sang his praises, especially after he landed a \$9.3-million grant from the NIMH for a trial on depression. The trial, as laid out, aimed to compare the effectiveness of two drugs against cognitive behavioral therapy – 'talk therapy' – in people with major depression, Wadman explained. Investigators, including Nemeroff, planned to employ genetic scans and brain imaging to see whether there are factors that predispose some people to respond well to different treatments, thus creating a rational alternative to the "try-this-pill-and-see-if-it-works" approach that is common practice in psychiatry today, she continued. Nemeroff and other researchers were planning to enroll 400 people but participants weren't so easy to find because they had to sign on to a demanding fourteen-week course of investigations and treatment and they couldn't participate if they had ever been previously treated for major depression. By August 2008, two years into their funding, Wadman reported they had only enrolled a handful of patients. But by then, despite any promise the study might have had, the NIH had become painfully aware of Grassley's investigation and as part of the agency's own inquiry, uncovered substantive discrepancies between what Nemeroff had disclosed to Emory and what he had actually earned.

Zerhouni suspended Nemeroff's study in mid-August, and the university stopped enrolling patients.[245] Grassley sent a letter to Emory about a month later that documented the discrepancies. In it, he informed the university's administrators that of the \$2 million Nemeroff had collected from drug companies between 2000 and 2007, Nemeroff had failed to report at least \$1.2 million. After that, it was clear to Emory, as Wadman wrote later, "any hope of keeping those disclosure lapses private disappeared in October when the story broke in the media. At the same time, reporters learned that the NIH had suspended the study." Zerhouni commented, in part, "We said: 'This can't continue [under this principal investigator].'"[246] Zerhouni had done exactly what Grassley had wanted to see happen all along, Wadman observed. "All NIH should have to do is pull back one grant or refuse to give a grant to a university that's not [policing conflicts properly] and they all get in line," the Iowa senator told Wadman for her October 2008 *Nature Medicine*[247] story about his committee investigations into biomedical researcher misconduct. Nemeroff has the distinction of being the only study suspended due to Grassley's inquiries.

While Grassley had his detractors, as did Dingell before him, others, like Zerhouni concur that something had to be done. "People flouted the rules, didn't disclose, and did it for years on end, repeatedly," Zerhouni told Wadman. "That tells you the problem is not Grassley. The problem is our current system of managing conflicts."[248] Ultimately, Emory did vet Nemeroff, prompted by Thacker's work for Grassley. The university announced in December 2008 that Nemeroff had received – and failed to declare – $800,000 in payments from GlaxoSmithKline for giving more than 250 talks to psychiatrists between January 2000 and January 2006. Nemeroff told the university that he didn't consider the talks promotional and thought they were "off-the-books" under Emory's rules for declaring outside income. While the evidence that Emory uncovered supported Nemeroff's description of the speaking engagements, he was nonetheless removed as chairman of the department and the university precluded him from applying for, or being involved with, any NIH grants for at least two years or to receive any outside compensation without prior review and approval by the dean's office. At the same time, the university issued a public statement to the effect it found that Nemeroff's outside speaking engagements did not adversely affect persons enrolled in clinical trials, and that there was no evidence that his activities biased scientific research in which he was engaged.[249] Given these limitations, Nemeroff left Emory for the University of Miami in 2009.

More recently, William A. Wilson, a software engineer in the San Francisco Bay area, took on the subject of big science gone wrong in a way no one else had done heretofore. "The problem with science," he wrote, "is that so much of it simply isn't. Last summer (in 2015), the Open Science Collaboration (OSC) announced that it had tried to replicate [there's that word again] one hundred published psychology experiments sampled from three of the most prestigious journals in the field. Scientific claims rest on the idea that experiments repeated under nearly identical conditions ought to yield approximately the same results," he continued, "but until very recently, very few had bothered to check in a systematic way whether this was actually the case. The OSC was the biggest attempt yet to check a field's results, and the most shocking. In many cases," Wilson explained, "they had used original experimental materials, and sometimes even performed the experiments under the guidance of the original researchers. Of the studies that had originally reported positive results, an astonishing sixty-five percent failed to show statistical significance on replication, and many of the remainder showed greatly reduced effect sizes."[250] In other words, a lot of published research is false. But that's not even the worst part. "Advocates of the existing scientific research paradigm," wrote Pascal-Emmanuel Gobry, "usually smugly declare that while some published conclusions are surely false, the scientific method has 'self-correcting mechanisms' that ensure that, eventually, the truth will prevail."[251] Certainly, this is also the point Dr. Sylvia McClain iterated in the Hauser story. "Unfortunately for all of us," Gobry continued, "Wilson makes a convincing argument that those self-correcting mechanisms are broken." Wilson directed much attention to the same issue raised by Hill in her allegations against Roger Howell and Anupam Bishayee in that there is a replication crisis in science.

The failure to replicate isn't just a problem in the field of experimental psychology. Wilson discovered that there's an unspoken rule in the pharmaceutical industry that half of all academic biomedical research will ultimately prove false, and in 2011, researchers at Bayer decided to put that to the test. "Looking at sixty-seven recent drug discovery projects based on preclinical cancer biology research, they found that in more than seventy-five percent of cases the published data did not match up with their in-house attempts to replicate. These were not studies, Wilson explained, by fly-by-night oncology journals but blockbuster research featured in *Science*, *Nature*, *Cell*, and the like.[252] "The Bayer researchers were drowning in bad studies," Wilson wrote, "and it was to this, in part, that they attributed the mysteriously declining yields of drug pipelines. Perhaps so many of these new drugs fail to have an effect because the basic research on which their development was based isn't valid."

So, to what can we attribute this failure to replicate? While there is the obvious innocent human error and what many would dub "everything in-between," there's also unabashed, intentional fraud. In a 2011 survey of 2,000 research psychologists [back to them again], over half admitted to selectively reporting those experiments that gave the result they were after. The survey also concluded that around ten percent of research psychologists have engaged in outright falsification of data, and more than half have engaged in what was categorized as less brazen but still fraudulent behavior such as reporting that a result was statistically significant when it was not, or deciding between two different data analysis techniques after looking at the results of each and choosing the more favorable.[253]

What Wilson discovered was that the interpretation that the original finding was false is likelier than an innocuous mistake by the researchers. The ability to parse out the fraud was first articulated by John P. A. Ioannidis, M.D., D.Sc., the F. Rehnborg professor in disease prevention in the Stanford University School of Medicine and professor of health research and policy (epidemiology), an argument based on the simple application of Bayesian statistics. "A tremendous amount depends on the proportion of possible hypotheses which turn out to be true, and on the accuracy with which an experiment can discern truth from falsehood. Ioannidis shows that for a wide variety of scientific settings and fields, the values of these two parameters are not at all favorable," Wilson observed. He described, for example, a team of molecular biologists investigating whether a mutation in one of the countless thousands of human genes is linked to an increased risk of Alzheimer's disease. Ioannidis would argue that the probability of a randomly selected mutation in a randomly selected gene having precisely the effect is quite low, and a positive finding is thus likelier than not to be spurious, unless the experiment is unbelievably successful at sorting the wheat from the chaff.[254] "Ioannidis," Wilson continued, "finds that in many cases, approaching even fifty percent true positives requires unimaginable accuracy. Hence, the eye-catching title of his paper – 'Why most published research findings are false.'" This goes to the paucity of accuracy.

Wilson pointed to the competitive nature of journals, those generally more interested in publishing research with an "impact factor," that new, exciting finding that reads better than what he called a "killjoy failure" in their association. But both of these – the exciting and the killjoy failures – can be quantified. Since the majority of all investigated hypotheses are false, if positive and negative evidence were written up and accepted for publication in equal proportions, he concluded that the majority of articles in scientific journals would report no findings. When tallies are actually made, though, the precise opposite turns out to be true: nearly every published scientific article reports the presence of an association. There must be massive bias at work – or outright fraud.[255] Many forms of statistical falsification are devilishly difficult to catch, or close enough to a genuine judgment call to provide plausible deniability, Wilson noted. Data analysis is very much an art, and one that affords even its most scrupulous practitioners a wide degree of latitude.

"One creative attempt to estimate how widespread such dishonesty really is involves comparisons between fields of varying 'hardness,'" Wilson offered. He reprised the work of Daniele Fanelli to make his point, noting that Fanelli theorized that the farther from physics one gets, the more freedom creeps into one's experimental methodology, and the fewer constraints there are on a scientist's conscious and unconscious biases.[256] "If all scientists were constantly attempting to influence the results of their analyses, but had more opportunities to do so the 'softer' the science, then we might expect that the social sciences have more papers that confirm a sought-after hypothesis than do the physical sciences, with medicine and biology somewhere in the middle," Wilson deduced. "This is exactly what the study discovered: a paper in psychology or psychiatry is about five times as likely to report a positive result as one in astrophysics. This is not necessarily evidence that psychologists are all consciously or unconsciously manipulating their data – it could also be evidence of massive publication bias – but either way, the result is disturbing."

But as it turned out, Wilson concluded that even physics wasn't all that it was cracked up to be, citing the Sokal hoax that demonstrated, in the worst way, that the peer review process doesn't work. Physicist Alan Sokal, Ph.D., a professor of mathematics at University College London and professor of physics at New York University, who works in statistical mechanics and combinatorics but is best known publicly for his criticism of postmodernism, gained greater notoriety after the Sokal affair in 1996 when his deliberately nonsensical paper was published by Duke University's *Social Text*. Sokal also works to counter faulty scientific reasoning, as seen with his involvement in criticizing the critical positivity ratio concept in positive psychology. This purposeful act to hoodwink the journal to publish the work to prove his point that junk science was creeping into the mainstream scientific "think" made the front pages of the May 18, 1996 *New York Times*.

Similarly, Wilson addressed the less famous but similar hoodwinking of the very prestigious *British Medical Journal*, to which a paper with eight major errors was submitted. He pointed to the fact that not a single one of the 221 scientists who reviewed the paper caught all the errors in it, and only thirty percent of reviewers recommended that the paper be rejected. Amazingly, the reviewers who were warned that they were in a study and that the paper might have problems found no more flaws than the ones who were blinded to this important fact. "If peer review is good at anything, it appears to be keeping unpopular ideas from being published," Wilson concluded. "Consider the finding of another (yes, another) of these replicability studies, this time from a group of cancer researchers. In addition to reaching the now unsurprising conclusion that only a dismal eleven percent of the preclinical cancer research they examined could be validated after the fact," he continued, "the authors identified another horrifying pattern: the 'bad' papers that failed to replicate were, on average, cited far more often than the papers that did! As the authors put it, 'some non-reproducible preclinical papers had spawned an entire field, with hundreds of secondary publications that expanded on elements of the original observation, but did not actually seek to confirm or falsify its fundamental basis.'" This is a serious finding.

"What they do not mention," Wilson opined, "is that once an entire field has been created – with careers, funding, appointments, and prestige all premised upon an experimental result which was utterly false due either to fraud or to plain bad luck – pointing this fact out is not likely to be very popular." In this case, Wilson writes that peer review switches from merely useless to actively harmful. "It may be ineffective at keeping papers with analytic or methodological flaws from being published," he offered, "but it can be deadly effective at suppressing criticism of a dominant research paradigm. Even if a critic is able to get his work published, pointing out that the house you've built together is situated over a chasm will not endear him to his colleagues or, more importantly, to his mentors and patrons."

Interestingly, Wilson identified that older scientists contribute to the propagation of scientific fields in ways that go beyond educating and mentoring a new generation. "In many fields, it's common for an established and respected researcher to serve as 'senior author' on a bright young star's first few publications, lending his prestige and credibility to the result, and signaling to reviewers that he stands behind it." In Hill's case, the question was raised regarding Roger Howell listing Anupam Bishayee's name first among all coauthors, including himself, most especially on the two publications in *Radiation Research* that became such a source of controversy in her *qui tam* naming Howell and Bishayee as defendants. During depositions in *Hill v. UMDNJ et al.* Bishayee admitted that he did not write either paper, and when asked by counsel who did, he identified Howell.

"In the natural sciences and medicine, senior scientists are frequently the controllers of laboratory resources – which these days include not just scientific instruments, but dedicated staffs of grant proposal writers and regulatory compliance experts – without which a young scientist has no hope of accomplishing significant research," Wilson concluded from his inquiry. "Older scientists control access to scientific prestige by serving on the editorial boards of major journals and on university tenure review committees. Finally," he continued, "the government bodies that award the vast majority of scientific funding are either staffed or advised by distinguished practitioners in the field." But all this tells us what? Certainly, from Wilson's findings it is troubling that older scientists are the most likely to be invested in what he dubbed "the regnant research paradigm," whatever it is, "even if it's based on an old experiment that has never successfully been replicated." He offered that the quantum physicist Max Planck famously quipped: "A new scientific truth does not triumph by convincing its opponents and making them see the light, but rather because its opponents eventually die, and a new generation grows up that is familiar with it."[257] Wilson thinks Planck may have been too optimistic. He cited a recent paper from the National Bureau of Economic Research that studied what happens to scientific subfields when star researchers die suddenly and at the peak of their abilities, and finds that while there is considerable evidence that young researchers are reluctant to challenge scientific superstars, a sudden and unexpected death does not significantly improve the situation, particularly when, he quotes: "key collaborators of the star are in a position to channel resources (such as editorial goodwill or funding) to insiders."[258]

Gobry asks why is it, then, that our scientific process is so structured as to reward the old and the prestigious? "Government funding bodies and peer review bodies are inevitably staffed by the most hallowed (read: out of touch) practitioners in the field," he opined. "The tenure process ensures that in order to further their careers, the youngest scientists in a given department must kowtow to their elders' theories or run a significant professional risk. Peer review isn't any good at keeping flawed studies out of major papers, but it can be deadly efficient at silencing heretical views." Read further: it has also bred younger scientists, postdoctoral fellows and doctoral students who won't speak truthfully, including under oath in a deposition, if it means going on record against a senior scientist, university or research institution, making them risk averse to being labeled a problem or worse, the one who broke ranks and blew the whistle on wrongdoing. All of this suggests, Gobry wrote, that the current system isn't just showing cracks, but is actually broken, and in need of major reform. "There is very good reason to believe that much scientific research published today is false," he concluded, "[and that] there is no good way to sort [what Wilson called the] wheat from the chaff, and, most importantly, that the way the system is designed ensures that this will continue being the case."

Faked and fabricated data and peer review in published research has become the norm, no longer the exception. "Made-up identities assigned to fake e-mail addresses. Real identities stolen for fraudulent reviews. Study authors who write glowing reviews of their own research, then pass them off as an independent report," wrote Sarah Kaplan in the August 18, 2015 *Washington Post*. "These are the tactics of peer review manipulators, an apparently growing problem in the world of academic publishing." But this isn't how it is supposed to go. Journals are discovering that the "authorities" tasked with peer review aren't exactly who they thought they were, she opines, and, in truth, the authoritative checks are being rigged. The episode behind Kaplan's article was, just then, the latest in a string of fake peer review phenomenon. One of the world's largest academic publishers, Springer, retracted 64 articles from ten of its journals after discovering that their reviews were linked to fake e-mail addresses. This announcement came just nine months after 43 studies were retracted by BioMed Central – one of Springer's imprints – for the same reason.

In 2013, the editor of the *Journal of Vibration and Control* (JVC) and publisher SAGE became aware of a peer review ring involving hundreds of fraudulent assumed and fabricated identities that appeared to boil down to Chen-Yuan "Peter" Chen, an associate professor and computer scientist (also called a physicist in some publications) at the National Pingtung University of Education (NPUE) in Taiwan, who orchestrated the scam made easier by the magazine's lackadaisical editorial policies. After a fourteen-month investigation that ended in 2014, and following what it called "an unsatisfactory response" from Chen, his university was notified. SAGE retracted a total of sixty research articles published over a four-year period in *JVC* publication. National Pingtung University of Education investigated Chen, and he resigned. Of note, Taiwan's education minister, Chiang Wei-ling, resigned on July 14, 2014, over his ties to Chen whose papers, some of which Chiang coauthored, were retracted from an international scientific journal due to what was, at that point, alleged fraud.

The common thread among the peer review scandals was that they were exposed by outsiders. The SAGE story broke on July 8, 2013, at *Retraction Watch*, but the first hint of a conspiracy emerged that May, observed Daniel Sherman, the publisher's head of public affairs. It was learned from Sherman that "an author (later confirmed to be an innocent party) contacted SAGE after receiving two suspicious e-mails from individuals related to a paper he had submitted to *JVC*." The senders claimed, according to media reporting, to be university-based scientists but were using Google g-mail accounts. By directly contacting the scientists via their official university e-mail accounts, SAGE investigators discovered that the identity of at least one of the scientists had been stolen – that researcher did not have a g-mail account. SAGE did not reveal the names of the researchers involved.[259]

Reporting for *Science*, John Bohannon, documented that SAGE investigators discovered that the assumed identity and g-mail accounts had been used many other times in ScholarOne, SAGE's online manuscript submission system, and the reviewers and coauthors for those papers were also attached to suspicious e-mail addresses. Sherman further explained that they checked the wording of reviews written by those individuals, as well as the time it took to complete the review, which in some cases amounted to just "a few minutes."[260] The network of *JVC* papers that emerged was dubbed "incestuous," with the same small group of authors reviewing each other's work and appearing together as coauthors, Bohannon noted. By the end of the year, the investigators had a list of 130 e-mail addresses associated with the sixty papers that were subsequently retracted, with one scientist as coauthor on all of them and that was Peter Chen. The big giveaway of the wrongdoing that had been secretly going on over a protracted period of time: when SAGE sent an e-mail to all 130 e-mail addresses requesting that the authors confirm their identity, none responded. "The authors were contacted again by SAGE in May 2014 to inform them that their papers would be retracted in the July 2014 issue," says Sherman, but again none responded.[261]

With the Springer incident, *Retraction Watch*, which monitors and reports on retractions of fraud, plagiarism and other forms of misconduct in academic publishing, told Kaplan that the tally of withdrawn papers for faked reviews was up to 230 for the period 2012 to 2015. Those papers made up only a fraction of the hundreds of thousands of studies published each year, but they are the canary in the coal mine to publishers of academic publications. Oransky didn't know of any cases of faked peer reviews prior to 2012 but since then, he told Kaplan that the practice has accounted for fifteen percent of all retractions logged by his website. "It's like a virus that maybe was lying dormant for decades or centuries and all of a sudden, it's coming out," Oransky told Kaplan. "What's not clear is, are we better at finding it? Or is it actually a new phenomenon?" Springer, unlike SAGE, didn't reveal which papers were retracted nor in what publication they'd appeared. Further, Springer didn't identify the source of the fake reviews, though, the publisher did inform *Retraction Watch* that it believed "third party agencies" were likely involved.[262]

Certainly, it is also not clear in all cases who is responsible for the fraudulent reviews. "In a report for the journal *Nature* [published in the fall 2014], Oransky and his colleagues told the story of a Korean medicinal plant researcher who wrote peer reviews for 28 of his own papers," Kaplan wrote. "In July, the publishing company Hindawi found that three of its own editors had subverted the process by creating fake peer reviewer accounts and then using the accounts to recommend articles for publication [all 32 of the affected articles were thereafter re-reviewed]." These investigations have turned up a number of other insidious practices, including services that sell names and contact information for made-up experts guaranteed to provide an expedited, positive review.[263]

In an effort to spread the word on this problem, the Committee on Publication Ethics (COPE) reported in February 2015 that some agencies were – and are – selling services, ranging from authorship of prewritten manuscripts to providing fabricated contact details for peer reviewers during the submission process and then supplying reviews from these fabricated addresses. From COPE's website: "Some of these peer reviewer accounts have the names of seemingly real researchers but with e-mail addresses that differ from those from their institutions or associated with their previous publications, others appear to be completely fictitious."[264] Asked why scientists take risky shortcuts to be published, Oransky observed, "Publishing papers is the coin of the realm when it comes to academic advancement. You're going to do whatever you have to do to do it."[265]

Beyond peer review fraud, *Retraction Watch* has reported that one in twenty-five papers contain inappropriately duplicated images. Citing the work of Elisabeth "Elies" Bik, Ph.D., a microbiologist at Stanford University and a force in scientific integrity, the retraction tracking site reported that she and two editors at microbiology journals conducted a massive study looking at image duplication and manipulation in 20,621 published papers. Bik and her coauthors, Arturo Casadevall, M.D., Ph.D., a microbiologist at the Johns Hopkins Bloomberg School of Public Health, and Ferric C. Fang, M.D., director of clinical microbiology at Harborview Medical Center and a University of Washington professor of laboratory medicine, microbiology and medicine (infectious diseases), discovered 782 occurrences of inappropriate image duplication, including 196 published papers containing duplicated figures with alterations.[266] Much of Bik's work has focused on detection of manipulation of western blots.

Bik's findings are not isolated. In 2013, *Nature* reported findings from a previous screen of published papers, reported *Retraction Watch*. The screening was done by Italian bioinformatics startup BioDigitalValley and focused largely on gel-electrophoresis in Italian studies; it was discovered that one in four papers had inappropriate duplications of images to include repeated use of the same image and also copying and pasting gel bands.[267]

None of the papers Bik studied had been retracted at the time she screened them for problems. According to *Retraction Watch*, she has since submitted over 700 reports to journal editors showing the duplications, and written to about ten institutions where she found what she called "clusters," three to six papers from the same group containing duplications. "This has resulted in six retractions," according to the retraction site, "four of them in Fang's journal, *Infection and Immunity*, much lower than the 42 percent retraction rate she has received in the past from reporting plagiarism she found through Google Scholar searches. About 60 inappropriately duplicated images have been corrected since Bik first began reporting these findings to editors in the spring of 2014. She estimates an average of six months between her reporting and the retractions or corrections over the course of [her] project."[268] Bik, like Helene Hill, has struggled with journals in a similar way – falling short of getting misconduct research and case information published. Since it is not in journals' best interest to retract papers, they've declined repeatedly to publish Bik's paper about her findings. "I expect this to be a controversial paper," she told *Retraction Watch*, "no journal wants to hear a percentage of their papers is considered very bad and for this reason she believes her paper has been rejected. "One reviewer said, oh, this has to be published. Most of the others said it's very controversial, that it's not novel."

David Vaux, Ph.D., a cell biologist at the Walter and Eliza Hall Institute of Medical Research in Melbourne, Australia, told *Retraction Watch* that he believed the study from Bik, Casadevall and Fang provided strong evidence supporting the idea that a major reason for the lack of reproducibility of science papers is deliberate falsification of data. "They completed the Herculean task of visually inspecting the figures in over 20,000 papers, looking for image duplications," he told the retraction site. "They then looked at suspicious images more closely using image processing software. In this way," he continued, "they detected duplicated images in 3.8 percent of the papers. Although they could not distinguish accidental duplications, they did sub-categorize the papers into less worrying classes such as 'cuts,' 'beautification,' and 'simple duplications' and more worrying classes in which the duplicated images were repositioned or further altered such as by stretching or rotating." Vaux shared further that he believed the strength of their evidence should be enough "to convince everyone that there is a major problem with how research is being conducted. Now we need," he offered, "to determine what to do with this information. Should journals implement similar visual screens, should they use computerized screens, or some sort of combination? What is the best way to handle the suspicious images that are detected?"[269]

Vaux and a growing number of research ethicists have reached the same conclusion regarding big science gone bad: the lack of reproducibility associated with published science papers is deliberate falsification of data included to support the premise of the study. According to a spring 2016 article in *Nature*, more than seventy percent of researchers have tried and failed to reproduce another scientist's experiments, and more than half have failed to reproduce their own experiments. Those are some of the telling figures that emerged from *Nature*'s survey of 1,576 researchers who took a brief online questionnaire on reproducibility in research.[270] Data gleaned from *Nature*'s survey pointed to often contradictory attitudes toward reproducibility. Although fifty-two percent of those surveyed agreed that there is a "significant crisis" of reproducibility, wrote *Nature*'s Monya Baker, less than thirty-one percent think that failure to reproduce published results means that the result is probably wrong, and most stated that they still trust the published literature. Baker reported that data on how much of the scientific literature is reproducible are rare and generally "bleak." She cited the best-known analyses from psychology and cancer biology, which found rates of around forty and ten percent, respectively. "Our survey respondents were more optimistic," wrote Baker, "seventy-three percent said that they think that at least half of the papers in their field can be trusted, with physicists and chemists generally showing the most confidence."

Casadevall observed that the results of *Nature*'s survey have captured a confusing snapshot of pervasive attitudes and perceptions regarding the reproducibility issue. "At the current time there is no consensus on what reproducibility is or should be," he told Baker. "The next step may be identifying what is the problem and to get a consensus." Baker reiterated that although the vast majority of researchers in the *Nature* survey had failed to reproduce an experiment, less than twenty percent of respondents stated that they had ever been contacted by another researcher unable to reproduce their work. She noted further that *Nature*'s findings regarding researcher-to-researcher conversations regarding reproducibility – lack of reproducibility – were "strikingly similar" to another online survey of nearly 900 members of the American Society of Cell Biology. Baker attributed those similarities to the difficulty of conversations reproducibility. "If experimenters reach out to the original researchers for help, they risk appearing incompetent or accusatory, or revealing too much about their own projects," she wrote. Further, the number of respondents to the survey who reported ever having tried to publish a replication study is very small. "When work does not reproduce," she continued, "researchers often assume there is a perfectly valid reason. What's more, incentives to publish positive replications are low and journals can be reluctant to publish negative findings; in fact, several respondents who had published a failed replication said that editors and reviewers demanded that they play down comparisons with the original study."

A study out of the Center for Open Science, a nonprofit based in Charlottesville, Virginia, and headed by Brian Nosek, Ph.D., the center's executive director and a University of Virginia psychology professor, was the subject of an August 27, 2015, *Washington Post* article. Though science's reproducibility problem isn't limited to psychology, Nosek's study, featured in a paper titled "Estimating the reproducibility of psychological science," was published in the journal *Science*, was the result of a widespread effort that took four years and 270 researchers who attempted to reproduce the results of 100 experiments (of the 1.5 million scientific studies published per year) that had been published in three prestigious psychology journals.[271] The impetus for the study was the long-held belief that much of what gets published in these well-respected journals is fundamentally "squishy – that the results tell a great story but can't be reproduced when the experiments are run a second time," wrote the *Post*'s Joel Achenbach. Nosek focused on psychology only because most of the center's researchers are psychologists. The phenomenon of irreproducible results in psychology comes on the heels of well publicized frauds, some of which have already been mentioned herein per the work of John Ioannidis and others but also include, wrote Achenbach, Dutch psychologist Diederik Stapel, who admitted in 2011 that he'd been fabricating his data for years.

"A more fundamental problem, say Nosek and other reform-minded scientists, is that researchers seeking tenure, grants or professional acclaim feel tremendous pressure to do experiments that have the kind of snazzy results that can be published in prestigious journals," Nosek explained to Achenbach. "They don't intentionally do anything wrong, but may succumb to motivated reasoning. That's a subtle form of bias, like unconsciously putting your thumb on the scale. Researchers see what they want and hope to see, or tweak experiments to get a more significant result. Moreover," continued Achenbach, "there's the phenomenon of 'publication bias.' Journals are naturally eager to publish significant results rather than null results. The problem is that, by random chance, some experiments will produce results that appear significant but are merely anomalies – spikes in the data that might mean nothing." Given the small sample size of Nosek's study, he acknowledged that his own study would itself be tricky to reproduce because there were subjective decisions made along the way and judgment calls about what, exactly, "reproduced" means. The very design of the review injected the possibility of bias, he explained to Achenbach, in that the volunteer scientists who conducted the replications were allowed to pick which experiments they wanted to do.[272]

The *Nature* survey asked scientists what led to problems in reproducibility in an effort to get at root causes of the problem. The results showed that sixty percent of respondents stated that each of two factors, including pressure to publish and selective reporting (cherry-picking an experiment's results to get a desired outcome), always or often contributed to irreproducibility. More than half of those admitted to insufficient replication in the lab, poor oversight or low statistical power, and a similar number pointed to obstacles such as variability in reagents or the use of specialized techniques that are often difficult to repeat.[273] But there are other factors that play leading roles in the irreproducibility problem, noted Judith Kimble, Ph.D., Vilas Professor, Howard Hughes Medical Institute investigator (also in the laboratory of molecular biology, department of medical genetics, and department of cell and regenerative biology) at the University of Wisconsin at Madison College of Agricultural and Life Sciences, to include competition for grants and positions, and what she called a growing burden of bureaucracy that takes away from time spent doing and designing research.[274] "Everyone is stretched thinner these days," she told *Nature*'s Baker, "and the cost extends beyond any particular research project. If graduate students train in labs where senior members have little time for their juniors, they may go on to establish their own labs without having a model of how training and mentoring should work. They will go off and make it worse." Still, these reasons are not an excuse for bad actors purposefully fabricating data.

"Publication of a flawed manuscript [based on flawed experiments] has significant consequences for the progress of science," according to an article by Sean P. Murphy et al. in the February 2014 *Journal of Neuroscience*. "When this proves to be intentional, science is brought into disrepute, and this puts even more pressure on the shrinking resources that society is prepared to invest in research. All scientific journals, including the *Journal of Neuroscience*, have witnessed a marked increase in the number of corrections and retractions of published papers over the last ten years, and uncovered a depressingly large number of fabrications amongst submitted manuscripts." Murphy et al. attributed the detection of these "spoiled" manuscripts to improved methods that journals employ to detect misconduct, which is responsible for more than two thirds of the 2,000 or more retracted articles indexed in *PubMed*.

Like Bik, Murphy et al. acknowledged that image manipulation has become increasingly sophisticated but also more readily identifiable by means of image software. In Murphy et al.'s field – and also Bik's – western blots have become highly questionable currency, as these can be cropped, reversed, duplicated, mislabeled, lightened and darkened, they explained.[275] "Journals with the highest visibility [the most prominent] clearly suffer from the most image manipulation. Journal editors," Murphy et al. explained, "resort increasingly to a request for original data to accompany manuscript submissions and insist that any form of 'improvement' be clearly described. On average," they continued, "three to four cases arise in Journal of Neurochemistry each year. Invariably, the corresponding author's institution is notified, which can result in dismissal, resignation, or nonrenewal of a contract."

But what about data fabrication? This is more difficult to detect, Murphy et al. opined, unless statistical analysis of the distribution of the data points reveals anomalies. "Ultimately, it is a 'failure to repeat' that throws doubt on the veracity of published data," they wrote. The reasons for fabricating data, many of which have been put forward in Hill's case and articulated by Fanelli, Wilson and others, fall in line with Murphy et al.'s belief that the answer can range from financial, predicated on securing a grant or in response to dubious practice by employers offering cash bonuses for publishing in high-profile journals, or personal advancement and reputation. From what prior misconduct cases have shown, and Murphy et al. confirm, more than half – fifty percent – of senior scientist authors keep their careers in science and continue to publish. But it's often a different outcome for junior laboratory scientists, who are almost universally more affected by a misconduct charge than the senior investigator, publishing an average of only one paper a year after a misconduct determination.[276] Ultimately, however, Murphy et al. observed that it is often the senior investigator who must take responsibility when former laboratory members have left the institution and cannot be contacted.

"The United States, which has far more biomedical research than any other country and where the problem [of scientific misconduct] has been taken seriously for two decades, has seen hundreds of cases, many of them serious," observed Richard Smith, M.D., in an article he wrote for the May 2006 *Journal of the Royal Society of Medicine*. Smith further noted that most misconduct cases, whether in the United States or abroad, are probably not publicized. "They are simply not recognized, covered up altogether, or the guilty researcher is urged to refrain, move to another institution, or retire from research. I have spoken perhaps a dozen times on research misconduct in several countries and often to audiences where people come from many countries," he continued, "I usually ask the members of these audiences how many know of a case of misconduct [consciously not offering a definition of misconduct]. Usually half to two-thirds of the audience put up their hands. I then ask whether those cases were fully investigated, people punished, if necessary, lessons learnt, and the published record corrected. Hardly any hands go up."[277] From his many years of research, much of which he has published in prestigious biomedical journals, Smith has come to the conclusion that it is the "cover ups" that explain why it is so hard to get good data on the prevalence of serious research misconduct.

Chapter 13 –
Who Does it Hurt?

When Charles Seife and his students exposed the Food and Drug Administration's purposeful cover-up of widespread misconduct, he also documented who gets hurt when the agency fails to notify the public of what it has uncovered. Such was the case with a RECORD 4 study, one of four large clinical trials that involved thousands of patients who were recruited at scores of clinical sites in more than a dozen countries around the world. The trial, Seife explained, was used as evidence that a new anti-blood-clotting agent, rivaroxaban[278], brand name Xarelto® and manufactured by Bayer, one of the world's largest pharmaceutical enterprises, was safe and effective. The FDA purportedly inspected or had access to external audits of sixteen of the RECORD 4 sites. "The trial was a fiasco," he wrote. "At Dr. Craig Loucks' [an M.D. and orthopedist] site in Colorado, the FDA found falsified data. At Dr. Ricardo Esquivel's [an M.D. and another orthopedist] site in Mexico, there was 'systematic discarding of medical records' that made it impossible to tell whether the study drug was given to the patients. Half of the sites that drew FDA scrutiny – eight out of sixteen – there was misconduct, fraud, fishy behavior, or other practices so objectionable that the data had to be thrown out. The problems were so bad and so widespread that," he continued, "contrary to its usual practice, the FDA declared the entire study to be 'unreliable.'" What bothered Seife were the medical journals. "The results from RECORD 4 sit quietly in The Lancet without any hint in the literature about falsification, misconduct, or chaos behind the scenes. This means that physicians around the world are basing life-and-death medical decisions on a study that the FDA knows is simply not credible."

More troubling, it wasn't just the one study. The FDA, Seife opined, had found major problems with sites involved in the other three clinical trials that were used to demonstrate rivaroxaban's safety and effectiveness as a prescription blood thinner. "RECORD 2, for example, was nearly as awful as RECORD 4: four out of ten sites that the FDA inspected showed evidence of misconduct, or other issues grave enough to render the site's data worthless – including clear evidence of data falsification at one site. In aggregate," Seife noted, "these problems raise serious doubts about the quality of all four key rivaroxaban studies – and, by extension, doubts about how seriously we should take the claim that rivaroxaban is safe and effective. The FDA is keeping mum," he iterated, "even as wrongful death lawsuits begin to multiply."

Seife has called the FDA's failure to notify the public more than a sin of omission. The FDA convened a committee of outside scientific experts in March 2009 to determine the robustness and what Seife couched as "meaningfulness" of the rivaroxaban trials, RECORDS 1, 2, 3 and 4. Of this decision to call the experts to weigh in the rivaroxaban study, he observed that the agency regularly calls in advisers to get advice or, in his words, "to get cover," about a decision the agency is being compelled to make about, in this case, a new drug. When the agency briefed its outside panel, the FDA wasn't entirely forthcoming with what it knew about rivaroxaban. "It was, to put it mildly, coy about the problems it was finding," Seife wrote later. "It said only that inspectors had found 'significant issues' at two clinical sites involved in the RECORD 4 study – and that data from one of them was included in the analysis. Inspections were still ongoing," he added just then, "so it's not easy to say precisely what the agency knew at that point, but it's clear that the FDA wasn't admitting to everything it knew." Clearly, according to Seife's findings, a significant number of inspections had been completed a month prior to the convening of the outside expert panel. "We know for certain," he documented after the fact, "that the agency was fully aware of major issues beyond the two it revealed at the advisory committee." To prove this to be true, he provided a memorandum dated three days before the advisory committee meeting met in which the FDA detailed falsification of data by a subinvestigator at a RECORD 2 site. The advisory committee was never informed of this information.[279]

Candidly, Seife reiterated that while all of this might seem to be a miscommunication or oversight, the FDA has a history of not notifying the public about the misconduct it finds. "About a decade ago, the agency got into trouble over a newly approved antibiotic, Ketek[280]," he wrote. "Inspectors had found extensive problems (including fraud) affecting key clinical trials of the drug. Yet the agency did its best to hide the problems from even its most trusted advisers." Congressman John Dingell's House Subcommittee on Oversight and Investigations, held hearings on the adequacy of the FDA efforts to assure the safety of the drug supply and in a report it issued on February 13, 2007, it documented the Ketek controversy:

> FDA staffers publicly complained that safety problems and data integrity issues were ignored prior to approval, and the House Committee on Energy and Commerce held hearings to examine these complaints. One doctor went to prison because she falsified data in her portion of the clinical trials because Ketek seemed to cause liver problems, including liver failure, to a greater extent than would be expected of a common-use antibiotic.

According to a July 19, 2006 *New York Times* article, it was revealed that in an internal review, a federal drug safety official concluded that the controversial antibiotic made by a French drug company Hoechst Marion Roussel (later Sanofi-Aventis) should be withdrawn, according to e-mail messages exchanged among top agency officials. The *Times* reported that the official, David Graham, M.D., part of the FDA's drug safety office, wrote in a message dated June 16 that the agency's approval of Ketek, an antibiotic made by Sanofi-Aventis that is also known as telithromycin, was "a mistake." He wrote: "It's as if every principle governing the review and approval of new drugs was abandoned or suspended where telithromycin is concerned." Referring to reports of negative, or adverse, drug reactions voluntarily submitted to the agency, he continued, "We don't really know if the drug works; no one is claiming it works better than other, safer drugs;

and we're flying blind as far as safety goes, except for our own A.D.R. [adverse drug reaction] data that suggests telithromycin is uniquely more toxic than most other drugs." Graham concluded that the agency should recommend the drug's "immediate withdrawal."

In writing about the Ketek controversy later, David B. Ross, M.D., Ph.D., put it bluntly:

> Three years ago, the Food and Drug Administration (FDA) approved the drug Ketek (telithromycin), lauding it as the first of a new class of antimicrobial agents that circumvent antibiotic resistance. Since then, Ketek has been linked to dozens of cases of severe liver injury, been the subject of a series of increasingly urgent safety warnings and sparked two Congressional investigations of the FDA's acceptance of fraudulent safety data and inappropriate trial methods when it reviewed the drug for approval. As a former FDA physician who was involved in the Ketek review, I believe there are lessons to be learned from an examination of the events surrounding the approval of this product.[281]

Though Ross also documented, in agreement with Seife's findings, that in January 2003, over reviewers' protests, "FDA managers hid the evidence of fraud and misconduct from the advisory committee, which was fooled into voting for approval."[282] But when the reports of misconduct at one clinical site began popping up in the press, along with stories of the liver damage to which Ross referred to in his reporting, and blurred vision associated with the new drug, that Congress stepped into the case, demanding information about the fraud that had taken place. While Dingell's subcommittee had gotten into the controversy, so did Senator Chuck Charles Grassley, who complained vigorously that even he couldn't get key information about the misconduct surrounding the drug – from the FDA. "Every excuse under the sun has been used to create roadblocks," he opined. "Even in the face of congressional subpoenas requesting information and access to FDA employees." FDA commissioner Andrew C. von Eschenbach, M.D. attempted to explain to Congress why his agency didn't inform its advisory committee about the problems with the Ketek study, peddling the story that since the investigation results were preliminary, the agency decided to hold the advisory committee meeting as planned, "without notifying" the committee of the potential problems.[283] But Democrat Michigan congressman Bart Stupak wasn't having it. "So either you are not being forthright with us, when I believe you are, but whoever is doing your work is trying to lead this committee down the wrong path."[284] As Seife observed, the correct path showed that site after site involved in study 3014, as well as other key Ketek studies, were tainted as well.

Von Eschenbach didn't mention to Congress that he'd warned FDA reviewers in a June 2006 meeting not to discuss Ketek outside the agency, in what could only be concluded was a circling the wagons maneuver. By the end of 2006, according to Ross, Ketek had been implicated in 53 cases of acute severe liver injury and 12 cases of acute liver failure. The FDA didn't relabel Ketek to indicate its possible severe heptotoxicity until 16 months after the first liver failure cases became public. The withdrawal of approval for two indications, acute bacterial sinusitis and acute exacerbation of chronic bronchitis, for which Ketek's efficacy had never been demonstrated, did not occur until February 12, 2007, only a day before the congressional hearing on Ketek.[285]

Certainly, to Seife and others, in the wake of the Ketek debacle, the FDA had learned little. "On occasion, the FDA has even actively approved and promoted statements about drugs that, according to its own inspectors, are based upon falsehoods. At the end of 2011," wrote Seife, "the FDA learned that an audit of a Chinese site involved in a key clinical trial of a different anti-clotting agent, apixaban, brand name Eliquis®, had turned up evidence of fraud." As he described it, personnel had apparently been fiddling with patient records. "Worse yet," he opined, "the fraud appeared to invalidate one key finding of the study. Just three months earlier, the researchers running the trial proudly announced in the *New England Journal of Medicine* that there was a 'significant reduction in mortality' among patients who took apixaban compared with those who took the old standby, warfarin. Alas," he continued, "the moment you exclude the data form the Chinese fraud site, as per standard FDA procedure, that statement went out the window." He drew attention to the apixaban – Eliquis – label, the one approved by the FDA after the fraud was discovered – "and you read that 'treatment resulted in a significantly lower rate of all-cause death

[...] than did treatment with warfarin, backed up by the data set with the Chinese site included. In other words," he continued, "the label is carrying a claim that the FDA knows is based upon fraud." When Seife queried the FDA on this subject, the agency replied: "The FDA extended the drug's review period to address the concerns. However, the review team did conclude concluded [*sic*] that the date at that site and other sites in China did reflect meaningful clinical information; that was not what was considered unreliable." But is that really the way it happened?

The FDA found out about data irregularities prior to denying approval of Bristol-Myers Squibb and Pfizer's Eliquis in a June 2012 complete response letter; this was before approval for use of the drug in the United States. Prior to the rejection, Bristol told the agency, according to a July 9, 2013 *FierceBiotech* article, that some patients got the wrong medicine, records were changed on the sly and serious adverse reactions were missing, *Bloomberg*'s Drew Armstrong noted. As Elsevier's *Pharmaceutical Approvals Monthly* first reported, concerns about the integrity of the data contributed to the nine-month delay in the FDA's approval, which came in December and went into effect in early 2014. There are several troubling issues exposed in the articles. One is whether the alleged fraud in the Eliquis study in China marks an isolated case or a pattern of misconduct in drug research in the country with the world's largest population. This story comes fast on the heels of GlaxoSmithKline firing its head of research and development in Shanghai after allegations of unscrupulous treatment of pre-clinical study data, prompting the company to slam the brakes on an early-stage human trial.[286] Each of these developments is largely out of sight, out of mind to the drug consumer. *Bloomberg*'s article also highlighted how the FDA cleared the death benefit on Eliquis' label over the objections of FDA investigator Thomas Marciniak, M.D., an outspoken critic of how clinical drug trials are conducted and overseen, after his independent review of the trial data.[287] Seife shared Marciniak's concerns as the FDA cleared the therapy, which is an alternative to generic warfarin, without the public's knowledge about its deep concerns about the data irregularities in China. Shortly before Eliquis was approved, at the beginning of December 2013, Marciniak went on record in the *BMJ*, opining that "the clinical trial system is broken." He didn't hold back. "Drug companies have turned into marketing machines," he said. "They've kind of lost sight of the fact that they're actually doing something which involves your health. You've got to take away the key components of the trials from drug companies."[288]

Seife, again, cautions that what happened with the Eliquis approval isn't an isolated incident. "I had previously encountered bogus data on FDA-approved labels when a colleague and I were looking into a massive case of scientific misconduct – a research firm named Cetero had been caught faking data from more than 1,400 drug trials. That suddenly worthless data had been used to establish the safety or effectiveness of roughly 100 drugs," he continued, "mostly generics, that were being sold in the United States. But even after the agency exposed the problem, we found fraud-tainted data on FDA-approved drug labels." The FDA, he added, still maintains its silence about the Cetero affair. "To this day, the agency refuses to release the names of the 100-odd drugs whose approval data were undermined by fraud." But the FDA covers up drug-related misconduct, according to Seife's findings, in other, perhaps subtler ways. "For example, the agency publishes the canonical listing of generic drugs in the United States known as the 'Orange Book.' Prescription drugs in this book are often given what's called a 'therapeutic equivalence code.' This code," he explained, "is a two-letter designation that signals the quality of the scientific evidence that a generic is 'bioequivalent' to the name-brand drug. The code 'AB,' for example, tells pharmacists and physicians that there are solid scientific studies proving that bioequivalence. Another code, 'BX,' signals that there isn't sufficient data to prove the generic is bioequivalent to the name brand."

Seife observed that when the Cetero misconduct was uncovered, key bioequivalence studies for scores of generic drugs turned out to be "worthless." But some of those drugs should have been downgraded to BX. "[E]ven though the FDA updates the Orange Book monthly," he explained, "there was no rash of drugs losing their AB rating in the months after the Cetero affair broke. In the year-and-a-half after the Cetero fraud was first announced, I was able to identify a grand total of four generic drugs (in various dosages) that were downgraded to BX, none of which," he iterated, "appeared to be linked to the Cetero problem. On the other hand, the one prescription drug that I knew for sure had been hit hard by the Cetero fraud – both key studies to keep its AB badge for months without any valid data backing the drug's bioequivalence. When asked, point blank, whether the agency had downgraded the bioequivalence code of any products due to the Cetero affair, officials promptly dodged the question." When the FDA ultimately responded to him, Seife got this response: "If the data were not provided within six months or the data provided did not support a finding of bioequivalence, FDA said it would consider changing the generic product's therapeutic equivalence rating in the Orange Book from AB to BX." The agency offered no indication that the bioequivalence rating had actually been changed. Seife considered the Cetero affair yet another indicator of a pattern of bad behavior.

Following on the heels of the Cetero troubles came the GVK Biosciences, a Hyderabad, India company, scandal. The FDA's European equivalent, the European Medicines Agency, found that one thousand drugs in various dosages were affected by GVK's data manipulations and suggested pulling 700 off the market. But the FDA didn't respond in kind to GVK drugs in the American market. According to the reporting, there were approximately 40 or more drugs whose approval rested on GVK testing. But, like the Cetero affair, no one knows which ones.[289]

Seife asked the important questions: why does the FDA remain silent about fraud and misconduct in scientific studies of pharmaceuticals? Why would the agency, he postulated, allow claims that have been undermined by fraud to appear on drug labels? Why on earth, he further iterated, would it throw up roadblocks to prevent the public, the medical community, its advisory panels, and even Congress from finding out about the extent of medical misconduct? "The answers the FDA gives are fascinating," he observed, "they show how an agency full of well-meaning people can do intellectual backflips to try to justify secrecy." He pointed to an agency that falls back to one excuse when challenged: exposing the details about scientific wrongdoing – naming the trials that were undermined by research misconduct, or revealing which drugs' approvals relied upon tainted data – would compromise what it called "confidential commercial information" (proprietary data on the drug) that might hurt drug companies if revealed. But as he iterated, this claim "falls apart under scrutiny."[290] Seife wrote that the courts have ruled that when information is provided by companies involuntarily, such as the information that an FDA inspector finds, commercial confidential information refers to propriety material that causes substantial, specific harm when it falls into the hands of a competitor. "It doesn't cover the embarrassing peccadilloes – or misconduct that might cause bad publicity when word gets out." Certainly, too, the American public isn't the competitor but too often the victim of a bad drug.

Seife and others have also come to know another excuse heard too often, one framed around the belief that if the FDA tells the public about the problems with a drug, especially when, in the FDA's judgment, the misconduct doesn't pose an immediate risk to public health, why tell them? "When my colleague and I asked the director of FDA's Center for Drug Evaluation and Research why the agency wouldn't name the drugs affected by the Cetero fraud," Seife reported, "she told us that the matter 'did not rise to the level where the public should be notified. We felt it would result in misunderstanding and inappropriate actions.' But even the most paternalistic philosophy of public health can't explain why the FDA would allow drug companies to put data on its labels that the agency knows are worthless, or to fail to flag bioequivalence problems in a publication that is specifically designed for the purpose of flagging those very problems." The FDA logic flies in the face of the very purpose of the agency – to protect the public health. "Yet, over and over again, the agency has proven itself willing to keep scientists, doctors, and the public in the dark about incidents when those scientific trials turn out to be less than reputable. It [the FDA] does so not only by passive silence, but by active deception," wrote Seife. "And despite being called out numerous times over the years for its bad behavior, including some very pissed-off members of Congress, the agency is stubbornly resistant to change. It's a sign," he concluded, "that the FDA is deeply captured, drawn firmly into the orbit of the pharmaceutical industry that it's supposed to regulate. We can no longer hope that the situation will get better without firm action from the legislature." In his farewell address to the nation on January 17, 1961, President Dwight D. Eisenhower warned the American people to keep a careful eye on what he called the rise of the "military-industrial complex" that had rapidly developed in the post-World War II years. In the spirit of Eisenhower's warning, perhaps the American public should be equally concerned of the rise of "Big Pharma" and the risk to the public health, a threat worthy of Eisenhower's presidential cautionary. The rise of "Big Pharma" is the military-industrial complex threat of 1961.

Dingell[291] had seen all of this before. He went on record on June 3, 1993, regarding the case of Robert C. Gallo, M.D., the NIH's world-famous AIDS researcher; in fact, the American co-discoverer of what causes the AIDS virus. "Because the subcommittee has not yet completed its investigation [just then]," he wrote, "I cannot tell you what our conclusion will be. I can say, however, that one of the things that puzzles and troubles us is Dr. Gallo's entanglement in a large number of unusual circumstances that he claims are misunderstandings and coincidences." Dingell explained that after Syed Zaki Salahuddin, Ph.D., one of Gallo's long-time laboratory scientists was convicted of a felony in connection with his activities at Gallo's lab, Gallo explained that he had been unaware of Salahuddin's activities. Shortly thereafter, Dingell opined, Prem S. Sarin, Ph.D., Gallo's deputy laboratory chief, was indicted for activities unrelated to those of Salahuddin but also stemming from work in the laboratory. Gallo explained, again, that he had no knowledge of his deputy laboratory chief's misconduct and that the two separate criminal cases involving his laboratory scientists were merely "unfortunate coincidences." But then Dingell's subcommittee learned that two subjects described in an article in *Lancet* coauthored by Gallo and French researcher Daniel Zagury, M.D., Ph.D., had died, but that Gallo had failed to report these deaths to the NIH as was required by grant regulations and that he had further erroneously reported in *Lancet* that he had observed no adverse reactions in the human subjects who were part of the study.[292] Predictably, Gallo had an explanation ready and waiting for someone to ask him why he'd failed to comply with NIH's grant requirements. He told investigators that the statement in *Lancet* was an inadvertent error on his part and that his failure to comply with NIH procedure was a result of his unfamiliarity with the regulations – this, Dingell, observed, despite Gallo's some twenty years of employment at the NIH.

"More recently," wrote Dingell just then, "in the controversy over the AIDS blood test, Dr. Gallo is under investigation because of, among other things, allegations that statements he made in the patent application and thereafter in the patent dispute were deliberately misleading. Dr. Gallo," he continued, "first stated that the virus he used was definitely different from that used by the competing French team. When genetic sequencing proved that the viruses were identical, he suggested that the French must have taken his virus. When that claim was challenged, Dr. Gallo explained that there must have been an inadvertent contamination in his laboratory." But that wasn't the worst of it. There were also serious questions about the cell line in which Gallo grew his viruses, similar in nature to the questions posed in the Helene Hill case. What eventually came to light was that Gallo's cell line belonged to Adi F. Gazdar, M.D., a researcher at another NIH institute. Gallo explained that this was a "misunderstanding" and that he had never intended to deprive Gazdar of credit but merely renamed the cell line for convenience.[293]

Dingell declared the entire affair a very troubling case. "An eminent scientist who heads one of the most important laboratories in the world is embroiled, directly or indirectly, in many different serious situations, which" Dingell opined, "Dr. Gallo himself invariably characterizes as inadvertent errors, miscommunications, and unfortunate coincidences." At that time, Dingell noted that NIH's then defunct Office of Scientific Integrity (not yet reconstituted as the Office of Research Integrity) had found Gallo innocent of fraud or serious misconduct, a finding about which Dingell had serious questions. "We understand," he wrote, that the NIH itself receives millions of dollars each year in royalties from the blood test, but that should be no impediment to its assessing the charges against Dr. Gallo objectively."

Gallo didn't get off on the misconduct charge. After three years of investigations, in December 31, 1992 *New York Times* reported that the Office of Research Integrity (ORI) found Gallo had, indeed, committed scientific misconduct. The investigators found that he had falsely reported a critical fact in a paper that appeared in the May 1984 *Science* journal in which he described isolating the virus that causes AIDS.[294] The ORI report stated that Gallo had intentionally misled colleagues to claim credit for himself and diminish credit due his French competitors, led by Luc Montagnier, M.D. The report further noted that his false statement had impeded potential AIDS research progress by diverting scientists from potentially fruitful work with the French researchers. Gallo had faced serious questions about his work since the paper was published. Many in the scientific community argued he'd taken credit for work performed by French scientists and that he may even have taken the virus the French were studying and claimed it was his own – the crux of Dingell's findings. Also charged with misconduct was Mikulas Popovic, M.D., Ph.D., a

Czechoslovak immigrant who'd actually performed the important AIDS experiments in Gallo's laboratory.

Gallo claimed he wasn't guilty and had every intention to appeal the ORI decision. "After reviewing everything I and my colleagues have ever published on the discovery of the AIDS virus and the development of the AIDS blood test, the Office of Research Integrity could only take issue with a few trivial mistakes and a single sentence written by me," he opined in the *Times* article. He took issue with the federal investigators' conclusions, calling them "utterly unwarranted." Gallo further offered this:

> On a broader level, this endless and incompetent Government investigation should be of concern to everyone seeking to advance medical knowledge. My laboratory's contributions to the advancement of medical science are undisputed. For the past three years, however, I have spent a substantial amount of my time responding to issues [raised in the investigations].[295]

Beyond the ORI's finding of Gallo's guilt, the report determined that Gallo intentionally misled scientific colleagues by informing them he had grown an AIDS virus in his laboratory for study and that he had not grown or studied a similar French strain of the virus; in truth, Gallo himself had grown the French virus and used it in furthering his own research, the report said. While searching for the cause of AIDS, Gallo had received a sample of a virus being studied by French researchers and had worked extensively with it to extend his own discoveries, the ORI report concluded. Gallo left little credit for the French scientists in his 1984 paper because he claimed he had not been able to grow enough of the French AIDS virus, explaining later that it "has not been transmitted to a permanently growing cell line for true isolation and therefore has been difficult to obtain in quantity."[296] Gallo subsequently argued that this wording had been misinterpreted and that what he meant was that the virus was difficult for anyone to grow, not that he himself had failed to grow it; in fact, investigators, according to the *Times*, showed that the French virus had been grown in cell lines in Gallo's own laboratory, and worked with there. When referring to that sentence, the new ORI report stated this:

> Dr. Gallo falsely reported the status of L.A.V. [lymph-adenopathy-associated virus] research when he wrote the statement, and this constitutes scientific misconduct. L.A.V. refers to the French strain of the AIDS virus. [The report went on:] The explanations that Dr. Gallo proffered for the statement are neither credible when the evidence is considered, nor do they vitiate the impropriety of falsely reporting the status of L.A.V. research.

Additionally, the *Times* reported that the ORI found that Gallo warranted censure on these four other counts: Firstly, referring to his role as a referee for a different article submitted to a journal by his French competitors, in which he altered several lines to favor his own hypothesis about the AIDS virus, the report said the revisions were "gratuitous, self-serving and improper." Secondly, as to the many errors in the 1984 paper, which was coauthored with Popovic, the report concluded: "In light of his role as senior author, Dr. Gallo must bear substantial responsibility for the numerous discrepancies, including four instances of scientific misconduct attributed to Dr. Popovic." Thirdly, on the standards of Gallo's laboratory recordkeeping, the report stated: "Especially in light of the groundbreaking nature of this research and its profound public health implications, ORI believes that the careless and unacceptable keeping of research records reflects irresponsible laboratory management that has permanently im-paired the ability to trace the important steps taken." Fourthly, Gallo, the report said, also failed to determine in a timely way the exact origin of some of the crucial cells in which he grew the finicky virus. Like the viruses themselves, the cells were also found to have been borrowed from another scientist without giving him due credit in the paper. Later, Gallo also refused to share the cells freely with other scientists trying to duplicate the important work, the report stated.

Dingell had every reason to be frustrated with the Gallo case and the original federal finding that had absolved him of guilt. This was a complicated, protracted case that had begun in the early 1980s and took nearly a decade to adjudicate. The controversy had become so heated and inextricably tethered to the scientific integrity of the United States and France, that the presidents of both nations tried to bring the matter to a close in 1987 when it was agreed that credit and payment of patent royalties from work with the AIDS virus and the blood test to detect it would be split fifty-fifty.[297] Still, the issue didn't die.

Reporting on the case[298], Janine Roberts wrote that Gallo had also gotten into trouble in Africa. "His laboratory had developed a vaccine based on transplanting into the shell of another virus a putative part of HIV," she wrote. "It seems this was easier than using HIV itself as it was difficult to find. This vaccine was injected into a few Congolese in Africa and Paris and three of them died. It was then discovered that his vaccine had only been approved for use in animals. But Gallo escaped with only a mild reprimand." But in May 1991, knowing that he was about to implicated by the OSI, Gallo wrote to *Nature*, reported Roberts, "confessing that he now realized that the French virus and his own were the same. He blamed his error on inadvertent laboratory contamination. Then," she continued, "a similar confession appeared in the United Kingdom from a leading British virologist and colleague of Robert Gallo, Dr. Robin Weiss, the scientist I had first met when he was chairing the NIH workshop on SV40[299], and again when he was chairing the Royal Society debate on the polio vaccine and HIV." Weiss claimed, too, that his mistake was attributable to inadvertent laboratory contamination. "He also, like Gallo, had used the French virus to secure a UK patent for the HIV virus."[300]

At stake in this controversy between the United States and France were millions of dollars the two governments received annually in royalties from the manufacturers of the blood tests. Lawyers for the Institut Pasteur in Paris, where the French work was performed, had previously asked the United States to turn over its half of the profits from the blood test – about $50 million since 1985. Gallo earned about $100,000 per year from the royalties on the blood test, as did his French colleague, Montagnier. After the *Chicago Tribune*'s John Crewdson[301], a major player in the story's unraveling, penned a fifty-thousand-word exposé laying out most of the charges against Gallo and his laboratory in November 1989, a federal investigation ensued. When the French found out about the renewed allegations against Gallo, largely through the media, including the *Tribune* and the *Times*, Michael Epstein, an attorney for the Institut Pasteur, called for the United States to renegotiate its agreement giving equal credit to each country, a plan that would assign a larger share to France going forward. "This ought to move the United States government to action," he said.

"Dr. Gallo has always told us that he was never able to grow L.A.V. One of the most important reasons why the Pasteur had settled the dispute two years before – in 1987 – was that Dr. Gallo told us that and said there was no evidence to the contrary. Now even the United States government is saying that he knowingly lied."[302] According to the *Times*, the new report reversed the findings of Bernadine P. Healy, M.D., then director of the National Institutes of Health and Gallo's superior. After receiving the report from the Office of Scientific Integrity, she had concluded in September 1991 – despite evidence to the contrary – that Gallo did a number of things wrong but was not guilty of the most serious charge, that of scientific misconduct; it was Healy's actions Dingell had found so questionable. Investigators at National Institutes of Health and in Congress disagreed, among them Dingell, as did a panel of independent scientists and Gallo was charged with scientific misconduct. He appealed. But his case, like the Imanishi-Kari debacle, wasn't decided on the merits but under the veil of politics, pressure and the chasm that had opened up between the Department of Health and Human Services and its own watchdog office. The perception and manipulation of the ORI findings process was being attacked from all sides, compromising the office's ability to do the work of policing bad science for which it was established.

By April 1992, Roberts wrote that an ABC *Prime Time* investigation stated the Gallo case might be "the greatest scientific fraud of the twentieth century." The televised segment ran over a portrait of Gallo: "Eight years ago this man was hailed as a genius who discovered the AIDS virus." But by then, it had become a story "of how a fight for wealth and glory can interfere with the desperate attempt to conquer a deadly disease."[303] By the time this program aired, federal investigators had already zeroed in on Gallo's laboratory staff, eventually finding one of them guilty that July of a federal crime.

According to the November 12, 1993, *New York Times*, until August the integrity office had expected that it would not be called upon to prove what Gallo intended in writing the 1984 paper. "But in two recent cases," read the *Times* article, "the appeals board at the Department of Health and Human Services used strong language in throwing aside the work of the integrity office as inadequate, saying a more stringent standard of evidence was necessary. In light of those other cases, the integrity office abandoned hope of winning the hearing on Dr. Gallo's appeal. In his official statement, Lyle W. Bivens, Ph.D., ORI director, offered that his office "maintains that the standards applied by the panel [the appeals review board] reflect a fundamental disagreement with ORI as to the importance of clarity, accuracy and honesty in science. However, because ORI is bound by the panel's decisions, it will not continue its proceeding against Dr. Gallo." The ORI officially dropped its charges against Gallo on November 12, 1993. But for his former laboratory scientists it was a different story.

None of what transpired during the various investigations of Gallo and his laboratory seemed to permanently tarnish anyone's reputation to the point they couldn't move on to other positions of importance. On May 3, 1993, months before the ORI opted to drop the Gallo case once and for all, the CEL-SCI Corporation and Alpha 1 Biomedicals, Inc. announced the appointment of Prem Sarin, Ph.D., as vice president of research for Viral Technologies, Inc., a joint venture of CEL-SCI Corporation and Alpha 1 Biomedicals, Inc. Sarin, it noted, served as deputy chief, National Cancer Institute Laboratory of Tumor Cell Biology, at Bethesda, Maryland, under Gallo from 1975 to 1991, where he'd been actively engaged in research on AIDS drugs and AIDS vaccines as well as working on different approaches to the treatment of AIDS.

Full disclosure, Sarin left Gallo's laboratory for a reason[304]; he'd been found guilty in federal court on July 7, 1992, of embezzling $25,000 from the Germany company Homburg Degussa Pharma intended to pay for testing a possible AIDS drug in Gallo's laboratory at the NIH. Sarin was the second Gallo assistant to be convicted of a federal crime involving misappropriation of funds. Sarin's guilty verdict, on three counts of embezzlement and making false statements to the NIH was handed down by a federal jury; he was acquitted of a fourth charge of illegal supplementation of income. Another Gallo aide, Syed Zaki Salahuddin, Ph.D., had already pleaded guilty two years before for accepting illegal gratuities from a company that supplied goods and services to the Gallo laboratory. The NIH had dismissed Sarin before his conviction in federal court. Though he faced a maximum of twenty years in prison and $750,000 in fines, Sarin's attorney appealed, and he subsequently landed the job with Viral Technologies.

As for the Homburg Degussa Pharma drug, it proved ineffective against the AIDS virus. Sarin and others associated with the Degussa study testified, according to the July 8, 1992 *Chicago Tribune*, that they had discussed it with Gallo on several times while it was under way. But Gallo was much more vague in his recollections of the same, testifying in federal court with the ubiquitous and familiar: "If I knew, I don't remember it today." Scientific personnel working on projects like the Degussa drug-testing experiment just then were hired at NIH through an adjunct entity called the Foundation for the Advancement of Education in the Sciences (FAES). Dale Kelberman, the federal prosecutor on the Sarin case, produced evidence showing that the personal bank account into which Sarin deposited the $25,000 Degussa check was also titled FAES. When asked about that, Sarin testified that it stood for "Family Account for the Education of the Sarin Children" to which Kelberman told the jury Sarin's explanation of what the acronym stood for was, in his words, "an insult to your intelligence."[305]

A third member of Gallo's laboratory, Mikulas Popovic, M.D., Ph.D., was found guilty by the NIH of falsifying and misrepresenting data in Gallo's highly controversial 1984 article[306] reporting the isolation of the AIDS virus. But this finding, too, met with pushback. Popovic was employed by the National Institutes of Health (NIH) between 1980 and 1989 as a scientist. He worked in the Laboratory for Tumor Cell Biology (LTCB). Principally, he worked with Gallo, the chief of the laboratory, in an effort to find a cure for AIDS. Popovic and Gallo made a breakthrough discovery by finding a way to isolate the AIDS virus and by proving that it was a retrovirus. Moreover, the doctors succeeded in growing amounts of the virus sufficient to facilitate the development of a test for humans. The results of the research were then published in *Science* magazine by Popovic, Gallo and other doctors.

Based on his success at NIH, Popovic left the laboratory there in 1989 to head a laboratory at New Mexico State University, which received some funding from the National Cancer Institute. When that laboratory began to suffer difficulties the same year, Popovic called Gallo in hopes of returning to his old position at NIH.

Gallo was unable to bring Popovic back to NIH because of the government investigation of his laboratory, which essentially got underway in October 1990. Eight months later, in June 1991, the Office of Scientific Integrity (OSI) (the ORI precursor) sent a report to Popovic detailing its findings and requesting his response. Popovic did respond the following September, and disputed the allegations contained in the agency's report. OSI then amended the report, and in March 1992 forwarded it to the Office of Scientific Integrity Review (OSIR), which reviews all final reports of investigations to assure that proposed findings or recommendations are sufficiently documented.

While that process was pending, ORI, which had by then succeeded OSI, prepared another report; it issued its final report on December 29, 1992. That final report contained ORI's conclusion that Popovic had behaved improperly and recommended sanctions. The report stated that Popovic had falsified certain data and methods in reporting the research in the 1984 article. Despite its findings, however, ORI did state that its report should not bar Popovic from gaining employment as a scientist. Perhaps the softening of the watchdog's eventual finding was due to Popovic's cooperation with Suzanne Hadley, former OSI chief, and suggested by Janine Roberts[307] in her documentation of the case. She wrote that when Popovic found out the Hadley was going to rule he'd committed scientific misconduct, he provided Hadley a 1984 draft of the key – and highly controversial – *Science* paper that he'd kept hiding overseas. "Among other things," she documented, "this draft revealed that Robert Gallo had extensively changed this paper at the last moment to hide their use of the French virus. With this," Roberts continued, "it seemed that evidence was at hand to prove Gallo guilty of illegal use of the French virus and thus of scientific deceit." Hadley then crafted her OSI report with that information. She concluded, Roberts revealed: "Dr. Gallo has claimed credit for the Popovic et al. paper and the other 1984 papers, so much he bear responsibility for the falsehoods in the Popovic et al. paper. Accordingly, the OSI finds that Dr. Robert Gallo engaged in scientific misconduct." While this conclusion should have been forwarded up the chain at NIH for adjudication, this is where Healy intervened and removed Hadley from her duties at the OSI.

Hadley's damning indictment of Gallo was removed from the final report. Despite Healy's attempts to preserve Gallo's career, enter the *Chicago Tribune* exposé and all hell broke loose once again. With this article came renewed scrutiny and stiff consequences. Gallo and Popovic took much of the heat. Talk circulated that the NIH was covering up Gallo's guilt to protect his AIDS research and it was just then that Dingell got fully involved. Dingell, according to Roberts, had Gallo and Popovic's OSI files brought to his office. He also wanted the services of Hadley, who would continue her investigation with additional congressional powers. Roberts reported that an aide to Dingell decried that everything Hadley had told them had "checked out 100 percent" against documentation the committee had received from the NIH. The aide also noted that Hadley had "obviously been treated very shabbily" by the NIH.

What Dingell soon discovered was that he was missing some of the Gallo research documents and he wanted to know why, according to Roberts. This was late 1992. He wrote to the NIH on November 24: "We have received reliable information that documents from the Gallo/Popovic investigation were being shredded at the NIH's Office of Scientific Integrity. NIH's actions [...] show a clear pattern of obstruction and attempted deception [...] particularly when juxtaposed with the curious diligence the NIH showed in its efforts to seek out and destroy the person or persons suspected of blowing the whistle on the shredding."[308]

In late January 1993, Popovic appealed ORI's conclusions to the Department of Appeals Board (DAB). Later in the year, the DAB held a *de novo* hearing. On November 3, 1993, the DAB exonerated Popovic, finding that ORI had not established its claims of wrongdoing by a preponderance of the evidence. NIH's investigation ended with that finding. Popovic believed that the entire investigation was conducted because of ill will, and filed a complaint with NIH, as he is required to do under the Federal Tort Claims Act (FTCA) per 28 U.S.C. § 2675 and cited in *Mikulas Popovic v. United States, et al.* No. 98-1432 (4th Cir. 1999). When NIH did not grant him relief, he filed the instant action. The Maryland federal district court dismissed the action, holding that Popovic's claims essentially were defamation claims. Popovic filed a timely appeal.

The district court dismissed Popovic's claims of negligence and invasion of privacy, reasoning that in reality Popovic was claiming that he was defamed. Popovic's claims of negligence are also undermined because he appeared to claim that HHS acted intentionally, accusing the OSI/ORI of continually leaking information to media sources. But he didn't even allege that the information was negligently disseminated. He claimed, too, that OSI/ORI continued to pursue him after filing reports by raising new allegations at the DAB hearing. In more general allegations, he claimed that OSI/ORI refused to consider any exculpatory evidence and repeatedly denied him other due process rights. The appeals court denied Popovic's claims and affirmed the district court's ruling in favor of the defendants, reference *Popovic v. United States*, 997 F. Supp. 672 (D. Md. 1998). Popovic is an adjunct professor of medicine at the University of Maryland School of Medicine, Institute of Human Virology.

So, what happened to Syed Zaki Salahuddin, Ph.D., who'd been described by congressional investigators as turning the battle against AIDS into a personal profit mill?[309] Salahuddin had set up a company that did business with Gallo's (and his) laboratory at NIH. Further, he hid his relationship with the company as he helped it gain hundreds of thousands of dollars in federal contracts, according to investigators and reported in the May 1, 1990 *Philadelphia Inquirer*. Dingell called it "a sorry tale of gross conflict of interest" and blamed slip-shod conflict of interest standards at the NIH. He also laid blame directly at the door of Gallo, Salahuddin's boss and colleague, for not pursuing reports of a conflict in his lab more aggressively. Gallo's "rather cavalier attitude" is "a curious one" for a public servant, Dingell observed.[310] Salahuddin invoked the Fifth Amendment when told he would be asked to testify before Dingell's subcommittee.

Investigators found the following when they investigated Salahuddin's arrangement within Gallo's laboratory that when Pan-Data Systems was set up in 1984, it had a net worth of $921 and operated from a condominium apartment in suburban Maryland. By 1988, it had grown to a $3-million-a-year biomedical firm. Much of its business was with NIH, but nowhere in Salahuddin's financial disclosure forms was his link to the firm reported. The company, meanwhile, took care of the bill for painting Salahuddin's house and paid his wife fees for her work as a director, the investigators discovered. NIH acting director William F. Raub, Ph.D., asked for media comment, noted that Salahuddin acknowledged in the last week of April 1990 that the firm had made some of his mortgage payments.

Investigators for the subcommittee said they had found a paper trail linking Salahuddin and his wife, Firoza Salahuddin, to Pan-Data. The company's articles of incorporation listed his wife as a twenty-five percent owner. According to Pan-Data's records, the two resigned as directors in 1984. But investigators found bank records, Small Business Administration records and documents from a United States Army AIDS testing contract indicating that the Salahuddins continued their involvement with the firm in 1985 and 1986.

According to media accounts of the Salahuddins' conflict of interest, in 1986 Gallo confronted Syed Salahuddin about rumors that he was linked to Pan-Data. But investigators said Gallo accepted Salahuddin's explanation that his wife worked for the firm and that he would see to it that she resigned. In January 1990, NIH signed a three-year, $1 million research support contract with Pan-Data. NIH was aware then of the congressional investigation of Salahuddin. But when NIH officials double-checked Salahuddin's financial disclosure forms, they found no mention of Pan-Data.

Dingell's committee ultimately released an unofficial 267-page report of its findings in the Gallo case, the highlights of which the *Chicago Tribune* published in two articles, the first on January 1, 1995, and the other on January 6, 1995. Dingell's findings concluded: "The result was a costly, prolonged defense of the indefensible in which the LTCB [Gallo's laboratory of tumor cell biology] 'science' became an integral element for HIV research were severely damaging, leading, in part, to a corpus of scientific papers polluted with systematic exaggerations and outright falsehoods of unprecedented proportions." The report, Roberts noted, presented detailed evidence that destroyed Gallo's central claim in the *Science* papers, notably that he had isolated HIV in dozens of AIDS patients in experiments conducted in 1982 and 1983. Congressional investigators further stated that he did not have the tools needed to do this – and consequently couldn't have isolated or identified a single AIDS virus.[311]

It wasn't until May 25, 1995, that it was announced that Gallo would be leaving the National Cancer Institute after a thirty-year career. The circular further noted that Gallo would be setting up his own Institute of Human Virology in downtown Baltimore, Maryland. Unfortunately, the Science papers he'd authored, despite having been found to contain scandalously fraudulent material, were never retracted nor were they corrected thus creating a circumstance in which potentially thousands of researchers consult them without full disclosure of the flawed data used to write them. There are, in Roberts' estimation, "thousands of papers on HIV and AIDS [which today] refer back to them, and all medical authorities point to them, too. The United States Centers for Disease Control (CDC) still states on its website that the key foundation papers in AIDS research are 'four papers from Dr. Gallo's laboratory, demonstrating that HTLV-III retrovirus [HIV] was the cause of AIDS."

So, who gets hurt? Dingell's 267-page subcommittee treatise on the case, in the executive summary, provided a clear picture of those most affected. "One of the most remarkable and regrettable aspects of the institutional response to the defense of Gallo et al. is how readily public service and science apparently were subverted into defending the indefensible," the report read at the end of a stinging executive summary. "To comprehend the significance of the subversion of public service in the cover-up, it is useful to review the Guidelines of the Ethics Committee of the Washington, D.C. Bar Association concerning the roles of government attorneys."[312] According to the guidelines:

> A government lawyer in a civil action or administrative proceeding has the responsibility to seek justice and to develop a full and fair record, and he should not use his position or the economic power of the government to harass parties or to bring about unjust settlements or results" (EC 7-14; p. 51).

"These guidelines, although they focus on government attorneys, apply at least as well to the NCI/HHS science administrators who played such a crucial, 'make-or-break' role at the outset of the French/American dispute. The deliberately negligent 'fact-finding' conducted by these individuals, combined with their deliberate suppression of incriminating evidence," concluded Dingell's minority staff report, "set the stage for everything that happened thereafter. But the attorneys bear significant responsibility as well, for they clearly did not seek diligently to 'develop a full and fair record' of the facts about the claims of Gallo et al." Further, neither did HHS and Department of Justice officials and attorneys, once the dispute was under way, deal responsibly with the accumulating evidence that there were serious problems in the United States government's claims. "Instead, they pushed on with their 'litigation strategy,' all the while adding deception to deception, consuming untold resources and squandering scientific and international good will. The fraud became self-perpetuating." Most damning of all, the subcommittee's investigators reached the conclusion, to which Dingell was fully in agreement: "Defending the indefensible became a reflex, until ultimately, the cover-up was so burdened with falsehoods that its collapse was inevitable. HHS officials and attorneys should have recognized early on that the falsehoods could not be indefinitely sustained. But HHS sought only to 'defend the position.' HHS did not seek the truth. HHS did not honor the public trust."

The minority staff report summarized it all thusly:

> The violence to principles of responsible, ethical science was just as profound. At a crucial point early in the LTCB's HIV research, international politics and the technocrats committed to those politics virtually took over that research, claiming the laboratory's putative accomplishments as accomplishments of the United States administration and by extension, the United States itself. Once done, the LTCB's interests became the government's interests; defending the LTCB scientists' reputations and claimed accomplishments became necessary for defending the honor of the United States. The defense thus became a consuming effort for significant portions of the [United States] government.

Ultimately, the Gallo case cast a long shadow, the result of which was a costly, prolonged defense of the indefensible in which the LTCB's so-called science became an integral element of the United States government's public relations/advocacy efforts, a conclusion also expressed in the report. "The consequences for HIV research were severely damaging, leading, in part, to a corpus of scientific papers polluted with systematic exaggerations and outright falsehoods of unprecedented proportions."

To comprehend how far from the ideal the defense of Gallo et al. led the scientific community, it is useful to note the words of two esteemed scientists, John Forster Cairns, M.D., and Paul M. Doty, Ph.D., observed the Dingell report:

> ...science is the pursuit of a truth that is external to our wishes. This truth is quite unlike the verdict of a court of law because it does not depend on advocacy.[313]

> This challenge to readdress the fundamental tenets of acceptable behavior in science comes at a time when the traditions of the scientific enterprise are under new threats arising from new stresses and temptations ... As a result, the scientific community may already be experiencing a gradual departure from the traditional scientific standards ... In this way we risk sliding down toward the standards of some other professions where the validity of action is decided by whether one can get away with it. For science to drift toward such a course would be fatal – not only to itself and the inspiration which carries it forward, but to the public trust which is its provider.[314]

There could be no better description of the disastrous consequences of the federal government's defense of Gallo et al., concluded Dingell's unofficial final report on the Gallo scandal that rocked the NIH and the scientific community.

Certainly, Gallo wasn't hurt by the controversy, not in any lasting way. He and Montagnier published a series of articles, one of which they co-wrote, for the November 29, 2002 *Science*. In this article they acknowledged one another's role in the discovery of HIV. But six years later, in 2008, Montagnier and colleague Françoise Barré-Sinoussi, of the Institut Pasteur, were awarded half the Nobel Prize in Physiology or Medicine for their work on the discovery of HIV,[315] the other half going to Harald zur Hausen for his discovery of the human papilloma viruses that lead to cervical cancer. Gallo wasn't named as a co-recipient; he wasn't even mentioned. Montagnier later expressed his chagrin that Gallo hadn't been included.

Gallo had long since moved on from his NIH laboratory when the Nobel Prize was awarded to Montagnier and Barré-Sinoussi for their work with HIV. Gallo cofounded the Institute of Human Virology (IHV) in 1996, and in addition to his position as director of the IHV remains the codirector of IHV's Division of Basic Science and Vaccine Development, with William Blattner, M.D., associate director of the IHV and director of IHV's Division of Epidemiology and Prevention and Robert Redfield, M.D., associate director of the IHV and director of IHV's Division of Clinical Care and Research. The institute is part of the University of Maryland School of Medicine and affiliated with the University of Maryland Medical Center. Also that year, Gallo's discovery that a natural compound known as chemokines can block HIV and halt the progression of AIDS was hailed by *Science* magazine as one of that year's most important scientific breakthroughs. This also helped others identify CCR5[316] as the HIV co-receptor since these chemokines were known to bind to CCR5.[317]

According to Gallo's IHV webpage, IHV has cared for and treated more than one million HIV positive individuals in seven African and two Caribbean nations in addition to more than 6,000 HIV positive Baltimoreans. The Institute for Human Virology is internationally renowned for its basic science research, which included the tentative launch of clinical trials in 2015 on a promising preventive HIV vaccine candidate funded largely by the Bill and Melinda Gates Foundation. Additionally, in 2011 Gallo cofounded the Global Virus Network (GVN) with William Hall, Ph.D., M.D., D.T.M.H., of the University College Dublin and Reinhard Kurth[318], M.D., of the Robert Koch Institute, to position the world to rapidly respond to new or reemerging viruses that threaten mankind, to bring together and achieve collaboration amongst the world's leading virologists, and to support training of the next generation of medical virologists.

Beyond Seife's revelations and the lengthy investigation of Gallo, the number of cases of misconduct that affect health outcomes has spun out of control, most of these transgressions, missteps and deceptions unknown to the public. The range of misconduct is sweeping, touching a wide range of biomedical study, leaving no discipline untouched by fraud, falsification and, ultimately, the deep erosion of public trust and crushed hope that follows. Issues raised by Helene Hill's story are magnified in the many cases that preceded, took place concurrently, and have since followed her own. Though it didn't make major headlines in 2014, a federal judge ruled in favor of whistleblowers who accused big pharma giant Merck of lying about the efficacy of its mumps vaccine, presently only available in combination with measles and rubella as the "MMR" shot. Major mainstream media backburnered the judge's ruling – effectively burying the story alive – despite the fact that it covered the case back in 2012, before it was hit with Merck's bastion of attorneys who appealed and tried to get the filing thrown out of court.[319] They failed. Just then, *Forbes* contributor Gergana Koleva observed that anyone who failed on either side of the debate about vaccines' alleged potential to cause harm "is sure to have heard the big news this week – the unsealing of a whistleblower suit against Merck, filed back in 2010 by two former employees accusing the drugmaker of overstating the effectiveness of its mumps, measles and rubella vaccine."[320] Well, perhaps it was hopeful thinking that "most everyone" heard of the case.

The June 27, 2012 *Forbes* article made it clear what was at stake:

> The scientists claim Merck defrauded the [United States] government by causing it to purchase an estimated four million doses of mislabeled and misbranded MMR vaccine per year for at least a decade and helped ignite two recent mumps outbreaks that the allegedly ineffective vaccine was intended to prevent in the first place.

> 'As the single largest purchaser of childhood vaccines (accounting for more than 50 percent of all vaccine purchasers), the United States is by far the largest financial victim of Merck's fraud. But the ultimate victims here are the millions of children who every year are being injected with a mumps vaccine that is not providing them with an

adequate level of protection against mumps. And while this is a disease the CDC targeted to eradicate by now, the failure in Merck's vaccine has allowed this disease to linger with significant outbreaks continuing to occur,' the suit alleges.

The mumps outbreaks to which the *Forbes* article referred were the 2006 mumps outbreak in the Midwest, in which 6,500 cases were reported among a highly vaccinated population, and another in 2009, in which 5,000 cases were confirmed. By comparison, the annual average of mumps cases in the United States in the two decades preceding the 2006 outbreak was 265; before the introduction of the single-shot Mumpsvax vaccine in 1967, there were approximately 200,000 cases of the disease, according to the fifty-five-page document.[321]

The whistleblowers who brought the Merck lawsuit were two virologists –Stephen A. Krahling and Joan A. Wlochowski – medical doctors who worked for the drug giant and soon found themselves in the inevitable and unenviable position of accusing the company of lying about the effectiveness of the mumps vaccine. Krahling and Wlochowski filed a *qui tam*[322] action in the name of the United States on August 27, 2010, spotlighting their belief that Merck had overstated the effectiveness of its mumps, measles and rubella vaccine. Further, they claimed the drugmaker had defrauded the federal government by causing it to purchase an estimated four million doses of mislabeled and misbranded MMR vaccine per year for at least a decade and helped ignite the two major outbreaks previously cited. Merck denied all wrongdoing. But Koleva observed just then that if the accusations proved true, the *qui tam* brought forward by Krahling and Wlochowski would most assuredly "lend credence to the perception held by many that pharmaceutical companies are more interested in pursuing profits and preserving their market share than in protecting consumers' health." Merck had everything to lose. Specifically, in *Krahling v. Wlochowski et al. v. Merck*, the plaintiffs cited the drug maker's effort to maintain its Food and Drug Administration (FDA) approval and exclusive license to sell the vaccine, Merck used improper testing techniques and falsified test data to fabricate a vaccine efficacy rate of ninety-five percent or higher. This is the efficacy threshold on which the FDA insists for its licensing and approval of the vaccine. In truth, according to the complaint, the efficacy rate of Merck's mumps vaccine had been, since at least 1999, significantly lower than this requisite threshold. Krahling and Wlochowski were employed by Merck in the laboratory that performed the efficacy testing. They witnessed firsthand the improper testing and data falsification in which Merck engaged to artificially inflate the vaccine's efficacy findings, and, in fact, they were pressured by their Merck superiors and senior Merck management to participate in the fraud and subsequent cover-up.

As the direct result of Merck's fraudulent scheme, the United States had paid the pharmaceutical giant hundreds of millions of dollars for a vaccine that did not, according to case documents, provide adequate immunization. The argument was made that had the government known the true efficacy of the vaccine, the government's decision to purchase the product surely would have been different, either purchasing the vaccine from another source, requiring that Merck produce a new vaccine with the requisite immunizing effect, or re-negotiating the contract for the existing product. This is significant. As the single largest purchaser of childhood vaccines, accounting for more than fifty percent of all vaccine purchases, the United States was – and remains – the largest financial victim of Merck's – or any other drugmaker – when fraud is committed. Yet the ultimate victims in the case against Merck remained the millions of children who every year were being injected with a mumps vaccine that was not providing them with an adequate level of protection. Mumps was a disease, according to the Centers for Disease Control (CDC), that was supposed to be eradicated by now. Per the suit, the failure of Merck's vaccine had allowed this disease to linger with significant outbreaks continuing to occur.

What was this Merck whistleblower case all about? On background, the suit explained that for more than thirty years [up to the 2010 filing of the case in federal court], Merck had an exclusive license from the FDA to manufacture and sell a mumps vaccine in the United States. The FDA first approved the vaccine in 1967; it was developed by Dr. Maurice Hilleman at Merck's West Point research facility, from the mumps virus that infected his five-year-old daughter Jeryl Lynn. Merck continued to use this "Jeryl Lynn" strain of the virus for its vaccine today. Merck's original mumps vaccine was delivered to patients in a single, stand-alone injection. But in 1971, Merck developed a combination vaccine that delivered Merck's vaccines for measles, mumps and rubella in one injection. The same year, the FDA handed Merck the exclusive United States license to manufacture and sell this MMR vaccine. Seven years later, in 1978, the FDA approved and gave Merck the exclusive United States license for the manufacture and sale of what it dubbed "MMRII," a replacement for MMR containing a different strain of the rubella virus. Since that time, Merck had sold more the 450 million doses worldwide, with approximately 200 million doses sold in the United States alone. Merck, per the suit, was still selling more than seven million doses of the vaccine in the United States annually.

To get its original FDA approval and license and to sell the mumps vaccine, Merck conducted tests that demonstrated the vaccine had an efficacy rate of ninety-five percent or higher. This meant that ninety-five percent of those taking the vaccine would be immunized against mumps. The FDA insists on such a high efficacy rate because only then can the disease ultimately be eradicated through what is commonly known as "herd immunity." Short of that, there remains a real risk of continued outbreaks of the illness. When outbreaks of mumps occur in vaccinated populations, the disease afflicts older children who are at greater risk of complications. The disease also presents greater risks for infants. Merck's mumps vaccine originally seemed well on its way to achieving the herd immunity threshold. Before the introduction of vaccine, the disease affected tens of thousands of Americans annually. This number dropped off precipitously after the widespread administration of Merck's vaccine. According to *Krahling v. Wlochowski et al. v. Merck*, in the 1980s, outbreaks of mumps still occurred but these too petered out for a while with the requirement beginning in 1989 that children receive two doses of the MMRII vaccine at twelve to eighteen months, and again at four to five years of age. The CDC projected that by 2010, mumps would be completely eradicated. Unfortunately, that did not happen. Beginning in 2006, there was a resurgence of mumps outbreaks. How could this happen? As it turned out, the continued outbreaks boiled down to Krahling's and Wlochowski's claim that the Merck vaccine no longer bore a ninety-five percent efficacy rate.

The vaccine may have had a ninety-five percent efficacy rate when it was originally licensed in 1967, but the vaccine virus had been waning as it continued to be passaged to create more vaccine virus for distribution. Vaccine propagation further attenuates the virus, a problem that is compounded with each additional passage of the virus to create more vaccine. This is especially evident in the case of Merck's mumps vaccine because the vaccine strain was established more than forty years before the case came to the federal court, and had already been used to manufacture hundreds of millions of doses. Rather than develop a new mumps vaccine with the requisite efficacy rate, Merck had instead gone to great lengths before the FDA and the public to push the narrative that the vaccine continued to have a ninety-five percent efficacy rate or better. This obfuscation was easy to pull off for a while because Merck was able to rely on the efficacy testing it conducted in connection with the FDA's original granting of the company's exclusive license. But in 1997 the FDA required Merck to conduct renewed efficacy testing of the mumps vaccine in MMRII. This was a problem. The FDA's demand also coincided with Merck's development and quest for FDA approval of a new vaccine called "ProQuad" that would effectively combine its vaccine against varicella (chickenpox) with MMRII. Without demonstrating that its mumps vaccine continued to be ninety-five percent effective, Merck would lose its exclusive license to manufacture and sell its MMRII vaccine, and if the company lost the license for MMRII, it would also be unable to secure FDA approval for its ProQuad vaccine. Thus, Merck set out to conduct testing of its mumps vaccine that would guarantee an efficacy of ninety-five percent or higher and it did this through manipulating the testing procedures and falsifying the test results, according to documentation provided in *Krahling v. Wlochowski et al. v. Merck*. Krahling and Wlochowski participated on the Merck team that conducted this testing and witnessed the fraud in which Merck engaged to reach the desired results. Merck internally referred to the testing as Protocol 007.

Suzanne Humphries, M.D., wrote of this testing on June 25, 2012, that Merck got the old virus to pass its new tests by doing three things.[323] First, she explained, the Merck laboratory tested efficacy on the vaccine strain virus and not the wild virus as had previously been done. The testing is done by vaccinating children and testing the ability of their blood to neutralize the virus before and after the vaccine. There should be a marked difference between the two tests. For the new testing method, the children's blood was tested for its ability to neutralize the virus, using the vaccine strain virus, instead of the wild-type strain that is much more infective, and the one that children would most likely catch. By using a weaker virus, the old vaccine strain virus allowed the neutralization to occur much more easily. But still it was not ninety-five percent effective.[324] To get blood to pass the test, Merck added antibodies from rabbits. The addition of rabbit antibody increased efficacy to one hundred percent. But that was not the end of it because the test had to be done, Humphries pointed out, on pre-vaccine blood and post-vaccine blood. Just the addition of rabbit antibody made the pre-vaccine blood go from ten percent positive to eighty percent positive and that was such an obvious sign of foul play that yet another manipulation had to be made to get the FDA to buy in. "The desired end result is to have very low pre-vaccine viral neutralization and ninety-five percent or more post-vaccine viral neutralization," she iterated. How was the Merck laboratory going to get past this problem? Another change in procedure was made. The pre-vaccine positive tests were all recounted. None of the post-vaccine tests and none of the pre-vaccine negatives were recounted, she revealed. Just the pre-vaccine positives were reported. According to Krahling and Wlochowski they did this by fabricating the plaque counts on the pre-vaccine blood samples, counting plaques that were not there. The test is called a plaque reduction neutralization (PRN) assay.

Plaques are areas in cell culture dishes where virus attacks cells, demonstrating viral activity and lack of immunity. Because the rabbit antibodies were used in the pre-vaccine tests, Humphries observed that there was positive PRN meaning that the pre-vaccine blood looked immune, or that there was plaque reduction. "This is not desired when wanting to show that the vaccine is what led to the immunity, where you would only want the post-vaccine samples to have positive PRN- or a reduction in plaques," she continued. "So, the pre-vaccine positive PRN assays were recounted because there were not enough plaques to show the desired result – that virus was active in the pre-vaccine tests." Krahling and Wlochowski reported that plaques were counted where there were none. This allowed a mathematical dilution of the pre-vaccine positive PRN tests, turning them negative. The complaint says that forty-five percent of the pre-vaccine samples recounted were revised.[325]

There was much more to the absence of wild-type strain in the Merck testing. While Merck's PRN test was modeled after the efficacy test generally accepted in the industry, it diverged from this gold standard test when it did not test the vaccine for its ability to protect against the wild-type mumps virus. The wild-type virus is a strain of the virus as it exists in nature and would confront a person in the real world – that is the type of real-life virus against which vaccines are generally tested. Instead, Merck tested the children's blood for its capacity to neutralize the same Jeryl Lynn mumps strain with which the children were vaccinated. The children's vaccine response was not tested for its capacity to neutralize virulent, disease-causing mumps virus. The use of the Jeryl Lynn strain, as opposed to a virulent wild-type strain, subverted the fundamental purpose of the PRN test that was to measure the vaccine's ability to provide protection against a disease-causing mumps virus with which a child would actually face in real life. The end result of this deviation from the PRN gold standard test was that Merck overstated the vaccine's effectiveness.

Interestingly, even with a deviation that could only overstate vaccine effectiveness, the results from Merck's preliminary testing yielded a seroconversion rate[326] of only 79.5 percent and this was more than 15 percent lower than the 95 percent efficacy threshold on which Merck's original FDA approval and exclusive license was based and which the FDA still required of the pharmaceutical giant. Merck knew that a seroconversion rate so far below ninety-five percent would not be acceptable to the FDA, nor would it support Merck's continued license to exclusively sell the mumps vaccine and, certainly, its new ProQuad. During the testing process, Merck senior investigator David Krah on several occasions emphasized to his staff, to include Krahling and Wlochowski, that if Merck could not show a minimum of ninety-five percent seroconversion rate in conducting the company's mumps efficacy tests, the FDA would rescind Merck's exclusive licensing rights to MMRII. Significantly, Merck abandoned the PRN test and the unsatisfactory results it yielded and worked towards what Humphries described as the desired results involving the use of rabbit antibodies in what amounted to an enhanced PRN test.

The second Merck test employed under Protocol 007 was formally called anti-IgG[327] enhanced mumps plaque reduction neutralization assay; it was started in 2000 and again led by Krah and his staff at Merck's West Point facility. Krahling and Wlochowski participated on the team that conducted this supposedly enhanced test. Each of them witnessed firsthand the falsification of the test data in which Merck engaged to reach its ninety-five percent efficacy threshold; in fact, each was significantly pressured by Krah and his deputy, Mary Yagodich, as well as other senior Merck personnel to participate in the fraud. Notably, Merck's executive director of vaccine research, Alan Shaw, was also named as having "approved the testing methodology Krah and Yagodich employed."[328] From the outset, the objective of this newly devised procedure was clear. It was not to measure the actual seroconversion rate of Merck's mumps vaccine. Rather, it was to come up with a methodology that would yield a minimum ninety-five percent efficacy threshold regardless of what the vaccine's true efficacy might actually show. The first page of the October 2000 Merck presentation on the newly devised efficacy test states just that:

> Objective: Identify a mumps neutralization assay format…
> that permits measurement of a ≥ 95% seroconversion rate
> in M-M-R®II vaccine.

Notably, nowhere in this presentation or anywhere else did Merck provide any kind of justification or explanation for abandoning its original PRN test and the unsatisfactory efficacy results the test yielded, according to the plaintiff's filing. Then there was the use of those rabbit antibodies, the anti-IgG, added to both the pre- and post-vaccination blood samples. The use of rabbit antibodies is not uncommon. Significantly, in those experiments where rabbit antibodies are added as an enzyme to identify human antibodies, the rabbit antibodies do not alter the outcome of the experiment. But not so with Merck, which added the rabbit antibodies for the singular purpose of altering the outcome of the test by increasing the virus neutralization count. In a laboratory setting, rabbit antibodies can combine with human antibodies to cause virus neutralization that would not otherwise occur from the human antibodies alone. Without applying proper control to the process, there is no way to isolate whether virus neutralization is caused by the human antibodies alone or in combination with rabbit antibodies. Merck did not employ this kind of control; it included in its seroconversion measure all virus neutralizations regardless of whether they resulted from human antibodies or by their combination with the rabbit antibodies. This enhanced PRN procedure thus allowed Merck to increase dramatically the recordable instances of mumps virus neutralization and to count those neutralizations toward seroconversion and its measure of the vaccine's success. Merck knew that the neutralizations attributable to the rabbit antibodies would never exist in the real world, and this is because the human immune system, even with the immunity boost provided by an effective vaccine, could never produce rabbit antibodies. Further, adding rabbit antibodies as a supplement to a vaccine was not an option because it could result in serious complications to a human, even death. Thus, the uncontrolled boost to neutralization Merck designed using rabbit antibodies in its laboratory did not in any way correspond to, correlate with, or represent real life – also called *in vivo* – virus neutralization in vaccinated people. Krah defended the use of rabbit antibodies in the enhanced PRN test by pointing to the FDA's purported approval of the process; however, whatever approval Merck might have received from this testing, the FDA was not fully aware of the extent of Merck's manipulation, including Merck's wholesale fabrication of test data to reach its preordained ninety-five percent efficacy threshold.[329]

Most troubling, Krah and Yagodich and other members of Krah's staff falsified the test results to ensure a pre-positive neutralization rate of below ten percent. They did this by fabricating their plaque counts on the pre-vaccination blood samples, counting plaques that were not actually there. Humphries was quick to point this out. With these inflated plaque counts, Merck was able to count as pre-negative those blood samples that would have otherwise been counted as pre-positive because of the increased neutralization caused by the rabbit antibodies. Merck's falsification of the pre-vaccination plaque counts was performed in a broad-based and systematic manner. In the suit, Krahling and Wlochowski documented that Krah stressed to his staff that the high number of pre-positives they were finding was a problem that needed to be fixed. Krah directed his staff to re-check any sample found to be pre-positive to see if more plaques could be found to convert the sample to a pre-negative. Krah and deputy

Yagodich falsified plaque counts to convert pre-positives to pre-negatives, and directed other staff scientists to do the same. Krah appointed Yagodich and two others to audit the testing that other staff scientists had performed. These audits were limited to finding additional plaques on pre-positive samples thereby rendering them pre-negatives. Krah instituted several measures to isolate the pre-positive samples, facilitate their re-count and consequent conversion to pre-negatives, and minimize the chances of detection. These included destroying test results, substituting original counting sheets with "clean" sheets, and entering and changing test results directly onto electronic – Excel program – spreadsheets that left no paper trail. Further, Merck cancelled a planned outsource of the efficacy testing to a laboratory in Ohio because the outside lab was unable to replicate the seroconversion results Krah was obtaining in his lab. Krah and his staff conducted all the testing instead. Unsurprisingly, none of the recounting and retesting that Merck performed as part of its enhanced PRN testing was performed on any post-vaccination samples or on any pre-vaccination samples that were pre-negative. This additional "rigor" was only applied to the pre-positive samples, the very samples that were keeping Merck from achieving the requisite ninety-five percent seroconversion threshold.

In July 2001, Krahling and Wlochowski conducted their own test to confirm statistically what they already knew to be true. They reviewed approximately twenty percent of the data that Merck had collected as part of the company's enhanced PRN test. In this sampling, they found that forty-five percent of the pre-positive data had been altered to make it pre-negative. No pre-negatives were changed to pre-positives. No post-positives were changed to post-negatives. No post-negatives were changed to post-positives. All changes were in one direction – reducing the incidence of pre-positives. The statistical probability of so many innocent changes occurring in just the pre-positive data and in no other data was more than a trillion to one – and that was a conservative measure given the likelihood that an even greater number of pre-positives were changed but remained undetected because the changes were not recorded in Merck's files.

Krahling v. Wlochowski et al. v. Merck implicated Merck's senior management in the mumps vaccine deception, naming Emilio Emini, vice president of the company's vaccine research division. In April 2001, Emini held a meeting with Krah and his staff in which he directed them to follow

Krah's orders to ensure the enhanced PRN testing would be successful. Emini also told the staff that they had earned very large bonuses for their work up to that point on the project, and that he was going to double the bonuses and pay them once the testing was complete. Then, in July 2001, Krahling met with Alan Shaw, Merck's executive director of vaccine research, and complained to him about the fraudulent testing. Krahling presumed that Shaw already knew about the fraud since he frequently visited Krah's laboratory and almost certainly would have witnessed the changing of pre-positive data that Krah was opening directing. Nevertheless, according to the suit, Krahling wanted to put Shaw on formal notice of the fraud and told him of the falsification of the pre-positive data. He further complained about the improper use of the rabbit antibodies to inflate the post-vaccine neutralization counts. Shaw responded that the FDA permitted the use of rabbit antibodies and that that should be good enough for Krahling. Shaw refused to discuss anything further about the matter. Instead, Shaw talked about the significant bonuses that Emini had promised to pay once the testing was complete. Krahling then met with Bob Suter, his human resources representative at Merck. Krahling told Suter about the falsification of testing data and Shaw's refusal to get involved. Krahling then tipped his hand, telling Suter that he was going to report the activity to the FDA. Suter informed Krahling that he would go to jail if he contacted the FDA. Suter's consolation prize was a weak proffer to set up a private meeting with Emini in which Krahling could discuss his concerns.

Shortly after Krahling's meeting with Suter, Emini agreed to meet with him. Krahling brought to the meeting actual testing samples and plaque counting sheets to demonstrate to Emini the fraudulent testing that Krah was directing. Emini agreed that Krah had misrepresented the data. Krahling also complained about the use of rabbit antibodies to inflate the seroconversion rate. Emini responded that the rabbit antibodies were necessary for Merck to achieve the project's objective. Krahling proposed a scientific solution to lower the pre-positive rate and end the need to falsify data – stop using rabbit antibodies. When Emini declined, Krahling asked him what scientific rationale justified using rabbit antibodies. Emini explained that Merck's choice to use these antibodies was a business decision. But to assuage Krahling's concerns, Emini promised to conduct an internal audit of the Protocol 007 testing. Krahling countered that the FDA should be contacted since only the FDA could perform an audit that was truly independent. Emini ordered Krahling not to call the FDA. Immediately after the meeting, Suter reappeared and again threatened Krahling with jail if he made the notification to the FDA. The next morning, Krah arrived early to the lab and packed up and destroyed evidence of the ongoing Protocol 007 efficacy testing – this included garbage bags full of the experimental plates that would have – and should have – been maintained for review until the testing was complete and final. Despite the threats he received from Suter and Emini, Krahling called the FDA to report this activity and Merck's ongoing fraud.

On August 6, 2001, in response to Krahling's call, an FDA investigator visited Merck to question Krah and Shaw. Krahling was able to situate himself near the interview and listen to the agent's questions and to what Krah and Shaw said in response. The FDA investigator's questions were largely focused on Merck's process for counting plaques in the enhanced PRN test. Not surprisingly, Krah and Shaw misrepresented the process that Merck was actually conducting and the fact that Merck was falsifying the pre-positive test data. The FDA investigator, in fact, asked Krah if it was typical laboratory procedure to recheck the original plaque counts. Krah replied that plaque counts were being rechecked only in the control plates, and only in order to verify the results. Krah also told the FDA investigator that data was not changed once it is entered into the Excel spreadsheet. When the FDA investigator pressed Krah on what criteria he used to alter data on the counting sheets, Krah left the room without giving her an answer. Shaw stepped in and told the FDA investigator that a memorandum would be added to the experimental procedure to explain the data alterations. When the FDA investigator asked Shaw why this had not been done before the project started, Shaw replied that Krah had identified problems and trends with the original counts that only became noticeable after the results were analyzed. Krah reentered the room and told the FDA investigator that no revisions had been made to the experimental plates. These responses were, according to *Krahling v. Wlochowski et al. v. Merck*, patently false and kept the FDA investigator from finding out what was really going on with Merck's manipulation of the testing procedure to reach its targeted seroconversion rate. The entire interview with Krah and Shaw was brief, perhaps less than half an hour. She did not question Krahling, Wlochowski or other members of Krah's staff in order to corroborate what Krah and Shaw told her. From what Krahling and Wlochowski witnessed, she did not attempt to substantiate Krah's or Shaw's responses by reviewing any of the testing samples or backup data that had escaped destruction. She also did not address the actual destruction of evidence that Krah had already facilitated. The FDA, the public's neutered gumshoe, did what it always did and issued a one-page deficiency report identifying a few relatively minor shortcomings in Merck's testing process. These "shortcomings" principally related to flaws in Merck's record-keeping and in its validation and explanation of changes to the test data. The report failed to address nor censure Merck for any issues related to Merck's improper use of rabbit antibodies or the company's wide-scale falsification of pre-positive test data. The FDA failed, too, to uncover fraudulent activity in the course of the agency's perfunctory visit, due largely to Krah's and Shaw's misrepresentations to the FDA investigator.

After it received the FDA's deficiency report, Merck made minor adjustments to its testing procedure relating to its heretofore *ad hoc* procedure for counting plaques. The new, more formalized procedure explicitly provided for supervisory oversight and review of plaque counts in pre-vaccinated blood samples and where plaques were difficult to read because of the condition of the sample. In other words, under the "new" procedure, Merck continued to falsify the test data to minimize the level of pre-positives and inflate the seroconversion rate. Merck simply used the deficiency report as a vehicle to legitimize the scheme. After the FDA visit, Krahling was barred from any further participation in the Protocol 007 project. He was also prohibited from accessing any data related to the project. Shortly thereafter, he was given a poor performance review and barred from continuing to work in Krah's lab on any project. Krahling was offered a position in a different laboratory within Merck's vaccine division, but it involved work for which Krahling had no prior experience or interest. At that point, he believed his only option was to resign from the company, which he did in December 2001. Wlochowski had continued to work in Krah's lab until she, too, was transferred out of Krah's lab at the end of September 2001. She spent an additional year working at Merck in a lab overseen by Thomas J. Palker[330], Ph.D., in the department of virus and cell biology before she also left Merck.

Merck went on to complete Protocol 007 testing in late summer/ early fall 2001. Unsurprisingly, the results Merck reported fell within the ninety-five percent seroconversion target Merck had provided to the FDA since the company's first mumps vaccine came to market. This is the result Merck provided to the FDA and the public, in general. But what no one knew outside of Merck – not the FDA, the CDC or any other government agency – was that this result was the product of Merck's improper use of rabbit antibodies and the wide-scale falsification of test data to conceal the inflated seroconversion numbers these antibodies generated.

After Protocol 007 was put to bed and up to the filing of *Krahling v. Wlochowski et al. v. Merck* nine years later, Merck continued to represent that its mumps vaccine had at least a ninety-five percent efficacy rate. The company did so even though it was well aware that the efficacy rate was far lower and that the FDA would rescind Merck's exclusive license if it became aware of the true efficacy rate of the vaccine. Revealed in the court filing, Merck principally made false representations in the package insert that accompanies each dose of a Merck vaccine. This is the product material that the FDA requires, among other things, for the purpose of informing the government, healthcare providers and the public of the composition of the vaccine and its overall efficacy at immunizing the recipient from contracting mumps. Merck's mumps vaccine insert had changed over the years, but at least one thing that remained constant was the company's reporting of at least a ninety-five percent efficacy rate. The insert at the time of the suit provided that "a single injection of the vaccine induced mumps neutralizing antibodies in ninety-six percent of susceptible persons." As support of this representation, Merck cited the studies it conducted nearly forty years before to obtain its original license and vaccine approval. Merck's insert had included this exact language and backup support material since at least 1999. The insert cited in the filing also provided that "following vaccination, antibodies associated with protection can be measured by neutralization assays." The citation for this statement was "unpublished data from the files of Merck Research Laboratories." While the source for this unpublished data was not specified, it appeared to have come from the falsified results obtained during the Protocol 007 efficacy testing. Thus, the product insert was a clear misrepresentation of the efficacy of the mumps vaccine. Clearly, it relied on outdated studies that were far removed from the vaccine's actual efficacy at the time the suit was filed in federal court. Further, it ignored the unfavorable seroconversion results from the company's 1999 PRN test that Merck ultimately abandoned. Worse still, it ignored the fraud and manipulation that was intrinsic to the enhanced PRN test. In short, as Merck was well aware, the efficacy rate of its mumps vaccine was nowhere close to ninety-five percent, and had not been for many years. Yet, Merck continued to misrepresent the vaccine's effectiveness as it continued vigorous sales of the same in the United States and Europe, where the company was approved to sell the MMRII®, and in the United States and Europe, where it was given the green light to sell the MMRII®/Varicella combination vaccine, referred to as ProQuad. To get these permissions, from the FDA and the European Medicines Agency (EMA), Merck, according to the complaint, misrepresented the efficacy of the mumps vaccine.

The EMA approved Merck's sale of the MMRII® analogue, which they marketed as MMR Vaxpro®, through a joint venture with Sanofi Pasteur MSD in 2006. Merck rolled out the falsified results of its enhanced PRN test to get into the European market. The EMA actually cited Protocol 007 test results in support of its decision to grant the approval. After the agreement was reached, Merck manufactured MMR Vaxpro at its West Point facility for Sanofi Pasteur MSD to sell in Europe. Around the same time, the EMA also approved Sanofi Pasteur MSD's application for sale of Merck's ProQuad® in Europe. As with MMR Vaxpro, Merck's joint venture submitted the falsified results of Protocol 007 to the EMA as supportive clinical information in its vaccine application. Relying on this information, the EMA found "no major concern" about the efficacy of the mumps component of the vaccine. Thus, by 2006, Merck had exclusive licenses to sell MMRII® and ProQuad® in the United States, as well as licenses to sell MMRII® and ProQuad® in Europe. During this time, Merck's misrepresentation of the mumps vaccine effectiveness and deficient testing that it used to support the application amounted to outright fraud.

Without the requisite ninety-five percent efficacy of the Merck mumps vaccine, the door was swung wide open for the resurgence of the disease in the United States and Europe. That is exactly what Krah, who was well aware of the vaccine's failings – predicted would happen. In a conversation Krah had with Krahling in the middle of the enhanced PRN testing, Krah acknowledged that the efficacy of Merck's vaccine had declined over time, explaining that the constant passaging of virus to make more vaccine for distribution had degraded the product. Krah predicted just then that because of this, mumps outbreaks would continue. Krah made this statement in an effort to *justify* Merck's falsification of the test data because, according to him, Merck's vaccine was still the safest one available. Krah's predilection came true.

In 2006, more than 6,500 cases of mumps were reported in the Midwest. This was the largest mumps outbreak in nearly twenty years – since the two-dose MMRII® requirement was implemented – and a significant spike from the annual average of 265 cases that had been reported for the years leading up to the 2006 outbreak, which began in Iowa with a group of college students and ultimately spread to the states of Kansas, Illinois, Nebraska, Missouri, South Dakota, Pennsylvania, and Wisconsin. The CDC, FDA and Merck publicly worked together to determine the cause of the 2006 outbreak; of course, only Merck knew that the primary cause was the insufficient efficacy of its vaccine. But Merck continued to maintain its inflated efficacy rate and the government continued to believe that there was no problem with the vaccine. During the investigation of the 2006 outbreak, CDC director Julie Gerberding, M.D., Ph.D., reaffirmed the CDC's position, no doubt fed by Merck's fabricated scientific studies and continued misrepresentations – that there was no problem with the vaccine:

> We have absolutely no information to suggest that there is any problem with the vaccine. [...] What is going on here in the context of the outbreaks is a number of people who have not received both doses, coupled together with people who have received the vaccine but are susceptible anyway because it is not perfect, living in crowded conditions like college dormitories or mixing up with other students at spring break or during holidays, and setting off a cascade of transmission that is going to take a while to curtail.

Gerberding and the CDC emphasized that "[t]he best protection against the mumps is the vaccine." In late 2009, after seven years as director of the CDC, Gerberding became president of Merck's vaccines division, where she's expanded the company's product reach beyond the United States and Europe.

Contrary to the CDC's blame-shifting in the wake of the 2006 outbreak, the scientific community was not as accepting of Merck's vaccine or that it had nothing to do with the Midwest mumps outbreak. Scientists and public health officials worldwide have continued researching the 2006 outbreak to understand the origins of such a large epidemic among a highly vaccinated population. One of the leading studies was led by Gustavo H. Dayan, M.D., working just then at the CDC, who published finding in 2008 in the *New England Journal of Medicine* that concluded "[a] more effective mumps vaccine or changes in vaccine policy may be needed to avert outbreaks and achieve elimination of mumps."[331] Perhaps most notable about this study was that it scientifically questioned Merck's stated efficacy based solely on Merck's use of the vaccine strain instead of the wild-type virus to test efficacy. But what the study could not account for was Merck's concealed effort to mask the vaccine's efficacy using improperly applied rabbit antibodies nor did was Dayan's research privy to the company's falsification of test data.

At the time Krahling and Wlochowski filed the *qui tam*, Emory University was conducting a clinical trial of its university students in yet another attempt to explain the cause of the 2006 mumps outbreak among largely college-age students who had received both doses of the vaccine. Certainly, this study became suspect when it was learned that Merck was a collaborator on the Emory study, thus continuing to position the company to perpetuate fraudulent efficacy findings. Merck's continuing misrepresentations of this testing precluded any scientist or public health official from getting to the cause of the 2006 outbreak, which Krahling and Wlochowski contended was Merck's vaccine failure and an efficacy rate far below the accepted ninety-five percent. In their suit, Krahling and Wlochowski stated that it was unlikely that Dayan would pursue his conclusion that perhaps it was time for a new vaccine, or to conduct future studies to help evaluate national vaccine policy. Dayan left the CDC to become director of clinical development at Sanofi

Pasteur, the vaccine division of Sanofi Aventis Group, Merck's partner in manufacturing and sale of the mumps vaccine in Europe. As of the August 2010 filing of their *qui tam*, the CDC had also not acted on Dayan's conclusions.

In his 2008 study, Dayan also predicted another mumps outbreak would follow three years after the 2006 epidemic in the Midwest. His prediction followed from the three-year cycles in which outbreaks occurred in the pre-vaccine era: "In the pre-vaccine era, mumps activity followed three-year cycles, so the current low activity rate [at the time of his 2008 study] may be transient while another crucial mass of susceptible persons accrues."[332] Dayan was correct. In August 2009, roughly three years after the 2006 outbreak and just as Dayan had stated, another mumps outbreak began, despite high vaccination coverage among the United States' children's population. By August 2010 there were more than 3,700 cases reported to the CDC.

Importantly, it was because of the 2006 and 2009 outbreaks that the CDC pushed back its target date for eradicating mumps from its original 2010 goal to no earlier than 2020. Beyond the scope of the Krahling and Wlochowski *qui tam*, there have been further mumps outbreaks. As of December 16, 2016, in fact, Arkansas had 2,200 cases of mumps, and the CDC reported that Arkansas led the nation in mumps cases during the outbreak.

Krahling v. Wlochowski et al. v. Merck made a compelling case that Merck violated the False Claims Act. In the first decade of the twenty-first century, Merck's fraudulent scheme to misrepresent the efficacy of its mumps vaccine cost the United States taxpayer hundreds of millions of dollars through the federal government's annual purchases of the vaccine under the National Vaccine Program (NVP). The NVP was created by then National Childhood Vaccine Injury Act in 1986 to coordinate all federal activities related to vaccines and immunization programs and is operated by the United States Department of Health and Human Services (HHS). The CDC plays the critical role of identifying and recommending which vaccines should be administered as part of the NVP. The CDC recommended Merck's mumps vaccine for more than thirty years, a recommendation premised on the CDC's belief that the vaccine had an efficacy rate of ninety-five percent or higher.

The CDC also negotiates and contracts for the government's purchase of vaccines. Federal funding for the NVP goes back to the 1962 Vaccination Assistance Act that established the Section 317 Program to support immunization programs. By 2010, the CDC was spending approximately $3.4 billion each year on federal and state programs to provide vaccines for free. This amount represented approximately fifty-two percent of all spending for childhood vaccines in the United States. The two government programs for which the CDC principally purchases vaccines are the Section 317 Program, which provides federal grants to state and local health departments to pay for vaccines in support of mass immunization campaigns, and the Vaccines for Children Program that provides vaccines to children who are uninsured, are on Medicaid, are Native Americans, or who may have insurance but which does not cover the cost of vaccines. Additionally, certain states participate in universal purchaser programs that provide free vaccines to all children who do not otherwise qualify for the two federal programs. The CDC coordinates these state programs but they are funded by the participating states.

The CDC contracts for the purchase of vaccines directly from the license holder. In the case of the MMRII® and ProQuad®, the CDC directly contracts with and purchases from Merck. The CDC purchases vaccines in batches of varying size throughout the year for administration to the public. As a matter of practice, for many years and as negotiated, Merck sent – and presumably still sends – the mumps vaccines to the CDC's designated repositories together with the relevant product information to be disseminated to the doctors and health clinics responsible for administering the vaccine. Between 2000 and 2010, the CDC had paid Merck more than $600 million for its MMRII® vaccine. The amount of sales is likely conservative because they do not account for the CDC's purchase of ProQuad®, which is significantly more expensive than MMRII®, and purchases of adult doses of MMRII® and ProQuad®, which Merck also sold to the CDC. Over this period, the United States had thus paid between a half and three-quarters of a billion dollars for a vaccine that does not provide adequate immunization.

Krahling v. Wlochowski et al. v. Merck was unsealed by the federal court in June 2012. This suit was promptly followed by a putative class action in which Chatom Primary Care, P.C., an Alabama health provider, alleged that Merck's behavior resulted in artificially high vaccine prices by effectively barring competitors from manufacturing their own vaccines. *Chatom Primary Care et al. v. Merck* was filed in the United States District Court for the Eastern District of Pennsylvania on June 25, 2012, as an antitrust case. The Sherman Antitrust Act, made law by the United States Congress in 1890, attempts to prevent the artificial raising of prices by restriction of trade or supply. Acts by a monopolist to artificially preserve that status, or nefarious dealings to create a monopoly, are illegal. The purpose of the Sherman Act is not to protect competitors from harm from legitimately successful businesses, nor to prevent businesses from gaining honest profits from consumers, but rather to preserve a competitive marketplace to protect consumers from abuses.

Chatom Primary Care alleged Merck had clearly violated antitrust law and brought appropriate legal action against the company for unlawfully monopolizing the United States market for mumps vaccine. "As with the market for any product, a potential competitor's decision to enter a market hinges on whether its product can compete with those products already being sold in the market," read the complaint. "If an existing vaccine is represented as safe and at least ninety-five percent effective, as Merck has falsely represented its vaccine to be, it would be economically irrational for a potential competitor to bring a new mumps vaccine to the relevant market." The complaint laid out the anticompetitive effects of Merck's unlawful monopolization of the mumps vaccine market. "Through its false representations of the mumps vaccine's efficacy rate and its efforts to conceal the significantly lower efficacy rate that the Protocol 007 confirmed, Merck has unlawfully monopolized the relevant market and foreclosed potential competitors from entering the market with a new mumps vaccine," read the suit. "No manufacturer is going to sink the time, energy and resources into developing the vaccine for sale in the United States with the artificially high bar Merck has unlawfully devised."

But for Merck's anticompetitive conduct, the Chatom Primary Care suit laid bare, including its fraud and other misconduct, one or more competing manufacturers would have entered this lucrative market – with its guaranteed sales of almost eight million doses a year – with a competing mumps vaccine. "For example, GlaxoSmithKline, a manufacturer of numerous FDA approved vaccines, has an MMR vaccine, Priorix®, that is widely sold in Europe, Canada, Australia and other markets. Priorix® is not licensed or sold in the United States, even though the company has a United States patent covering the vaccine and, according to an industry journal, had plans to enter the United States market with it." The complaint emphasized that there are no legitimate pro-competitive efficiencies that justify Merck's anticompetitive and/or otherwise unlawful conduct or outweigh its substantial anticompetitive effects. Merck had effectively harmed competition by foreclosing other manufacturers from entering the market, a claim made clear in the Chatom Primary Care filing.

The case originally brought to federal court by Krahling and Wlochowski had been slow to move through the legal process. Oral arguments stemming from the whistleblowers' *qui tam* did not take place until the summer of 2013 – approximately three years after the claim was filed with the court. The Chatom Primary Care antitrust case certainly helped their case. The two lawsuits accusing Merck of lying about efficacy of its mumps vaccine to keep competitors from bringing their own versions of the vaccine to market was moved forward after the Pennsylvania federal judge overseeing both cases ruled in favor of the whistleblowers and direct purchasers on September 5, 2014.[333] United States District Judge C. Darnell Jones II ruled that the whistleblowers had sufficiently pled that Merck might have provided false statements to the government and that the direct purchasers – Chatom Primary Care – had shown enough evidence to establish that these falsehoods could have helped the company gain a monopoly. The judge did throw out breach of contract and unjust enrichment against Merck but allowed the rest of the suit to be carried forward. Importantly, Judge Jones found that the plaintiffs' – Krahling's and Wlochowski's – first-hand experience in the Merck laboratories, where they witnessed supervisors and managers instructing staff persons to withhold information from the government regarding the diminished efficacy of the mumps vaccine was enough to establish the defendant knew the claims about the product were false. "For the purposes of this stage of litigation, these allegations provide defendant with sufficient notice of the claims at issue. Relators' [Krahling and Wlochowski] claim alleging a violation of [False Claims Act] § 3729(a)(1)(A) is well pled."

Chapter 14 –
The Whistleblower Catch-22

W ere it not for the whistleblowers and watchdogs, the majority of cases of scientific misconduct would go under- and unreported. The public would never get to read headlines like the following:

"Authors retract study with contaminated cell lines"

"Former UT [University of Texas]-Southwestern cancer researchers faked data in 10 papers: ORI"

"Seventh retraction for Ohio researcher who manipulated dozens of figures"

"Columbia [University] investigation reveals researcher faked data – and a degree"

"Big Pharma criminality no longer a conspiracy theory: bribery, fraud, price fixing now a matter of public record"

"USDA finds 'evidence of manipulation' in vaccine study"

"Millions of children put at risk by Merck"

"Vaccine researcher charged with felony crimes for research fraud; may spend 20 years in prison over faked AIDS vaccine"

"'Dr. Drew' was paid by Glaxo [SmithKline]"

"Heart researcher faked 70-plus experiments, 100-plus images"

"Re-analysis of controversial Paxil study shows drug 'ineffective and unsafe' for teens"

"Stunner: researchers retract paper because company complains it's hurting profits"

These cases are among hundreds of recent reports, investigations and court filings tied to big science gone bad. The spike in reporting over the past ten years has been palpable. But just as remarkable is the role of the whistleblower and the catch-22 in which they are often left when the misconduct they have reported is put to bed.

"As an unsuccessful whistleblower," wrote Helene Hill for the February 11, 2014 *Scientist Magazine*, "I know first-hand that the system to deal with misconduct in science often fails. [...] I advocate that, to minimize scientific misconduct, we establish a better system for reporting and adjudicating it."

In recapping her case, she iterated that it had started when she witnessed what she believed – and were supported by a second observer – to be data fabrication. In short, numbers were recorded for samples that they believed did not exist. "I followed the guidelines promulgated by my university for reporting perceived misconduct as closely as I could," she continued. "I was referred to the campus committee on research integrity (CCRI), a standing committee of representatives of the various schools on the Newark campus of the University of Medicine and Dentistry of New Jersey, of which my employer, the New Jersey Medical School, was a part." To her astonishment – still – her employer, like too many other colleges and universities with similar investigative bodies, appointed investigators with no prior expertise in the research being scrubbed for misconduct. "Some were administrators," she recalled. "There was [also] a nurse, a midwife, a lawyer, and the vice president for research." But then there was the wife of the questioned principal investigator who had been, she opined, "a postdoctoral fellow working down the hall from the CCRI chairman. The guidelines mandated a two-stage process, but the CCRI assumed both roles and did not call on an independent expert in the field in question."

The CCRI subsequently determined there was insufficient evidence of scientific misconduct. "I disagreed. I was not given the final report, and there was no mechanism for me – the whistleblower – to appeal, although the respondents would have been granted that privilege were the case not to have gone their way." After exhausting options via the United States Department of Health and Human Services' Office of Research Integrity (ORI), she filed the *qui tam*. "For those unfamiliar with such suits," she explained, "*qui tam* cases arose in the Civil War and involved [cases of] bilking the federal government in various ways – none of which realistically apply to scientific research. And, indeed, I maintain that the judge who ruled against me paid no attention to the scientific evidence of fraud. His decision was based on his belief," she added, "that the principal investigator did not know of the fraud the day he submitted the grant application to the National Institutes of Health, even though, at that time, I believed he had plenty of cause for suspicion."

In hindsight, Hill observed, although the *qui tam* case, though in many respects a waste of time and money, did prove enormously useful in that one of her expert witnesses, Joel Pitt, and Hill gained access to data from the accused postdoctoral fellow and nine other members of the same laboratory. "We [then had] compelling evidence for manipulation of results that were used to obtain the original grant, its renewal, and were published in up to eight journal articles," Hill explained. "We analyzed several hundred datasets, finding that the alleged fraudster's results were extremely far from [expected] random compared to the consistently random results of the nine other investigators in the same laboratory."

Hill was a whistleblower who refused to give up. After losing the *qui tam* and subsequent appeal, she returned to the CCRI and ORI with what she considered extremely compelling evidence of data fabrication but was stonewalled again by both investigative bodies. Then she tried, with Pitt, to publish their data analysis from the case. They submitted and were summarily rejected. No one, it seemed just then, wanted to publish statistical analysis of allegedly fraudulent data nor, time would tell, correct the scientific record. "In the meantime, though, there are some things that policy makers, administrative officials, journal editors, and we scientists can do to make correcting the record much more straightforward," wrote Hill, as she outlined what those things should be:

1. Scientists should make raw data publicly available.
2. Scientists should be obligated to report failed replication attempts to the journal that published the original results.
3. All misconduct proceedings involving state and federally funded research should be open to the public.
4. The CCRI [or equivalent] ought to be *ad hoc*, composed of individuals unfamiliar with the complainant or respondent.
5. The CCRI [or equivalent] should employ at least one expert in the field under examination.
6. The final report of the CCRI [or equivalent] and/or the ORI ought to be provided to both the complainant and the respondent.
7. Both complainant and respondent should have the right to appeal, and the appeal board ought not overlap with the original committee.
8. Journals ought to take a more active role in retracting papers containing false information.
9. The scientific community should establish a board of last resort for appeal and adjudication of unresolved scientific questions raised regarding data fabrication and falsification.[334]

Hill's detractors called her efforts quixotic. But she survived as a whistleblower in a climate in which others faced punitive backlash and legal repercussions: the whistleblower's catch-22. She bucked the odds. She did the right thing.

As it turned out, Hill was not the only whistleblower at her medical school. Though her case concluded unsuccessfully, Rutgers University quietly resolved several major whistleblower cases inherited with the merger of the University of Medicine and Dentistry of New Jersey (UMDNJ), agreeing to pay nearly $2 million in settlements with former high-level administrators.[335] When reported on April 26, 2015, the never-disclosed confidential agreements brought an end to two long-running lawsuits that charged UMDNJ, indeed the state's troubled former medical university, with wrongful termination over alleged fraud and illegal bidding practices. Media reports indicate that in one case, Rutgers reached a $1.2 million settlement with Edward Burke, the UMDNJ Newark hospital's former chief financial officer, who made it known that he was fired after he accused top administrators of systematically defrauding Medicaid. Documents show administrators at the UMDNJ inflated medical expenditures by at least $21 million a year, boosting Medicaid and Medicare reimbursement rates. The abuses were allowed to continue even after a consultant repeatedly warned them about the problem.[336] In 2014, Rutgers reached a $700,000 settlement with Ellen Casey, a UMDNJ purchasing officer, who claimed she was terminated after discovering that telecommunications contracts worth millions of dollars were being awarded without public bids.[337] Few knew about the cases because Rutgers imposed strict restrictions on the disclosure of any terms of the settlements on lawyers engaged by Burke and Casey. But NJ Advance Media obtained copies of both settlement documents through Open Public Records Act requests filed with Rutgers, a public university and required by law to comply. Adam Henick[338], M.B.A., M.P.H., one-time vice president of ambulatory care at UMDNJ's university hospital who oversaw Medicaid billing for outpatient services and first discovered evidence of overbilling in 2002 before being forced out, opined to NJ Advance Media writer Ted Sherman that Rutgers "is probably trying to put UMDNJ's sordid past to rest."[339] Sherman's article laid bare that past, much of it exposed earlier in this narrative. In review, UMDNJ's history included "a culture of waste, fraud and abuse, including lucrative consulting contracts that went to political insiders, double-billing Medicare and Medicaid by millions of dollars, and an illegal kickback scheme that gave doctors no-show jobs in return for referring patients to the university's faltering cardiac surgery program."[340]None of this included, at any point, an admission of any wrongdoing by Rutgers.

The Rutgers administrative whistleblower settlements were low-hanging fruit when contrasted to charges of fraud, fabrication and misconduct that were taking place in the UMDNJ's laboratories over a protracted period of time, to include the same time frame in which the Burke and Casey controversies raged, and Helene Hill reported the same in her own department. In the two cases that follow, one involving a UMDNJ neuroscientist and the other a pharmacologist, it is clear that UMDNJ investigative bodies were more than familiar with a number of concerning cases on campus of falsifying and fabricating research data and the inevitable cover-ups that ensued.

The Office of Research Integrity published findings on June 28, 2012, that found a neuroscientist who studied the effects of pesticides on a mouse model of Parkinson's disease, made up data. The researcher in question, Mona Thiruchelvam, Ph.D., a UMDNJ assistant professor in the department of environment and occupational health sciences, faked cell counts in two grant applications and a number of papers that claimed to show how the pesticides paraquat, maneb, and atrazine might affect parts of the brain involved in Parkinson's.[341] Based on the report of the investigation conducted by the UMDNJ and additional analysis conducted by ORI in its oversight review, ORI found Thiruchelvam, engaged in research misconduct in research supported by National Institute of Environmental Health Sciences (NIEHS), National Institutes of Health (NIH), grants P30 ES05022, P30 ES01247, and R01 ES10791 and the intramural program at the National Institute on Drug Abuse (NIDA) at the National Institutes of Health.

ORI found that Thiruchelvam engaged in research misconduct by falsifying and fabricating cell count data that she claimed to have obtained through stereological methods in order to falsely report the effects of combined exposure of the pesticides paraquat and maneb on dopaminergic neuronal death and a neuroprotective role for estrogen in a murine model of Parkinson's disease. Thiruchelvam further provided to the institution corrupted data files as the data for stereological cell counts of nigrostriatal neurons in brains of several mice and rats by copying a single data file from a previous experiment and renaming the copies to fit the description of 13 new experiments composed of 293 data files when stereological data collection was never performed for the questioned research. The fabricated data, falsified methodology, and false claims based on fabricated and falsified data were reported in two NIEHS,

NIH, grant applications, two publications, a poster, and a manuscript in preparation. The ORI further determined that Thiruchelvam falsified cell count data published in two papers in 2005 in *Environmental Health Perspectives* and *Journal of Biological Chemistry*, both of which she agreed to retract. The now-retracted papers investigated the neurological response to the combined pesticides paraquat and maneb and suggest the pesticide atrazine also has a role in disrupting dopamine pathways. The false data were used, according to *The Scientist*'s June 29, 2012 reporting, to create several summary bar graphs, which Thiruchelvam modified to support the hypothesis that proteasomal dysfunction is higher in males than females with Parkinson's, and that exposure to paraquat and maneb enhances this effect.[342]

As a purveyor of bad data, Thiruchelvam's misconduct may long outlive her punishment. *The Scientist* exposé on the case, Hayley Dunning quoted Emory University professor of environmental health Gary W. Miller, Ph.D., who cited the *Environmental Health Perspectives* paper in one of his own. Miller wrote that his laboratory had always been skeptical about the association between certain herbicides and Parkinson's. "There is strong evidence of an association between pesticides and PD, but figuring out exactly which compounds are driving this has been difficult," he told *The Scientist* by email. "I suspect some laboratories have pursued studies based on these findings, which is unfortunate. The retraction of these papers doesn't help the field." According to the ISI Web of Science, Thiruchelvam's Environmental Health Perspectives articles have been cited thirty-six times. Deborah Cory-Slechta, Ph.D., a coauthor on the same paper, informed Dunning that she was "both shocked and disappointed" by the news. Both papers have had influence in the field, Dunning iterated, with the *Journal of Biological Chemistry* study being cited seventy-three times according to ISI Web of Science.

None of what Thiruchelvam had done would have been reported had a collaborator at UMDNJ first brought the matter to the attention of the university research integrity committee a few years after Thiruchelvam joined the university in 2003. The whistleblower realized she was publishing cell density data without using his laboratory as she had done before, reported *The Scientist*. "An initial inquiry was launched, for which Thiruchelvam provided the name of a researcher in California who she said had provided her with data," Dunning continued. "The witness, who Thiruchelvam said by that point had moved to England, was called and confirmed the story, but further investigation by UMDNJ revealed that this was a false witness. When investigators got in touch with the actual person Thiruchelvam had named, they learned she still resided in California and that she denied providing any data to Thiruchelvam."[343] Thiruchelvam's deception did not end there. As Dunning documented:

> Thiruchelvam, in her defense, then produced [the aforementioned] 293 data files she said were the product of a confocal microscope system manufactured by the company Micro Bright Field (MBF). When UMDNJ investigators gave MBF the data to interpret, the company concluded that the files were corrupted and could not be verified as real or false; however, when the case was passed to the ORI for oversight review, agents used forensic computing software to determine that many of the files, despite having different file names and dates, were identical in content. This suggested to the ORI that the corruption had been intentional on the part of Thiruchelvam, and not due to damage by a computer virus. The identical files were then sent back to MBF for further analysis, which subsequently discovered that all the files had come from one single file created in 2002 during one of Thiruchelvam's previous investigations at the University of Rochester, before she joined UMDNJ.

The *Journal of Biological Chemistry* paper was the subject of a 2008 correction, in which coauthor Eric Richfield, M.D., Ph.D., was removed. When *Retraction Watch* asked him why his name was taken off the paper, he refused comment. "It's not clear," wrote Ivan Oransky on July 2, 2012, "if Richfield is the collaborator referred to in *The Scientist* story."[344] Thiruchelvam left UMDNJ in February 2010, per *The Scientist*, and was barred by the ORI from receiving federal grants for seven years.

The UMDNJ's spotty past also included the case of Bernd Hoffmann, Ph.D., a postdoctoral fellow and adjunct assistant professor, in the department of pharmacology. Based on two inquiry/investigation reports from the UMDNJ and additional analysis provided by the ORI, the United States Public Health Service published findings on February 24, 2004, that determined Hoffmann engaged in scientific misconduct by falsifying and fabricating research data in a manuscript entitled "LIS1/ NUDF and CLIP-170 are required for dynein-mediated vesicle transport on microtubules" that had been submitted to the *Journal of Cell Biology*, but was withdrawn before publication. Specifically, the Public Health Service investigation found that Hoffmann had:

1. falsified data values on the second line from the bottom of Table IV; for example, the correct number under "Bound" in the first column was only one-third of that shown (325) in the manuscript.
2. falsified data by erasing a band of approximate molecular weight 15KD from Figure 5A in the manuscript, and
3. falsified a related movie film available on the Internet by altering the movement of the vesicles.

The Public Health Service also found that Hoffmann engaged in scientific misconduct by falsifying and fabricating research data in a published paper entitled "The LIS1-related protein nudF [nuclear distribution protein] of *Aspergillus nidulans* and its interaction partner nudE bind directly to specific subunits of dynein and dynactin and to alpha- and gamma-tubulin"[345] that had been published in the *Journal of Biological Chemistry* in 2001. Just as he had done in the prior paper, he falsified in Figure 5A left, Western blot with the alpha tubulin antibody for incubated proteins; the lower right band was reused twice in Figure 2A. In Figure 5A, it was used as gamma tubulin band for the coprecipitation experiment with nudF-Protein S and as nude for the coprecipitation experiments with nudG (CDLC[346]) – FLAG[347]. He also falsified Figure 5A left, nudF Western blot with the alpha tubulin antibody for incubated proteins; the lower left band was reused in Figure 2A as alpha tubulin in the coprecipitation experiment with nudF-Protein S. The inquiry further found that he falsified Figure 4A left, nudF and for the interaction between the two proteins nudA and nudF, pulled out with nudA-FLAG-agarose, had been used at several other places such as Figure 5A left, left gamma tubulin band, Figure 5B left, nudE band for the interaction E plus alpha, and Figure 5B right, nudE band for the interaction E plus K (ARP1[348]).

Hoffmann entered into a voluntary exclusion agreement in which he agreed for a period of three years, beginning on January 30, 2004, to exclude himself from any contracting or subcontracting with any agency of the United States government and from eligibility or involvement in non-procurement programs of the same, referred to as covered transactions as defined in debarment regulations set forth in 45 CFR Part 76. He also agreed to exclude himself from serving in any advisory capacity to the Public Health Service, including but not limited to, service on any of its advisory and/or peer review committees, and boards, or as a consultant. Hoffman lastly agreed to draft a letter of retraction to be forwarded to the ORI along with the signed agreement. The draft letter requested the retraction of the *Journal of Biological Chemistry* article and stated that he falsified and fabricated data in Figures 2A, 4A, 5A, and 5B. A final retraction letter was sent to the editor of the journal.

Michael Doran's article in the April 20, 2016 *Nature* offered a whistleblower's perspective not all that different than what Hill documented – at least at the beginning. "In 2012," he wrote, "a graduate student [Luke Cormack] came to me for advice." Doran, a stem cell scientist at Queensland University of Technology in Brisbane, Australia, listened. "He [Cormack] explained that his supervisor was traveling, and [he] needed images for a review article that he was writing. I suggested," he continued, "that he check his lab's folders for something suitable. When he came to me again, he was shaking." The student had stumbled upon images and data that he knew all too well but that had descriptions that did not match up. Cormack had not been the lead author of the paper, first published in February 2010, but the subject matter had formed the core of his doctoral study. Cormack subsequently submitted it to his doctoral committee, which rejected it because he did not have enough data nor, as it turned out, cells to do the necessary assays. The committee further believed he fell short on the literature review. Of interest, the university did think there were other findings he might pursue and gave Cormack a scholarship to write manuscripts. When he went back to the original data from the now-retracted paper, the media had been developed in June 2007, before he came to the laboratory, and this is how he came to discover that the paper was so full of errors.[349]

Thus, as Doran explained further, Cormack was worried that findings he was unable to replicate from a previous publication – that were the foundation, of his doctoral project – were false. "Misleading" was the word Doran used. But he looked at the data with the student and on his own, asked a more senior colleague about it, and then decided to work with the student to submit a complaint. "Ultimately, I advised that he take his concerns to the vice chancellor," Doran explained. "Although we did not know it at the time, doing so mandated that the complaint be formally investigated. It also gave the student some protection from the possible repercussions of being a whistleblower."

Like Hill's case, a university investigation ensued. That is where the similarity ends. In the case to which Doran was privy, there was a retraction, repayment of a major national grant and an external inquiry that found research misconduct and multiple misrepresentations. "The vice chancellor violated state policy by disclosing my name to a reporter," Doran opined. "Although he self-reported the breach, I wondered what else was being said about me and to whom. The difficult situation was new to us all. It affected my health, productivity and relationships; I lost countless hours that I could have devoted to lab work. The student, who has still not graduated, is taking a break from research."[350] The graduate student's response and subsequent predicament – the catch-22 – should not have been unexpected. Several well documented *qui tam* cases have involved a graduate student – or two, or three – reporting a faculty member for fraud, fabrication or misconduct. In nearly every instance of such reporting, the graduate student became the object of backlash, and few went on to complete their respective degree program. As reported by *Retraction Watch*'s Ivan Oransky on December 27, 2012, "a contested retraction in *Stem Cells and Development* has left the Queensland University of Technology graduate student who fought for it in limbo, uncertain if he will earn his Ph.D. And many of those who didn't want the paper retracted have a significant financial interest in a company whose work was promoted by research, despite any lack of disclosure in the now-retracted paper."[351]

Doran offered that there is no handbook that describes what to do in these situations. "If you decide to be a whistleblower, you must realize that it will be stressful. And because it is so stressful, you want to ensure that any investigations that are carried out will be robust," he wrote. "Every case needs to be considered on an individual basis, but I hope that sharing my recommendations will help others who find themselves in a similar position."[352]

Like Hill, Doran compiled a series of recommendations for future whistleblowers:

1. *Don't confront potential misconduct alone.* Although postdocs and Ph.D. students are the most likely to identify inconsistencies in the previously published data of their groups, they are often the least equipped to highlight serious problems. Be ready to give your supervisor the benefit of the doubt but also be aware that raising concerns directly could provide an opportunity to obscure evidence of misconduct. In fact, the co-founders of the blog *Retraction Watch* recommend against contacting authors first if no one else knows of issues in the research. [Good advice.] Doran also remarked that before submitting a complaint, ask a technical expert, preferably from outside your university, to corroborate your assessment. In his case, he asked others to verify his evaluation of data without disclosing exactly why. You must be absolutely confident in the veracity of the complaint, he advised, and be ready for your expertise to be called into question.

2. *Make your case clearly and keep detailed records.* Describe inconsistencies thoroughly, and ensure that any correspondence is recorded electronically, complaints with a digital trail are less likely to disappear. Maintain careful records, and retain all of the original data, if possible. Do not relinquish these records, and neither deny nor admit that you have them. Doing so could mean that authorities might either dismiss your assertions or accuse you of holding unauthorized copies of data.

3. *Submit your complaint to the highest authority.* A stifled complaint will degrade your credibility. Take steps to maximize the chance that your concerns will be investigated and that you will be protected from retribution [of any kind]. In Doran's case, an initial internal inquiry did not result in what he and his graduate student

believed were the necessary corrective actions. Frustrated, as he described it in *Nature*, the student then sent his complaint to *Retraction Watch*, as well as to the journal that had published the original paper.

4. *Avoid public disclosure.* 'In my view, it is not appropriate to make public statements about such cases until they are resolved. It would have been much easier,' he opined, 'for me if our case had not been discussed openly until its outcome had been decided.' That noted, Cormack discovered that talking to the press [*Retraction Watch*] helped move the case forward with the university involved.[353]

Would Doran blow the whistle again – yes. He encourages other scientists to report misconduct if they find it. "As more cases emerge," he wrote, "it becomes easier for other whistleblowers to come forward: this incident prompted the university to implement progressive policies concerning responsible research practices."[354]

While the *Stem Cells and Development* journal ultimately retracted the article in question, a subsequent external inquiry by Australia's National Health and Research Council concluded that although there had been misconduct by one researcher, there was no misconduct by the traveling supervisor referred to in Doran's article, and there was no intentional wrongdoing, thus protecting the financial interest vested in the research, and documented through a series of *Retraction Watch* articles on the case. Here is the journal's retraction notice:

> Concerns were raised by one of the co-authors, Luke Cormack [a Ph.D. candidate], and after an investigation was conducted by the authors' institution, The Queensland University of Technology, Stem Cells and Development is officially retracting the paper, 'A Chimeric Vitronectin: IGF-1 Protein Supports Feeder-Cell-Free and Serum-Free Culture of Human Embryonic Stem Cells,' by Manton et al., from Volume 19, Issue 9 (pages 1298-1305).

> The concerns center primarily on whether the images provided in Figure 1 show colonies of the stem cell line

described at the passage stated stained for the marker listed. There were also related concerns with the PCR data included.

It is important to note that the corresponding author, K. J. Manton [Kerry J. Manton, Ph.D.], has denied deliberate wrongdoing. Stem Cells and Development acknowledges and appreciates the thoroughness of the investigation undertaken by Queensland University of Technology.

Stem Cells and Development believes that had the peer reviewers of the paper been aware of the extent and nature of the mistakes in the paper that apparently went undetected by the authors through an initial submission, two revisions, as well as galley proof, they would not have deemed the paper acceptable.

Stem Cells and Development is dedicated to the highest ethical standards of scientific publishing.

Cormack won in the sense he got the retraction. Certainly, his coauthors wanted only an erratum. The erratum would have been insufficient given the gross errors discovered in the paper. Graham Parker, the journal's editor, told *Retraction Watch*: "There were problems with the paper that I believe had the reviewers known they would not have accepted as it stood. Those problems must have been known to at least one of the authors. After revisions and galley proof approvals," he continued, "no one except Cormack alerted the journal to those problems. Following an investigation by the host institution those problems were acknowledged yet no explanation was offered as to why the publishers had not been previously informed."[355] According to Institute of Scientific Information's (ISI's) Web of Science, the now-retracted paper has been cited six times. Oransky iterated that what caught *Retraction Watch*'s attention about this particular case was that senior author Zee Upton, Ph.D., a biochemist, inventor and tissue engineer known for her research in growth factors, extracellular matrix proteins and wound healing, is consulting chief scientific officer of Tissue Therapies, which Queensland University of Technology spun off to develop technologies based on vitronectin and other compounds. Vitronectin is a glycoprotein of the hemopexin family that is abundantly found in serum, the extracellular matrix and bone. Upton and other investors cofounded the company in 2003, approached the university for a license on the intellectual property, and had Tissue Therapies listed on the Australian Stock Exchange in 2004.[356] An e-print[357] of another paper by Upton and coauthors, Oransky discovered[358], noted that several authors had bought stock in the company. The notice reads: "The authors have purchased shares in Tissue Therapies Ltd., an enterprise spun-out from the Queensland University of Technology, Brisbane, to commercialize some of the technology described in the manuscript."

Doran missed a piece of advice that might serve all would-be whistleblowers well as they boldly put their careers on the line to report misconduct in science:

Follow the money.

The Queensland University of Technology became the definition of the fox guarding the henhouse, the most likely to exploit information or resources that it had been charged to protect or control, in this case the integrity of the research in the university's purview. What transpired was a conflict of interest for the authors and the university charged to investigate the research and determine, if any, wrongdoing had occurred. The university was forced to repay a $275,000 research grant after it was determined one of the school's academics knowingly provided misleading and incorrect information on her stem cell research over five years. *The Australian*'s Julie Hare reported on August 30, 2014, "also found her 'failures to correct were made with gross and persistent negligence.'"[359] The inquiry by the National Health and Research Council focused on bioscientist Kerry Manton, Ph.D., who was a junior researcher with the university's Institute for Health and Biomedical Innovation at the time of the 2009 grant application, and senior researcher Zee Upton, Ph.D. The council determined that Manton "failed to fulfill (her) responsibilities in relation to the responsible dissemination of research findings and that this, coupled with a failure to correct the errors, constituted research misconduct." But it found that while Upton failed in her duties as supervising researcher, it was due to "extraordinary work commitments and high expectations of performance" that she should not be censured or punished in any other manner. "Her failures," the report concluded and as addendum to Doran's *Nature* article laid out, "were not deliberate or intentional, nor were they reckless."[360]

Not all whistleblowers want their names publicly known, largely due to concern – and a host of well documented cases – of retribution. What's the phrase oft repeated? No one likes a snitch. Paul S. Brookes, Ph.D., a professor in the department of anesthesiology, pharmacology and physiology at the University of Rochester Medical Center in Rochester, New York, was outed as a whistleblower when his identity as the founder of a blog called Science-Fraud.org made headlines. Brookes had begun the site over his own frustration with the scientific establishment's handling of irregularities that popped up with increasing frequency in published papers. Science-Fraud.org was only up and running for six months but in that time, it spotlighted 275 papers with apparent problems, to include everything from undisclosed but recognizable slicing of gels, duplication of bands, to unacknowledged reuse of images. Brookes posted to the blog anonymously, often based on tips he received from equally anonymous scientists.[361] Then he got outed. Someone unhappy with the light Brookes shone of those committing fraud, fabrication and scientific misconduct – and under the guise of "anonymous" – sent an e-mail on January 2, 2013, to many scientists and university administrators, to include the president of Brookes' university, disclosing his name and calling for a lawsuit. This "someone" may well have been one of the many academics Brookes tagged for wrongdoing on his blog. After he was exposed, Brookes was forced to close down the site amid multiple legal threats. In an extensive interview with *Science*, Brookes discussed just how serious the legal threats became, and also how his institution responded to the entire ordeal – Rochester was not happy, making it clear that his blog was outside his role as a university faculty member. "Nothing I did," he said later, "was protected by any legal protections from the university, so when I began to receive legal threats, I had to hire an attorney, paid from my own pocket. This is one of the downsides," he continued, "to this kind of activity [whistleblowing]: it requires you to draw a line between your job – your academic career – and this kind of activity, which you are doing as a private citizen."[362]

Brookes' case is illustrative of the great risks to whistleblowers. The question-and-answer he subsequently did with *Science*'s Elisabeth Pain was revealing not only for what he offered about Rochester's response but everything that followed the anonymous e-mail that shut him down. When asked if he ever thought his identity might be revealed after he set up the blog, he admitted the possibility of his unmasking was certainly a consideration. "I think it was always in the back of my mind that it was a possibility, so I took certain precautions, but as we now know from the NSA [National Security Agency] scandal, there is no such thing as privacy on the Internet," he told Pain. "There are several people who are now friends of mine who are also blogging anonymously, and I think there is a fear, especially for junior people [in academia]: 'What happens if my university or department chair finds out about this?' If you are a grad student, what happens if your mentor finds out? It is still a little bit dangerous," he continued. "We are still in the first decade of social media, and we have yet to work out all the rules about what is permissible." Those "junior people" in academia were responsible for many of the problem reports passed to his blog by their faculty. "So, if you are too junior or you are too scared, then you can always ask somebody who is more senior or who is more experienced with this if they would like to do it for you."[363] Brookes also had advice for early-career scientists tempted to commit misconduct:

> You have to decide not to do it. Obviously, people are put under a lot of pressure. I know of a case where the lead author of a retracted paper was a postdoc and wrote a very long letter detailing the working conditions in the lab of the mentor and how they were harassing them and telling them, 'Get the data, get the data, get the data.' And the postdoc felt they had no other option than to comply. If you are in a situation where you are asked to do something that you don't want to do, you have to speak out; you have to refuse to do it.

Brookes' university may not have been pleased with his blog, but it was at least open to discussion of Brookes blog and the implications of what he had been doing and why. One of the major concerns of his university was how much it might distract from the job Rochester hired him to do. "To run a lab, be a faculty member, teach, do research [he listed them off]," he explained. "I would say that at the peak in 2012 I was spending 10 or 12 hours per week on this, but much of this was done after work, at home. Last year, we [he and the university] reached an understanding. This activity was now considered to be unpaid outside consulting."[364] To make sure he was clear of conflicts with the university – after he was outed – Brookes bought a separate laptop out of his own funds to continue his scrutiny of problem publications. With his blog and subsequent engagement on additional academic peer review sites, Brookes filled and continues to engage in discussion of flawed publications and suspected cases of misconduct. He has had to do this because while medical centers like his own and in academia have established boards, special committees, and offices to review and redress breaches of ethical behavior, they are inevitably flawed. "[T]hese mechanisms repeatedly prove themselves ineffective in addressing research misconduct within the institutions of academic medicine," in short. "[I]nstitutional design systematically ignores serious ethical problems, makes whistleblowers into institutional enemies and punishes them, and thereby fails to provide an ethical environment."[365] This is a dark prognosis for academic medicine but an even darker one for whistleblowers who are vigorously dismissed by their own institutions. At the end of the day, as Patricia A. Patrick, Ph.D., C.F.E., C.P.A., C.G.F.M., reported: "Anyone can report wrongdoing, but the level of protection an employee will receive will differ depending on whether they're public or private, to whom they report, the manner in which they report, the type of wrongdoing they report, and the law under which they report."[366]

The scope and seriousness of unethical behavior and misconduct in academic medicine, most especially, is a serious problem. When academic medicine abdicates responsibility in cases of fraud and scientific misconduct, failing to hold wrongdoers accountable, everyone loses. Asked by Pain if he believed his blog might cost him his job and career, Brookes confessed that he was worried about that possibility. "I have to continue to get grants, publish papers, and obviously if there are people out there who are upset with me," he opined to her, "then maybe they will review my grants badly, maybe they will review my papers badly. The potential for retaliation is there; there is not a way to get around this." While he was threatened with a total of six possible legal actions, none came to fruition. But the threats did affect him in other ways. "Some of the places I was scheduled to speak were in other states where they have different legal rules, and, in fact, in one case I was scheduled to speak [at] the university of a person who'd threatened to sue me, so I decided not to go on that trip because I didn't want to be served. Fighting a legal challenge at home," he added, "is difficult and expensive, but having to travel to another state would be just impossible."[367]

Brookes observed that in some sense that his outing did make him aware – long term – of what you can and cannot believe in the course of research. "[W]e are becoming increasingly less willing to publish Western blot data, in particular. There are so many limitations to this method, and I don't think we can rely on it very much. For example," he offered, "there are many papers published using commercially available antibodies, all showing that a particular potassium channel is expressed in certain tissues. We have not been able to repeat any of this. We can get a band, but it's still there in the knockout mouse." Thus, he concluded, "[...] I can publish a Western blot that shows what I want to show, but ethically I know that would be wrong. And this has caused problems for us with publishing some things, because reviewers are asking for Western blot data and we have to tell them, 'Sorry, we don't have it. We tried. It didn't work.' And that's frustrating."[368]

Asked if he might have done anything different with this blog, Brookes told Pain he should have been more careful about his anonymity and also security. "One of the major problems was the name of the blog; it was a little bit harsh, and the rhetoric was maybe in some cases a little bit harsh. I am British; we swear a lot. So, you will see from the things I am posting, more recently on PubPeer and other places," he continued, "that I am being more careful with the language." Nothing that he went through with this blog and the dissolution of the same has deterred Brookes' interest in continuing to expose problems in scientific literature. What's at stake? Pain asked. "First of all, fairness. It is just absolutely not fair for people to build careers and get publications and grants and promotion and publicity and fame and fortune on the basis of poor science," he iterated. "We have to live in a meritocracy, and the system of rewards has to be based on reality." Brookes spoke of those who criticized him, suggesting that discussion of possible scientific misconduct [and other problems] in public might be damaging in the long term because it leaves the public with the impression that the majority of scientific literature is wrong, leading to refusal to continue to fund studies. "This is a catch-22, because if we do nothing, then the bad science will be perpetuated. If we speak out, then we risk the public will not believe in science anymore, so we need to be careful."[369]

What about outcomes? Did Brookes achieve any forward progress exposing misconduct with this blog? What should happen next, in his view? When asked these questions of his blog's effectiveness, he rattled off data. "Out of the 275 papers discussed, there have so far been 16 retractions and 47 corrections," which Brookes thought was good. "[B]ut it's not enough. [T]here are many potential problems still unresolved." He had a paper about to be published, in fact, that analyzed the outcomes of his blog work, data on the papers about which he blogged and what happened to them. "I have also kept in private files another set of more than 200 papers that contain the same kinds of problems but were never publicized. I found," he emphasized, "that the public papers have a seven- to eight-fold higher level of corrections and retractions. So, the fundamental idea that I am trying to get across is that discussing this stuff in public produces results, and it appears that when stuff is kept private, less is done about it."[370] Brookes told *Science*'s Pain he wants everything to happen faster with investigations of misconduct:

> Many cases at the Office of Research Integrity (ORI) take 3 or 4 years [to make it through the inquiry]. The ORI needs more money and more people. Every journal needs to have a staff member who is dedicated to dealing with this kind of issue [of weeding out problem papers].
>
> As an example of how things can drag on, I have dealt with cases where it took 7 to 8 months for the investigator to hand over original data to the journal. In my lab, we have a very open data policy: every single piece of data is stored on a common open lab drive, so anybody in the lab can access any of the data from any experiment that

we did for the past 10 years. If somebody e-mails me and says, 'Can you give me the original Western blot from this paper?' within a couple of days I can do that. I don't see why it should take 8 months to produce original data when you get an e-mail from a journal editor or from somebody reading the paper. You either have the data or you don't, and too many people use the excuse that 'the postdoc who did the experiment has left the country and they did not leave behind good records.' That's just unbelievable to me. The prevailing standards of science, which always existed but are now more important than ever, require everybody to keep good records.[371]

Brookes was not the only whistleblower to report anonymously. Tom Reller, the vice president of global corporate relations at Elsevier, the world's largest provider of scientific, technical and medical information products and services, observed that of the three techniques most commonly employed by individuals who report scientific fraud, anonymity is often used. Reller cited an individual working under the pseudonym "Clare Francis." "Clare" is a highly likely a "he"[372] who has flagged hundreds of suspected cases of potential fraud since 2010. There is also another person writing under the name "Julie H. Miller," equally adept at catching those who have committed misconduct.

More recently, the March 8, 2017 *New York Times* called "Clare" both legendary and loathed in biomedical circles, a so-called scientific gadfly. "Whoever Clare Francis actually is, he or she has an uncanny knack for seeing improperly altered images, as well as smaller flaws that some editors are inclined to ignore," according to the *Times*. "Clare Francis has a particular, high-strung style. 'You misunderstand that Carlo [M.] Croce is a great scientist,' Clare Francis wrote in a note copied to John Dahlberg, then the deputy director at the research integrity office. 'My reply was that he has over a thousand papers. That is not the same thing. You simply misunderstand science. Please stop talking about reputations and look at the reality!'" Carlo Croce, M.D., a professor and chair of the department of cancer biology and genetics at the Ohio State University College of Medicine, was the subject of the *Times* article. In 2013, an anonymous critic ["Clare"] contacted Ohio State and the federal authorities with allegations of falsified data in more than thirty of Croce's papers. Further, since 2014, another critic, David A. Sanders, Ph.D., a virologist who teaches at Purdue University, has made claims of falsified data and plagiarism directly to scientific journals where more than twenty of Croce's papers have been published.[373] "Clare" specifically targeted improperly manipulated Western blots and other data. "Despite the lashing criticisms of his work," according to the *Times*, "Dr. Croce has never been penalized for misconduct, either by federal oversight agencies or by Ohio State, which has cleared him in at least five cases involving his work or the grant money he receives."

Sanders made a strong case for further inquiry. "It's a reckless disregard for the truth," Sanders told the *Times*. "As a result of complaints by Dr. Sanders and others, journals have been posting notices of problems with Dr. Croce's papers at a quickening pace. From just a handful of notices before 2013 – known as corrections, retractions and editors' notices – the number has ballooned to at least twenty, with at least three more on the way, according to journal editors. Many of the notices," it was emphasized, "involve the improper manipulation of a humble but universal lab technique called Western blotting [also raised by "Clare"], which measures gene function in a cell and often indicates whether an experiment has succeeded or failed."

Asked if the fact that Clare Francis is a pseudonym made a difference in how Reller perceived his allegations, he told *Nature*'s Richard Van Noorden:

> For some, it's not that Clare Francis is a pseudonym; it's that the pseudonym is Clare Francis. Editors and publishers are as interested in cleaning up the scientific record as anyone, but like any other facet of life, there's something to be said about how you work with or make requests of people. For example, I don't think repeated threats to expose people's names to the media every time they disagree with the interpretation of a particular case helps Clare's cause.[374]

Some, Reller add, would question the classification of "Clare" as a whis-tleblower, "if you believe the term 'whistleblower' suggests someone who reveals confidential knowledge or information." There are those who want the public to believe – wrongly – that "Clare" is the product of running a software application over publicly available articles, that he's not really providing substantive intelligence, just pointing out flaws in the record, Reller offered. In patronizing fashion, "Clare's" naysayers would rather academia – and the public – believe they would "prefer to spend [their] time with people telling [them] things about scientific record [they] can't otherwise know."[375] Adam Marcus and Ivan Oransky, who run *Retraction Watch*, suggest that editors should stop ignoring anonymous whistleblow-ers. "Clare Francis is a thorn in many journal editors' sides," they write. "His [...] modus operandi is simple: he uses plagiarism detection software to compare various papers, then sends his findings to journal editors, often with a request for a retraction."[376] Yes, editors can and do ignore "Clare." But there are others who cull through what he sends and respond. "Those [who] do reply often dismiss the alleged overlap, saying the paper might have been the full version of a conference abstract published earlier, or that it was simply a review," they continued. "Others respond by thanking him for pointing out a paper that, in fact, should be retracted."

Marcus and Oransky honed in on editors who respond to "Clare" as they would to any anonymous whistleblower that "we'd be happy to look into this if you tell us who you are and where you work."[377] They pulled a thread from an exchange about Francis that transpired between *Retraction Watch* and Eric Murphy, the editor in chief of the journal *Lipids* in which Murphy wrote:

> I think it is critical that people identify themselves for the EIC to know who is bringing about the accusation with regards to misconduct. As my students [...] can tell you, misconduct, including self-plagiarism, can and has ruined careers. Hence, it is imperative that these cases are handled appropriately and often this includes contacting the accusing party.

I think a recent misconduct case in Germany [of Silvia Bulfone-Paus] highlights how blogs and tipsters can disrupt the investigative process and people are then tried on-line and the damage lasts forever. Whether there was guilt or not has been and will continue to be irrelevant in some people's minds. Hence, I think the motive is important. Perhaps we have someone who merely wants to ruin someone else's career for personal reasons or for a perceived 'greater good' that they themselves believe gives them carte blanche to launch whatever accusation on the web via a blog. Without the individuals responsible for examining this accusation involved, the falsely accused are then potentially damaged for life via actions of someone who is not in a position to make that decision. In this case 'Clare' did contact the EIC [editor in chiefs] of both journals but identifying themselves would have been helpful for follow-up via a phone call.[378]

The aforementioned Silvia Bulfone-Paus, Ph.D., was a researcher at Germany's Research Center Borstel, who had thirteen papers retracted[379] amid investigations into allegations of image manipulation. Certainly, "Clare's" information had an effect. "In a perfect world," Marcus and Oransky iterated, "we'd love to know who Francis and these other whistleblowers are." Like Reller, they concur that understanding someone's motivations for speaking out can add a great deal to the reporting. No one condones the use of anonymity to press *ad hominem* attacks to avoid reprisal. "In the case of industry-funded medical research, it's critical [to get to and beyond the motivation]. But we're baffled as to why editors and institutions ignore private e-mails from anonymous whistleblowers. Unless, of course," the *Retraction Watch* founders suggest, "they're trying to find ways not to do the work of investigating the claims – work that, one way or another, is their responsibility."[380]

David Cyranoski wrote in the January 28, 2014 *Nature*[381] about the South Korean researcher who blew the whistle on cloning fraudster Woo Suk Hwang – a major scandal – and the backlash he faced for doing the right thing. Cyranoski explained that Young-Joon Ryu, a key figure in

Hwang's laboratory at Seoul National University for several years, maintained his silence for eight years, holding back that he was the one who had played a key role in exposing Hwang, until December 2013 and a follow-on interview with *Nature*. Ryu initiated the investigation that uncovered one of the biggest frauds in science, but he had done it with the promise his identity would be kept from public discovery. "The nature of the Hwang scandal is the abuse of other people's sacrifice and other people's lives for personal success," Ryu, now in the pathology department at Kangwon National University in Chuncheon, told *Nature*.[382] To summarize from *Nature*, Ryu, who joined Hwang's laboratory in 2002, was the scientist who led the team that attempted to create cloned human embryos and stem cell lines from them. He wrote the first draft of the *Science* paper published in February 2004. Ryu's achievement sparked Hwang's public claim that his laboratory had cloned a human embryo and produced stem cells from it, potentially opening the way for new disease treatments. Despite the positive spotlight, Ryu was uncomfortable working with Hwang and left the university in April 2004 for a new position at the Korea Cancer Center Hospital.

Though he had left Hwang's laboratory, Ryu took note when Hwang's group published what Cyranoski described as "a dazzling follow-up" [to Ryu's article] the following year suggesting that the previous proof-of-principle was almost ready for the clinic. "Ryu was suspicious. He knew that important lab members had left, yet the team had pumped out 11 embryonic stem cell lines in a short time," Cyranoski wrote. Ryu added, "I knew how difficult it was. It wasn't logical."[383] What upset Ryu the most was the idea that Hwang was set to initiate a clinical trial using a ten-year-old with a spinal cord injury, a boy Hwang promised to make walk again. Ryu knew the boy and, importantly, that the trial could inflict further injury. But he also lacked evidence to stop Hwang. Adding to his frustration, he also worried that he would be unmasked as the whistleblower, a concern that led him to summarily reject suggestions that he approach Seoul National University or the police. "Instead, on June 1, 2005, he e-mailed television network Munhwa Broadcasting Corporation (MBC) to recommend an investigation," *Nature* reported. "MBC producers were initially intimidated by Hwang's star status but decided to work with Ryu to develop their case. A first program on the subject, about ethical violations in the way that

Hwang recruited egg donors, aired on November 22, 2005, and forced a confession from [Hwang]."[384] He admitted fabrication of his findings. Interestingly, *Nature* also reported that MBC's second program, which concerned Hwang's fraudulent research, was postponed after sponsors withdrew support for the television network and producers faced legal and physical threats. The tide turned in MBC's favor, however, as suspicion of Hwang grew. "Posts on the website of the Biological Research Information Center (BRIC), in which volunteers noted errors in papers, helped to force Seoul National University to open an investigation. By the time the second show aired [that] December 15, Hwang's fate was sealed."[385]

Hwang's fall from grace in 2005 and 2006, was covered around the globe, and featured two high-profile retractions from Science and convictions, subsequently appealed, on charges he embezzled government funds and broke South Korea's bioethics law, according to *Retraction Watch*'s Ivan Oransky.[386] But then some of the same publications that panned Hwang began to write favorably about his biotech research foundation. "Despite his legal troubles – and the widespread belief that his career was over – Hwang continued to work, thanks to the supporters who amassed $3.5 million to launch Sooam[387]. About fifteen scientists followed Hwang from Seoul National University, and around half of those remain today among Sooam's forty-five staff. His team now creates some three hundred cow and pig embryos per day, and delivers about fifteen cloned puppies per month."[388] According to *Retraction Watch*[389], the United States Patent and Trademark Office (USPTO) awarded Hwang a patent based on a cell line described in his retracted Science papers, and confirmed by the *Korea Times*.[390] "The USPTO acknowledged the technological edge of Hwang's team, which means something in consideration of the global scientific leadership of the U.S.," said Sang-Hwan Hyun, D.V.M., Ph.D., of Chungbuk National University, and one of Hwang's closest aides.[391] The *New York Times* had a different take: "Despite all that [he had done wrong in the past], Dr. Hwang has just been awarded an American patent covering the disputed work, leaving some scientists dumbfounded and providing fodder to critics who say the Patent Office is too lax."[392] Few could understand it.

"Shocked, that's all I can say," said Shoukhrat Mitalipov, Ph.D., an American biologist, chief of the Center for Embryonic Cell and Gene

Therapy at Oregon Health and Science University, and who appears to have actually accomplished what Hwang claims to have done. "I thought somebody was kidding, but I guess they were not."[393] Equally perplexed was Jeanne F. Loring, Ph.D., a professor of developmental neurobiology and director of the Center for Regenerative Medicine in the department of molecular biology at the Scripps Research Institute in San Diego. Loring, a stem cell scientist, told the *New York Times* that her first reaction was "You can't patent something that doesn't exist." But, she said, she later realized that "you can." But even those outside biomedical sciences panned the USPTO's move. Daniel B. Ravicher, executive director of the Public Patent Foundation, which challenges patents it believes are invalid and obstruct innovation, iterated strongly that the issuance of the patent to Hwang was more evidence that the patent office was nothing more than "a rubber-stamp, fee-motivated government agency."[394] The *Times* did note that Mitalipov published a paper in *Cell*[395] in mid-2013, ahead of Hwang's patent, that established Mitalipov as the first to create human embryonic stem cells using cloning, even though it was also subject to extensive corrections that were figure-related and typographical errors. None of the errors affected the conclusions of the paper.

The dysfunctionality of Hwang's interaction with the USPTO boiled down to the simple fact that the patent cites both of Hwang's retracted *Science* papers without noting that both were retracted. This stark fact did not seem to trouble the patent office. The *New York Times* spotlighted the issue, incredulous that something as significant as two full retractions did not matter to the United States government agency handing out patents. "But a spokesman for the United States Patent and Trademark Office, and some outside patent lawyers, said the system operates on an honor code and that patent examiners cannot independently verify claims." In the eyes of the issuing agency, "[t]he patent is 'definitely not an assertion by the United States government that everything he is claiming is accurate,' the USPTO spokesman, Patrick Ross, said of Hwang. He said the agency was aware of Hwang's history and took steps to make sure the claimed invention complied with patent statutes."[396]

Back to Ryu. How did the public find out about Ryu? MBC leaked his identity in the first program it aired on the case. Then the retribution started, most of it coming from Hwang's supporters. Ryu told *Nature* that

Hwang's supporters hacked his blog and sent threatening e-mails to him, his employer and his wife, another former researcher in Hwang's laboratory. On December 6, 2005, Ryu resigned from his hospital job under pressure to do so. Shortly thereafter, Ryu, his wife and infant daughter went into hiding for six months. "It was 2007 before the ostracized Ryu could find paid employment, as a pathology resident at Korea University in Seoul."[397]

Ryu's story has a bittersweet ending. On December 23, 2013, he decided to post a note on the BRIC website to express his thanks to all who had supported him and signed it with his real name. Cyranoski wrote that some eight thousand people viewed the post, and a few dozen sympathetic comments showed up on the page. "But then the story was picked up by a local newspaper and the tone changed. Of more than one thousand comments on the popular Daum news-aggregator website, ninety percent [were] negative. Online commenters," he added, "said that by 'revealing a petty truth,' Ryu caused South Korea to 'fall behind in the stem-cell business.' Another accuse[d] him of 'satisfying his arrogance' while 'seriously injuring the nation' as the 'entire project was stolen by other nations.'"[398] Ryu had no regrets about being a whistleblower. Despite the backlash, he told Cyranoski that the Hwang scandal did not destroy his faith in science. He did finish his doctoral degree in bioethics and safe research in 2011 and later returned to Seoul National University to pursue another doctorate in animal reproductive biology.

Ryu's experience demonstrates how whistleblowing carries substantial risks, especially for junior researchers, observed Bernd Pulverer, head of scientific publications at the European Molecular Biology Organization in Heidelberg, Germany, who weighed in on the case for *Nature*. "The Hwang case was a wake-up call for many journals to police [fraud] more seriously," he emphasized. But then he added this missive: "Little has formally changed regarding the protection and encouragement of constructive whistleblowing."[399] Beryl Lieff Benderly would agree. "Science is said to be a search for truth, and that search can sometimes exact a heavy personal price from individuals who take it seriously. Especially," she added, "at risk are junior scientists who dare to confront revered elders."[400] Benderly referred back to *Science*'s 2010 story on what she dubbed "the heroes in the furor that unmasked the deceptions of the once-admired Harvard University psychologist Marc Hauser, whose upright actions did not benefit their academic careers." No, instead, the whistleblowers in that case paid a price – a heavy one. Then, of course, came the plight of Ryu, "the whistleblower in a much bigger scandal" and the woes he experienced after tagging the massive fraud perpetuated by Hwang. Was it fair treatment? No – but it can be worse.

Take the case of Wyn Ellis, Ph.D., a British agricultural researcher and environmental consultant working in Thailand as the coordinator of the United Nations Sustainable Rice Platform. After Ellis blew the whistle on Supachai Lorlowhakarn, the former director of Thailand's National Innovation Agency (NIA), for ethical violations, and plagiarism of his thesis, he was stopped at Bangkok's Suvarnabhumi Airport and incarcerated in a nearby detention center over allegations that he was now a threat to Thai national security for four days starting on September 3, 2015. "I was told that I had been blacklisted as a danger to Thai society,"[401] he told *The Telegraph* from the immigration holding center just then. They even showed him the 2009 letter that Lorlowhakarn wrote in which no doubt marked Ellis for detention and prosecution by Thai immigration – all for having reported Lorlowhakarn for his deception. After the trouble with Lorlowhakarn started, Thai officials previously tried to deport Ellis, who had lived in the country for thirty years, to Norway, from which he arrived, but he refused to leave. British media reporting at the time noted that his detention came after he won a long-running plagiarism battle with the Thai government, which raised questions about the country's position on intellectual property and whistleblowers. Ellis was on the Thai immigration blacklist.

As backstory, Ellis complained to Thailand's elite Chulalongkorn University after he noticed that a dissertation submitted by Lorlowhakarn had been copied from a report Ellis had written for the country's international trade center. Ellis further called for the retraction of a 2008 paper based on Lorlowhakarn's thesis published in the *Thai Journal of Agricultural Science* that also listed the official as first author. Despite the complaint of plagiarism, Lorlowhakarn was still allowed to graduate. What ensued was a protracted legal fight that drew a solid line in the sand between Ellis and the NIA, which filed nine lawsuits against him, including a serious charge of criminal defamation, which carries a possible prison sentence. Much happened across the continuum of time and that line in the sand between Ellis and the NIA. Nearly two years of death threats, smashed windows and monitoring of his home by the Thai military took place before Chulalongkorn University did what it should have done from the beginning: investigate Lorlowhakarn. The Thai official had plagiarized eighty percent of his thesis. Still, it took an exposé in the *Times Higher Education* magazine – in 2012 – for the university to strip Lorlowhakarn of his doctoral degree, which it did in a very public way. Lorlowhakarn paid a significant price for his deception.[402] As one further twist to the story, it was also reported that Lorlowhakarn was later convicted of forging Ellis' employment contract.

In an *Independent* story on the case, Ellis did win seven of the nine lawsuits filed against him, two of which were settled out of court, with the last case resolved in the Thai supreme court in May 2014. According to further reporting, in the meantime, Ellis and his wife, a Thai citizen, lived in a safe house under witness protection. In another extraordinary revelation, despite Lorlowhakarn's now widespread wrongdoing, the Thai government continued to employ him until February 2015. "For me, as the original whistleblower who first alerted authorities to the problems with Lorlowhakarn's Ph.D. thesis," wrote Ellis four years after Chulalongkorn University revoked Lorlowhakarn's doctoral degree, "the knowledge that justice was eventually served is far from cause for celebration. Indeed, the Byzantine twists and turns, the lawsuits, surveillance, physical attacks, and even death threats over the past nine years have – without a doubt – taken their toll on my family and I, and should serve as a salutary lesson to anyone harboring naïve notions of civic duty. This was certainly my own motivation back then, as an advocate and passionate supporter of Thai science and innovation."[403] Invited the *Retraction Watch* to tell his story from the whistleblower's perspective, Ellis got specific and in doing so, illustrated what others have long believed should be stiffer penalties that fit the crimes being committed under the header of "scientific misconduct." Here are some of the threats Ellis encountered, in his own words:

> Listening to a surreal, disembodied voice on the line, yet again informing me of my own address, and how he intends to abduct and kill my family and myself; the

shock of a large rock smashed into my car window[404] on two occasions as my wife and I drove to court hearings. I experienced repeated 'investigations' of my tax and immigration status; attempts to have me kicked out of my job, my Ph.D. studies, even my own adopted country. And of course, the nine lawsuits and police reports, which could have landed me in a Thai prison for years. Looking back at such systemic and long-term intimidation, it seems incredible that anyone would continue to pursue such a cause, given the very real prospect of rocks being replaced by bullets. The pressure cost Erika Fry, the *Bangkok Post* investigative journalist who famously broke the story in 2009, her job; facing criminal defamation charges while those against her employer were dropped, she jumped bail and returned home to the USA.[405]

After all of this, Ellis would still be a whistleblower again. "Corruption requires complicity," he explained, "whether from peers, superiors or subordinates. However, something in this case does not fit. The scale and coordination of the harassment over the years can surely not be justified by a simple case of plagiarism; the obsession with face – both personal and institutional – guaranteed a united front against a transparent investigation by any of the institutions with a duty of care. The obvious exception to this is the National Anti-Corruption Commission (NACC), whose importance as Thailand's last bastion for truly independent investigation cannot be underestimated."[406] Beyond the institutional corruption and draconian cultural impetus to "save face," Ellis opined of the active and long-running harassment, of which his blacklisting and subsequent 2015 detention at the Bangkok airport was only the most recent chapter. "Having lived harmoniously for over 30 years in my adopted country, the very real prospect of deportation as *persona non-grata* brought my whole life crashing down around me," he continued. Lorlowhakarn's letter labeling him as a "danger to national security," written in 2009, was still being used against Ellis and resulted, in fact, in his being grabbed up at the airport. The memory of that detention and the retelling of it has become a lesson that all would-be whistleblowers need to read:

I was held in a windowless detention room with lights on 24 hours a day, with up to 30 other unfortunates. It was an unnerving experience to have guards bursting in at any hour to summon some poor inmate from his slumber and usher him off for summary deportation.

Fortunately, I was able to use my phone and laptop and resolving not to 'go quietly' to an uncertain fate, turned to Facebook, and called everyone I knew as the seriousness of my predicament began to take hold. I was completely unprepared for the scale of the response. Within two days, over the weekend, the story made global news, and was covered by *The Guardian*, BBC, Reuters as well as *Retraction Watch* (among others). Mark Kent, the British ambassador, called personally, and the embassy played a large part in facilitating my ultimate release from detention on the night of Monday, September 7, 2015.

My wife Arada is, of course, the real hero of this tale; she fought untiringly to face down a very entrenched bureaucracy and finally harangued the immigration department into submission by sheer force of logic, supported by a heap of court verdicts exonerating me from any of the wrongdoing alleged by Lorlowhakarn.

Though the immigration department staff treated me as best they could (they even called me 'professor'), to this day I am constantly reminded of the fragility of our daily life, and truly, deeply hate making that long walk through the airport.[407]

Of note, Ellis never received an apology from Lorlowhakarn's NIA successor, Pun-arj Chairatana, Ph.D., for his considerable troubles. As a result, Ellis filed a police report against Lorlowhakarn and the NIA. "NIA's continuing obduracy in refusing to recognize my authorship of a [United

Nations] study and its continuing online distribution of its own republished and expunged version of the report has finally obliged me to file suit at Thailand's intellectual property and international trade court," Ellis wrote. "[...] But in Thailand it never really ends: one of the defendants – a retired senior civil servant- has already counter-sued, demanding [fifty thousand dollars] in damages for defamation."[408] Lorlowhakarn did not get off with a slap on the wrist. According to Ellis, Thailand's national anti-corruption court (NACC) recommended Lorlowhakarn's criminal indictment based on his use of state budget funds for unauthorized publication and sale of the academic works of others, of conflict of interest in hiring a third party to conduct field research that ended up in his own thesis, and of wrongful registration of copyright. These charges are small consolation for Ellis. Lorlowhakarn used his position to do far greater harm to Ellis.

Ellis remains unconvinced that on the other side of his ordeal, the lessons are being learned at an academic level, despite the introduction, he added, of anti-plagiarism software in most Thai universities following the scandal that enveloped him. "The editor of the *Thai Journal of Agricultural Science* has yet to comply with my requests to officially retract a paper co-authored by Lorlowhakarn and his academic supervisors at Chulalongkorn University, and drawn from his cancelled thesis, despite the resignation of several members of the [journal's] editorial board and the journal's relegation by Scopus to Group 3 over its handling of the matter."[409] One commenter on Ellis' *Retraction Watch* article remarked that the penalty should fit the crime. "Grad student plagiarizes dissertation – take away the Ph.D.; scientist fabricates data that lands big grants that pay their executive salaries – jail time, [and most importantly] let the threats that the scientific thugs [and their surrogates] make against whistleblowers become a crime, too, or part of the court's or judge's decision for penalty." In large measure, it is *that* simple. The punishment needs to fit the crime. Make no mistake that what many of those who commit scientific misconduct are prosecutable criminal acts.

Decades ago, Brian Martin wrote that the narrow definition of scientific fraud that existed just then – and largely persists today – was convenient to the groups in society that included scientific elites, and powerful government and corporate interests that have maintained a stronghold on priorities in science.[410] In a letter from Charles W. McCutchen, Ph.D., a physicist with extensive experience with the United States Public Health Service, to Martin dated December 12, 1989, McCutchen wrote:

> In protesting scientific fraud, the whistleblower soon realizes that he or she will have few allies. The biomedical science establishment has taken the position that fraud is very rare and will use almost any means to maintain that illusion. Its response is sufficiently savage to make whistleblowing professionally suicidal if the accused is either important or can involve someone important. This means that one cannot in good conscience ask for support, because one has no right to get the career of an innocent third party destroyed.

> The *de facto* alliance between perpetrators of scientific fraud and the biomedical science establishment is reflected in the response of the scientific journals. I know of only one, *Neurology*, that has published a direct exchange between accuser and accused. *Nature* has been ambivalent, and all other journals I know of have either avoided the issue, or, like *Science*, been captives of the nothing-is-wrong establishment. [411]

Fast forward to April 16, 2014, McCutchen wrote to Helene Hill that though he had never been involved with a *qui tam* suit, so he did not know what was needed to win one. "But I suspect that the atmosphere, pro- or anti-whistleblower, is more important than legal details," he advised her. "It may be vital to win one's ORI case, which in turn would depend on the atmosphere because," as McCutchen knew all too well from Lyle W. Bivens, Ph.D., a neuropsychologist formerly of the United States Department of Health and Human Services' National Institute for Mental Health (NIMH) who had headed the ORI, "They have to compromise." Bivens had issued in 1993 an ORI policy statement that authorship disputes would not be handled by ORI as "misconduct" but instead be left to the collaborators and their institutions to resolve. That policy remains in effect today.[412]

In his communication with Hill, McCutchen pulled no punches. "Do you note a resemblance to the [old] Soviet government – and the resulting growth of *samizdat*[413] publications, now easy via the Internet? I do not expect," he observed, "our bioscience establishment to experience a Gorbachev-like transformation, but in a rational world one would ask why Alan Price overruled Kay Fields' advice to look into your statistical arguments. That an informed forger could defeat forensic statisticians is no reason not to use statistics to catch sloppy crooks. How about putting [Terence] Speed's review of your paper on your *samizdat*, along with his criticizing the evidence in the Imanishi-Kari trail by modifying it? And maybe ask why the establishment is so anxious to suppress statistical evidence? Could it be because it takes Speed-like tactics to dispute it?"

Why was ORI so unwilling to look into statistical evidence in fraud cases? "Because it is hard to refute," McCutchen wrote Hill on May 18, 2014. "Exoneration becomes brazen if data numbers exist and are fishy. If the presence of fishy numbers is acknowledged it takes a smooth, authoritative confuser like Terry Speed to provide a smoke screen for authority to hide behind. In the Imanishi-Kari case Speed admitted that the oddities in Imanishi-Kari's hand-copied data must have resulted from human intervention, but he did not propose any sort of intervention that would replace real data numbers with those that had an uneven distribution of digits. I think," McCutchen opined, "no one asked him to do so, thus the unlikeliness of such an intervention did not come to light." McCutchen's letters to Hill take this story back to the beginning – where this narrative started. "Suppose that in the Imanishi-Kari case Mosimann had satisfied Speed's demands, trolled others' data sets and found one like Imanishi-Kari's. Speed would no doubt have declared Imanishi-Kari innocent, but the other data set could, as likely as Imanishi-Kari's, have been forged, the trolling having found another crook. I have never," he continued, "seen Speed, or anyone opposed to him, mention this possibility."

McCutchen was spot on. "Ever since the bioscience establishment defeated John Dingell and associated whistleblowers by smearing them, and flaunted its victory by obviously rigging the departmental appeals board's handling of Imanishi-Kari's appeal, it has feared no one. With confidence in its power, it enforces the appearance that all is well in bioscience. Yes," he continued, "the resulting cover-ups damage scientists, their work, victims of disease and the taxpayer, but these sacrifices are the price of defending the great enterprise that is science."[414]

About Amy Yarsinske

A my Waters Yarsinske is the author of several best-selling, award-winning nonfiction books, most recently *An American in the Basement: The Betrayal of Captain Scott Speicher and the Cover-up of His Death* (Trine Day, 2013), which won the Next Generation Indie Book Award for General Non-fiction in 2014. To those who know this prolific author, it's no surprise that this Renaissance woman became a writer. Amy's drive to document and investigate history-shaping stories and people has already led to publication of over 75 nonfiction books, most of them spotlighting current affairs, the military, history and the environment. Amy graduated from Randolph-Macon Woman's College in Lynchburg, Virginia, where she earned her Bachelor of Arts in English and Economics, and the University of Virginia School of Architecture, where she earned her Master of Planning and was a DuPont Fellow and Lawn/Range resident. She also holds numerous graduate certificates, including those earned from the CIVIC Leadership Institute and the Joint Forces Staff College, both headquartered in Norfolk, Virginia. Amy serves on the national board of directors of Honor-Release-Return, Inc. and the National Vietnam and Gulf War Veterans Coalition, where she is also the chairman of the Gulf War Illness Committee. She is a member of the American Society of Journalists and Authors (ASJA), Investigative Reporters and Editors (IRE), Authors Guild and the North Carolina Literary and Historical Association (NCLHA), among her many professional and civic memberships and activities.

If you want to know more about Amy and her books, go to www.amywatersyarsinske.com.

Bibliography

Primary Sources
Government Documents

Steneck, Nicholas H. (HHS/ORI). *Introduction to the Responsible Conduct of Research.* Washington, D.C.: United States Government Printing Office, 2007. https://ori.hhs.gov/sites/default/files/rcrintro.pdf

Steneck, Nicholas H. and Mary D. Scheetz (HHS/ORI). *Investigating Research Integrity, Proceedings of the First ORI Research Conference on Research Integrity,* Washington, D.C.: United States Government Printing Office, 2002. https://ori.hhs.gov/documents/proceedings_rri.pdf

United States Environmental Protection Agency (EPA). *Improvements to EPA Policies and Guidance Could Enhance Protection of Human Study Subjects.* Report No. 14-P-0154, March 31, 2014. https://www.epa.gov/sites/production/files/2015-09/documents/20140331-14-p-0154.pdf

United States Food and Drug Administration (FDA). *Wesley A. McQuerry: Debarment Order. Federal Register* Document 2016-06104, filed March 17, 2016 and posted March 18, 2016. https://www.federalregister.gov/articles/2016/03/18/2016-06104/wesley-a-mcquerry-debarment-order

United States Department of Health and Human Services (HHS), Office of Research Integrity (ORI). *Findings of Research Misconduct. Federal Register* Document 2013-31160, filed December 27, 2013 and posted December 30, 2013. https://www.federalregister.gov/articles/2013/12/30/2013-31160/findings-of-research-misconduct

—. "Case summaries," *Office of Research Integrity Newsletter*, Volume 21, Number 4, September 2013. https://ori.hhs.gov/images/ddblock/sept_vol21_no4.pdf

—. "Case summaries," *Office of Research Integrity Newsletter*, Volume 21, Number 3, June 2013. https://ori.hhs.gov/images/ddblock/june_vol21_no3.pdf

—. "Case summaries," *Office of Research Integrity Newsletter*, Volume 21, Number 2, March 2013.

—. Case Summary: Han, Dong-Pyou, December 23, 2013. http://ori.hhs.gov/content/case-summary-han-dong-pyou

—. *Case Summary: Sheehy, Timothy*, December 5, 2013. http://ori.hhs.gov/content/case-summary-sheehy-timothy

—. *Case Summary: Wang, Hao*, November 20, 2013. http://ori.hhs.gov/content/case-summary-wang-hao
https://ori.hhs.gov/images/ddblock/march_vol21_no2.pdf

—. *Case Summary: Aggarwal, Nitin*, October 17, 2013. http://ori.hhs.gov/content/case-summary-aggarwal-nitin

—. *Case Summary: Francis, Peter J. Federal Register*, Volume 77, Number 72, April 13, 2012, pages 22320-22321. https://grants.nih.gov/grants/guide/notice-files/NOT-OD-12-104.html

—. *Findings and Consequences of Research Misconduct.* n.d. 2008. https://ori.hhs.gov/education/products/RIandImages/misconduct_cases/findings_of_misconduct.pdf

—. *Findings of Scientific Misconduct*, NOT-OD-10-130, August 30, 2010. http://grants.nih.gov/grants/guide/notice-files/NOT-OD-10-130.html

—. *ORI Oversight Report, University of Medicine and Dentistry of New Jersey,* ORI 2001-28, August 2002.

—. *Public Health Service Policies on Research Misconduct; Final Rule* (42 CFR Parts 50 and 93). *Federal Register,* Volume 70, Number 94, May 17, 2005. https://ori.hhs.gov/sites/default/files/42_cfr_parts_50_and_93_2005.pdf

—. *Handling Misconduct.* http://ori.hhs.gov/handling-misconduct

—. *Sample Policy and Procedures for Responding to Research Misconduct Allegations.* http://ori.hhs.gov/sample-policy-procedures-responding-research-misconduct-allegations

—. Press release. "ORI concludes investigation at University of Wisconsin-Madison, April 2, 1996.

—. *Integrity and Misconduct in Research: Report of the Commission on Research Integrity.* Washington, D.C.: United States Government Printing Office, 1995. http://ori.hhs.gov/sites/default/files/report_commission.pdf

Academic Documents

Columbia University. *Responsible Conduct of Research: Research Misconduct.* Curriculum, 2003-2004. http://ccnmtl.columbia.edu/projects/rcr/rcr_misconduct/foundation/

Pitt, Joel H. and Helene Z. Hill, "Statistical detection of potentially fabricated data: a case study," submitted November 21, 2013, https://arxiv.org/ftp/arxiv/papers/1311/1311.5517.pdf

Rutgers University Libraries. Health Sciences—Scientific Misconduct, 2013. http://rbhs.rutgers.edu/rwjlbweb/posters/scimisconduct.html

Stasik, Justine, "Examples of research misconduct," Office of the Research Integrity Officer, Baruch College, updated October 3, 2011. http://www.baruch.cuny.edu/rio/research_misconduct_examples.htm

"The cloning scandal of Hwang Woo-Suk," module supplement in Stem Cells: Biology, Bioethics and Applications course. http://stemcellbioethics.wikischolars.columbia.edu/The+Cloning+Scandal+of+Hwang+Woo-Suk

Weill Cornell Medical College and Weill Cornell Graduate School of Medical Sciences. Policy and Procedures Governing Research Integrity. July 2007. http://researchintegrity.weill.cornell.edu/pdf/RIP_062707_FinalJuly20200_1.pdf

Secondary Sources
Books

Angell, Marcia. *The Truth About the Drug Companies: How They Deceive Us and What to Do About It.* New York: Random House, 2005.

Barry, Kevin. *Vaccine Whistleblower: Exposing Autism Research Fraud at the CDC.* New York: Skyhorse Publishing, 2015.

Bass, Alison. *Side Effects: A Prosecutor, a Whistleblower, and a Bestselling Antidepressant on Trial.* New York: Algonquin, 2008.

Bell, Robert. *Impure Science: Fraud, Compromise and Political Influence in Scientific Research.* Hoboken, New Jersey: Wiley, 1992.

Hill, Helene Z. *Hidden Data—The Blind Eye of Science.* West Orange, New Jersey: CreateSpace, 2016.

Howell, B. Lindsay, ed. *Foreign Temporary Workers in America: Policies that Benefit the U.S. Economy: Policies that Benefit the U.S. Economy.* Westport, Connecticut: Quorum Books, 1999.

Lock, Stephen, Frank Wells and Michael Farthing, eds. *Fraud and Misconduct in Biomedical Research*. London: BMJ Books, 2001.

Journals, Newspapers, Magazines, and Papers

Achenbach, Joel, "Many scientific studies can't be replication. That's a problem," *Washington Post*, August 27, 2015. https://www.washingtonpost.com/news/speaking-of-science/wp/2015/08/27/trouble-in-science-massive-effort-to-reproduce-100-experimental-results-succeeds-only-36-times/

Adams, Mike, "Vaccine researcher charged with felony crimes for research fraud; may spend 20 years in prison over faked AIDS vaccine," *NaturalNews.com*, June 25, 2014. http://www.naturalnews.com/045726_research_fraud_aids_vaccine_science-based_medicine.html

—. "Big Pharma criminality no longer a conspiracy theory: bribery, fraud, price fixing now a matter of public record," *NaturalNews.com*, July 9, 2012. http://www.naturalnews.com/z036417_Glaxo_Merck_fraud.html

—. "Merck vaccine fraud exposed by two Merck virologists; company faked mumps vaccine efficacy results for over a decade, says lawsuit," *NaturalNews.com*, June 28, 2012. http://www.naturalnews.com/036328_Merck_mumps_vaccine_False_Claims_Act.html

Ahmed, Muhammed Z., "Opinion: the postdoc crisis," *The Scientist*, January 4, 2016. http://www.the-scientist.com/?articles.view/articleNo/44874/title/Opinion—The-Postdoc-Crisis/

Associated Press. "AIDS researcher suspended over finances," May 2, 1990. http://www.nytimes.com/1990/05/02/us/aids-researcher-suspended-over-finances.html

Associated Press Science. "Faked research results on rise?" *Wired.com*, July 10, 2005. http://www.wired.com/2005/07/faked-research-results-on-rise/

Baker, Monya, "1,500 scientists lift the lid on reproducibility," *Nature*, May 25, 2016. http://www.nature.com/news/1-500-scientists-lift-the-lid-on-reproducibility-1.19970

Barbash, Fred, "Scientist falsified data for cancer research once described as 'holy grail,' feds say," *Washington Post*, November 9, 2015. https://www.washingtonpost.com/news/morning-mix/wp/2015/11/09/scientist-falsified-data-for-cancer-research-once-described-as-holy-grail-feds-say/

Bavel, Jay Van, "Why do so many studies fail to replicate?" *New York Times*, May 27, 2016. http://www.nytimes.com/2016/05/29/opinion/sunday/why-do-so-many-studies-fail-to-replicate.html?_r=0

Benderly, Beryl Lieff, "Paying a price for truth," *Science*, June 29, 2014. http://www.sciencemag.org/careers/2014/01/paying-price-truth

Benson, Jonathan, "Merck senior management tried to pay off its own vaccine scientists to remain silent about scientific fraud," *Global Research*, February 9, 2015. http://www.globalresearch.ca/merck-senior-management-tried-to-pay-off-its-own-vaccine-scientists-to-remain-silent-about-scientific-fraud/5430364

Birch, Douglas M. and Gary Cohn, "How a cancer trial ended in betrayal," *Baltimore Sun*, June 24, 2001. http://www.baltimoresun.com/bal-te.research24jun24-story.html

Boffey, Philip M., "U.S. study finds fraud in top researcher's work on mentally retarded," *New York Times*, May 24, 1987. http://www.nytimes.com/1987/05/24/us/us-study-finds-fraud-in-top-researcher-s-work-onmentally-retarded.html

Bohannon, John, "Updated: lax reviewing practice prompts 60 retractions at SAGE journal," *Science*, July 14, 2014. http://www.sciencemag.org/news/2014/07/updated-lax-reviewing-practice-prompts-60-retractions-sage-journal

Broad, William J., "A tempest in a test tube, 10 years later," *New York Times*, March 23, 1999. http://www.nytimes.com/library/national/science/032399sci-cold-fusion.html

Budd, John M., MaryEllen Sievert and Tom R. Schultz, "Phenomena of retraction: reasons for retraction and citations to the publications," *Journal of the American Medical Association* (JAMA), Volume 280, Number 3, July 15, 1998, pages 296-297. DOI: 10.1001/jama.280.3.296. http://jama.jamanetwork.com/article.aspx?articleid=187739

Carpenter, Tim, "KSU researcher contests firing amid research feud," *cjonline.com*, the home of the *Topeka Capital-Journal* on the web, May 1, 2015. http://cjonline.com/news/state/2015-05-01/ksu-researcher-contests-firing-amid-research-feud

—. "KSU severs ties to author of controversial prairie-burn research article," *cjonline.com*, the home of the *Topeka Capital-Journal* on the web, October 11, 2014. http://cjonline.com/news/business/2014-10-11/ksu-severs-ties-author-controversial-prairie-burn-research-article

Castillo, Stephanie, "The FDA underreports scientific misconduct in peer-reviewed articles: the benefits of negative science," *MedicalDaily.com*, February 10, 2015. http://www.medicaldaily.com/fda-underreports-scientific-misconduct-peer-reviewed-articles-benefits-negative-321548

CBS. "Second whistleblower Donna Busche fired at troubled Washington State Hanford nuke plant," February 19, 2014. http://www.cbsnews.com/news/second-whistleblower-donna-busche-fired-at-troubled-wash-state-hanford-nuke-plant/

Chang, Kenneth, "Columbia chemistry professor is retracting 4 more papers," *New York Times*, June 15, 2006. http://www.nytimes.com/2006/06/15/science/15chem.html

Chicago Tribune. "Ex-Gallo aide guilty of pocketing $25,000, July 8, 1992. http://articles.chicagotribune.com/1992-07-08/news/9203010832_1_gallo-laboratory-dr-mikulas-popovic-aids-virus

Chronicle of Higher Education. "Stanford researcher, accused of conflicts, steps down as NIH principal investigator, August 1, 2008. http://chronicle.com/article/Stanford-Researcher-Accused/41395

Clarke, Toni, "Alzheimer's research fraud case set for trial," *Reuters*, May 11, 2012. http://www.reuters.com/article/us-science-fraud-idUSBRE84A0O820120511

Consoli, Luca, "Scientific misconduct and science ethics: a case study based approach," *Science and Engineering Ethics*, 2006, Volume 12, Issue 3, pages 533-541. http://ltc-ead.nutes.ufrj.br/constructore/objetos/5%20-%20Scientific%20misconduct%20and%20science%20ethics.pdf

Corbyn, Zoe, "Misconduct is the main cause of life-sciences retractions," *Nature*, Volume 490, October 4, 2012, page 21.

Couzin-Frankel, Jennifer, "Scientist turned in by grad students for misconduct pleads guilty," *Science*, June 28, 2010. http://www.sciencemag.org/news/2010/06/scientist-turned-grad-students-misconduct-pleads-guilty

Cox, Harold C. and Drummond Rennie, "Research misconduct, retraction, and cleansing the medical literature: lessons from the Poehlman Case," *Annals of Internal Medicine*, Volume 144, Number 8, April 18, 2006, pages 609-613. DOI:10.7326/0003-4819-144-8-200604180-00123. http://annals.org/article.aspx?articleid=722445

Crewdson, John, "The great AIDS quest: science under the microscope," *Chicago Tribune*, November 19, 1989. http://archives.chicagotribune.com/1989/11/19/page/101/article/the-great-aids-quest

Cyranoski, David, "Whistle-blower breaks his silence," *Nature*, January 28, 2014, Volume 505. http://www.nature.com/news/whistle-blower-breaks-his-silence-1.14598

—. "Cloning comeback," *Science*, January 14, 2014. http://www.nature.com/news/cloning-comeback-1.14504

Davis, Mark S., Michelle Riske-Morris, and Sebastian R. Diaz, "Causal factors implicated in research misconduct: evidence from ORI case files," *Science and Engineering Ethics*, December 2007, Volume 13, Issue 4, pages 395-414. DOI: 10.1007/s11948-007-9045-2. https://www.researchgate.net/profile/Sebastian_Diaz9/publication/5809908_Causal_factors_implicated_in_research_misconduct_evidence_from_ORI_case_files/links/5512b1ea0cf268a4aaeacd0c.pdf

Dayan, Gustavo H., "Recent resurgence of the mumps," *New England Journal of Medicine*, Volume 358, Number 15, April 10, 2008. http://www.nejm.org/doi/full/10.1056/NEJMoa0706589

Dingell, John D., "Shattuck lecture—misconduct in medical research," *New England Journal of Medicine*, Volume 328, Number 22, June 3, 1993, pages 1610-1615.

Dubois, James M., D.Sc., Ph.D., Emily E. Anderson, Ph.D., and John Chibnall, Ph.D., "Assessing the need for a research ethics remediation program," *Clinical and Translational Science*, June 2013; 6 (3): 209-213, first published online February 11, 2013. DOI 10.111/cts.12033.

Dunning, Hayley, "Parkinson's researcher fabricated data," *The Scientist*, June 29, 2012. http://www.the-scientist.com/?articles.view/articleNo/32305/title/Parkinson-s-Researcher-Fabricated-Data/

Doran, Michael, "How to survive as a whistleblower," *Nature*, Volume 532, April 20, 2016, page 405. DOI:10.1038/nj7599-405a http://www.nature.com/naturejobs/science/articles/10.1038/nj7599-405a

Durso, Thomas, "Dismissal of false claims suit shows scientific sophistication, experts say," *The Scientist*, February 19, 1996. http://www.the-scientist.com/?articles.view/articleNo/17792/title/Dismissal-Of-False-Claims-Suit-Shows-Scientific-Sophistication—Experts-Say/

Editorial. "Judge: lawsuit against Merck's MMR vaccine fraud to continue," *Health Impact News*, 2014. http://healthimpactnews.com/2014/judge-lawsuit-against-mercks-mmr-vaccine-fraud-to-continue/

Editorial. "A painful remedy," *Nature*, Volume 468, November 4, 2010, page 6. DOI:10.1038/468006b http://www.nature.com/nature/journal/v468/n7320/full/468006b.html

—. "Trust, but verify," *Nature*, Volume 461, page 315, September 17, 2009. DOI:10.1038/461315a http://www.nature.com/nature/journal/v461/n7262/full/461315a.html

Ellis, Wyn, "Broken windows, threats, and detention: is whistleblowing worth it?" *Retraction Watch*, July 12, 2016. http://retractionwatch.com/2016/07/12/broken-windows-threats-and-detention-is-whistleblowing-worth-it/

Engber, Daniel, "Ask us anything: how common is scientific fraud? *Popular Science*, March 30, 2015. http://www.popsci.com/ask-us-anything-how-common-scientific-fraud

Faigman, David L., "Putting scientific peer review in the courtroom," *Scientific American*, December 18, 2015. http://www.scientificamerican.com/article/putting-scientific-peer-review-in-the-courtroom/

Fanelli, Daniele, "How many scientists fabricate and falsify research? A systematic review and meta-analysis of survey data," *PLOS One*, May

29, 2009. DOI: 10.1371/journal.pone.0005738. http://journals.plos. org/plosone/article?id=10.1371/journal.pone.0005738

Fang, Ferric C., R. Grant Steen and Arturo Casadevall, "Misconduct accounts for the majority of retracted scientific publications," *Proceedings of the National Academy of Sciences* (PNAS), Volume 109, Number 42, pages 17028–17033. DOI: 10.1073/pnas.1212247109.

http://www.pnas.org/content/109/42/17028.full

Ferguson, Cat, "One in 25 papers contains inappropriately duplicated images, screen finds," *Retraction Watch*, April 19, 2016. http:// retractionwatch.com/2016/04/19/one-in-25-papers-contains-inappropriately-duplicated-images-screen-finds/

—. "RW cited in scientist's $8 million suit against university," *Retraction Watch*, February 11, 2015. http://retractionwatch.com/2015/02/11/ rw-cited-scientists-8-million-suit-university/

—. "Former Vanderbilt scientist faked nearly 70 images, will retract 6 papers: ORI," *Retraction Watch*, November 20, 2014. http://retractionwatch. com/2014/11/20/former-vanderbilt-scientist-faked-nearly-70-images-will-retract-6-papers-ori/

FierceBiotech. "Fraud found at China site for key study of Pfizer and Bristol's Eliquis, July 9, 2013. http://www.fiercebiotech.com/r-d/ fraud-found-at-china-site-for-key-study-of-pfizer-and-bristol-s-eliquis

Foster, Hope, et al., "Mintz Levin health care *qui tam* update—recent developments and unsealed cases," *Mintz.com*, April 2016. https:// www.mintz.com/newsletter/2016/Newsletters/5730-HL/index. html?_cldee=a3Nsb3ZpdGNoQG1pbnR6LmNvbQ%3d%3d

—. "Mintz Levin health care *qui tam* update," *Mintz.com*, June 1, 2015. https:// www.mintz.com/legal-insights/industry/articletype/articleview/ articleid/2768/mintz-levin-health-care-emqui-tamem-update

Foster, Hope and Kevin M. McGinty, "Claims, awards, enforcement action: *qui tam* update—recently unsealed whistleblower cases," *The National Law Review*, April 13, 2016. http://www.natlawreview.com/article/claims-awards-enforcement-action-qui-tam-update-recently-unsealed-whistleblower

—. "Health care *qui tam* update: recently unsealed whistleblower cases," *The National Law Review*, October 8, 2015. http://www.natlawreview.com/article/health-care-qui-tam-update-recently-unsealed-whistleblower-cases

Foster, Hope and Kevin M. McGinty and Sean Grammel, "Mintz Levin health care *qui tam* update—recent developments and unsealed cases," *Mintz.com*, March 2015. https://www.mintz.com/newsletter/2015/Newsletters/4751-0315-NAT-HL/

Friedly, Jock, "After 9 years, a tangled case lurches toward a close," *Science*, Volume 272, May 17, 1996, pages 947-948.

Gander, Kashmira, "Dr. Wyn Ellis: British academic detained in Thailand after exposing official as plagiarist," *Independent*, September 6, 2015. http://www.independent.co.uk/news/uk/home-news/dr-wyn-ellis-british-academic-detained-in-thailand-after-exposing-official-as-plagiarist-10489026.html

George, Stephen L., "Research misconduct and data fraud in clinical trials: prevalence and causal factors," *International Journal of Clinical Oncology*, Volume 21, Issue 1, August 20, 2015. DOI: 10.1007/s10147-015-0887-3.

Gerashchenko Bogdan I., Akira Yamagata Akira, Ken Oofusa, Katsutoshi Yoshizato, Sonia M. de Toledo, Roger W. Howell, "Proteome analysis of proliferative response of bystander cells adjacent to cells exposed to ionizing radiation," *Proteomics*, Volume 7, Number 12, 2007, pages 2000-2008. DOI:10.1002/pmic.200600948.

Gerashchenko, Bogdan I. and Roger W. Howell, "Bystander cell proliferation is modulated by the number of adjacent cells that were exposed to ionizing radiation," *Cytometry Part A*, Volume 66, Issue 1, 2005, pages 62-70. DOI:10.1002/cyto.a.20150.

— ."Proliferative response of bystander cells adjacent to cells with incorporated radioactivity," *Cytometry Part A*, Volume 60, Issue 2, 2004, pages 155-164. DOI: 10.1002/cyto.a.20029 https://www. researchgate.net/publication/8419266_Proliferative_response_of_ Bystander_cells_adjacent_to_cells_with_incorporated_radioactivity

—. "Cell proximity is a prerequisite for the proliferative response of bystander cells co-cultured with cells irradiated with gamma-rays," *Cytometry Part A*, Volume 56, Issue 2, 2003, pages 71-80. DOI:10.1002/ cyto.a.10092.

—. "Flow cytometry as a strategy to study radiation-induced bystander effects in co-culture systems," *Cytometry Part A*, Volume 54, Issue 1, 2003. DOI:10.1002/cyto.a.10049.

Gerashchenko, Bogdan I, Edouard I. Azzam and Roger W. Howell,. "Characterization of cell-cycle progression and growth of WB-F344 normal rat liver epithelial cells following gamma-ray exposure," *Cytometry Part A*, Volume 61, Number 2, 2004, pages 134-141. DOI:10.1002/cyto.a.20065.

Glanz, James, and Agustin Armendariz, "Years of ethics charges, but star cancer researcher gets a pass," *New York Times*, March 8, 2017. https:// www.nytimes.com/2017/03/08/science/cancer-carlo-croce.html?_r=0

Gobry, Pascal-Emmanuel, "Big science is broken," *The Week*, April 18, 2016. http://theweek.com/articles/618141/big-science-broken

Goldberg, Daniel, "Research fraud: a sui generis problem demands a sui generis solution (plus a little due process), *Thomas M. Cooley Law Review*, Hilary Term 2003, Volume 20, Number 47, pages 47-69.

Goldberg, Paul, "Prominent Duke scientist claimed prizes he didn't win, including Rhodes Scholarship," *The Cancer Letter*, Volume 36, Number 27, July 16, 2010.

Graham, Troy and Jennifer Moroz, "UMDNJ monitor says fraud, failures now up to $243 million, the 'disheartening' tally could rise, the school said the figure included issues it had already addressed," *Philadelphia Inquirer*, July 21, 2006. http://articles.philly.com/2006-07-21/news/25404571_1_potential-fraud-medicaid-investigators

Guerin, Lisa, "Workplace retaliation: What are your rights?" *NOLO*, n.d. http://www.nolo.com/legal-encyclopedia/workplace-retaliation-employee-rights-30217.html

Hare, Julie, "Queensland University of Technology forced to repay cells grant," *The Australian*, August 30, 2014. http://www.theaustralian.com.au/higher-education/queensland-university-of-technology-forced-to-repay-cells-grant/news-story/ec642f24b1fe9d9a881011546a222c44

Harris, Gardiner, "Approval of antibiotic worried safety officials," *New York Times*, July 19, 2006. http://www.nytimes.com/2006/07/19/health/19fda.html

Herper, Matthew, "Face of the year: David Graham," *Forbes*, December 13, 2004. http://www.forbes.com/2004/12/13/cx_mh_1213faceoftheyear.html

Heyboer, Kelly, "Gov. Christie helps celebrate Rutgers-UMDNJ merger," NJ Advance Media for *NJ.com*, July 1, 2013. http://www.nj.com/news/index.ssf/2013/07/gov_christie_helps_celebrate_rutgers-umdnj_merger.html

Hill, Helene Z., "Opinion: reducing whistleblower risk," *The Scientist*, February 11, 2014. http://www.the-scientist.com/?articles.view/articleNo/39139/title/Opinion—Reducing-Whistleblower-Risk/

Hilts, Philip J., "U.S. drops misconduct case against AIDS researcher," *New York Times*, November 12, 1993. http://www.nytimes.com/1993/11/13/us/us-drops-misconduct-case-against-an-aids-researcher.html?pagewanted=all

—. "Federal inquiry finds misconduct by a discoverer of the AIDS virus," *New York Times*, December 31, 1992. http://query.nytimes.com/gst/fullpage.html?res=9F0CEFDA103DF932A05751C1A9649 58260&pagewanted=all

—. "Biologist who disputed a study paid dearly," *New York Times*, March 22, 1991. http://www.nytimes.com/1991/03/22/us/biologist-who-disputed-a-study-paid-dearly.html?pagewanted=all

Holloway, James P., "The pitfalls of settling *qui tam* lawsuits," *Health Law Alert Newsletter*, Ober-Kaler Attorneys at Law, 2015, Issue 5. http://www.ober.com/publications/2889-the-pitfalls-settling-qui-tam-lawsuits

Howell, Roger W. and Dandamudi V. Rao, "In memoriam: Kandula S. R. Sastry (1935-2001)," ***Journal of Nuclear Medicine*, Volume** 43, Issue 1, January 2002: N33. http://search.proquest.com/openview/099ddd 09684249d54c35360784713e41/1?pq-origsite=gscholar&cbl=40808

Howell, Roger W., Prasad Venkata S.V. Neti, Massimo Pinto, Bogdan I. Gerashchenko, Venkat R. Narra, Edouard I. Azzam, "**Challenges and progress in predicting biological responses to incorporated radioactivity,**" *Radiation Protection Dosimetry*, Volume 122, Issues 1-4, 2006, pages 521-527. DOI:10.1093/rpd/ncl448.

Humphries, Suzanne, "Scientists sue Merck: allege fraud, mislabeling, and false certification of MMR vaccine," International Medical Council on Vaccination, June 25, 2012. http://www.vaccinationcouncil.org/2012/06/25/scientists-sue-merck-allege-fraud-mislabeling-and-false-certificaion-of-mmr-vaccine-suzanne-humphries-md/

India West. "Former department chair sues George Washington University for $8M," February 10, 2015. http://www.indiawest.com/news/global_indian/former-department-chair-sues-george-washington-university-for-m/article_9bcdb624-b165-11e4-8a7f-c365ca7943c1.html

Ingham, Janis Costello, "Ethics and research," *The ASHA Leader*, March 2004, Volume 9, 10-25. DOI:10.1044/leader.FTR6.09052004.10

Johnson, Carolyn Y., "Harvard report shines light on ex-researcher's misconduct," *Boston Globe*, May 30, 2014. https://www.bostonglobe.com/metro/2014/05/29/internal-harvard-report-shines-light-misconduct-star-psychology-researcher-marc-hauser/maSUowPqL4clXrOgj44aKP/story.html

Kaiser, Jocelyn, "Top U.S. scientific misconduct official quits in frustration with bureaucracy," *Science*Insider, March 12, 2014. http://www.sciencemag.org/news/2014/03/top-us-scientific-misconduct-official-quits-frustration-bureaucracy

Kevles, Daniel J., "Scientific fraud and misconduct in American political culture: reflections on the Baltimore case," *Engineering and Science*, No. 3 (1998). http://calteches.library.caltech.edu/3951/1/Fraud.pdf

Kintisch, Eli, "Poehlman sentenced to 1 year in prison," *Science*, June 28, 2006. http://www.sciencemag.org/news/2006/06/poehlman-sentenced-1-year-prison

Kocieniewski, David, "Latest twist in a scandal hits a medical school when it's down," *New York Times*, November 24, 2006. http://www.nytimes.com/2006/11/24/nyregion/24hosp.html

Koleva, Gergana, "Merck whistleblower suit a boon to vaccine foes even as it stresses importance of vaccines," *Forbes*, June 27, 2012. http://www.forbes.com/sites/gerganakoleva/2012/06/27/merck-whistleblower-suit-a-boon-to-anti-vaccination-advocates-though-it-stresses-importance-of-vaccines/#5d5dfd00caf7

Korpela, K. M., "How long does it take for the scientific literature to purge itself of fraudulent material?: the Breuning case revisited," *Current Medical Research and Opinion*, Volume 26, Issue 4, 2010; DOI: 10.1185/03007991003603804.

Krimsky, Sheldon, "The funding effect in science and its implications for the judiciary," *Journal of Law and Policy*, Volume 13, Number 1, 2005, pages 45-51. http://brooklynworks.brooklaw.edu/jlp/vol13/iss1/4

Lacetera, Nicola and Lorenzo Zirulia, "The economics of scientific misconduct," *Journal of Law, Economics, and Organization*, first completed March 30, 2008, printed in the journal 2011, Volume 27, Number 3, pages 568-603. DOI: 10.1093/jleo/ewp031 https://www.researchgate.net/profile/Lorenzo_Zirulia/publication/5010502_Erratum_The_Economics_of_Scientific_Misconduct/links/0912f50db1b1a7eb4f000000.pdf

LaFollette, Marcel C., "The evolution of the 'scientific misconduct' issue: an historical overview," *Experimental Biology and Medicine*, Volume 224, Number 4, pages 211-215. http://ebm.sagepub.com/content/224/4/211.full.

Lawrence III, Cleveland, ed., "False Claims Act liability," *False Claims Act and Qui Tam Quarterly Review*, Volume 56, April 2010.

—. "Jurisdictional issues," *False Claims Act and Qui Tam Quarterly Review*, Volume 56, April 2010.

—. "False Claims Act retaliation claims," *False Claims Act and Qui Tam Quarterly Review*, Volume 56, April 2010.

—. "Common defenses to FCA allegations," *False Claims Act and Qui Tam Quarterly Review*, Volume 56, April 2010.

—. "Federal rules of civil procedure," *False Claims Act and Qui Tam Quarterly Review*, Volume 56, April 2010.

—. "Litigation developments," *False Claims Act and Qui Tam Quarterly Review*, Volume 56, April 2010.

—. "Judgments and settlements," *False Claims Act and Qui Tam Quarterly Review*, Volume 56, April 2010.

Lestch, Corinne, "Iowa professor cops to faking results of big-bucks AIDS vaccine research," *New York Daily News*, December 27, 2013. http://www.nydailynews.com/news/national/professor-cops-faking-aids-vaccine-research-data-article-1.1559245

Leys, Tony, "Ex-ISU scientist pleads guilty of AIDS vaccine fraud," *Des Moines Register*, February 25, 2015. http://www.desmoinesregister.com/story/news/2015/02/25/dong-pyou-han-iowa-state-university-guilty-plea/23996449/

—. "Grassley wants more specifics about response to ISU scientist's multimillion-dollar research fraud," *Des Moines Register*, March 11, 2014. http://blogs.desmoinesregister.com/dmr/index.php/2014/03/11/grassley-wants-more-specifics-about-response-to-isu-scientists-multimillion-dollar-research-fraud

Loikith, Lisa and Robert P. Bauchwitz, "The essential need for research misconduct allegation audits," *PeerJ* Preprints 4:e1577v4 https://doi.org/10.7287/peerj.preprints.1577v4

Lubalin, James S. and Jennifer L. Matheson, "The fallout: what happens to whistleblowers and those accused of scientific misconduct," *Science and Engineering Ethics*, Volume 5, Issue 2, 1999, pages 229-250. http://link.springer.com/article/10.1007%2Fs11948-999-0014-9#page-1

Lutz, Holley Thames, "Scientific misconduct—revised and revisited by ORI," program paper, American Health Lawyers Association (AHLA), January 26, 2006. https://www.healthlawyers.org/Archive/Program%20Papers%202/2006_AMC/[2006_AMC]%20Scientific%20Misconduct%20-%20Revised%20and%20Revisited%20By%20ORI.pdf

Macagnone, Michael, "1ˢᵗ Circ. rules $12M NIH fraud suit had its chance," *Law360.com*, March 18, 2015. http://www.law360.com/articles/632509/1ˢᵗ-circ-rules-12m-nih-fraud-suit-had-its-chance

Marcus, Adam, "Misconduct dissolves paper on possible clot-busters," *Retraction Watch*, March 10, 2015. http://retractionwatch.com/2015/03/10/misconduct-dissolves-paper-on-possible-clot-busters/

—. "Faking data earns stem cell researcher a ban on federal funding," *Retraction Watch*, December 9, 2014. http://retractionwatch.com/2014/12/09/faking-data-earns-stem-cell-researcher-ban-federal-funding/

—. "Senator 'unsatisfied' with ORI's response on recovery of tainted granted money," *Retraction Watch*, March 11, 2014. http://retractionwatch.com/2014/03/11/senator-unsatisfied-with-oris-response-on-recovery-of-tainted-grant-money/

—. "Office of Research Integrity (ORI) head David Wright leaves agency," *Retraction Watch*, March 7, 2014. http://retractionwatch.com/2014/03/07/office-of-research-integrity-ori-head-david-wright-leaves-agency/

—. "Nature yanks controversial genetics paper whose coauthor was found dead in lab in 2012," *Retraction Watch*, November 6, 2013. http://retractionwatch.com/2013/11/06/nature-yanks-controversial-genetics-paper-whose-co-author-was-found-dead-in-lab-in-2012/

—. "ORI says Case Western skin scientist falsified data," *Retraction Watch*, February 1, 2013. http://retractionwatch.com/2013/02/01/ori-says-case-western-skin-scientist-falsified-data/

Marcus, Adam and Ivan Oransky, "Crack down on scientific fraudsters," *New York Times*, July 10, 2014. http://nyti.ms/1rZUEnU

—. "Who are you? Editors should stop ignoring anonymous whistleblowers," *Lab Times*, July 2011. http://www.labtimes.org/labtimes/issues/lt2011/ lt07/lt_2011_07_39_39.pdf

Margolin, Josh, "Oversight didn't halt abuses at UMDNJ," *NJ.com*, October 26, 2008. http://www.nj.com/news/index.ssf/2008/10/oversight_ didnt_halt_abuses_at.html

Marshall, Elliot, "Scientific misconduct: how prevalent is fraud? That's a million dollar question," *Science,* Volume 290, December 1, 2000, http://www.sciencemag.org/careers/2000/12/how-prevalent-fraud-thats-million-dollar-question

Martin, Brian, "Scientific fraud and the power structure of science," *Prometheus*, Volume 10, Number 1, June 1992. https://www.uow. edu.au/~bmartin/pubs/92prom.html

Martinson, Brian C., Melissa S. Anderson and Raymond de Vries, "Scientists behaving badly," *Nature*, Volume 435, June 9, 2005, pages 737-738. http://pages.stolaf.edu/ross/files/2014/05/ScientistsBehavingBadly.pdf

Marusic, Ana, Vedran Katavic and Matko Marusic, "Role of editors and journals in detecting and preventing scientific misconduct: strengths, weaknesses, opportunities, and threats," *Medicine and Law*, September 2007, Volume 26, Number 3, pages 545-566.

Mathews, Anna Wilde, "Infected data: fraud, errors taint key study of widely used Sanofi drug despite some faked results, FDA approves antibiotic; one doctor's cocaine use; company defends safety," *Wall Street Journal*, May 1, 2006.

Mathews, David, "Secret dossier on research fraud suggests government concern over science," *Times Higher Education*, December 3, 2015. https://www.timeshighereducation.com/news/secret-dossier-research-fraud-suggests-government-concern-over-science

Mattiuzzi, Paul G., "What is workplace retaliation? It's about making people afraid," *Everyday Psychology*, March 26, 2012. http://www. everydaypsychology.com/2012/03/what-is-workplace-retaliation-its-about.html#.Vz4eJ-SJdv8

Mayhew, Ruth, "Can my employer force me to sign a letter of resignation? The Nest, n.d. http://woman.thenest.com/can-employer-force-sign-letter-resignation-18537.html

McClain, Sylvia, "Scientific fraud: a sign of the times?" *The Guardian*, October 12, 2012. https://www.theguardian.com/science/occams-corner/2012/oct/12/scientific-fraud

McCook, Alison, "FDA bans trial coordinator who pocketed patient funds and went to prison," *Retraction Watch*, March 21, 2016. http://retractionwatch.com/2016/03/21/fda-bans-trial-coordinator-who-pocketed-patient-funds-and-went-to-prison/

—. "Kansas ecology prof loses whistleblower protection after alleging misconduct," *Retraction Watch*, September 15, 2015. http://retraction watch.com/2015/09/15/kansas-ecology-prof-loses-whistleblower-protection-after-alleging-misconduct/

—. "AIDS vaccine fraudster sentenced to nearly 5 years in prison and to pay back $7 million," *Retraction Watch*, July 1, 2015. http://retraction watch.com/2015/07/01/aids-vaccine-fraudster-sentenced-to-nearly-5-years-in-prison-and-pay-back-7-million/

McCutchen, Charles W. "The Baltimore case misrepresents a major piece of evidence," *Journal of Information Ethics*, Volume 11, Number 1, Spring 2002.

—. "The departmental appeals board and the Imanishi-Kari case," http://conductinscience.com/ August 4, 2000.

McGinty, Kevin, et al. "Mintz Levin health care *qui tam* update—recent developments and unsealed False Claims Act cases," *Mintz.com*, October 2014. https://www.mintz.com/newsletter/2014/Newsletters/ 4338-1014-NAT-HL/

Michalek, Arthur M., et al., "The costs and underappreciated consequences of research misconduct: a case study, *PLOS Medicine*, Volume 7, Number 8, August 17, 2010. DOI: 10.1371/journal.pmed.1000318. http://www.plosmedicine.org/article/info:doi/10.1371/journal.pmed. 1000318

Murphy, Sean P., et al., "Submitting a manuscript for peer review—integrity, integrity, integrity," *Journal of Neurochemistry*, Volume 128, Number 3, February 2014, pages 341-343. DOI:10.1111/jnc.12644. http://www. ncbi.nlm.nih.gov/pmc/articles/PMC3926655/pdf/nihms553048.pdf

National Whistleblowers Center. "Harvard teaching hospital to face trial for research fraud," *Whistleblowers.org*, May 8, 2012. http://www. whistleblowers.org/index.php?option=com_content&task=view&id= 1462&Itemid=208

New York Times. "The fraud case that evaporated," June 25, 1996. http://www.nytimes.com/1996/06/25/opinion/the-fraud-case-that-evaporated.html

Norton, Christopher, "Mass. Hospitals seek quick win in research grant suit," *Law360.com*, September 10, 2010. http://www.law360.com/ articles/192055/mass-hospitals-seek-quick-win-in-research-grant-suit

Oransky, Ivan, "Weekend reads: peer review, troubled from the start; how to survive as a whistleblower," *Retraction Watch*, April 23, 2016. http://retractionwatch.com/2016/04/23/weekend-reads-peer-review-troubled-from-the-start-how-to-survive-as-a-whistle-blower/

—. "It's official: Anil Potti faked data, say Feds," *Retraction Watch*, November 7, 2015. http://retractionwatch.com/2015/11/07/its-official-anil-potti-faked-data-say-feds/

—. "Nature issues expression of concern for paper by author who threatened to sue Retraction Watch," *Retraction Watch*, December 1, 2014. http://retractionwatch.com/2014/12/01/nature-issues-expression-of-concern-for-paper-by-author-who-threatened-to-sue-retraction-watch/

—. "Australian university to repay $275,000 grant because of 'misleading and incorrect' information," *Retraction Watch*, August 29, 2014. http://retractionwatch.com/2014/08/29/australian-university-to-repay-275k-grant-because-of-misleading-and-incorrect-information/

—. "Former UT-Southwestern cancer researchers faked data in 10 papers: ORI," *Retraction Watch*, September 18, 2014. http://retractionwatch.com/2014/09/18/former-ut-southwestern-cancer-researchers-faked-data-in-10-papers-ori/

—. "'Crack down on scientific fraudsters'—our op-ed in today's *New York Times*," *Retraction Watch*, July 11, 2014. http://retractionwatch.com/2014/07/11/crack-down-on-scientific-fraudsters-our-op-ed-in-todays-new-york-times/

—. "SAGE Publications busts 'peer review and citation ring,' 60 papers retracted," *Retraction Watch*, July 8, 2014. http://retractionwatch.com/2014/07/08/sage-publications-busts-peer-review-and-citation-ring-60-papers-retracted/

—. "STAP stem cell papers officially retracted as *Nature* argues peer review couldn't have detected fatal problems," *Retraction Watch*, July 2, 2014. http://retractionwatch.com/2014/07/02/stap-stem-cell-papers-officially-retracted-as-nature-argues-peer-review-couldnt-have-detected-fatal-problems/

—. "Stunner: researchers retract paper because company complains it's hurting profits," *Retraction Watch*, May 19, 2014. http://retractionwatch.com/2014/05/19/wow-researchers-retract-paper-because-company-complains-its-hurting-profits/

—. "In sharp resignation letter, former ORI director Wright criticizes bureaucracy, dysfunction," *Retraction Watch*, March 13, 2014. http://retractionwatch.com/2014/03/13/in-sharp-resignation-letter-former-ori-director-wright-criticizes-bureaucracy-dysfunction/

—. "Nature paper retracted following multiple failures to reproduce results," *Retraction Watch*, February 27, 2014. http://retractionwatch.com/2014/02/27/nature-paper-retracted-following-multiple-failures-to-reproduce-results/

—. "Fraud, retractions no barrier to U.S. cloning patent for Woo-Suk Hwang," *Retraction Watch*, February 16, 2014. http://retractionwatch.com/2014/02/16/fraud-retractions-no-barrier-to-us-cloning-patent-for-woo-suk-hwang/

—. "Weekend reads: MIT professor accused of fraud, biologist who retracted paper suspended, and more," *Retraction Watch*, February 15, 2014. http://retractionwatch.com/2014/02/15/weekend-reads-mit-professor-accused-of-fraud-biologist-who-retracted-paper-suspended-and-more/

—. "Fraud, retractions no barrier to U.S. cloning patent for Woo-Suk Hwang," *Retraction Watch*, February 16, 2014. http://retractionwatch.com/2014/02/16/fraud-retractions-no-barrier-to-us-cloning-patent-for-woo-suk-hwang/

—. "Psychology researcher explains how retraction-causing errors led to change in her lab," *Retraction Watch*, January 15, 2014. http://retractionwatch.com/2014/01/15/psychology-researcher-explains-how-retraction-causing-errors-led-to-change-in-her-lab/

—. "Weekend reads: Most scientific fraudsters keep their jobs, random acts of academic kindness, and more," *Retraction Watch*, January 4, 2014. http://retractionwatch.com/2014/01/04/weekend-reads-most-scientific-fraudsters-keep-their-jobs-random-acts-of-academic-kindness-and-more/

—. "Former NIH scientist falsified images in hepatitis study: ORI," *Retraction Watch*, December 29, 2013. http://retractionwatch.com/2013/12/29/former-nih-scientist-falsified-images-ori/

—. "Case Western dermatology department hit with second ORI sanction within 6 months," *Retraction Watch*, August 9, 2013. http://retractionwatch.com/2013/08/09/case-western-dermatology-department-hit-with-second-ori-sanction-within-6-months/

—. "Stem cell retraction leaves grad student in limbo, reveals tangled web of industry academic ties, *Retraction Watch*, December 27, 2012. http://retractionwatch.com/2012/12/27/stem-cell-retraction-leaves-grad-student-in-limbo-reveals-tangled-web-of-industry-academic-ties/

—. "Anil Potti resurfaces with job at North Dakota cancer center," *Retraction Watch*, August 20, 2012. http://retractionwatch.com/2012/08/20/anil-potti-resurfaces-with-job-at-north-dakota-cancer-center/

—. "ORI finds Parkinson's-pesticides researcher guilty of faking data, two papers retracted," *Retraction Watch*, July 2, 2012. http://retractionwatch.com/2012/07/02/ori-finds-parkinsons-pesticides-researcher-guilty-of-faking-data-two-papers-to-be-retracted/

—. "The Anil Potti retraction record so far," *Retraction Watch*, February 14, 2012. http://retractionwatch.com/2012/02/14/the-anil-potti-retraction-record-so-far/

Packel, Dan, "Antitrust, FCA claims on Merck mumps vaccine to advance," Law360.com, September 5, 2014. https://www.law360.com/articles/574389/antitrust-fca-claims-on-merck-mumps-vaccine-to-advance

Pain, Elisabeth, "Paul Brookes: Surviving as an outed whistleblower," *Science*, March 10, 2014. http://www.sciencemag.org/careers/2014/03/paul-brookes-surviving-outed-whistleblower

Palus, Shannon, "Columbia investigation reveals researcher faked data—and a degree," *Retraction Watch*, June 13, 2016. http://retractionwatch.com/2016/06/13/columbia-investigation-reveals-researcher-faked-data-and-a-degree/

—. "DC court allows part of lawsuit against GW to proceed," *Retraction Watch*, April 13, 2016. http://retractionwatch.com/2016/04/13/dc-court-allows-part-of-lawsuit-against-gw-to-proceed-million-lawsuit-in-part/

—. "Danish neuroscientist sentenced by court for lying about faked experiments," *Retraction Watch*, October 1, 2015. http://retractionwatch.com/2015/10/01/danish-neuroscientist-sentenced-by-court-for-lying-about-faked-experiments/

—. "Re-analysis of controversial Paxil study shows drug 'ineffective and unsafe' for teens," *Retraction Watch*, September 16, 2015. http://retractionwatch.com/2015/09/16/re-analysis-of-controversial-paxil-study-shows-drug-ineffective-and-unsafe-for-teens/

Parrish, Debra M., "Scientific misconduct and correcting the scientific literature," *Academic Medicine*, Volume 74, Number 3, March 1999.

Patrick, Patricia A., "Be prepared before you blow the whistle," *Fraud*, September/October 2010. http://www.fraud-magazine.com/article.aspx?id=4294968656

Perry, Nick, and Carol M. Ostrom, "UW: Researcher faked AIDS data, altered images," *Seattle Times*, November 28, 2007. http://www.seattletimes.com/seattle-news/education/uw-researcher-faked-aids-data-altered-images/

Perry, Susan, "'Lone wolf' investigator says 'clinical trial system is broken," *Minnpost*, December 10, 2013. https://www.minnpost.com/second-opinion/2013/12/lone-wolf-fda-investigator-says-clinical-trial-system-broken

Pollack, Andrew, "Disgraced scientist granted U.S. patent for work found to be fraudulent," *New York Times*, February 14, 2014. https://www.nytimes.com/2014/02/15/science/disgraced-scientist-granted-us-patent-for-work-found-to-be-fraudulent.html?partner=rss&emc=rss&_r=0

Poon, Peter, "Legal protections for the scientific misconduct whistleblower," *Journal of Law, Medicine and Ethics*, Volume 23, Issue 1, March 1995, pages 88-95.

Powers, Scott, "Posey looking at whistleblower's CDC autism documents," *Orlando Sentinel*, "September 9, 2014. http://www.orlandosentinel.com/news/politics/political-pulse/os-us-rep-bill-posey-looking-at-immunizationautism-20140909-post.html

Price, Alan R., "Research misconduct and its federal regulation: the origin and history of the Office of Research Integrity—with personal views by ORI's former associate director for investigative oversight," original manuscript of an article ultimately published in *Accountability in Research*, Volume 20, Issue 5-6, September 18, 2013. http://researchmisconductconsultant.com/Alan%20Price%20paper%20on%20ORI%20History%20for%20Accountability%20in%20Research%20-%20Sept%202013.pdf

Price, Michael, "Sins against science: data fabrication and other forms of scientific misconduct may be more prevalent than you think," *Monitor*, July/August 2010, Volume 41, Number 7. http://www.apa.org/monitor/2010/07-08/misconduct.aspx

Qualters, Sheri, "Can an NIH grant application form the basis of a False Claims Act case?" *The National Law Journal*, September 15, 2011. http://rose-law.net/pdf/NLJ_FalseClaimsArticle.pdf

Reardon, Sara, "US vaccine researcher sentenced to prison for fraud," *Nature*, Volume 523, pages 138-139, July 9, 2015. DOI: 10.1038/nature.2015.17660. http://www.nature.com/news/us-vaccine-researcher-sentenced-to-prison-for-fraud-1.17660

Redman, Barbara K. and Arthur Caplan, "No one likes a snitch," *Science and Engineering Ethics*, Volume 21, Issue 4, June 17, 2014, pages 813-819. DOI: 10.1007/s11948-014-9570-8.

Redman, Barbara K. and Jon F. Merz, "Scientific misconduct: do the punishments fit the crime?" *Science*, Volume 321, August 8, 2008. DOI: 10.1126/science.1158052

Reich, Eugenie Samuel, "Report on alleged scientific misconduct to remain secret, court rule," *Nature* News Blog, August 24, 2011. http://blogs.nature.com/news/2011/08/report_on_alleged_scientific_m.html

Reiss, Dorit Rubenstein, "Merck whistleblowers—mumps and motions—updated," *Skeptical Raptor's Blog*, January 31, 2016. http://www.skepticalraptor.com/skepticalraptorblog.php/merck-mumps-motions-whistleblowers-the-actual-story/

Reller, Tom, "It's not that Clare Francis is a pseudonym; it's that the pseudonym is Clare Francis," *Elsevier.com*, December 6, 2013. https://www.elsevier.com/connect/its-not-that-clare-francis-is-a-pseudonym-its-that-the-pseudonym-is-clare-francis

Resnik, David B., "From Baltimore to Bell Labs: reflections on two decades of debate about scientific misconduct," *Accountability in Research: Policies and Quality Assurance*, Volume 10, Number 2, 2003, pages 123-135. DOI: 10.1080/08989620390199890. http://dx.doi.org/10.1080/08989620300508

Reynolds, Sandra M., "ORI findings of scientific misconduct in clinical trials and publicly funded research, 1992-2002," *Clinical Trials*, Volume 1, Number 6, December 2004, pages 509-516. DOI: 10.1191/1740774504cn048oa.

Rhodes, Rosamond and James J. Strain, "Whistleblowing in academic medicine," *Journal of Medical Ethics*, 2004, Volume 40, pages 35-39. DOI: 10.1136/jme.2003.005553 http://www.ncbi.nlm.nih.gov/pmc/articles/PMC1757136/pdf/v030p00035.pdf

Roberts, Janine, "Fraud found in AIDS research," excerpt from *Fear of the Invisible*, August 22, 2008. http://reaids.com/fearoftheinvisible.com/aidsresearch.html

Robins, Richard B., "Anatomy of a federal investigation and trial for alleged Stark and Anti-Kickback violations," *Brach Eichler L.L.C. Newsletter*, n.d. http://www.bracheichler.com/C3F493/assets/files/News/AnatomyofaFederalInvestigation.pdf

Rockey, Sally, "Postdoctoral researchers—facts, trends, and gaps," *Extramural Nexus*, June 29, 2012, a publication of the National Institutes of Health Office of Extramural Research. https://nexus.od.nih.gov/all/2012/06/29/postdoctoral-researchers-facts-trends-and-gaps/

Ross, David B., "The FDA and the case of Ketek," *New England Journal of Medicine*, Issue 356, April 19, 2007, pages 1601-1604. DOI: 10.1056/NEJMp078032 http://www.nejm.org/doi/full/10.1056/NEJMp078032#t=article

Ruben, Adam, "Sins of the principal investigator," *Science*, April 20, 2016. DOI: 10.1126/science.caredit.a1600065 http://www.sciencemag.org/careers/2016/04/sins-principal-investigator

Ryan, Kenneth J., "Commission on Research Integrity report sparks debate on science and ethics," *Professional Ethics Report*, Volume 9, Number 2, Spring 1996. http://www.aaas.org/sites/default/files/migrate/uploads/per5.pdf

Schachman, Howard K., "What is misconduct in science?" *Science*, Volume 261, July 9, 1993. http://www.albany.edu/~scifraud/data/sci_fraud_3175.html

Scudellari, Megan, "Third retraction for GWU biologist as university seeks to dismiss his $8 million lawsuit," *Retraction Watch*, May 13, 2015. http://retractionwatch.com/2015/05/13/third-retraction-for-gwu-biologist-as-university-seeks-to-dismiss-his-8-million-lawsuit/

Seife, Charles, "Research misconduct identified by the US Food and Drug Administration," *JAMA Internal Medicine*, Volume 175, Issue 4, April 2015, pages 567-577. DOI: 10.1001/jamainternmed.2014.7774. http://archinte.jamanetwork.com/article.aspx?articleID=2109855

—. "Are your medications safe? The FDA buries evidence of fraud in medical trials. My students and I dug it up," *Slate*, February 9, 2015. http://www.slate.com/articles/health_and_science/science/2015/02/fda_inspections_fraud_fabrication_and_scientific_misconduct_are_hidden_from.html

Shepard, Ray M., "First-to-file bar held inapplicable to *qui tam* suits," *Health Law Alert Newsletter*, Ober-Kaler Attorneys at Law, Spring 2006. http://www.ober.com/publications/971-first-to-file-bar-held-inapplicable-qui-tam-suits

Sherman, Ted, "UMDNJ whistleblower cases cost Rutgers nearly $2M in settlements," NJ Advance Media for *NJ.com*, April 26, 2015. http://www.nj.com/news/index.ssf/2015/04/umdnj_whistleblower_cases_cost_rutgers_nearly_2m_i.html

Silbergeld, Ellen K., "Annotation: protection of the public interest, allegations of scientific misconduct, and the Needleman Case," *American Journal of Public Health*, February 1995, Volume 85, Number 2, pages 165-166. http://ajph.aphapublications.org/doi/pdf/10.2105/AJPH.85.2.165

Smith, Jane, and Fiona Godlee, eds., "Investigating allegations of scientific misconduct," *BMJ* (formerly the *British Medical Journal*), Volume 331, July 30, 2005, pages 245-246. DOI: 10.1136/bmj.331.7511.245. http://www.ncbi.nlm.nih.gov/pmc/articles/PMC1181252/pdf/bmj33100245.pdf

Smith, Richard, "If Volkswagen staff can be criminally charged so should fraudulent scientists," *The BMJ* (formerly the *British Medical Journal*), September 28, 2015. http://blogs.bmj.com/bmj/2015/09/28/richard-smith-if-volkswagen-staff-can-be-criminally-charged-so-should-fraudulent-scientists/

—"Science and journalism threatened in the high court," *The BMJ*, July 30, 2015. http://blogs.bmj.com/bmj/2015/07/30/richard-smith-science-and-journalism-threatened-in-the-high-court/

—. "Research misconduct: the poisoning of the well," *Journal of the Royal Society of Medicine*, May 2006, Volume 99, Number 5, pages 232-237. http://www.ncbi.nlm.nih.gov/pmc/articles/PMC1457763/pdf/0232. pdf

— ed., "The need for a national body for research misconduct: nothing less will reassure the public," *BMJ* (formerly the *British Medical Journal*), June 6, 1998, pages 1686-1687; *BMJ* 1998; 316 (7146): 1686-1687. http://www.ncbi.nlm.nih.gov/pmc/articles/PMC1113269/

Solomon, Lawrence, "Merck has some explaining to do over its MMR vaccine claims," *Huffington Post*, April 18, 2016. http://www.huffingtonpost.ca/lawrence-solomon/merck-whistleblowers_b_5881914.html

Soltis, Andy, "Professor admits faking AIDS vaccine to get $19M in grants," *New York Post*, December 26, 2013. http://nypost.com/2013/12/26/professor-admits-faking-aids-vaccine-to-get-19m-in-grants/

Sovacool, Benjamin K., "Exploring scientific misconduct: isolated individuals, impure institutions, or an inevitable idiom of modern science? *Journal of Bioethical Inquiry*, December 2008, Volume 5, Issue 4, pages 271-282. DOI 10.1007/s11673-008-9113-6.

—. "Using criminalization and due process to reduce scientific misconduct," *American Journal of Bioethics*, Volume 5, Number 5: W1-W7, 2005. DOI: 10.1080/15265160500313242

Steen, R. Grant, "Retractions in the scientific literature: is the incidence of research fraud increasing?" *Journal of Medical Ethics*, December 24, 2010, DOI: 10.1136/jme.2010.040923.

Stein, Callan, "8 things you might not know about research misconduct proceedings: guest post," *Retraction Watch*, August 13, 2015. http://retractionwatch.com/2015/08/13/guest-post-8-things-you-might-not-know-about-research-misconduct-proceedings/

Stokes, Trevor L., "A grad student was caught in the crossfire of fraud—and fought back," *Retraction Watch*, November 22, 2016. http://retraction watch.com/2016/11/22/a-grad-student-was-caught-in-the-crossfire-of-fraud-and-fought-back/

Sullivan, Charles A., "Suing whistleblowers," *Health Reform Watch.com*, analysis and commentary from Seton Hall Law School's Center for Health and Pharmaceutical Law and Policy, February 20, 2015. http://www.healthreformwatch.com/2015/02/20/suing-whistleblowers/

Swinton, William Elgin, "Physician contributions to nonmedical science: Sir Grafton Elliot Smith and Piltdown man," *Canadian Medical Association Journal*, November 20, 1976, Volume 115, Number 10, pages 1047-1053. PMCID: PMC1878881.

Tae-gyu, Kim, "Hwang wins U.S. patent for human stem cells," *Korea Times*, February 11, 2014. http://www.koreatimes.co.kr/www/news/nation/2014/02/116_151430.html

Taubes, Gary, "The strange case of chimeraplasty," *Science*, January 2003, DOI: 10.1126/science.298.5601.2116. https://www.researchgate.net/publication/10992026_Gene_therapy_The_strange_case_of_chimeraplasty

Tavare, Aniket, "Managing research misconduct: is anyone getting it right?" *BMJ* (formerly the *British Medical Journal*), December 29, 2011; *BMJ* 2011; 343: d8212. DOI: http://dx.doi.org/10.1136/bmj.d8212

University of Miami Ethics Programs. "The Baltimore Case." http://www.miami.edu/index.php/ethics/projects/timelines_project/baltimore/

Van Noorden, Richard, "Publishers withdraw more than 120 gibberish papers," *Nature*, February 24, 2014. DOI: 10.1038/nature.2014.14763. http://www.nature.com/news/publishers-withdraw-more-than-120-gibberish-papers-1.14763?utm_content=buffer95c78&utm_medium=social&utm_source=twitter.com&utm_campaign=buffer

Wade, Nicholas, "Harvard finds scientist guilty of misconduct," *New York Times*, August 20, 2010. http://www.nytimes.com/2010/08/21/education/21harvard.html

Wadman, Meredith, "Money in biomedicine: the senator's sleuth," *Nature*, Volume 461, September 16, 2009, pages 330-334. DOI:10.1038/461330a http://www.nature.com/news/2009/090916/full/461330a.html

Wallace, Clementine, "Biologist charged with more fraud," *The Scientist*, June 12, 2006. http://www.the-scientist.com/?articles.view/articleNo/24077/title/Biologist-charged-with-more-fraud/

Weaver, David, Moema H. Reis, Christopher Albanese, Frank Costantini, David Baltimore and Thereza Imanishi-Kari, "Altered repertoire of endogenous immunoglobulin gene expression in transgenic mice containing a rearranged mu heavy chain gene," *Cell*, Volume 45, Issue 2, April 25, 1986, pages 247-259.

Weber, Patricia, "Suspect something. Say something? Several methods to report scientific misconduct," Association for Women in Science (AWIS) New Jersey, n.d. 2014. (hard copy) (reference November 23, 2013 *Nature* article "3 ways to blow the whistle") http://www.awisnj.org/

Weed, Douglas L., "Preventing scientific misconduct," *American Journal of Public Health*," Volume 88, Number 1, January 1998. http://ajph.aphapublications.org/doi/pdf/10.2105/AJPH.88.1.125

Weiss, Rick, "Many scientists admit to misconduct," Washington Post, June 9, 2005. http://www.washingtonpost.com/wp-dyn/content/article/2005/06/08/AR2005060802385.html

Wessler, Seth Freed, "Alabama whistleblower fired and sued after protesting conditions," NBCNews.com, April 15, 2015. http://www.nbcnews.com/feature/in-plain-sight/alabama-whistleblower-fired-sued-after-protesting-conditions-n341296

Whalen, Jeanne, "'Dr. Drew' was paid by Glaxo," Wall Street Journal, July 3, 2012. http://www.wsj.com/articles/SB10001424052702303933404577505032006855076

Willingham, Emily, "A congressman, a CDC whistleblower and an autism tempest in a trashcan," Forbes.com, August 6, 2015. https://www.forbes.com/sites/emilywillingham/2015/08/06/a-congressman-a-cdc-whisteblower-and-an-autism-tempest-in-a-trashcan/#24fcbf215396

Wilson, William A., "Scientific regress," First Things, May 2016. http://www.firstthings.com/article/2016/05/scientific-regress

Winter, Suzanne, "Former Wisconsin researcher sentenced for misconduct," BioTechniques, September 17, 2010. http://www.biotechniques.com/news/Former-Wisconsin-researcher-sentenced-for-misconduct/biotechniques-302891.html

Xu, Baoyan, et al., "Hybrid DNA virus in Chinese patients with seronegative hepatitis discovered in deep sequencing," Proceedings of the National Academy of Sciences (PNAS) USA, June 18, 2013, Volume 110, Number 25, pages 10264–10269. Published online May 28, 2013. DOI: 10.1073/pnas.1303744110.

Yang, Debra Wong, Nick Hanna and Alexander H. Southwell, "Health care compliance in 2009 and going forward: Part 1," Westlaw Journal Health Care Fraud, Volume 15, Issue 9, March 2010. http://www.gibsondunn.com/publications/Documents/YangHannaSouthwell-HealthCareCompliancePart1.pdf

Yong, Ed, "Why a new case of misconduct in psychology heralds interesting times for the field," *Discover*, June 26, 2012. http://blogs. discovermagazine.com/notrocketscience/2012/06/26/why-a-new-case-of-misconduct-in-psychology-heralds-interesting-times-for-the-field/#.VxWBiXr3Nv8

Yong, Ed, Heidi Ledford and Richard Van Noorden, "3 ways to blow the whistle," *Nature*, Volume 503, November 28, 2013. http://www. nature.com/polopoly_fs/1.14226!/menu/main/topColumns/topLeft Column/pdf/503454a.pdf

Zaldivar, R. A., "Government aids scientist accused of secretly steering work to his firm," *Philadelphia Inquirer*, May 1, 1990. http://articles.philly. com/1990-05-01/news/25885670_1_nih-investigators-william-f-raub

Websites

http://www.law360.com

http://www.amerares.com (Amerandus Research)

https://www.hzhill.net (Helene Z. Hill website)

http://www.iThenticate.com

http://www.mintz.com (Mintz Levin Cohn Ferris Glovsky and Popeo, P.C.)

http://www.ori.hhs.gov (Office of Research Integrity)

http://www.popehat.com

http://www.publicationethics.org/ (COPE—Committee on Publication Ethics)

http://www.PubPeer.com

http://www.RetractionWatch.com

http://www.Turnitin.com

http://www.whistleblowers.org (National Whistleblowers Center)

Endnotes

1   Hill, Helene Z. Hidden Data – The Blind Eye of Science. West Orange, New Jersey: CreateSpace, 2016.

2   Ibid., iThenticate.com

3   www.popehat.com

4   Hill, Hidden Data, ibid.

5   Ibid.

6   Ibid.

7   A cause of action arises under the False Claims Act if an individual or institution knowingly presents a false or fraudulent claim for payment or makes a false statement to get a false or fraudulent claim paid or approved by the government per 31 U.S.C. § 3729(a). A private party may assert a claim under the False Claims Act in the name of the United States government, but the complaint must be filed in camera and under seal and served on the government so that the government has the opportunity to intervene. 31 U.S.C. § 3730(b)(2). The private party may elect to pursue the case even if the government declines to intervene. 31 U.S.C. § 3730(c)(3). In Milam's original 1990 filing the United States notified the court that it would not intervene in the pending case. See 31 U.S.C. § 3730(b)(4)(B).

8   United States, ex rel. Kathryn M. Milam, v. The Regents of the University of California, et al., 912 F. Supp. 868 (D. Md. 1995), and also earlier, United States, ex rel. Kathryn M. Milam, v. The Regents of the University of California, et al., 961 F.2d 46 (4th Cir. 1992); United States, ex rel. Kathryn M. Milam, v. The Regents of the University of California, et al., No. B-90-523 (D. Md. 1992); United States, ex rel. Kathryn M. Milam, v. The Regents of the University of California, et al. No. B-90-523 (D. Md. 1990).

9   Thereza Imanishi-Kari, Ph.D., DAB No. 1582 (1996). http://www.hhs.gov/dab/decisions/dab1582.html

10   University of Miami Ethics Programs. "The Baltimore Case." http://www.miami.edu/index.php/ethics/projects/timelines_project/baltimore/

11   Winter, Suzanne, "Former Wisconsin researcher sentenced for misconduct," BioTechniques, September 17, 2010. http://www.biotechniques.com/news/Former-Wisconsin-researcher-sentenced-for-misconduct/biotechniques-302891.html

12   Couzin-Frankel, Jennifer, "Scientist turned in by grad students for misconduct pleads guilty," Science, June 28, 2010. http://www.sciencemag.org/news/2010/06/scientist-turned-grad-students-misconduct-pleads-guilty

13   Ibid.

14   Robert Bauchwitz received his undergraduate degree in biochemistry and molecular biology from Harvard University, and his M.D. and Ph.D. in molecular genetics from Cornell University.

15   United States, ex rel. Robert Bauchwitz, M.D., Ph.D. v. William K. Holloman, Ph.D. et al., No. 04-cv-2892 (E.D. Pa. June 30, 2004).

16   Ibid.

17   Kenneth Berns, M.D., Ph.D. is a distinguished professor emeritus in the University of Florida Department of Molecular Genetics and Microbiology College of Medicine.

18   In March 1989, the PHS created the Office of Scientific Integrity (OSI) in the Office of the Director, NIH, and the Office of Scientific Integrity Review (OSIR) in the Office of the Assistant Secretary for Health (OASH). The sole purpose of these offices was to deal with research misconduct; the creation of OSIR also began the process of removing responsibility for research misconduct from the funding agencies. In May 1992, OSI and OSIR were consolidated into the Office of Research Integrity (ORI) in the OASH. Later that year, HHS established a hearing opportunity before the Research Integrity Adjudications Panel of the Departmental Appeals Board, HHS, for all scientists formally charged with research misconduct.

19   Bauchwitz, ibid., Taubes, Gary, "The strange case of chimeraplasty," Science, January 2003, DOI: 10.1126/science.298.5601.2116. https://www.researchgate.net/publication/10992026_Gene_therapy_The_strange_case_of_chimeraplasty

20   http://www.amerares.com/About_Robert_Bauchwitz_120515.html (Bauchwitz)

21   Ibid.

22   Ibid.

23   Ibid, http://publicationethics.org/

24   Bauchwitz, ibid.

25   Loikith, Lisa and Robert P. Bauchwitz, "The essential need for research misconduct allegation audits," PeerJ Preprints 4:e1577v4 https://doi.org/10.7287/peerj.preprints.1577v4

26   Bauchwitz, ibid.

27   Ibid, Mathews, David, "Secret dossier on research fraud suggests government concern over science," Times Higher Education, December 3, 2015. https://www.timeshighereducation.com/news/secret-dossier-research-fraud-suggests-government-concern-over-science

28   The False Claims Act creates civil liability for individuals or entities that make false or fraudulent claims for payment to the federal government. See

31 U.S.C. § 3729(a). An individual violates the Act when he "knowingly presents, or causes to be presented, a false or fraudulent claim for payment or approval" to an officer or employee of the United States government. 31 U.S.C. § 3729(a)(1)(A). An action may be filed by the Attorney General or by a private individual, called a relator, as an assignee of the government. 31 U.S.C. § 3730(a)-(b). In order to state a claim under the Act, an individual must allege that the defendant: "(1) knowingly presented or caused to be presented, (2) a false claim, (3) to the United States government, (4) knowing its falsity, (5) which was material, (6) seeking payment from the federal treasury." United States ex rel. Hutcheson v. Blackstone Medical, Inc., 694 F. Supp. 2d 48, 61 (D. Mass. 2010). Here, the "false claim or statement," "materiality," and "knowingly" elements are at issue with respect to the relator's claims. There are three theories under which a claim may be "false or fraudulent" under the Act. These are: (1) factual falsity; (2) legal falsity under an express certification theory; and (3) legal falsity under an implied certification theory, meaning that misrepresentations regarding the allegedly falsified data, blinded methodologies, and reliability tests arise under the factual falsity theory. Alleged failure to comply with the responsibilities of applicants' grounds liability under both the express and implied certification theories. Finally, the alleged failure to comply with the post-award requirements regulation appears to be grounded upon an implied certification theory. A claim is deemed "factually false" when "the goods or services provided are either incorrectly described, or make claim for a good or service never provided." Hutcheson, 694 F. Supp. 2d at 62 (citing United States ex rel. Mikes v. Straus, 274 F. 3d 687, 697 (2nd Circ. 2001)). A claim is legally false under an express certification theory when the party making the claim expressly but falsely states that it has complied with any precondition of payment. Hutcheson, 694 F. Supp. 2d at 62 (citing United States ex rel. Conner v. Salina Regional Health Center, Inc., 543 F. 3d 1211, 1217 (10th Circ. 2008)). In Hutcheson, the court recognized that a claim is legally false under an implied certification theory when "a claimant makes no express statement about compliance with a statute or regulation, but by submitting a claim for payment implies that it has complied with any preconditions to payment (citing Conner, 543 F. 3d at 1218)." In adopting this definition, the court restricted liability under an implied certification theory to "compliance with expressly stated preconditions of payment found in the relevant statute or regulations. (citing Mikes, 274 F. 3d at 700)."

29   Law Offices of Fell and Spalding to Robert Bauchwitz, M.D., Ph.D., November 13, 2013, regarding whether a court ever decided if the defendants in United States v. Holloman et al. had or had not submitted grant applications containing false scientific information. http://www. amerares.com/Law_Firms_Evidence_Use_Letter_res_jud_2013.pdf

30   Ibid.

31   Kenneth James Jones earned his bachelor of science at Massachusetts Institute of Technology, and his master of arts and doctor of education at Harvard University.

32   United States ex rel. Kenneth James Jones, Plaintiff v. Brigham and Women's Hospital, et al., No. 07-cv-11481 (D. Mass. June 14, 2006); United States ex rel. Kenneth James Jones, Plaintiff v. Brigham and Women's Hospital, et al., No. 07-cv-11481-amended (D. Mass. November 10, 2010); United States ex rel. Kenneth James Jones, Plaintiff v. Brigham and Women's Hospital, et al., No. 10-2301 (1st Cir., May 7, 2012);

33   NIH Grants Policy Statement (present and past); United States ex rel. Kenneth James Jones, Plaintiff v. Brigham and Women's Hospital, et al., No. 13-1973 (1st Cir. March 16, 2015).

34   Dr. Killiany received his master's degree in psychology from the University of Hartford and completed doctoral training in psychology at Northeastern University. He completed postdoctoral fellowship training in neuroanatomy, neurobiology, and neuropsychology at the Boston University School of Medicine and joined its faculty in 2001.

35   Killiany, Ronald J. et al., "Use of structural magnetic resonance imaging to predict who will get Alzheimer's disease," Annals of Neurology, April 2000, Volume 47, Number 4 [hereinafter Killiany, et al. Structural MRI] ; United States ex rel. Kenneth James Jones v. Brigham and Women's Hospital, et al., No. 07-cv-11481 (D. Mass. June 14, 2006).

36   Keith A. Johnson, M.D. is a professor of radiology and neurology at the Harvard Medical School. He is also an associate radiologist and the director of molecular neuroimaging in the Division of Nuclear Medicine and Molecular Imaging, Department of Radiology, Massachusetts General Hospital (MGH). Johnson also serves as an associate physician and staff neurologist in the Memory Disorders Unit at the Brigham and Women's Hospital as well as a clinical associate in neurology at the MGH. Johnson is co-director of the neuroimaging program of the Massachusetts Alzheimer's Disease Research Center and its Dominantly Inherited Alzheimer Network (DIAN) research initiatives. He oversees the Clinical Brain Positron Emission Tomography (PET) Service at the MGH and also practices as a neurologist that specializes in neurodegenerative disorders.

     Dr. Johnson maintains an Internet teaching atlas of neuroimaging known as the Whole Brain Atlas: www.med.harvard.edu/AANLIB/home.html

37   Dr. Moss received his doctorate in psychology from Northeastern University and completed postdoctoral training at Beth Israel Hospital, Harvard Medical School in neuroanatomy and neuropsychology. He joined the Department of Anatomy and Neurobiology at Boston University School of Medicine in 1982 and has served as its chairman since 1998. Together with Dr. Douglas Rosene, Moss is codirector for the Laboratory for Cognitive Neurobiology.

38   United States ex rel. Kenneth James Jones v. Brigham and Women's Hospital, et al., No. 07-cv-11481-opinion (D. Mass. November 10, 2010).

39   Ibid.

40   Ibid.

41   Ibid.

42   Ibid.

43   Ibid.

44   Norton, Christopher, "Mass. Hospitals seek quick win in research grant suit," Law360.com, September 10, 2010. http://www.law360.com/articles/192055/mass-hospitals-seek-quick-win-in-research-grant-suit

45   Harvard Medical School, Harvard University, and Marie F. Kijewski, D.Sc., were also named as defendants in this action. The parties stipulated to voluntary dismissal without prejudice of the case against Harvard Medical School and Harvard University on March 17, 2009. Dr. Kijewski moved separately to dismiss the case against her. After a hearing on the matter, the district court granted her motion to dismiss on July 10, 2009.

46   Whistleblowers Protection Blog. "Whistleblower Kenneth Jones wins appeal and forces Harvard to trial for research fraud," May 8, 2012. http://www.whistleblowersblog.org/2012/05/articles/false-claims/whistleblower-kenneth-jones-wins-appeal-and-forces-harvard-to-trial-for-research-fraud/

47   Saykin held these positions as of August 4, 2010, the date of his expert report.

48   United States, ex rel. Kenneth James Jones, et al. v. Brigham and Women's Hospital, et al., Case No. 10-2301 (1st Cir. 2012).

49   Schuff held these positions as of the date of his expert report, August 2, 2010.

50   Dávila-Garcia held this position as of the date of her expert report, September 15, 2010.

51   United States, ex rel. Kenneth James Jones, et al. v. Brigham and Women's Hospital, et al., Case No. 10-2301 (1st Cir. 2012).

52   Ibid.

53   Foster, Hope and Kevin M. McGinty, "Claims, awards, enforcement action: qui tam update – recently unsealed whistleblower cases," The National Law Review, April 13, 2016. http://www.natlawreview.com/article/claims-awards-enforcement-action-qui-tam-update-recently-unsealed-whistleblower

54   Ibid.

55   Foster, Hope, et al., "Mintz Levin health care qui tam update," Mintz.com, June 1, 2015. https://www.mintz.com/legal-insights/industry/articletype/articleview/articleid/2768/mintz-levin-health-care-emqui-tamem-update

56   Of the 65 cases, 28 were unsealed in April, 18 in May, and 19 in June 2014.

57   McGinty, Kevin, et al. "Mintz Levin health care qui tam update – recent developments and unsealed False Claims Act cases," Mintz.com, October 2014. https://www.mintz.com/newsletter/2014/Newsletters/4338-1014-NAT-HL/

58   Associated Press Science. "Faked research results on rise?" Wired.com, July 10, 2005. http://www.wired.com/2005/07/faked-research-results-on-rise/

59 Ibid.

60 Ibid.

61 Ibid.

62 Ibid.

63 Ibid.

64 Ibid. Debbi Gilad, J.D., has been, more recently, executive director, Office of Research Compliance and Integrity, Perelman School of Medicine, University of Pennsylvania, and director of the Bon Secours Health System Office of Research.

65 Associated Press, ibid.

66 Ibid.

67 Ibid.

68 Ibid.

69 Ibid.

70 http://rbhs.rutgers.edu/ Rutgers Biological and Health Sciences (RBHS) website.

71 http://rbhs.rutgers.edu/rwjlbweb/posters/scimisconduct.html Robert Wood Johnson Library of the Health Sciences website.

72 Yong, Ed, Heidi Ledford and Richard Van Noorden, "3 ways to blow the whistle," Nature, Volume 503, November 28, 2013. http://www.nature.com/polopoly_fs/1.14226!/menu/main/topColumns/topLeftColumn/pdf/503454a.pdf

73 Ibid.

74 Ibid.

75 Ibid.

76 On July 1, 2013, the New Jersey Medical and Health Sciences Education Restructuring Act went into effect, integrating Rutgers, The State University of New Jersey, with all units of the University of Medicine and Dentistry of New Jersey (UMDNJ), except University Hospital in Newark and the School of Osteopathic Medicine in Stratford. Rutgers Biomedical and Health Sciences serves as an umbrella organization for legacy UMDNJ schools and clinical units, several pre-existing Rutgers units with key health-related missions, and two research units that historically were jointly operated by Rutgers and UMDNJ.

77 Hill, Hidden Data, ibid.

78 http://csu-cvmbs.colostate.edu/pages/ueno-akiko-radiation-biology-fund.aspx

79 Ibid.

80 Today, Roger W. Howell is a professor of radiology at Rutgers New Jersey Medical School. He received his bachelor's degree in physics in 1982 and a doctorate in 1987 from the University of Massachusetts – Amherst.

He is chief of the Division of Radiation Research and chair of the Rutgers Radiation Safety Committee. Dr. Howell has authored over 100 scientific publications on radiation dosimetry and radiobiology of internal radionuclides, including two books and two patents. Howell serves on the Society of Nuclear Medicine's Medical Internal Radiation Dose Committee. He has served on committees for NCRP Report 167 Potential impact of individual genetic susceptibility and previous radiation exposure on radiation risk for astronauts, ICRU Report 67 Absorbed dose specification in nuclear medicine, ICRU Report 86 Quantification and reporting of low-dose and other heterogeneous exposures, and he serves on the International Commission on Radiation Units and Measurements. He is the recipient of the 2004 Loevinger-Berman Award from the Society of Nuclear Medicine and New Jersey Medical School's 2009 Basic Science Faculty of the Year Award. http://njms.rutgers.edu/resource_locator/find_people/profile. cfm?mbmid=rhowell#tab-bio

81   Strudler is the executive secretary of the NIH Radiation Study Section.

82   Sastry was born in the state of Andhra Pradesh, India, in 1935. He received his bachelor of science and master of science from Andhra University, Waltair, India, in 1955 and 1956, respectively. He came to the United States two years later, in 1958, to pursue a degree from Indiana University, where he received his Ph.D. in nuclear physics in 1962. The following year he accepted an assistant professorship in the department of physics at the University of Massachusetts-Amherst, where Dandamudi Rao and Roger Howell would also receive their doctorate degrees. He remained at the University of Massachusetts for the duration of his career.

83   Dandamudi Vishnuvardhana Rao received his Ph.D. from the University of Massachusetts in 1972 and is today a professor emeritus, Rutgers New Jersey Medical School.

84   Howell, Roger W. and Dandamudi V. Rao, "In memoriam: Kandula S. R. Sastry (1935-2001)," Journal of Nuclear Medicine, Volume 43, Issue 1, January 2002: N33. http://search.proquest.com/openview/099ddd09684249 d54c35360784713e41/1?pq-origsite=gscholar&cbl=40808

85   Azzam received his bachelor of science from the University of Calgary, Canada in microbiology, in 1973; his master of science in physiology in 1989 from the University of Manitoba, Canada, and his Ph.D. in radiation biology in 1995 from the University of Ottawa, Canada. From 1998-2000 Azzam was a research associate at the Harvard School of Public Health in Boston. He was named a full professor at Rutgers-New Jersey Medical School in 2008.

86   Anupam Bishayee, Ph.D. received his bachelor of pharmacy (B.Pharm.) in 1989; a master of pharmacy (M.Pharm.) in biomedical pharmacology in 1991, and his doctorate in tumor biology in 1996. All of his degrees were earned at Jadavpur University, Calcutta, India.

87   Howell, NIH grant application.

88 From 1983 to 2006, Denny was at the New Jersey Medical School, Department of Pathology, Laboratory Medicine and Pediatrics. Denny is presently a professor of medicine at Duke University School of Medicine, Duke University Medical Center (since 2010); chief operating officer, Duke Human Vaccine Institute and Center for HIV/AIDS Vaccine Immunology (since 2006), and director, Duke University Human Vaccine Institute – Immunology Quality Assessment Center (since 2006).

89 The first Newark CCRI's report along with the minutes of the various meetings were not made available until several years later during discovery proceedings as part of Hill's qui tam law suit. The retaliation was fully documented in the initial charges of this case.

90 Yong, ibid.

91 www.ori.dhhs.gov/statistical-forensics-check

92 Bishayee, Anupam, Dandamudi V. Rao, and Roger W. Howell, "Evidence for pronounced bystander effects caused by nonuniform distributions of radioactivity using a novel three-dimensional tissue culture model," Radiation Research, Volume 152, 1999, pages 88-97 (Attachment 3), cited in grant application 1 RO1 CA83838-01A1, pages 2, 26 and 48 (as referenced 66) (Attachment 1).

93 Bishayee, Anupam, Helene Z. Hill, Dana Stein, Dandamudi V. Rao, and Roger W. Howell, "Free-radical initiated and gap junction-mediated bystander effect due to nonuniform distribution of incorporated radioactivity in a three-dimensional tissue culture model," Radiation Research, Volume 155, 2001, pages 335-344 (Attachment 4), also reported in grant application RO1 CA83838-02, page 6 (Attachment 2).

94 United States Department of Health and Human Services (HHS), Office of Research Integrity (ORI). ORI Oversight Report, University of Medicine and Dentistry of New Jersey, ORI 2001-28, August 2002. The questioned research was reported as preliminary data in a National Institutes of Health (NIH) grant application or was supported by the following NIH grant: 1 RO1 CA83838-01A1 and -02, "Effects of non-uniform distributions of radioactivity," bearing the name of Roger W. Howell, principal investigator, submitted October 21, 1999, and awarded on July 1, 2000, to June 30, 2005.

95 FACS is a form of flow cytometry. In biotechnology, flow cytometry is a laser- or impedance-based, biophysical technology employed in cell counting, cell sorting, biomarker detection and protein engineering, by suspending cells in a stream of fluid and passing them by an electronic detection apparatus. It allows simultaneous multiparametric analysis of the physical and chemical characteristics of up to thousands of particles per second. Flow cytometry is routinely used in the diagnosis of health disorders, especially blood cancers, but has many other applications in basic research, clinical practice and clinical trials. A common variation is to physically sort particles based on their properties, so as to purify populations of interest.

96 Bishayee, et al. Radiation Research, 1999.

97   The John B. Little Center for Radiation Sciences is named for John B. Little, M.D., the James Stevens Simmons Professor of Radiobiology, Emeritus in the Department of Genetics and Complex Diseases at the Harvard T. H. Chan School of Public Health. http://www.hsph.harvard.edu/jbl-center/a-living-legacy/

98   NIH Grant No. 5P01CA049062-23 with component studies from February 1, 1997, to June 30, 2015.

99   President George W. Bush appointed Christie the United States Attorney for New Jersey, a position he held from 2002 to 2008.

100 On March 3, 2005, the Somerset County Board of Freeholders honored thirteen outstanding women for public service as follows, including Susan Schlow, senior special agent, United States Department of Health and Human Services, Office of the Inspector General. A graduate of Rutgers University, she was employed by that time with the law enforcement component of the state Department of Health and Senior Services, currently assigned to the Fairfield office covering Somerset County. Her work included, up to that time, conducting criminal and civil investigations into potential Medicare fraud.

101 Wallace, Clementine, "Biologist charged with more fraud," The Scientist, June 12, 2006. http://www.the-scientist.com/?articles.view/articleNo/24077/title/Biologist-charged-with-more-fraud/

102 Davidson, Keay, "Berkeley lab, Stanford researchers retract study – 1997 article based on data allegedly fabricated by North Carolina scientist," SFGate.com, June 17, 2005. http://www.sfgate.com/bayarea/article/Berkeley-lab-Stanford-researchers-retract-study-2661753.php

103 Ibid.

104 Graham, Troy and Jennifer Moroz, "UMDNJ monitor says fraud, failures now up to $243 million, the 'disheartening' tally could rise, the school said the figure included issues it had already addressed," Philadelphia Inquirer, July 21, 2006. http://articles.philly.com/2006-07-21/news/25404571_1_potential-fraud-medicaid-investigators

105 Kocieniewski, David, "Latest twist in a scandal hits a medical school when it's down," New York Times, November 24, 2006. http://www.nytimes.com/2006/11/24/nyregion/24hosp.html

106 Robins, Richard B., "Anatomy of a federal investigation and trial for alleged Stark and Anti-Kickback violations," Brach Eichler L.L.C. Newsletter, n.d. http://www.bracheichler.com/C3F493/assets/files/News/AnatomyofaFederalInvestigation.pdf

107 Ibid.

108 Discovery, as a reminder, is the entire efforts of a party to a lawsuit and that party's attorneys to obtain information before trial through demands for production of documents, depositions of parties and potential witnesses, written interrogatories (questions and answers written under oath), written requests for admissions of fact, examination of the scene and the petitions

and motions employed to enforce discovery rights. The theory of broad rights of discovery is that all parties will go to trial with as much knowledge as possible and that neither party should be able to keep secrets from the other (except for constitutional protection against self-incrimination). Often much of the fight between the two sides in a suit takes place during the discovery period, which was decidedly the case in Hill's qui tam.

109 Chang, Kenneth, "Columbia chemistry professor is retracting 4 more papers," New York Times, June 15, 2006. http://www.nytimes.com/2006/06/15/science/15chem.html

110 Ibid.

111 Dingell, John D., "Shattuck lecture – misconduct in medical research," New England Journal of Medicine, Volume 328, Number 22, June 3, 1993, pages 1610-1615.

112 Ibid.

113 Ibid.

114 Ibid.

115 Ibid.

116 Ibid.

117 Boffey, Philip M., "U.S. study finds fraud in top researcher's work on mentally retarded," New York Times, May 24, 1987. http://www.nytimes.com/1987/05/24/us/us-study-finds-fraud-in-top-researcher-s-work-onmentally-retarded.html

118 Hilts, Philip J., "Biologist who disputed a study paid dearly," New York Times, March 22, 1991. http://www.nytimes.com/1991/03/22/us/biologist-who-disputed-a-study-paid-dearly.html?pagewanted=all

119 Dingell, ibid.

120 Ibid.

121 McCutchen to Hill, letter, December 8, 2011.

122 Hilts, ibid.

123 Ibid.

124 Ibid.

125 Dingell, ibid.

126 Dennis M. Cavanaugh's page on the McElroy, Deutsch, Mulvaney and Carpenter website: http://www.mdmc-law.com/attorneys/Dennis_Cavanaugh/

127 Hall is presently Higgins Professor Emeritus of radiation biophysics, special research scientist, special lecturer in radiation oncology, Center for Radiological Research, College of Physicians and Surgeons of Columbia University.

128 Synchronization forces all the cells in a population to progress through the cell cycle in lock step. That way, all the cells will be in S (DNA synthesis) phase at the same time.

129   Certification of Dr. Helene Z. Hill in Opposition to Motion to Quash Subpoena issued to Dr. Thomas [Tom] Hei, dated August 1, 2008.

130   Edwin H. Goodwin, Ph.D., is currently a member of KromaTiD™'s Scientific Advisory Board. As a graduate student, Goodwin participated in an NIH-funded initiative to evaluate charged-particle radiations as a therapy for cancer. Using the cyclotrons at Lawrence Berkeley Laboratory, Goodwin analyzed chromosome fragmentation patterns in an effort to understand the biological effects of these exotic radiations. His studies of radiation continued at Los Alamos National Laboratory in research funded by the Department of Energy's radon and low-dose radiation risk assessment programs. While at Los Alamos, Goodwin invented Chromosome Orientation Fluorescence in situ Hybridization (CO-FISH). Application of CO-FISH reveals the orientation of DNA sequences in chromosomes, information that has proven to be useful in studies of repetitive DNA sequences and telomere biology. He received B.S. and M.S. degrees in physics from Fairleigh Dickinson University, Teaneck, New Jersey, and a Ph.D. in biophysics from the University of California, Berkeley, California. http://www.kromatid.com/

131   Passannante is presently a full professor in the UMDNJ-turned-Rutgers University New Jersey Medical School Department of Preventive Medicine and Community Health.

132   Holland is director, Educational Evaluation and Research at UMDNJ-turned-Rutgers University New Jersey Medical School.

133   Pogozelski is chair and State University of New York (SUNY) distinguished teaching professor of chemistry at SUNY Geneseo. She has been a member of the university's faculty since 1996.

134   Gerashchenko, Bogdan I., and Roger W. Howell, "Proliferative response of bystander cells adjacent to cells with incorporated radioactivity," Cytometry Part A, Volume 60, Issue 2, July 2004, pages 155-164. DOI: 10.1002/cyto.a.20029 https://www.researchgate.net/publication/8419266_Proliferative_response_of_Bystander_cells_adjacent_to_cells_with_incorporated_radioactivity

135   See bibliography.

136   Lange is reviewing articles for more than twenty journals, including Radiation Research, International Journal of Radiation Biology, and International Journal of Radiation Oncology Biology Physics.

137   Press release. http://www.24-7pressrelease.com/press-release/sir-christopher-s-lange-honored-for-excellence-in-radiation-oncology-402879.php

138   Today @ Colorado State. "John Terrence Lett, 75, retired radiation biology professor, died Oct. 6," November 15, 2009. http://today-archive.colostate.edu/story2402.html?id=2671

139   As of March 17, 2010, Round Table Group, Inc. operates as a subsidiary of Thomson Reuters Corporation.

140 When he died November 23, 2012, Michael "Mike" E. Robbins, Ph.D., had gained a national and, indeed, international, reputation as a leader in the field of radiation-induced normal tissue effects. He had been at Wake Forest Medical Center since 2001, an institution where he had developed a world-class research program looking at the prevention and treatment of radiation-associated tissue injury, with a specific interest in radiation-induced brain injury.

141 When Lenarczyk left Howell's lab in 2002 he went home to Poland for roughly six months then returned to the United States to work at Colorado State University until the spring of 2005 as a research associate four. He left Colorado for a brief stint as lab support at the University of Arkansas Medical School in Little Rock before taking a postdoctoral fellowship at the Medical College of Wisconsin; he remained there until 2012. Lenarczyk left Milwaukee, and after a short hiatus, took a position as a research associate at the University of Tennessee Health Science Center in 2013. The majority of Lenarczyk's positions were relatively short-lived because he was hired under grant funding that would eventually run out. Lenarczyk would state under oath during his deposition that he didn't realize how hard it was to keep his family in the United States. "And we only had one income, which was my income, so I couldn't stay over there [Little Rock]. I tried to stay in science, but not necessarily for that kind of money which they offered me. So, it was absolutely impossible to stay and keep my family. So, I decided to move out."

142 Biphasic means having two phases.

143 The period of growth of a population of cells (as of a microorganism) in a culture medium during which numbers increase exponentially and which is represented by a part of the growth curve that is a straight line segment if the logarithm of numbers is plotted against time – called also logarithmic phase.

144 MEMB is the acronym for Modified Eosin Methylene Blue Agar (cell culture medium).

145 Note that the bystander effect is not the same as the abscopal effect. The abscopal effect is a phenomenon where the response to radiation is seen in an organ/site distant to the irradiated organ/area, that is, the responding cells are not juxtaposed with the irradiated cells. T-cells and dendritic cells have been implicated to be part of the mechanism, according to Demaria, Sandra, Bruce Ng, Mary Louise Devitt, et al. (March 2004), "Ionizing radiation inhibition of distant untreated tumors (abscopal effect) is immune mediated," International Journal of Radiation Oncology Biology Physics, Volume 58, Number 3, March 2004, pages 862-70. DOI:10.1016/j.ijrobp.2003.09.012.

146 Web of Science [previously known as (ISI) Web of Knowledge] is an online subscription-based scientific citation indexing service maintained by Thomson Reuters that provides a comprehensive citation search. It gives access to multiple databases that reference cross-disciplinary research, which allows for in-depth exploration of specialized sub-fields within an academic or scientific discipline This is the world's most trusted citation index.

147   Anupam Bishayee, B.Pharm., M.Pharm, Ph.D. has been professor and chair of the Department of Pharmaceutical Sciences, College of Pharmacy, Larkin Health Sciences Institute, in Miami, Florida, since February 2015.

148   http://unemploymenthandbook.com/state-unemployment-directory/69-new-jersey/940-new-jersey-unemployment-if-you-quit

149   http://lwd.dol.state.nj.us/labor/ui/aftrfile/quit.html

150   http://www.lsnjlaw.org/Jobs-Employment/Unemployment-Insurance/Voluntary-Quit-Misconduct/Pages/Voluntary-Quit.aspx#.VzpjhOSJdv8

151   Ibid.

152   Ibid.

153   In addition to being chairman of the radiology department, Baker, M.Phil. and M.D., was also just then the association dean for graduate medical education in the UMDNJ. He had been chair of the department for 18 years and associate dean for six years at the time of the deposition. Baker had been a full professor for 22 years at that point in time.

154   Robert A. Saporito, D.D.S., served as senior vice president, academic affairs of the University of Medicine and Dentistry of New Jersey (UMDNJ) and as a faculty practitioner of the UMDNJ-New Jersey Dental School in Newark, with courtesy staff appointment at University Hospital in Newark. Dr. Saporito held various administrative and academic appointments within UMDNJ, including that of dean of the New Jersey Dental School. His significant professional activities revolve around teaching, research, and service on both a local, national, and international level. He is actively involved in community service. He is a trustee at Cooper University Hospital and Cooper University Health Care. Dr. Saporito received a Bachelor of Arts degree in Sociology from Franklin and Marshall College in Lancaster, Pennsylvania; his Doctor of Dental Science from the School of Dental and Oral Surgery of Columbia University, and a Specialty Certificate in Prosthodontics from the College of Dentistry, New York University.

155   Vincent Lanzoni, M.D., was New Jersey Medical School dean from 1975 to 1987. He died on June 28, 2007. A graduate of Tufts University, Lanzoni served in the United States Air Force from 1954 to 1956 and received his medical degree from Boston University in 1960.

156   Yang, Debra Wong, Nick Hanna and Alexander H. Southwell, "Health care compliance in 2009 and going forward: Part 1," Westlaw Journal Health Care Fraud, Volume 15, Issue 9, March 2010. http://www.gibsondunn.com/publications/Documents/YangHannaSouthwell-HealthCareCompliancePart1.pdf

157   Ibid.

158   Rockey, Sally J., "Postdoctoral researchers – facts, trends , and gaps," Extramural Nexus, June 29, 2012, a publication of the National Institutes of Health Office of Extramural Research. https://nexus.od.nih.gov/all/2012/06/29/postdoctoral-researchers-facts-trends-and-gaps/

159 Howell, B. Lindsay, ed. Foreign Temporary Workers in America: Policies that Benefit the U.S. Economy: Policies that Benefit the U.S. Economy. Westport, Connecticut: Quorum Books, 1999.

160 Ibid.

161 Mayhew, Ruth, "Can my employer force me to sign a letter of resignation? The Nest, n.d. http://woman.thenest.com/can-employer-force-sign-letter-resignation-18537.html

162 "Retaliation in the workplace," http://topics.hrhero.com/retaliation-in-the-workplace/#

163 Mattiuzzi, Paul G., "What is workplace retaliation? It's about making people afraid," Everyday Psychology, March 26, 2012. http://www.everydaypsychology.com/2012/03/what-is-workplace-retaliation-its-about.html#.Vz4eJ-SJdv8

164 Guerin, Lisa, "Workplace retaliation: What are your rights?" NOLO, n.d. http://www.nolo.com/legal-encyclopedia/workplace-retaliation-employee-rights-30217.html

165 Ibid.

166 The Krebs cycle is a series of enzymatic reactions that catalyzes the aerobic metabolism of fuel molecules to carbon dioxide and water, thereby generating energy for the production of adenosine triphosphate (ATP) molecules.

167 Mosimann, James, et al., "Terminal digits and the examination of questioned data," chapter in Investigating Research Integrity Proceedings of the first ORI research conference on research integrity, ed. Nicholas H. Steneck, Ph.D., and Mary D. Scheetz, Ph.D., 2002, pages 269-290. https://ori.hhs.gov/documents/proceedings_rri.pdf

168 Persaud, Rudranath, Hongning Zhou, et al., "Assessment of low linear energy transfer radiation-induced bystander mutagenesis in a three-dimensional culture model," Cancer Research, Volume 65, 2005, pages 9876-82.

169 Previous studies have shown that the inclusion of dCyd in the medium prevents the 3H-TdR from blocking cell movement through the cell cycle leading to an exponential decrease in cell survival.

170 According to the IUPAC-IUB Commission on Biochemical Nomenclature, the 1965 revision of Abbreviations and Symbols for Chemical Names of Special Interest in Biological Chemistry was completed and published in 1965 and 1966, almost coincident with the elucidation of the first complete nucleic acid sequence and with the development of methods for the synthesis of specific polynucleotide sequences. The proscribed acronyms have been in place for decades. http://www.chem.qmul.ac.uk/iupac/misc/naabb.html

171 "Grantee misconduct: Dr. Eric T. Poehlman," https://www.nih.gov/news-events/grantee-misconduct-dr-eric-t-poehlman; Office of Research Integrity, "Case summary: Eric T. Poehlman," http://www.ori.dhhs.gov/poehlman_notice

172　United States Attorney for the District of Vermont, "Press release – Dr. Eric T. Poehlman," March 17, 2005. http://ori.hhs.gov/press-release-poehlman

173　Ibid.

174　Ibid.

175　Kintisch, Eli, "Poehlman sentenced to 1 year in prison," Science, June 28, 2006. http://www.sciencemag.org/news/2006/06/poehlman-sentenced-1-year-prison

176　March 17, 2005 press release, ibid.

177　http://ori.hhs.gov/statistical-forensics-check

178　McCann, Joyce C., Mark L. Hudes, and Bruce N. Ames, "Unusual clustering of coefficients of variation in published articles from a medical biochemistry department in India," FASEB Journal, Volume 23, 2009, pages 698-708. http://www.fasebj.org/content/23/3/706.full.pdf

179　On July 1, 2013, Rutgers University took over seven of the University of Medicine and Dentistry of New Jersey's schools, including two of its medical schools, as part of a statewide higher education restructuring.

180　McCutchen to Hill, letter, April 4, 2014.

181　MDPI AG Publications, special issue submission requirements, http://www.mdpi.com/journal/publications/special_issues/scientific-publishing

182　Engber, Daniel, "Ask us anything: how common is scientific fraud? Popular Science, March 30, 2015. http://www.popsci.com/ask-us-anything-how-common-scientific-fraud

183　Ibid.

184　Ibid.

185　Hwang Woo-Suk was a highly regarded and well funded South Korean veterinarian, researcher and professor of theriogenology and biotechnology at Seoul National University who achieved international fame for his work on embryonic stem cells and the promises his findings offered. Until November 2005, he was regarded as one of the pioneering experts in the field, best known for two articles published in the journal Science in 2004 and 2005, where he reported to have succeeded in creating human embryonic stem cells by cloning. But this reputation quickly unraveled, according to multiple media reports, and his research activities were halted when his success in somatic cell nuclear transfer (SCNT) became mired in scandal, particularly when it emerged that many of his data on SCNT were fabricated. He was dismissed from his university position on March 20, 2006, and his two important papers on embryonic stem cell research had to be retracted from Science. On May 12, 2006, Hwang was charged with embezzlement and bioethics law violations after it emerged much of his stem cell research had been faked. The Korea Times reported on June 10, 2007, that Seoul National University had, indeed, fired him, and the South Korean government canceled his financial support and barred him from engaging in stem cell research. While being charged with fraud and embezzlement, he kept a relatively low profile at the Sooam Bioengineering Research Institute in Yongin, Gyeonggi Province, where he continued research efforts on creating cloned pig embryos and using them to make embryonic stem-cell lines. Since the controversy died down, despite the history and his lost credibility as a scientist, Hwang's lab has been actively publishing manuscripts, many of which have appeared on PubMed, the online database for biomedical research. In February 2011, Hwang visited Libya as part of a $133 million project in the North African country to build a stem cell research center and transfer relevant technology; however, the project was canceled when civil war broke out there. Hwang continued to land on his feet, finding new outlets for his work. On November 2015, a Chinese biotech company Boyalife Group announced that it would partner with Hwang's laboratory, Sooam Biotech, to open the world's largest animal cloning factory in Tianjin as early as 2016. The factory was set up to produce up to one million cattle embryos per year to meet the increasing demand for quality beef in China.

186 The Schön scandal was yet another example of a scientist of great prominence who committed misconduct who made national headlines. Schön, then a physicist at Bell Laboratories, had briefly risen to prominence after a series of apparent breakthroughs with semiconductors that were subsequently discovered to be fraudulent. Before he was exposed, Schön had received the Otto-Klung-Weberbank Prize for physics and the Braunschweig Prize in 2001 as well as the Outstanding Young Investigator Award of the Materials Research Society in 2002, both of which were later rescinded. He had also fabricated or falsified data in 17 published papers, some of which appeared in highly prestigious journals like Science, Nature, and Applied Physics Letters. The scandal provoked discussion in the scientific community about the degree of responsibility of coauthors and reviewers of scientific papers. The debate centered on whether peer review, traditionally designed to find errors and determine relevance and originality of papers, should also be required to detect deliberate fraud.

187 Fanelli, Daniele, "How many scientists fabricate and falsify research? A systematic review and meta-analysis of survey data," PLOS One, May 29, 2009. DOI: 10.1371/journal.pone.0005738. http://journals.plos.org/plosone/article?id=10.1371/journal.pone.0005738

188 Martinson, Brian C., Melissa S. Anderson and Raymond de Vries, "Scientists behaving badly," Nature, Volume 435, June 9, 2005, pages 737-738. http://pages.stolaf.edu/ross/files/2014/05/ScientistsBehavingBadly.pdf

189 Ibid.

190 Ibid.

191 Ibid.

192 Ibid.

193 Weiss, Rick, "Many scientists admit to misconduct," Washington Post, June 9, 2005. http://www.washingtonpost.com/wp-dyn/content/article/2005/06/08/AR2005060802385.html

194 George, Stephen L., "Research misconduct and data fraud in clinical trials: prevalence and causal factors," International Journal of Clinical Oncology, Volume 21, Issue 1, August 20, 2015. DOI: 10.1007/s10147-015-0887-3.

195 Ibid.

196 Martinson et al., ibid.

197 Weed, Douglas L., "Preventing scientific misconduct," American Journal of Public Health," Volume 88, Number 1, January 1998. http://ajph. aphapublications.org/doi/pdf/10.2105/AJPH.88.1.125

198 United States Department of Health and Human Services (HHS). Integrity and Misconduct in Research: Report of the Commission on Research Integrity. Washington, D.C.: United States Government Printing Office, 1995. http://ori.hhs.gov/sites/default/files/report_commission.pdf

199 Ingham, Janis Costello, "Ethics and research," The ASHA Leader, March 2004, Volume 9, 10-25. DOI:10.1044/leader.FTR6.09052004.10

200 Smith, Richard, "Research misconduct: the poisoning of the well," Journal of the Royal Society of Medicine, May 2006, Volume 99, Number 5, pages 232-237. http://www.ncbi.nlm.nih.gov/pmc/articles/PMC1457763/pdf/0232. pdf

201 Ingham and Horner, ibid.

202 Lock, Stephen, Frank Wells and Michael Farthing, eds. Fraud and Misconduct in Biomedical Research. London: BMJ Books, 2001.

203 Drummond Rennie is a nephrologist and high altitude physiologist who today remains a contributing deputy editor of the Journal of the American Medical Association (JAMA) and an adjunct professor of medicine at the University of California, San Francisco. He is an editor of JAMAevidence, a project for education related to evidence-based medicine sponsored by the American Medical Association. He is known for involvement in reform of scientific publishing and for advocating improvements in reporting standards for clinical trials. He is the director of the International Congress on Peer Review and Biomedical Publication. In 2008, the American Association for the Advancement of Science awarded him its Award for Scientific Freedom and Responsibility.

204 Fanelli, ibid.

205 Ibid.

206 Ibid.

207 Ibid.

208 Ibid.

209 Ibid.

210 Resnik, David B., "From Baltimore to Bell Labs: reflections on two decades of debate about scientific misconduct," Accountability in Research: Policies and Quality Assurance, Volume 10, Number 2, 2003, pages 123-135. DOI: 10.1080/08989620390199890. http://dx.doi.org/10.1080/08989620300508

211 Ibid.

212 McClain, Sylvia, "Scientific fraud: a sign of the times?" The Guardian, October 12, 2012. https://www.theguardian.com/science/occams-corner/2012/oct/12/scientific-fraud

213 Ibid.

214 Johnson, Carolyn Y., "Harvard report shines light on ex-researcher's misconduct," Boston Globe, May 30, 2014. https://www.bostonglobe.com/metro/2014/05/29/internal-harvard-report-shines-light-misconduct-star-psychology-researcher-marc-hauser/maSUowPqL4clXrOgj44aKP/story.html

215 Ibid.

216 McClain, ibid.

217 Castillo, Stephanie, "The FDA underreports scientific misconduct in peer-reviewed articles: the benefits of negative science," MedicalDaily.com, February 10, 2015. http://www.medicaldaily.com/fda-underreports-scientific-misconduct-peer-reviewed-articles-benefits-negative-321548

218 Seife, Charles, "Are your medications safe? The FDA buries evidence of fraud in medical trials. My students and I dug it up," Slate, February 9, 2015. http://www.slate.com/articles/health_and_science/science/2015/02/fda_inspections_fraud_fabrication_and_scientific_misconduct_are_hidden_from.html

219 Ibid.

220 Castillo, ibid.

221 Seife, ibid.

222 Ibid.

223 Wright began his term as ORI director in January 2012. Previously, he served as an expert consultant to ORI from 2001 until 2011, working with both the division of investigative oversight and the division of education and integrity. In that capacity he was the architect of the RIO Boot Camp program. From 1993 to 2004, Wright served as Michigan State University's (MSU) assistant vice president for research ethics and standards, as well as its intellectual integrity officer, overseeing most of MSU's research regulatory compliance activity. He also chaired the university's committee on research involving human subjects for eleven years. From 2005 to 2011 he was professor and chairperson of MSU's department of community sustainability where he taught and wrote on the history of science and technology and the responsible conduct of research. Wright has a bachelor's degree from Princeton University and a Ph.D. in American Studies from Michigan State University.

224 Kaiser, Jocelyn, "Top U.S. scientific misconduct official quits in frustration with bureaucracy," ScienceInsider, March 12, 2014. http://www.sciencemag.org/news/2014/03/top-us-scientific-misconduct-official-quits-frustration-bureaucracy

225 Ibid.

226 Leys, Tony, "Grassley wants more specifics about response to ISU scientist's multimillion-dollar research fraud," Des Moines Register, March 11, 2014. http://blogs.desmoinesregister.com/dmr/index.php/2014/03/11/grassley-wants-more-specifics-about-response-to-isu-scientists-multimillion-dollar-research-fraud

227 Ibid.

228 Marcus, Adam and Ivan Oransky, "Crack down on scientific fraudsters," New York Times, July 10, 2014. http://nyti.ms/1rZUEnU

229 Ibid.

230 Wadman, Meredith, "Money in biomedicine: the senator's sleuth," Nature, Volume 461, September 16, 2009, pages 330-334. DOI:10.1038/461330a http://www.nature.com/news/2009/090916/full/461330a.html

231 Nemeroff has been the Leonard M. Miller professor and chairman, Department of Psychiatry and Behavioral Sciences and Director, Center on Aging, at the University of Miami Leonard M. Miller School of Medicine since 2009. He received his M.D. and Ph.D. in neurobiology from the University of North Carolina at Chapel Hill.

232 Wadman, ibid.

233 Ibid.

234 Ibid.

235 DelBello joined the College of Medicine faculty in 2000 as an assistant professor of psychiatry and pediatrics. She has served in several leadership roles in the department of psychiatry and behavioral neuroscience, including vice chair of clinical research since 2007 and co-director of the division of bipolar disorders research since 2004. Prior to being named the chair of the department, she had also been codirector of the Mood Disorders Center at the University of Cincinnati Neuroscience Institute. A nationally recognized expert on child and adolescent mood disorders, she has lectured and published extensively on bipolar disorder and served as principal or co-investigator of several NIH grants. She has been a member of the Neural Basis of Psychopathology, Addictions and Sleep Disorders Study Section of the Center for Scientific Review at the National Institutes of Health since 2009, most recently serving as chairperson for a term running from July 2012 through June 2014. Since 2012, she has chaired the research committee of the American Academy of Child and Adolescent Psychiatry. At the University of Cincinnati Medical Center, she has served as medical director of the Resident Mood Medication Clinic since its launch in September 2013. She is codirector of the Mood Disorders Program at Cincinnati Children's Hospital Medical Center. Nationally, she is a fellow of the American Academy of Child and Adolescent Psychiatry and also holds membership in the American Psychiatric Association, International Society for Bipolar Disorders and American College of Neuropsychopharmacology.

236 Wadman, ibid.

237 Emilia DeSanto, now a former Grassley aide, is the deputy director of the United States Department of State Office of the Inspector General (OIG).

238 Wadman, ibid.

239 Ibid.

240 Chronicle of Higher Education. "Stanford researcher, accused of conflicts, steps down as NIH principal investigator, August 1, 2008. http://chronicle.com/article/Stanford-Researcher-Accused/41395

241 Zerhouni was the fifteenth director of the NIH, serving from May 2, 2002, to October 31, 2008. Zerhouni, a world renowned leader in the field of radiology and medicine, spent his career providing clinical, scientific, and administrative leadership. He is credited with developing imaging methods used for diagnosing cancer and cardiovascular disease. As one of the world's premier experts in magnetic resonance imaging (MRI), he had extended the role of MRI from taking snapshots of gross anatomy to visualizing how the body works at the molecular level. He pioneered magnetic tagging, a non-invasive method of using MRI to track the motions of a heart in three dimensions. He is also renowned for refining an imaging technique called computed tomographic (CT) densitometry that helps discriminate between non-cancerous and cancerous nodules in the lung.

242 Wadman, ibid.

243 Nemeroff, Charles B., et al., "VNS therapy in treatment-resistant depression: clinical evidence and putative neurobiological mechanisms,"

Neuropsychopharmacology, Volume 31, Number 7, July 2006, pages 1345–1355; published online April 19, 2006. DOI:10.1038/sj.npp.1301082. http://www.ncbi.nlm.nih.gov/pubmed/16641939

244 Wadman, ibid.

245 Ibid.

246 Ibid.

247 Wadman, Meredith, "Straight talk with … Charles Grassley," Nature Medicine, Volume 14, Number 10, October 2008, pages 1006-1007. DOI: 10.1038/nm1008-1006

248 Wadman, Nature (2009), ibid.

249 Ibid.

250 Wilson, William A., "Scientific regress," First Things, May 2016. http://www.firstthings.com/article/2016/05/scientific-regress

251 Gobry, Pascal-Emmanuel, "Big science is broken," The Week, April 18, 2016. http://theweek.com/articles/618141/big-science-broken

252 Wilson, ibid.

253 Gobry, ibid.

254 Wilson, ibid.

255 Ibid.

256 Ibid.

257 Ibid.

258 Ibid.

259 Bohannon, John, "Updated: lax reviewing practice prompts 60 retractions at SAGE journal," Science, July 14, 2014. http://www.sciencemag.org/news/2014/07/updated-lax-reviewing-practice-prompts-60-retractions-sage-journal

260 Ibid.

261 Ibid.

262 Kaplan, ibid.

263 Ibid.

264 Committee on Publication Ethics (COPE). "COPE statement on inappropriate manipulation of peer review processes."
   http://publicationethics.org/news/cope-statement-inappropriate-manipulation-peer-review-processes

265 Kaplan, ibid.

266 Ferguson, Cat, "One in 25 papers contains inappropriately duplicated images, screen finds," Retraction Watch, April 19, 2016. http://retractionwatch.com/2016/04/19/one-in-25-papers-contains-inappropriately-duplicated-images-screen-finds/

267 Ibid.

268 Ibid.

269 Ibid.

270 Baker, Monya, "1,500 scientists lift the lid on reproducibility," Nature, May 25, 2016. http://www.nature.com/news/1-500-scientists-lift-the-lid-on-reproducibility-1.19970

271 Achenbach, Joel, "Many scientific studies can't be replication. That's a problem," Washington Post, August 27, 2015. https://www.washingtonpost.com/news/speaking-of-science/wp/2015/08/27/trouble-in-science-massive-effort-to-reproduce-100-experimental-results-succeeds-only-36-times/

272 Ibid.

273 Baker, ibid.

274 Ibid.

275 Murphy, Sean P., et al., "Submitting a manuscript for peer review – integrity, integrity, integrity," Journal of Neurochemistry, Volume 128, Number 3, February 2014, pages 341-343. DOI:10.1111/jnc.12644. http://www.ncbi.nlm.nih.gov/pmc/articles/PMC3926655/pdf/nihms553048.pdf

276 Ibid.

277 Smith, Richard, "Research misconduct: the poisoning of the well," Journal of the Royal Society of Medicine, May 2006, Volume 99, Number 5, pages 232-237. http://www.ncbi.nlm.nih.gov/pmc/articles/PMC1457763/pdf/0232.pdf

278 There are extensive physician and patient cautionaries and warnings published for rivaroxaban. https://www.drugs.com/cdi/rivaroxaban.html

279 Seife, ibid.

280 Telithromycin is the first ketolide antibiotic to enter clinical use and is sold under the brand name of Ketek. It is used to treat community acquired pneumonia of mild to moderate severity. After significant controversy regarding safety and research fraud, the FDA sharply curtailed the approved uses of the drug in 2007. Telithromycin is a semi-synthetic erythromycin derivative; it is created by substituting a ketogroup for the cladinose sugar and adding a carbamate ring in the lactone ring. An alkyl-aryl moiety is attached to this carbamate ring. Furthermore, the carbon at position 6 has been methylated, as is the case in clarithromycin, to achieve better acid-stability.

281 Ross, David B., "The FDA and the case of Ketek," New England Journal of Medicine, Issue 356, April 19, 2007, pages 1601-1604. DOI: 10.1056/NEJMp078032 http://www.nejm.org/doi/full/10.1056/NEJMp078032#t=article

282 Seife, ibid.

283 Ibid.

284 Ibid.

285 Ross, ibid.

286 FierceBiotech. "Fraud found at China site for key study of Pfizer and Bristol's Eliquis, July 9, 2013. http://www.fiercebiotech.com/r-d/fraud-found-at-china-site-for-key-study-of-pfizer-and-bristol-s-eliquis

287 Ibid.

288 Perry, Susan, "'Lone wolf' investigator says 'clinical trial system is broken," Minnpost, December 10, 2013. https://www.minnpost.com/second-opinion/2013/12/lone-wolf-fda-investigator-says-clinical-trial-system-broken

289 Seife, ibid.

290 Ibid.

291 Dingell, ibid.

292 Ibid.

293 Ibid.

294 Hilts, Philip J., "Federal inquiry finds misconduct by a discoverer of the AIDS virus," New York Times, December 31, 1992. http://query.nytimes.com/gst/fullpage.html?res=9F0CEFDA103DF932A05751C1A964958260&pagewanted=all

295 Hilts, ibid.

296 Ibid.

297 Ibid.

298 Roberts, Janine, "Fraud found in AIDS research," excerpt from Fear of the Invisible, August 22, 2008. http://reaids.com/fearoftheinvisible.com/aidsresearch.html

299 SV40 is an abbreviation for Simian vacuolating virus 40 or Simian virus 40, a polyomavirus that is found in both monkeys and humans. It was named for the effect it produced on infected green monkey cells, which developed an unusual number of vacuoles. Like other polyomaviruses, SV40 is a DNA virus that has the potential to cause tumors in animals, but most often persists as a latent infection.

300 Roberts, ibid.

301 Crewdson, John, "The great AIDS quest: science under the microscope," Chicago Tribune, November 19, 1989. http://archives.chicagotribune.com/1989/11/19/page/101/article/the-great-aids-quest

302 Hilts, ibid.

303 Roberts, ibid.

304 Chicago Tribune. "Ex-Gallo aide guilty of pocketing $25,000, July 8, 1992. http://articles.chicagotribune.com/1992-07-08/news/9203010832_1_gallo-laboratory-dr-mikulas-popovic-aids-virus

305 Ibid.

306 This was one of four papers for Science by Gallo et al.

307 Roberts, ibid.

308 Ibid.

309 Zaldivar, R. A., "Government aids scientist accused of secretly steering work to his firm," Philadelphia Inquirer, May 1, 1990. http://articles.philly.com/1990-05-01/news/25885670_1_nih-investigators-william-f-raub

310 Ibid.

311 Roberts, ibid.

312 Subcommittee on Oversight and Investigations Committee on Energy and Commerce United States House of Representatives. Investigation of the Institutional Response to the HIV Blood Test Patent Dispute and Related Matters. Washington, D.C.: United States House of Representatives, unofficial minority staff report, 1994. http://www.virusmyth.com/aids/hiv/gallo/ExeSum.html

313 Doty, Paul, "Responsibility and Weaver et al.," Nature, July 18, 1991, Volume 352, pages 183-184. This commentary includes the comment from John Cairns, M.D.

314 Ibid.

315 The 2008 Nobel Prize in Physiology or Medicine: http://www.nobelprize.org/nobel_prizes/medicine/laureates/2008/ The Nobel Prize in Physiology or Medicine 2008 was divided, one half awarded to Harald zur Hausen "for his discovery of human papilloma viruses causing cervical cancer." the other half

jointly to Françoise Barré-Sinoussi and Luc Montagnier "for their discovery of human immunodeficiency virus."

316   C-C chemokine receptor type 5, also known as CCR5 or CD195, is a protein on the surface of white blood cells that is involved in the immune system as it acts as a receptor for chemokines. This is the process by which T cells are attracted to specific tissue and organ targets.

317   About Robert Gallo: http://www.ihv.org/about/robert_gallo.html

318   Kurth passed away on February 2, 2014.

319   Editorial. "Judge: lawsuit against Merck's MMR vaccine fraud to continue," Health Impact News, 2014. http://healthimpactnews.com/2014/judge-lawsuit-against-mercks-mmr-vaccine-fraud-to-continue/

320   Koleva, Gergana, "Merck whistleblower suit a boon to vaccine foes even as it stresses importance of vaccines," Forbes, June 27, 2012. http://www.forbes.com/sites/gerganakoleva/2012/06/27/merck-whistleblower-suit-a-boon-to-anti-vaccination-advocates-though-it-stresses-importance-of-vaccines/#5d5dfd00caf7

321   A total of 6,584 cases of mumps were reported in 2006, with seventy-six percent occurring between March and May. There were eighty-five hospitalizations, but no deaths were reported; eighty-five percent of patients lived in eight contiguous midwestern states. The national incidence of mumps was 2.2 per 100,000, with the highest incidence among persons 18 to 24 years of age (an incidence 3.7 times that of all other age groups combined). In a subgroup analysis, eight-three percent of these patients reported current college attendance. Among patients in eight highly affected states with known vaccination status, sixty-three percent overall and eighty-four percent between the ages of 18 and 24 years had received two doses of mumps vaccine. For the twelve years preceding the outbreak, national coverage of one-dose mumps vaccination among preschoolers was eighty-nine percent or more nationwide and eighty-six percent or more in highly affected states. In 2006, the national two-dose coverage among adolescents was eighty-seven percent, the highest in United States history. Citing Dayan, Gustavo H., et al., "Recent resurgence of the mumps," New England Journal of Medicine, Volume 358, Number 15, April 10, 2008. http://www.nejm.org/doi/full/10.1056/NEJMoa0706589

322   Krahling and Wlochowski et al. v. Merck (E.D. Pa., 2010).

323   Humphries, Suzanne, "Scientists sue Merck: allege fraud, mislabeling, and false certification of MMR vaccine," International Medical Council on Vaccination, June 25, 2012. http://www.vaccinationcouncil.org/2012/06/25/scientists-sue-merck-allege-fraud-mislabeling-and-false-certificaion-of-mmr-vaccine-suzanne-humphries-md/

324   Ibid.

325   Ibid.

326   The seroconversion rate is the scientific term for measuring the percentage of children that are successfully immunized from the vaccine. A seroconversion

occurs when the pre-vaccination blood sample is negative, meaning insufficient antibodies to neutralize the virus, and the post-vaccination sample is positive, meaning sufficient antibodies to neutralize the virus. For the purpose of its testing, Merck needed a seroconversion rate of ninety-five percent or higher. This was, of course, the FDA's established threshold.

327 IgG antibodies are involved in the secondary immune response (IgM is the main antibody involved in primary response). IgG can bind pathogens, like for example viruses, bacteria, and fungi, and thereby protects the body against infection and toxins.

328 Benson, Jonathan, "Merck senior management tried to pay off its own vaccine scientists to remain silent about scientific fraud," Global Research, February 9, 2015. http://www.globalresearch.ca/merck-senior-management-tried-to-pay-off-its-own-vaccine-scientists-to-remain-silent-about-scientific-fraud/5430364

329 Krahling v. Wlochowski et al. v. Merck.

330 More than a decade later, Palker became a program official in the innate immunity section, division of allergy, immunology and transplantation at the National Institutes of Health/National Institute of Allergy and Infectious Diseases (NIAID) in Rockville, Maryland.

331 Dayan, Gustavo H., et al., "Recent resurgence of the mumps," New England Journal of Medicine, Volume 358, Number 15, April 10, 2008. http://www.nejm.org/doi/full/10.1056/NEJMoa0706589

332 Ibid.

333 Packel, Dan, "Antitrust, FCA claims on Merck mumps vaccine to advance," Law360.com, September 5, 2014. https://www.law360.com/articles/574389/antitrust-fca-claims-on-merck-mumps-vaccine-to-advance

334 Hill, Helene Z., "Opinion: reducing whistleblower risk," The Scientist, February 11, 2014. http://www.the-scientist.com/?articles.view/articleNo/39139/title/Opinion – Reducing-Whistleblower-Risk/

335 Sherman, Ted, "UMDNJ whistleblower cases cost Rutgers nearly $2M in settlements," NJ Advance Media for NJ.com, April 26, 2015. http://www.nj.com/news/index.ssf/2015/04/umdnj_whistleblower_cases_cost_rutgers_nearly_2m_i.html

336 Margolin, Josh, "Oversight didn't halt abuses at UMDNJ," NJ.com, October 26, 2008. http://www.nj.com/news/index.ssf/2008/10/oversight_didnt_halt_abuses_at.html

337 Sherman, ibid.

338 Henick serves as the president of Beth Israel Ambulatory Care Services Corporation and the senior vice president of medical enterprise and ambulatory care for Continuum Health Partners in New York City, New York, where he has been for more than seven years. Previously, he held several executive positions in the metropolitan New York and New Jersey markets, to include his stint at UMDNJ, and as the executive director of Montefiore Medical Center, the senior vice president of primary care for Saint Barnabas

Health Care System, and a director of real estate for Mount Sinai Medical Center.

339 Sherman, ibid.

340 Ibid.

341 Dunning, Hayley, "Parkinson's researcher fabricated data," The Scientist, June 29, 2012. http://www.the-scientist.com/?articles.view/articleNo/32305/title/Parkinson-s-Researcher-Fabricated-Data/

342 Ibid.

343 Ibid.

344 Oransky, Ivan, "ORI finds Parkinson's-pesticides researcher guilty of faking data, two papers retracted," Retraction Watch, July 2, 2012. http://retractionwatch.com/2012/07/02/ori-finds-parkinsons-pesticides-researcher-guilty-of-faking-data-two-papers-to-be-retracted/

345 Tubulin in molecular biology can refer either to the tubulin protein superfamily of globular proteins, or one of the member proteins of that superfamily. α- and β-tubulins polymerize into microtubules, a major component of the eukaryotic cytoskeleton. Microtubules function in many essential cellular processes, including mitosis. Tubulin-binding drugs kill cancerous cells by inhibiting microtubule dynamics, which are required for DNA segregation and therefore cell division. The tubulin superfamily contains six families of tubulins, to include alpha- , beta- , gamma- , delta- , epsilon- and zeta-tubulins).

346 Cytoplasmic dynein light chain (CDLC) is required for nuclear migration and for dynein heavy chain localization in Aspergillus nidulans and other fungi.

347 FLAG is an affinity tag widely used for rapid and highly specific one-step protein purification.

348 Alpha-centractin (also called ARP1) is a protein that in humans is encoded by the ACTR1A gene.

349 Oransky, Ivan, "Stem cell retraction leaves grad student in limbo, reveals tangled web of industry academic ties, Retraction Watch, December 27, 2012. http://retractionwatch.com/2012/12/27/stem-cell-retraction-leaves-grad-student-in-limbo-reveals-tangled-web-of-industry-academic-ties/

350 Doran, Michael, "How to survive as a whistleblower," Nature, Volume 532, April 20, 2016, page 405. DOI:10.1038/nj7599-405a http://www.nature.com/naturejobs/science/articles/10.1038/nj7599-405a

351 Oransky, ibid.

352 Doran, ibid.

353 Ibid.

354 Ibid.

355 Oransky, ibid.

356 Ibid.

357 Link to e-print article: Leavesley, David I., et al. [to include Zee Upton], "Vitronectin – master controller or micromanager?" IUBMB Life, Volume 65, Number 10, October 2013. http://eprints.qut.edu.au/55378/

358 Oransky, Ivan, "Australian university to repay $275,000 grant because of 'misleading and incorrect' information," Retraction Watch, August 29, 2014. http://retractionwatch.com/2014/08/29/australian-university-to-repay-275k-grant-because-of-misleading-and-incorrect-information/

359 Hare, Julie, "Queensland University of Technology forced to repay cells grant," The Australian, August 30, 2014. http://www.theaustralian.com.au/higher-education/queensland-university-of-technology-forced-to-repay-cells-grant/news-story/ec642f24b1fe9d9a881011546a222c44

360 Ibid.

361 Pain, Elisabeth, "Paul Brookes: Surviving as an outed whistleblower," Science, March 10, 2014. http://www.sciencemag.org/careers/2014/03/paul-brookes-surviving-outed-whistleblower

362 Ibid.

363 Ibid.

364 Ibid.

365 Rhodes, Rosamond and James J. Strain, "Whistleblowing in academic medicine," Journal of Medical Ethics, 2004, Volume 40, pages 35-39. DOI: 10.1136/jme.2003.005553 http://www.ncbi.nlm.nih.gov/pmc/articles/PMC1757136/pdf/v030p00035.pdf

366 Patrick, Patricia A., "Be prepared before you blow the whistle," Fraud, September/October 2010. http://www.fraud-magazine.com/article.aspx?id=4294968656

367 Pain, ibid.

368 Ibid.

369 Ibid.

370 Ibid.

371 Ibid.

372 Per Retraction Watch's Ivan Oransky.

373 Glanz, James, and Agustin Armendariz, "Years of ethics charges, but star cancer researcher gets a pass," New York Times, March 8, 2017. https://www.nytimes.com/2017/03/08/science/cancer-carlo-croce.html?_r=0

374 Reller, Tom, "It's not that Clare Francis is a pseudonym; it's that the pseudonym is Clare Francis," Elsevier, December 6, 2013. https://www.elsevier.com/connect/its-not-that-clare-francis-is-a-pseudonym-its-that-the-pseudonym-is-clare-francis

375 Ibid.

376 Marcus, Adam and Ivan Oransky, "Who are you? Editors should stop ignoring anonymous whistleblowers," Lab Times, July 2011. http://www.labtimes.org/labtimes/issues/lt2011/lt07/lt_2011_07_39_39.pdf

377 Ibid.

378 Ibid.

379 Link to Silvia Bulfone-Paus, Ph.D.'s Retraction Watch file: http://retractionwatch.com/category/by-author/silvia-bulfone-paus-retractions/

380 Marcus and Oransky, Lab Times, ibid.

381 Cyranoski, David, "Whistle-blower breaks his silence," Nature, January 28, 2014, Volume 505. http://www.nature.com/news/whistle-blower-breaks-his-silence-1.14598

382 Ibid.

383 Ibid.

384 Ibid.

385 Ibid.

386 Oransky, Ivan, "Fraud, retractions no barrier to U.S. cloning patent for Woo-Suk Hwang," Retraction Watch, February 16, 2014. http://retractionwatch.com/2014/02/16/fraud-retractions-no-barrier-to-us-cloning-patent-for-woo-suk-hwang/

387 Sooam Biotech Research Foundation is a non-profit organization that focuses on research in advanced biotechnology for industrial and biomedical applications using animal cloning and pluripotent stem cells combined with transgenic technology.

388 Cyranoski, David, "Cloning comeback," Science, January 14, 2014. http://www.nature.com/news/cloning-comeback-1.14504

389 Oransky, Ivan, "Fraud, retractions no barrier to U.S. cloning patent for Woo-Suk Hwang," Retraction Watch, February 16, 2014. http://retractionwatch.com/2014/02/16/fraud-retractions-no-barrier-to-us-cloning-patent-for-woo-suk-hwang/

390 Tae-gyu, Kim, "Hwang wins U.S. patent for human stem cells," Korea Times, February 11, 2014. http://www.koreatimes.co.kr/www/news/nation/2014/02/116_151430.html

391 Ibid.

392 Pollack, Andrew, "Disgraced scientist granted U.S. patent for work found to be fraudulent," New York Times, February 14, 2014. https://www.nytimes.com/2014/02/15/science/disgraced-scientist-granted-us-patent-for-work-found-to-be-fraudulent.html?partner=rss&emc=rss&_r=0

393 Ibid.

394 Ibid.

395 Link to Mitalipov, Shoukhrat, et al., "Human embryonic stem cells derived by somatic cell nuclear transfer," Cell, Volume 153, Issue 6, June 6, 2013. http://www.cell.com/cell/fulltext/S0092-8674(13)00824-6

396 Pollack, ibid.

397 Cyranoski, Nature, ibid.

398 Ibid.

399 Ibid.

400 Benderly, Beryl Lieff, "Paying a price for truth," Science, June 29, 2014. http://www.sciencemag.org/careers/2014/01/paying-price-truth

401 Gander, Kashmira, "Dr. Wyn Ellis: British academic detained in Thailand after exposing official as plagiarist," Independent, September 6, 2015. http://www.independent.co.uk/news/uk/home-news/dr-wyn-ellis-british-academic-detained-in-thailand-after-exposing-official-as-plagiarist-10489026.html

402 Ibid.

403 Ellis, Wyn, "Broken windows, threats, and detention: is whistleblowing worth it?" Retraction Watch, July 12, 2016. http://retractionwatch.com/2016/07/12/broken-windows-threats-and-detention-is-whistleblowing-worth-it/

404 Link the Wyn Ellis video: https://www.youtube.com/watch?v=cXaKYyy931o of the rock put this his vehicle read window.

405 Ibid.

406 Ibid.

407 Ibid.

408 Ibid.

409 Ibid.

410 Martin, Brian, "Scientific fraud and the power structure of science," Prometheus, Volume 10, Number 1, June 1992. https://www.uow.edu.au/~bmartin/pubs/92prom.html

411 Ibid.

412 Price, Alan R., "Research misconduct and its federal regulation: the origin and history of the Office of Research Integrity – with personal views by ORI's former associate director for investigative oversight," original manuscript of an article ultimately published in Accountability in Research, Volume 20, Issue 5-6, September 18, 2013. http://researchmisconductconsultant.com/Alan%20Price%20paper%20on%20ORI%20History%20for%20Accountability%20in%20Research%20-%20Sept%202013.pdf

413 Samizdat is the clandestine copying and distribution of literature banned by the state, especially formerly in the communist countries of eastern Europe. Another way of phrasing it – "an underground publication."

414 Charles W. McCutchen to Helene Z. Hill, April 16, 2014.